Polarization Bremsstrahlung

PHYSICS OF ATOMS AND MOLECULES

Recent volumes in the series:

ATOMIC PHOTOEFFECT
M. Ya. Amusia

ATOMIC SPECTRA AND COLLISIONS IN EXTERNAL FIELDS
Edited by K. T. Taylor, M. H. Nayfeh, and C. W. Clark

ATOMS AND LIGHT: INTERACTIONS
John N. Dodd

COHERENCE IN ATOMIC COLLISION PHYSICS
Edited by H. J. Beyer, K. Blum, and R. Hippler

COLLISIONS OF ELECTRONS WITH ATOMS AND MOLECULES
G. F. Drukarev

ELECTRON-MOLECULE SCATTERING AND PHOTOIONIZATION
Edited by P. G. Burke and J. B. West

THE HANLE EFFECT AND LEVEL-CROSSING SPECTROSCOPY
Edited by Giovanni Moruzzi and Franco Strumia

INTRODUCTION TO THE THEORY OF X-RAY AND ELECTRONIC SPECTRA OF FREE ATOMS
Romas Karazija

MOLECULAR PROCESSES IN SPACE
Edited by Tsutomu Watanabe, Isao Shimamura, Mikio Shimizu, and Yukikazu Itikawa

POLARIZATION BREMSSTRAHLUNG
Edited by V. N. Tsytovich and I. M. Oiringel

PROGRESS IN ATOMIC SPECTROSCOPY, Parts A, B, C, and D
Edited by W. Hanle, H. Kleinpoppen, and H. J. Beyer

QUANTUM MECHANICS VERSUS LOCAL REALISM: The Einstein–Podolsky–Rosen Paradox
Edited by Franco Selleri

RECENT STUDIES IN ATOMIC AND MOLECULAR PROCESSES
Edited by Arthur E. Kingston

THEORY OF MULTIPHOTON PROCESSES
Farhad H. M. Faisal

ZERO-RANGE POTENTIALS AND THEIR APPLICATIONS IN ATOMIC PHYSICS
Yu. N. Demkov and V. N. Ostrovskii

A Continuation Order Plan is available for this series. A continuation order will bring delivery of each new volume immediately upon publication. Volumes are billed only upon actual shipment. For further information please contact the publisher.

Polarization Bremsstrahlung

Edited by
V. N. Tsytovich
Institute of General Physics
Moscow, Russia

and
I. M. Oiringel
Presidium of Irkutsk Scientific Center
Irkutsk, Russia

Translated from Russian by
D. H. McNeill

Springer Science+Business Media, LLC

Library of Congress Cataloging-in-Publication Data

Polarization bremsstrahlung / edited by V.N. Tsytovich and I.M. Oiringel ; translated from Russian by D.H. McNeill.
p. cm. -- (Physics of atoms and molecules)
Includes bibliographical references and index.
ISBN 978-1-4613-6329-3 ISBN 978-1-4615-3048-0 (eBook)
DOI 10.1007/978-1-4615-3048-0
1. Bremsstrahlung--Polarization. I. Tsytovich, V. N. (Vadim Nikolaevich), 1929- . II. Oĭringel′, I. M. (Isaak Mikhaĭlovich)
III. Series.
QC484.3.P65 1992
539.7'222--dc20 92-20860
CIP

ISBN 978-1-4613-6329-3

Originally published by Plenum Press, New York in 1992

CONTRIBUTORS

M. Ya. Amus'ya • Theoretical Department, A. P. Ioffe Physicotechnical Institute, St. Petersburg 194021, Russia.

V. A. Astapenko • Department of Physics and Quantum Electronics, Moscow Physicotechnical Institute, Moscow 141700, Russia.

V. M. Buimistrov • Department of Physics and Quantum Electronics, Moscow Physicotechnical Institute, Moscow 141700, Russia.

V. I. Gervids • Theoretical Department, Moscow Institute of Physical Engineering, Moscow 115409, Russia.

E. B. Kleiman • Department of Physics, Nizhnii Novgorod State University, Nizhnii Novgorod 603600, Russia.

V. I. Kogan • Theoretical Department, Russian Scientific Center, I. V. Kurchatov Institute, Moscow 123182, Russia.

A. V. Korol • Department of Physics, Naval Technical University, St. Petersburg 198262, Russia.

Yu. A. Krotov • Theoretical Department, Polus Scientific Research Institute, Moscow 117342 Russia.

A. B. Kukushkin • Theoretical Department, Russian Scientific Center, I.V. Kurchatov Institute, Moscow 123182, Russia.

V. S. Lisitsa • Theoretical Department, Russian Scientific Center, I. V. Kurchatov Institute, Moscow 12882, Russia.

A. M. Oiringel • Presidium of Irkutsk Scientific Center, Russian Academy of Sciences, Irkutsk 664033, Russia.

A. V. Solov'ev • Theoretical Department, A. F. Ioffe Physicotechnical Institute, St. Petersburg 194021, Russia.

V. N. Tsytovich • Theoretical Department, Institute of General Physics, Russian Academy of Sciences, Moscow 117942, Russia.

B. A. Zon • Department of Physics, Voronezh State University, Voronezh 394038, Russia.

FOREWORD

This book was written by a group of authors and provides a systematic discussion of questions related to bremsstrahlung in many-particle systems. A number of new results have recently been obtained in this area which require a fundamental revision of the previously existing traditional concepts of bremsstrahlung. This applies both to complicated atoms containing a large number of electrons and to the additional bremsstrahlung in a system of many particles forming a medium. In fact, the traditional approach was rigorously applicable only either to isolated "structureless" particles (e.g., to the emission of an electron on a proton) or to particles radiating in the limit of extremely high frequencies. Polarization effects (either polarization of an atom itself by an incident particle or polarization of the medium surrounding an atomic particle) have a significant effect in the practically important optical and x-ray frequency ranges and sometimes even predominate. The first effect has come to be known as polarization atomic (or dynamic) bremsstrahlung and the second, as polarization transition bremsstrahlung. The authors of this book use a single term: polarization bremsstrahlung. It seems that, in contrast to earlier ideas on the subject, bremsstrahlung during collisions of heavy incident particles with atoms is by no means small and is entirely caused by polarization effects.

The authors have been able to explain correctly an entire series of experiments on bremsstrahlung in many-electron systems, on laser breakdown in alkali metal vapor, and to use the results to examine the radiative properties of low-temperature plasmas. Ultimately, they have developed a general theoretical scheme for describing radiative processes during particle collisions in which the bremsstrahlung and polarization mechanisms enter as necessary and inseparable component parts. This justifies the claim of a new discipline with a general theoretical foundation and important applications.

The authors of this book have participated actively in creating this new discipline. It has been developed by several groups of researchers using different methods. Their joint efforts are the result of the coordinating activity of the Scientific Council of the Academy of Sciences of the USSR on the Physics of Low-Temperature Plasmas. In this way a good format was found for fruitful discussion of the pressing new problems that arose: topical working groups. These groups were created to provide an opportunity for deeper study of the research topics that arose. One of these groups was organized by the Section on collective phenomena in low-temperature plasmas together with the Scientific Council of the Academy of Sciences on the Physics of Electronic and Atomic Collisions. This working group was the organizational basis of the group of authors who wrote this book and reflects a new form of scientific communication that permits the testing of new results by means of discussions among the members of the group. The subsequent tasks of the group included a detailed joint discussion of each section of the book, writing collective introductions and conclusions, and collectively discussing and proofreading each section of the book. This method of working made it possible to create a collective work with a unified point of view of the problem that was common to the entire group of authors. This is a definite advantage of the book.

The quest for faster acquisition of scientific results and their generalization that is characteristic of our time requires that new types of communication and collective effort among scientists be found. Perhaps one such way will the method chosen by the authors of this book.

We would especially like to acknowledge the creative and organizational contributions of the chairman of the Scientific Council on Collective Phenomena in Low-Temperature Plasmas and leader of the working group, Prof. V. N. Tsytovich.

B. B. Kadomtsev
L. M. Biberman

PREFACE TO THE ENGLISH EDITION

This book is devoted to a new polarization mechanism for bremsstrahlung which has been neglected in most discussions of the bremsstrahlung problem in textbooks. Important new results have also been obtained very recently in the physics of the traditional mechanism for bremsstrahlung. In the English edition of this book we have included a final, thirteenth chapter, written by V. I. Gervids and V. I. Kogan and entitled *Nonpolarization Bremsstrahlung of Electrons in Static Potentials*. The polarization bremsstrahlung mechanism studied in this book acts against a background of traditional bremsstrahlung and competes with it. Thus, Chapter 13, which is concerned with some new approaches to the physics of traditional bremsstrahlung of electrons in both Coulomb and non-Coulomb fields, is an important addition to the material contained in the earlier Russian edition.

In terms of the coexistence with the polarization mechanism, bremsstrahlung on many-electron atoms or ions (Chapters 6–8) is of greatest interest here. As opposed to the case of a pure Coulomb potential, this does not admit of an exact analytic treatment and, in addition, does not obey the well-explored Born approximation at low energies. At low energies, the motion of electrons in atomic fields is quasiclassical. This opens up unexpectedly broad possibilities for describing their bremsstrahlung by the methods of classical electrodynamics, which are used extensively in Chapter 13.

With regard to classical electrodynamics, in the literature there is a large gap both in the analytic description of emission at the rather high frequencies produced by "sharp" electron trajectories in the framework of classical electrodynamics and in the treatment of the real, as opposed to intuitive (such as $\hbar\omega \ll E$), limits of its applicability in the framework of quantum electrodynamics. This gap is essentially filled by the material discussed in Chapter 13. It forms the basis of a new, basically classical approximate method for describing and calculating highly inelastic (and not only radiative; cf. Chapter 2) quantum mechanical transitions involving low-energy electrons.

This book is devoted to the modern theory of bremsstrahlung. As in many others areas, Hans Bethe has made a fundamental contribution to this topic. The papers presented here are distinguished by a detailed accounting for polarization bremsstrahlung owing to the dynamic response of the electronic structure of an atom. One of the first and most important papers in this area was written by Percival and Seaton.

The role of the electronic structure of atoms in bremsstrahlung has turned out to be much greater than assumed previously. It has become evident that a clear analogy exists between the bremsstrahlung of a charged particle on an atom and the bremsstrahlung on the unique classical atom formed by a plasma ion surrounded by its electron cloud (so-called transition bremsstrahlung). Radiation is produced by the polarization of either the atom itself or its surroundings and it makes sense to refer to this radiation as polarization bremsstrahlung.

It is significant that, in a number of cases, it is impossible to construct a correct physical picture of bremsstrahlung without taking electronic structure into account. If we include polarization bremsstrahlung simultaneously with the ordinary mechanism, then it is possible to explain easily and clearly the results that were not contained in the earlier theory of bremsstrahlung which relied on the shielding approximation. This also applies to the scattering of electrons on atoms in long-range collisions, when shielding generally ceases to occur, as well as to the bremsstrahlung of protons.

Polarization bremsstrahlung has been observed in the specially set-up experiments of Verkhovtseva, Gnatenko, and Pogrebnyak and in the experiments of Ishii and Morita. Some earlier experiments on laser breakdown in alkali metal vapors have been explained in terms of absorption through inverse polarization bremsstrahlung.

In the name of the authors of this book, I should like to express our deep appreciation of the fact that it will be available to the English speaking reader. I would also like to express my appreciation for discussions of the content of this book with Prof. D. ter Haar and for his comments and help in locating English translations of Russian articles.

V. N. Tsytovich

PREFACE

Several years ago the physics of bremsstrahlung underwent a rebirth: the traditional, seemingly stagnant, classical views were reexamined or refined and many new terms arose: "transition," "dynamic," and "atomic" bremsstrahlung. It turned out that the physical meanings of these effects were not only close, but were often identical for the same processes involving different objects. Many experiments were set up, and important applications were found in low-temperature plasma radiation, laser breakdown, and many other areas and were reported in Soviet and foreign publications. New experiments are still being set up.

Briefly stated, this is a new, rapidly developing discipline which is important both in its applications and in that it unifies previously differing approaches and physical phenomena. Different investigators have arrived at a unified understanding of the physical process in their own ways and this has led to a disparity in terminology.

In 1985 a meeting of the working group on dynamic and transition bremsstrahlung, a section of the Scientific Council of the Academy of Sciences of the USSR on the Physics of Low-Temperature Plasmas, and the Scientific Council on the Physics of Electronic and Atomic Collisions was held. About two dozen scientists who had been actively working in recent years on new problems in the physics of bremsstrahlung participated in this meeting. As a result of their discussions, the general physical picture of the phenomena became clearer and the conviction developed that this is indeed an important and new area of research. It was natural to present this physical theory in a rigorous way and combine the important results of research on this problem, along with their possible applications.

In writing this book a great deal of work was necessary to ensure consistency and unity in the presentation. Here an important, although not fundamental, point was the question of terminology.

We begin with a few words on the essence of this effect, without which any discussion of terminology would be unclear. When a charge is placed in a plasma, it is shielded by a Debye polarization "cloud." (Here a plasma is introduced as the simplest of examples; similar effects appear in any medium, as will be discussed in more detail below.) In a collision the charges in the cloud begin to oscillate and this leads to radiation. This has been referred to as transition radiation. It is important that the momentum and energy of the emitted photon originate in the momentum and energy of the colliding particles, so that until the cloud has transferred the energy and momentum to the charge that it is shielding, the cloud itself will be nonstationary and the emission process will not come to an end. Thus, the conservation laws applying to emission from the cloud are the same as for ordinary bremsstrahlung. The term "transition" bremsstrahlung is fully justified here, since scattering on a polarization charge is precisely transition scattering, while bremsstrahlung is known to reduce to scattering of the virtual field of the colliding particles into a real photon.

The terms "dynamic" and "atomic" bremsstrahlung have been used in papers devoted to bremsstrahlung on atoms and ions with bound electrons, including many-electron atoms. In this case, when a charged particle passes by atoms, the atomic electrons acquire an oscillating (dynamic) polarization which radiates. An atom or ion may remain in the same energy state after radiating, but they can also be excited or ionized. The energy and momentum of the emitted photon come from the energy and momentum of the incident particle or from the energy and momentum of the center of mass of the atom (the energy and momentum of the ejected electrons must also be taken into account during ionization). In this case the conservation law is the same (provided the atom does not change its energy state) as in ordinary bremsstrahlung, and the term "dynamic" or "atomic" bremsstrahlung is fully justified. It also emphasizes the facts that, as opposed to traditional bremsstrahlung, where the static polarization leading to shielding is important, here the dynamic polarization (the time variation in the polarization associated with the atomic electrons) plays a role and it is actually (in the sense of a change in the dipole moment) the atomic electrons which emit, rather than the incident charged particle. This sort of emission, of course, also occurs during collisions of two neutral polarizable particles.

The terms "transition," "dynamic," and "atomic" bremsstrahlung have been used extensively in recent scientific publications. Thus, it would be fairly painful to drop them. Nevertheless, the authors decided to unify the terminology for the new discipline. The general term "polarization bremsstrahlung" was proposed and such additional terms as "polarization dynamic radiation" or "polarization atomic radiation" can be used when necessary for clarity or to establish a connection to earlier work.

Polarization bremsstrahlung in general is inseparable from traditional bremsstrahlung caused by the slowing down of an incident particle, but it may predominate in certain frequency ranges. For example, it is extremely large when both colliding particles are heavy. For relativistic particles there is always a frequency range within which practically all the bremsstrahlung is polarization bremsstrahlung.

We shall not examine all the effects that arise since they form the subject of the later discussion. We have mentioned a few of them only to emphasize that the effect itself is significant and in many cases controls the radiation during collisions. It is, therefore, surprising that it has attracted almost no attention until recently. The papers written by the founders of the theory of bremsstrahlung on individual particles (Sommerfeld, Bethe and Heitler, Born) merely mention the fact that a rigorous statement of the problem would require consideration of an entire many-electron system, including both the incident electron (in electron–ion collisions) and atomic electrons.

In this book the presentation builds up from simple to complicated. It includes both a review of the problem and a considerable amount of new material.

Chapter 1 contains a qualitative "thumbnail sketch" of polarization radiation processes and has been written on the basis of results from independently developed research on atomic bremsstrahlung, dynamic bremsstrahlung, laser breakdown, and transition bremsstrahlung. The intuitive picture presented in this chapter had not actually existed previously and can now be used to explain the common characteristics of the different manifestations of polarization bremsstrahlung. The coherence of the emission from atomic electrons and electrons in a Debye cloud, the clear picture of the contributions from different impact parameters, and the so-called "stripping" of atoms with increasing bremsstrahlung frequency seem to be of particular importance.

Chapter 2 discusses the classical (nonquantum) theory of polarization bremsstrahlung in distant or long-range collisions. The classical concepts must be discussed both because they are more intuitive than the quantum mechanical ones and because the main contribution from polarization bremsstrahlung does indeed arise from long-range collisions, when the momentum transfers are small compared, for example, to the momentum of the incident particle.

The theory of polarization transition bremsstrahlung on Debye shielding clouds has been developed previously. It turns out that a classical theory of polarization radiation can also be constructed for bound electrons, i.e., for an atom, if we treat the atom as a polarizable dipole. Naturally, the polarizability itself must be calculated quantum mechanically. The result of the classical theory is the same as the quantum mechanical result if the quantum mechanical polarizability of the atom is used and the dimension of the atom is substituted in the argument of the logarithm (or, more precisely, the impact parameter is cut off at a distance on the order of the size of the atom). Thus, one can obtain the intensity of the polarization radiation of bound atomic electrons with logarithmic accuracy. This conclusion is new.

Chapter 3 contains a description of the fundamental physical questions relating to fluctuations and bremsstrahlung. A uniform medium does not radiate or scatter; thus, bremsstrahlung, like other types of radiation from media that are uniform on the average, is related to fluctuations. The simplest system for which the theory of fluctuations can be developed to completion is a plasma. On the other hand, the bremsstrahlung process is most complicated in plasmas. Every electron, as it simultaneously participates in the averaged and fluctuating motions, creates a Debye shield and is scattered by (collides with) other electrons and ions in the plasma. It appears that the picture of elementary excitations as a characteristic of the averaged motion, i.e., of the motion of particles "dressed" in clouds, is valid both for describing particle collisions and for describing bremsstrahlung. It is shown that polarization (transition) bremsstrahlung "automatically" follows from the theory of fluctuations and is an inseparable element of the general picture of our concepts of radiation in homogeneous media.

The general balance equations for bremsstrahlung including transition polarization bremsstrahlung are known from before, as is a proof that the frictional force of plasma particles on bremsstrahlung photons associated with stimulated bremsstrahlung and inverse bremsstrahlung is caused by the combined effect of traditional and polarization bremsstrahlung. Here we do offer a proof that the theory of fluctuations shows that polarization bremsstrahlung is necessarily present in a given volume of plasma. This material is also new.

Chapter 4 is devoted to a nonrelativistic quantum mechanical treatment of polarization bremsstrahlung on atoms. There it is shown that a rigorous analysis of bremsstrahlung (including polarization effects) leads to a correct physical picture which differs substantially from the shielding approximation within a certain range of frequencies. Sums over the intermediate states of an atom corresponding to virtual transitions appear and are calculated in this theory. The special role of polarization bremsstrahlung in heavy-particle collisions is emphasized, for example, when the incident particle is a proton, which is important both in plasmas and for bremsstrahlung on individual atoms. This radiation has been observed in experiments, some of which are discussed here.

Chapter 5 contains an analysis of polarization bremsstrahlung by the methods of quantum electrodynamics, which is the only rigorous theory for relativistic particles.

Chapter 6 uses the methods of quantum electrodynamics to calculate the polarization radiation of an ion with a core of bound electrons in a plasma, when the electrons lacking for quasineutrality are created by the Debye cloud. Here combined polarization bremsstrahlung of bound and free electrons takes place. Within a certain frequency range, the free and bound electrons can even radiate together (coherently). The exact results obtained here correspond to those obtained in Chapter 2 using a classical approach. The calculations are carried out simply by summing over all the bound and free electrons of the system using the methods and results of Chapter 5. Some of these results are also new.

Chapter 7 is devoted to the study of polarization bremsstrahlung by many-electron atoms. Here the diagram techniques of many-body theory are used. The interest in many-electron atoms is dictated primarily by the available experimental material and by the fact that polarization effects are extremely important for many-electron atoms and also show up at extremely high frequencies that may exceed the x-ray range. Here good agreement has been obtained between theory and experiment. The "stripping" of an atom with increasing bremsstrahlung frequency, when the various atomic shells are sequentially "shut off," is of some interest. Polarization bremsstrahlung enters here as an effective method for studying the structure of complicated atoms. Our investigation of polarization bremsstrahlung in collisions of two atoms and in collisions with such "exotic" systems as positronium is also original and new.

Chapter 8 is concerned with the theory of nuclear collisions from the standpoint of the concept of polarization radiation and is important for developing a wider view of the general physical problem of polarization radiation on the part of the reader.

Chapter 9 touches upon polarization bremsstrahlung of relativistic atoms and relativistic particles, including neutrinos, with atoms. Relativistic effects can determine the total bremsstrahlung cross section in a number of cases, even at nonrelativistic velocities, if the amplitudes cancel out when these effects are neglected. Apparently the emission from such "elementary" particles as protons, neutrons, and mesons, which are made up of quarks, can also be polarization radiation.

This ends the discussion of general concepts and we now begin to discuss a few simpler concepts, which are, nevertheless, important for direct applications of polarization bremsstrahlung effects. Among these, we must first include inverse polarization bremsstrahlung, when photons are absorbed in an "atom + incident particle" system and transfer their energy to the incident particle. If it acquires enough energy, it may cause ionization and then optical (laser) breakdown takes place. In order to explain the experiments in this area, one must draw upon the concept of inverse resonant polarization bremsstrahlung, which is two orders of magnitude greater than traditional inverse bremsstrahlung at the ruby laser wavelength. These questions are discussed in Chapter 10.

In Chapter 11 a number of previously known mechanisms are discussed from the standpoint of polarization bremsstrahlung. Here we are speaking primarily of elastic processes for resonant polarization radiation (when the atom remains in its initial state) and inelastic (not referring to polarization bremsstrahlung as such, but adjacent to it) processes such as dielectronic and polarization recombination. Resonant polarization radiation is also discussed in Chapters 4 and 12. Naturally, near atomic transition frequencies the dynamic polarization of an atom becomes resonantly large. Here, two aspects of this problem show up.

First, in the highly collisional case, when the incident particles "force" an atom to emit at frequencies close to a transition frequency but outside its natural linewidth, this radiation can be treated as collisional broadening of spectrum lines. This is done in Chapter 11, whereas Chapter 4 is concerned with resonant emission in a single elementary collision event. This interpretation of line broadening, as in Chapter 11 (or, more precisely, the contribution to line broadening attributable to fast electrons) does, of course, enrich our understanding of the nature of polarization bremsstrahlung and gives greater clarity to the old concept of collisional broadening. (In particular, it is clear that the velocity of the incident particle is important, rather than its mass, and broadening can be produced by collisions of atoms with heavy particles.)

All this makes it possible to extend further the range of problems which can be described in the language of polarization bremsstrahlung. Incidentally, under conditions such that several resonances (and not just one) make a significant contribution to the polarization, polarization bremsstrahlung no longer has a direct connection to line broadening and, therefore, represents a clearly more general phenomenon.

Chapter 12 is devoted to an examination of photon–plasmon transitions in plasmas as a mechanism for quenching metastable states of atoms. This question is akin to those discussed in Chapter 11. Here the incident particle is not a charged particle, but a quantum of the medium, namely a plasmon which polarizes the atom, while the latter begins to emit electromagnetic waves, and those excited states which were metastable are destroyed. Since plasmons of different types are very easily excited in plasmas and their level is generally high, this process is of interest both for astrophysical applications and for the general problem of the quenching of metastable states in weakly ionized low-temperature plasmas. In this, as in the preceding chapters, we discuss the polarization radiation of neutral incident particles.

The authors of this book previously worked on these problems independently of one another, so it is natural that, despite their overall agreement on the main points, they would have different opinions on a few particular questions. Hence, the main problem was to achieve a unified point of view on the fundamental questions. The solution of this problem began at the working group and then extended continuously over more than six months. Now the differences in viewpoint (for example, on resonant bremsstrahlung or broadening of spectrum lines) involve only the details. Our joint work on the Introduction and Conclusion, as well as on a number of the other chapters, was of great use for this purpose. For just this reason a few chapters have a large number of authors. The standardization of notation also required a lot of effort.

The authors considered one of the main purposes of this book to be the stimulation of more extensive experimental research on polarization bremsstrahlung. A discussion of several possible experiments of this sort is contained in the concluding part of Chapter 11, where experiments on bremsstrahlung in thermonuclear plasmas are also reviewed.

In conclusion, we note that this book by no means pretends to an exhaustive review of bremsstrahlung. The main emphasis has been on the new mechanism of polarization bremsstrahlung. Of course, in recent years many interesting and new results have been obtained even in the traditional approach. The traditional mechanism, however, is usually touched on here only to the extent that it competes with polarization bremsstrahlung (and in some cases almost completely "cancels" it).

A large part of the material presented here was reported at a Moscow seminar chaired by V. L. Ginzburg. The discussions at that seminar played an important role in deepening comprehension of the different approaches and in clarifying the common physical foundation of the processes which had not seemed so closely related before. It then became clear that polarization bremsstrahlung on bound electrons in atoms and ions also fits into the overall pattern of transition scattering of the virtual waves of colliding particles.

The authors sincerely thank their colleagues in the working group for useful discussions, especially A. K. Gailitis, who did the first work on the influence of polarization effects on scattering processes in plasmas. The authors also acknowledge a fruitful collaboration with A. V. Akopyan, N. B. Avdonina, A. S. Baltenkov, M. Yu. Kuchiev, L. K. Mikhailov, and L. I. Trakhtenberg, as well as useful discussions with their colleagues F. V. Bunkin, B. M. Bolotovskii, É. I. Rashboi, A. A. Rukhadze, G. V. Skrotskii, M. F. Stel'makh, and M. L. Ter-Mikaélyan. The authors are extremely grateful to I. P. Druzhinin for support in preparing the book.

V. N. Tsytovich
I. M. Oiringel

CONTENTS

BASIC NOTATION AND UNITS OF MEASUREMENT

s_i, s_f – complete set of quantum numbers of the initial and final states

$\varepsilon_i = \varepsilon_0$, $\mathbf{p}_i = \mathbf{p}_0$, $\mathbf{v}_i = \mathbf{v}_0$ – initial energy, momentum, and velocity of an incident particle

ε_f, $\mathbf{p}_f$, $\mathbf{v}_f$ – final energy, momentum, and velocity of an incident particle

E_i, E_f – initial and final energy states of a target atom

$\mathbf{q} = \mathbf{p}_f - \mathbf{p}_i + \mathbf{k}$ – momentum transferred from a target particle to an incident particle during bremsstrahlung

$\mathbf{q}_1 = \mathbf{p}_f - \mathbf{p}_i$ – with minus sign, is the momentum transferred from an incident particle to a target particle and bremsstrahlung photon

$-e = -|e|$, m – charge and mass of an electron

e_0, m_0 – charge and mass of an incident particle

$e_i = Z_i e$ – charge on an ion or nucleus (for $Z_i = Z$)

N – number of electrons in an atom ($Z_i = Z - N$)

N_i, I_i – number of electrons and ionization potential of the ith electron shell of an atom

$\alpha_i(\omega, q)$ – generalized nondipole polarizability of an atom in state i

$\alpha(\omega) = \alpha_i(\omega) = \alpha_d(\omega)$ – dipole polarizability of an atom in state i

$F(\mathbf{q})$ – atomic form factor

r_d – Debye radius

$\omega_{pe} = (4\pi ne^2/m)^{1/2}$ – plasma frequency

$\gamma = \varepsilon/mc^2$ – relativistic factor

$T_{i,f}$ – kinetic energy of nonrelativistic particles

$\mathbf{k}$ – momentum (wave number) of an emitted photon

$d\Omega_{\mathbf{k}}$ – element of solid angle for emitted photons with the angle taken relative to the initial velocity of the incident particle

$d\Omega_{\mathbf{p}f}$ – element of solid angle for an incident particle after a collision

$\mathbf{e} = \mathbf{e}_{\mathbf{k},\sigma}$ – unit polarization vector of an emitted photon
$\mathbf{n} = \mathbf{k}/k$ – unit propagation vector of a photon
$\mathbf{A}^{(0)}$, $\mathbf{E}^{(0)}$ – vector potential and field of an incident particle
$d\sigma$ – differential cross section
$d\sigma^{t}$ ($d\sigma^{st}$) – contribution of the traditional (static) part to the bremsstrahlung cross section
$d\sigma^{p}$ – contribution of the polarization part to the bremsstrahlung cross section
$d\sigma^{\text{int}}$ – interference component of the cross section
$d\sigma(\omega)/d\omega$ – cross section per unit frequency interval
$Q_{\omega} = n_i v \hbar\omega d\sigma(\omega)/d\omega$ – bremsstrahlung power per unit frequency interval
ρ, $\mathbf{j}$ – charge and current densities

In the *atomic units* employed here, $|e| = 1$, $m = 1$, $\hbar = 1$, $e = -1$, and $e_i = Z_i$. In atomic units velocity is measured in $e^2/\hbar$, length in Bohr radii $a_B = \hbar^2/me^2$, momentum in units of $\hbar/a_B$, electric field strength in $e/a_B{}^2$, frequency in $e^2/\hbar a_B = \omega_a$, and the velocity of light c is given by $\hbar c/e^2 \sim 137$.

In order to obtain the bremsstrahlung cross section in gaussian units from the bremsstrahlung cross section in terms of atomic units, one must make the substitution

$$\sigma_a = \frac{1}{c^3 p^2} \Lambda(\omega, q \ldots) \rightarrow \sigma_{Gs} = \frac{e^6}{\hbar c^3 p^2} \Lambda\left(\frac{\omega}{\omega_a}, q a_B \ldots\right).$$

The cross product of two vectors $\mathbf{a}$ and $\mathbf{b}$ is denoted by $[\mathbf{ab}]$ and the dot product by $(\mathbf{ab})$.

Chapter 1

POLARIZATION BREMSSTRAHLUNG EFFECT IN PARTICLE COLLISIONS

M. Ya. Amus'ya, V. M. Buimistrov, B. A. Zon, and V. N. Tsytovich

1.1. Introductory Remarks

The electromagnetic radiation from charged particles is a traditional topic of physical research, at least since the end of the last century. The production of radiation requires the interaction of a charged particle with matter or with an external field. Various kinds of radiation can result from these interactions, including bremsstrahlung, Cerenkov, transition, synchrotron, and undulator radiation [1]. Here we hope to emphasize one fundamental property of radiation in a medium, specifically the important role played in transition and Cerenkov radiation by the variable polarization induced by a charge in a medium. The question of the role of polarization effects in such fundamental phenomena as bremsstrahlung arises naturally.

Until quite recently, bremsstrahlung was a phenomenon that seemed to have been exhaustively studied, both theoretically and experimentally, and formed a rather extensive branch of modern physics. It is, therefore, all the more surprising that a whole class of effects associated with radiation in particle collisions, which appears simultaneously with "traditional" bremsstrahlung, has clearly been left out. In some cases, this new effect is inseparable from the traditional effect and may dominate it under certain conditions. This effect can be referred to as polarization bremsstrahlung in particle collisions.

1.2. Elementary Concepts of "Traditional" Bremsstrahlung in the Static Field of a Charge

Here we begin by recalling the basic features of the traditional mechanism for bremsstrahlung.

Particle collisions involve the slowing down (or acceleration) of the particles. During a collision between a light particle and a heavy one, it is primarily the light particle which is accelerated and emits radiation. If the incident particle is nonrelativistic, then the spectral distribution of the energy radiated in a single collision event is determined by the classical formula [2,3] for dipole radiation

$$W_\omega = \frac{8\pi e_0^2}{3c^3}(\ddot{\mathbf{r}}_\omega)^2 = \frac{8\pi e_0^4}{3c^3 m_0^2}(\mathbf{E}_\omega)^2 = \frac{8\pi e_0^4 e_i^2}{3c^3 m_0^2}\left[\left(\frac{\mathbf{r}}{r^3}\right)_\omega\right]^2,$$
$$\ddot{\mathbf{r}}_\omega = \frac{e_0}{m_0}\mathbf{E}_\omega, \quad v \ll c, \tag{1.1}$$

where W_ω is the energy radiated over the entire duration of the collision in the frequency interval $d\omega$; e_0 is the charge and m_0 is the mass of the light particle; $\ddot{\mathbf{r}}_\omega$ is the Fourier component of its acceleration; and $\mathbf{E}_\omega$ is the Fourier component of the field of the heavy particle, whose charge is given by $e_i = Ze$ (the charge on the electron is $-e$).

The simplest case is that in which the particles pass by one another at distances large enough that the trajectory of the incident particle undergoes little change. The shortest distance between an incident particle and a heavy particle, if the trajectory of the incident particle were not perturbed, is usually referred to as the impact parameter ρ. For long-range (distant) encounters, we can neglect the change in the trajectory when calculating the acceleration of the incident particle in the field of a heavy charge. Then $r^2 = \rho^2 + v_0^2 t^2$, where v_0 is the velocity of the incident particle. Evidently, when the particles pass by one another at a distance ρ, radiation at frequencies lower than v_0/ρ will be emitted. Then the Fourier component $(\rho/r^3)_\omega$ in Eq. (1.1) is constant and equal to $\rho/\pi v_0 \rho^2$. This represents radiation with a "white" spectrum (i.e., the spectral distribution of the radiant energy is independent of frequency).

The term "bremsstrahlung" appeared after the work of Stokes and Sommerfeld [4] who explained the continuous "white" x-ray spectrum observed when cathode rays were slowed down in the material of an anticathode. In this case, of course, we are speaking of radiation from a flux of particles interacting with a target, rather than of radiation in individual collision events. Each of the incident particles encounters a sequence of motionless heavy particles along its path. If the particle density in this target is n_i per cubic centimeter, then the spectral power of the radiation is given by the spectral energy density of a single particle, W_ω, multiplied by $n_i 2\pi\rho d\rho v_0$. Here it is assumed that the incident particle radiates at each heavy particle independently. This means that the scattering centers are positioned randomly and the incident particle is able to emit a bremsstrahlung photon before it collides with the next scattering center (when the latter condition is not satisfied, one speaks of the influence of multiple scattering on bremsstrahlung [5-7]). As a result, we obtain the following expression for the spectral power of the radiation:

$$Q_\omega = \int W_\omega n_i v_0 2\pi\rho \, d\rho = \frac{16 e_0^4 e_i^2 n_i}{3 m_0^2 v_0 c^3} \ln \frac{\rho_{max}}{\rho_{min}} . \tag{1.2}$$

From this derivation it is clear that $\rho_{max} \sim v_0/\omega$, while ρ_{min} is determined in accordance with the following considerations. The radiated power (1.2) for a single particle moving in a target has a weak logarithmic dependence on the frequency. Thus, we can speak of a "white" spectrum. The integral over all frequencies of such a spectrum diverges, but there is an obvious quantum mechanical limit $\hbar\omega \leq T_i$, where T_i is the kinetic energy of the incident particle. The criterion $\hbar\omega \ll T_i$ is equivalent to the assumption of long-range collisions.

The standard quantity used in the theory of bremsstrahlung is the differential (with respect to the frequency) bremsstrahlung cross section. In order to evaluate it, we must calculate the power radiated by a flux of particles incident on an isolated ion and divide that by the particle flux and the energy $\hbar\omega$ of a single photon:

$$Q_\omega = \hbar\omega n_i v_0 \frac{d\sigma}{d\omega}, \qquad \sigma = \int \frac{d\sigma}{d\omega} d\omega,$$
$$\frac{d\sigma}{d\omega} = \frac{16 e_0^4 e_i^2}{3 m_0^2 c^3 v^2 \hbar\omega} \ln \frac{\rho_{max}}{\rho_{min}} . \tag{1.3}$$

Note that Planck's constant appears here in a classical cross section because it refers, by definition, to the energy of a single photon.

This discussion of the bremsstrahlung effect is at the simplest level and serves only as a reminder of the well-known results of the theory which are currently in widespread use.

When bremsstrahlung takes place on a real atom, rather than on a "bare" atom, we must take the shielding of the field of the nucleus by the atomic electrons into account. The theory of bremsstrahlung initially developed in just this direction. If R_a is the shielding radius (the radius of an atom), then when $v_0/\omega > R_a$ instead of $\rho_{max} \sim v_0/\omega$, we have $\rho_{max} \sim R_a$ under the logarithm in Eq. (1.2). It should be noted that in the strictly quantum mechanical case, we cannot introduce an impact parameter. The uncertainty relation, however, can be used to estimate ρ_{min} [8].

If an ion is located in a plasma, then Debye shielding of the ion's field occurs and ρ_{max} must be replaced by the Debye radius r_d when $v_0/\omega > r_d$. It is extremely important that an ion in a plasma is actually a very good example of a real "classical atom." As for real atoms, of course, their atomic electrons must be described quantum mechanically.

We have derived the traditional bremsstrahlung formulas for distant collisions here because polarization bremsstrahlung can be very important precisely for distant collisions. Polarization bremsstrahlung can even make a greater contribution than the traditional mechanism discussed above.

Note the characteristic dependence of the radiated power on the charge e_0 and mass m_0 of the incident particle and the charge e_i of the scattering center. Usually, if the incident particle is an electron ($e_0 = -e$, $m_0 = m$) and the scattering center is a multiply charged ion ($e_i = Ze$), then the factor in front of the logarithm has a characteristic dependence of the form Z^2e^6/m^2.

Generally speaking, for polarization bremsstrahlung this factor is different and is independent of the mass of the incident particle. The qualitative difference between the two mechanisms for radiation by particles shows up in this way, although it is the same for a collision of an electron with a multiply charged ion.

In describing the traditional mechanism for bremsstrahlung, we have used classical concepts. This is permissible only when $|e_0e_i| \gg \hbar v_0$, i.e., for small velocities. Then ρ_{min} is determined by the characteristic impact parameter at which the deflection angle of the trajectory of the incident particle from a straight trajectory approaches unity. In the opposite case, ρ_{min} is determined from quantum mechanical considerations and, specifically, $\rho_{min} \sim \hbar/m_0v_0$, so that the term under the logarithm in Eq. (1.2) should then be

$$\rho_{max}/\rho_{min} \approx m_0v_0^2/\hbar\omega = 2T_i/\hbar\omega.$$

The quantum theory of bremsstrahlung was created by Sommerfeld [4], Bethe and Heitler [8], and Sauter [9]. The Sommerfeld formula in the Born approximation (the nonrelativistic limit of the Bethe–Heitler formula) for $\hbar\omega \ll T_i$ has the form

$$Q_\omega = \frac{16e_0^4e_i^2n_i}{3m_0^2v_0c^3}\left(\ln\frac{2T_i}{\hbar\omega} - \frac{1}{2}\right) \tag{1.4}$$

and actually differs from Eq. (1.2) only in containing a more accurate expression under the logarithm.

Recently, new and interesting results have been obtained even in the theory of traditional bremsstrahlung [see Chapter 13]. These are primarily concerned with very "steep" trajectories (that is, close collisions), non-Coulomb scattering centers, and the relationship between bremsstrahlung (free–free transitions) and recombination radiation (free–bound transitions). They also involve inelastic collisions and general ideas about quantum effects.

Here, at the beginning of this book, we wanted to focus the reader's attention on the basic problem to which the book is devoted, namely, polarization bremsstrahlung.

1.3. Polarization (Transition) Bremsstrahlung

We now consider the simplest example of polarization bremsstrahlung by an ion in a completely ionized plasma. As noted above, an ion in a plasma is an example of a "classical atomic system" in the sense that it has a "nucleus" (the ion) and surrounding electrons, which shield out the ion's field at a sufficiently large distance. As a whole, the charge of the screening electrons is equal to the charge of the ions; that is, this "atom" is quasineutral, as is an ordinary atom. The fact that quasineutrality occurs because of the statistically different plasma electrons is of no special significance in this case (the electrons are not distinguishable). The electrons have such high velocities that the shielding electrons move quasiclassically. We now discuss the bremsstrahlung of a "classical" atom of this type.

It will be clear from the discussion below (Chapter 3) that the emission from a uniform plasma, which, as in every uniform system, is caused by fluctuations, can be written as the sum of the emission from isolated particles shielded by a "cloud" of the opposite sign (ions shielded by a "cloud" of electrons and electrons shielded by a "cloud" of electron deficiencies or "holes").

By repeating the discussion of the previous section, it might seem that we could conclude that bremsstrahlung of fast electrons on a "classical" atom (an ion shielded by an electron "cloud") is given by Eq. (1.2) with $\rho_{max} \sim r_d$, where $r_d = v_{Te}/\omega_{pe} = (T_e/4\pi n_0 e^2)^{1/2}$ is the Debye radius, $v_{Te} = (T_e/m)^{1/2}$ is the average thermal velocity of the electrons, $\omega_{pe} = (4\pi n_0 e^2/m)^{1/2}$ is the plasma frequency, and n_0 is the plasma density.

This statement, however, is false. The error essentially lies in the implicit assumption that the polarization cloud surrounding the ion is static. This is not really so, since the electrons in the polarization cloud are brought into motion (made to oscillate) by the field of the incident fast electron, and this should cause additional radiation. This type of polarization is dynamic.

When $v_0 \gg v_{Te}$, different parts of the cloud may oscillate asynchronously if the wavelength of the emitted wave is less than the Debye radius ($\lambda < r_d$) and will oscillate synchronously if $\lambda > r_d$. Emission should take place in both cases, but its intensity should be greater when the cloud oscillates synchronously. Since the oscillating charge is equal to the charge of the cloud when $\lambda > r_d$ and that charge, in turn, is equal to the charge e_i on the ion, the radiated energy will be proportional to e_i^2, as in Eq. (1.2). The other factors, however, do change. The electrons in the cloud now experience an acceleration; that is, the square of the charge to mass ratio of the electron, $(e/m)^2$, now enters in the formula. The acceleration of the electrons in the cloud will be determined by the field of the incident particle; that is, a factor of e_0^2 will appear. Thus, for polarization bremsstrahlung the factor $e_0^4 e_i^2/m_0^2$ in

Eq. (1.2) must be replaced by $e_0^2 e^2 e_i^2/m^2$. If, however, the incident particle is an electron, then $e_0 = -e$ and $m_0 = m$ ($-e$ and m are the charge and mass of the electron) and these factors coincide. This means, in general, that in this case, the emission from a shielding electron in the cloud is of the same order of magnitude as the emission from the incident particle (1.2).

The logarithmic factor for the emission from the cloud should be determined by the ratio of the maximum and minimum impact parameters. It turns out to have the form $\ln v_0/\omega r_d$, which has a quite simple explanation.

When $v_0/\omega \gg \rho \gg r_d$ the physics of the radiation changes significantly as compared to the case of static shielding. The cloud acting on the particle now appears as a whole. That can happen only when $v_0/\omega > r_d$. That is, the Debye radius r_d is already serving as a minimum impact parameter. As for the maximum impact parameter, it equals v_0/ω as before. Here one mechanism for the emission is actually replaced by another: for $\rho > r_d$ the static mechanism does not operate (the field of the nucleus is shielded), but the polarization mechanism is activated. The claim that impact parameters $\rho > r_d$ do not contribute to the radiation is untrue. The above discussion shows that for polarization bremsstrahlung,

$$\ln(\rho_{max}/\rho_{min}) = \ln(v_0/\omega r_d).$$

Of course, all the above discussion is purely qualitative. An exact quantitative theory [10–13] will be presented below. Note that polarization bremsstrahlung can also be referred to as transition bremsstrahlung [12, 13], since it can be described by the transition scattering of the virtual field of the colliding particles into electromagnetic radiation (transition scattering [13, 14] is the radiation from the oscillations of a cloud of charges acted on by an incident wave).

The exact theory of polarization (transition) bremsstrahlung [10–13] yields the following result for the power radiated by a nonrelativistic fast electron on the cloud surrounding an ion at rest:

$$Q_\omega = \frac{16 e_0^2 e_i^2 e^2 n_i}{3 m^2 v_0 c^3}\left(\ln \frac{v_0}{r_d \omega} - \frac{1}{2}\right). \tag{1.5}$$

(The traditional bremsstrahlung mechanism is neglected. Strictly speaking, this is appropriate only for an infinitely heavy incident particle.) Compared with the qualitative estimate given above, this formula has a more accurate numerical coefficient in the logarithm. (The term $-1/2$ can be written in the form $e^{-1/2}$ under the logarithm.) The coefficient in front of the logarithm is the same in both the approximate and exact formulas.

The term "bremsstrahlung" for the emission from the cloud is fully justified, since the emission actually is bremsstrahlung of the electrons in the cloud. For heavy ions, when the charge of the cloud substantially exceeds that of a single electron, this type of radiation will be most effective when the coherence condition is satisfied (all the electrons are perturbed synchronously). If the coherence condition is not met, then this radiation is analogous to the emission from recoil electrons. The coherence condition has the form $\lambda > r_d$ or $\omega < \omega_{pe}c/v_{Te}$. This is also the range of validity of the formula given above for polarization (transition) bremsstrahlung.

It should, however, be emphasized that the emission of electromagnetic waves during particle collisions is a unified process, in which the radiation cannot always be separated into traditional and polarization bremsstrahlung (if one or the other of the two mechanisms does not predominate). In general, of course, they interfere. Actually, as usual, the amplitudes of processes, not their cross sections, are additive. Thus, for a light particle (an electron), the amplitude of polarization bremsstrahlung and the amplitude in the static field are partially compensated in the dipole approximation. At sufficiently high frequencies, radiation from the "bare" ion remains (cf. the "stripping" of an atom for a bremsstrahlung process in Sec. 1.4).

Although this mechanism for bremsstrahlung in plasmas is just as important as the traditional mechanism for bremsstrahlung on a static potential, it was first mentioned in the literature only in 1973 [10]. At that time [10] it was noted that separating these mechanisms was essentially inadmissible in many cases, especially when both mechanisms were important in a certain frequency range. This happens, for example, in the radiation of nonrelativistic electrons on ions.

The factors in front of the logarithms in Eqs. (1.2) and (1.5) for traditional and polarization bremsstrahlung are identical only when the incident particle and the shielding particle are both electrons. In general, the factor Λ in front of the logarithm in the formula for polarization bremsstrahlung looks different. It includes the product of the squares of the charges of the incident particle, scattering particle, and polarizing particles (in this case electrons), while the denominator includes the square of the mass of the polarizing particles (electrons):

$$\Lambda_p = e_0^2 e_i^2 e^2 / m^2. \tag{1.6}$$

Traditional bremsstrahlung is determined by the fourth power of the charge of the incident particle, the square of the charge of the scattering particle, and the mass of the incident particle. The factor in front of the logarithm has the form

$$\Lambda_T = e_0^4 e_i^2 / m_0^2. \tag{1.7}$$

Evidently, Λ_p and Λ_T are not the same in general, but can differ by orders of magnitude. This shows up most clearly for the case of an incident ion. Then $m_0 = m_i \gg m_e$ and Eq. (1.7) is less than Eq. (1.6) by roughly $(m/m_i)^2$ times. Ion collisions, therefore, involve substantial polarization emission and comparatively negligible emission in the static field of the shielded ionic charge. In addition, the dependence on the charge is different. If the charge of the incident ion is $Z_0 e$ and the charge of the scattering ion is $e_i = Ze$, then $\Lambda_p \propto Z_0^2 Z^2$, while $\Lambda_T \propto Z_0^4 Z^2$. All these points have important practical consequences.

Polarization radiation does not require the acceleration of the incident particle. Formally, the incident particle can have infinite mass; i.e., it can move along a straight line without deviating from its original direction. In that case, the emission is associated with oscillations of the cloud. The classical picture of polarization bremsstrahlung is of this type. We must, of course, also keep in mind that a small change in the particle velocity takes place during the emission of a photon of a given frequency, in accordance with the conservation of energy.

For relativistic particles ($v \to c$), the difference between the two mechanisms becomes even more striking. Traditional bremsstrahlung for ultrarelativistic particles is concentrated in a narrow cone of angles along the particle velocities, while polarization radiation, which is associated with the polarization of nonrelativistic particles, does not have this directionality. This shows that the two types of radiation differ sharply in their angular distribution. In addition, at low frequencies traditional bremsstrahlung is suppressed by the density effect [7, 12, 13] (cf. Chapter 2), while polarization bremsstrahlung is not [12, 13]. Thus, even for electron–ion collisions, there is a range of frequencies for which polarization bremsstrahlung predominates.

Finally, bremsstrahlung of an incident electron on the electrons of a plasma, each of which is surrounded by an electron cloud of the opposite sign, exists. (The cloud is made up of electrons, although it has a positive charge, since near each electron there is a deficiency of other electrons, which are repelled by the given electron; that is, so-called electron "holes" develop.) Here also, at electron velocities on the order of the thermal velocities, polarization bremsstrahlung is important, along with traditional bremsstrahlung. One further point: an incident electron acquires a polarization cloud as it enters a plasma and may produce additional radiation. True, this emission is weak for electron velocities v_0 much greater than v_{Te}.

Plasmas provide a special example of the appearance of polarization bremsstrahlung. A more general treatment has been given by Ginzburg and Tsytovich [13] based the general ideas of transition radiation [14]. There the field of the incident particle in a certain approximation can be replaced by the field of the equivalent photons. Scattering of these photons on the dynamic polarization yields polarization transition bremsstrahlung. Transition scattering refers to the scattering of electromagnetic waves in a medium on the dynamic polarization surrounding a charge [11, 13]. This dynamic polarization includes both the dynamic polarization of the individual atoms, molecules, ions, etc., and the modifications in it owing to the action of the "neighbors," including some which may not be nearest neighbors (as in the case of an ion shielded by a cloud in a plasma). The general concept of transition bremsstrahlung (as introduced in [13]) includes the scattering of equivalent photons on this complete dynamic polarization near a scattering center. In the special case where the atoms in the medium do not affect one another and the dynamic polarization of the medium is given by the sum of the dynamic polarizations of the individual atoms and molecules, we obtain so-called dynamic bremsstrahlung, the physics of which is discussed in the next section of this chapter. Here the important point is, of course, that for bound electrons, the Coulomb field of the nucleus is not a weak perturbation. Because of this fact, the theory of polarization bremsstrahlung on atoms developed independently of the theory of bremsstrahlung in plasmas, despite the relevance and importance of the preceding analysis.

Detailed discussions of polarization (transition) bremsstrahlung are contained in the book by Ginzburg and Tsytovich [13] and, below, in Chapters 2 and 3.

1.4. Polarization Bremsstrahlung on Atoms

We now examine polarization bremsstrahlung on neutral atoms and incompletely ionized ions, i.e., ions with an electronic structure (in articles devoted to this question, polarization bremsstrahlung on atoms has been referred to as dynamic and atomic bremsstrahlung). Atomic electrons, as opposed to plasma electrons, are bound in discrete electronic levels, so that the calculation of polarization bremsstrahlung on atoms is naturally a quantum mechanical problem. This problem must be solved by the methods of quantum mechanics and quantum electrodynamics (see Chapters 4–10.) We shall see below that the dynamic polarizability of an atom enters into the cross section for polarization bremsstrahlung. This is a quantum mechanical quantity and it is calculated by the methods of quantum mechanics. Even in this case, however, it is possible to use classical concepts in part to explain the radiation process itself (but not the polarizability of the atom) and they can help in better understanding the physical picture. Describing the radiation process itself as classical corresponds here to the case $\hbar\omega \ll T_i$ (where T_i is the kinetic energy of the incident particle). If this condition is satisfied, then one can replace the quantum mechanical transition flux by the classical particle flux, exactly as in the case of traditional bremsstrahlung.

In a plasma, as we saw above, polarization radiation occurs because the Debye cloud oscillates; in an atom, on the other hand, it occurs because of oscillations in the dipole moment of the atomic electrons induced by the field of the incident particle. On the basis of qualitative considerations, we can obtain a formula for polarization radiation on an atom at wavelengths greater than the characteristic size of the atom (the Bohr radius for the hydrogen atom), i.e., in the dipole approximation for the photon and in the dipole approximation for the incident particle. The dipole approximation for the incident particle means that $qR_a \ll 1$, where $\hbar q$ is the momentum transferred. Thus, in the classical picture this corresponds to impact parameters greater than the size of the atom $[\rho \sim (1/q) \gg R_a]$.

The physical picture described above can be used to derive a formula for polarization bremsstrahlung. The polarizability of the Debye electron cloud must be replaced by the polarizability of the atom, $\alpha(\omega)$; i.e.,

$$-\frac{e}{m}\frac{e_i}{\omega^2} \to \alpha(\omega). \tag{1.8}$$

Here $-e_i$ is the total charge of the electron cloud, which equals the charge of the fully ionized central ion (in the preceding discussion, we have assumed implicitly that an ion in a plasma does not have an electronic structure, that is, it is fully ionized). One must replace the Debye radius with the size of the atom (for hydrogen, the Bohr radius) in the logarithm.

Then, after making the substitution (1.8) and replacing the Debye radius by the size R_a of the atom, from Eq. (1.5) we obtain

$$Q_\omega = \frac{16\,|\,\omega^2\alpha\,(\omega)\,|^2\, e_0^2 n_i}{3c^3 v_0} \ln \frac{v_0}{\omega R_a}\,. \tag{1.9}$$

The corresponding formula for the (differential) bremsstrahlung cross section is obtained by evident steps from Eqs. (1.9) and (1.3):

$$\frac{d\sigma}{d\omega} = \frac{16\,|\,\omega^2\alpha\,(\omega)\,|^2\, e_0^2}{3c^3 v_0^2 \hbar\omega} \ln \frac{v_0}{\omega R_a}\,. \tag{1.10}$$

Strictly speaking, this formula is quantum mechanical. Planck's constant appears in it through the polarizability and the "size" R_a of the atom. Planck's constant also appears in a trivial way since the bremsstrahlung cross section has been evaluated for a single photon.

The exact theory of polarization bremsstrahlung on atoms for nonrelativistic incident particles has been developed in [15–19]. Equation (1.9) has been derived there with an exact expression under the logarithm. For many-electron atoms, evaluating the logarithm requires solution of a complicated many-electron quantum mechanical problem which actually gives the effective size of the atom for this problem. Thus, the exact value of the argument of the logarithm has so far only been determined for the hydrogen atom. At high frequencies (considerably above that corresponding to the ionization energy of all the electron shells of the atom), the electron polarizability $\alpha(\omega)$ transforms to the polarizability of free electrons. As a result, one obtains the following formula for the differential cross section for polarization bremsstrahlung:

$$\frac{d\sigma}{d\omega} = \frac{16 e_0^2 e^4 N^2}{3m^2 c^3 v_0^2 \hbar\omega} \left(\ln \frac{v_0}{\omega R_a} - \frac{1}{2} \right). \tag{1.11}$$

Here N is the number of electrons in the atom.

Although this formula differs from Eq. (1.5) only in the factor under the logarithm, the domains of applicability of Eqs. (1.5) and (1.11) are different. For Eq. (1.5) $\omega < v_0/r_d$ and for Eq. (1.11) $I/\hbar \ll \omega < v_0/R_a$. Since usually $R_a \ll r_d$, Eq. (1.11) describes radiation at substantially higher frequencies. It should be kept in mind, of course, that polarization radiation exists in a plasma for $\omega > v_0/r_d$, but it falls off with frequency as ω^{-4} (see Chapter 2). In exactly the same fashion, the polarization radiation from an atom falls off for $\omega > v_0/R_a$.

This comparison, of course, is somewhat abstract. A specific comparison will be offered in Chapter 6 for a partially ionized ion (an ion with electronic structure) in a plasma. The radiation described by Eq. (1.11) may lie in the soft x-ray range.

If we are considering an ion, rather than a neutral atom, then the charge of the electrons is not equal in magnitude to the nuclear charge and the polarization radiation is accordingly determined by the total charge of the electrons and not by the nuclear charge. Then the nuclear charge e_i in Eq. (1.10) must be replaced by the total charge of the electrons in the ion.

It should be emphasized that, for both bound atomic electrons and free plasma electrons, the traditional mechanism for bremsstrahlung does contribute to the total amplitude of the process (but not to the cross section for this process); that is,

$$A_{\text{total}} = A_p + A_T, \qquad d\sigma/d\omega \propto |A_p + A_T|^2. \tag{1.12}$$

Here A_T is the amplitude of traditional bremsstrahlung on the static field and A_p is the amplitude of the polarization bremsstrahlung. The exact formula for the bremsstrahlung cross section, therefore, contains an interference term, as well as the sum of the cross sections associated with these mechanisms. The case in which these amplitudes completely or partially cancel one another and the total cross section decreases sharply is of special interest (this is by no means an exotic case).

We now mention a number of features of polarization bremsstrahlung on atoms which make it distinct from bremsstrahlung on polarization clouds. The amplitude of polarization bremsstrahlung is resonant at the frequencies of atomic transitions. The cross section increases sharply near these frequencies. This explains some features of laser breakdown in gases (see [18] and Chapter 10). Resonance effects are also possible in a fully ionized plasma, but only at frequencies close to the plasma frequency, when the plasma "cloud" is at an effective resonance (see Chapter 2).

For an atom, both the frequency ranges $\hbar\omega \leq I$ and $\hbar\omega > I$, where I is the ionization energy of the atom, are important. For $\hbar\omega \gg I$ an extremely interesting physical effect arises in the total polarization radiation of a nonrelativistic electron on an atom. It has a simple intuitive interpretation: in the dipole approximation the amplitudes of the radiation in the static field (more precisely, the part associated with shielding) and the polarization radiation compensate one another (see Eq. (1.12)]. Thus, the radiation takes place as if the incident electron were being scattered on a "bare" nucleus. For large impact parameters, this completely contradicts the picture corresponding to shielding. In fact, when the impact parameter is greater than the radius R_a of the atom, the field of the nucleus (in the shielding approximation) is completely shielded but the bremsstrahlung actually takes place as if the electron were scattered on a "bare" nucleus. The interpretation of this effect is as follows: when $\hbar\omega \gg I$ the incident electron is actually scattered on the "free" atomic electrons. It is known from quantum and classical electrodynamics that dipole bremsstrahlung does not occur when the colliding particles have the same charge to mass ratio. Thus, in the dipole approximation, only the bremsstrahlung which originates in the scattering on the nucleus is left [16a]. Naturally, in a many-electron atom the condition $\hbar\omega \gg I_i$ (where I_i is the ionization energy of the ith shell) is met at different frequencies for different shells. Thus, as the frequency increases, there is a gradual "stripping" of the atom until, at a sufficiently high frequency ω (greater than that corresponding to the ionization energy for all the shells in the atom), bremsstrahlung takes place as if the electron were scattered on a completely unshielded, bare nucleus. Thus, the bremsstrahlung spectrum of fast electrons includes an effect owing to the "stripping" of the electronic shells in addition to the peaks associated with the dependence of the dipole dynamic polarizability of the target atoms on the photon frequency. This effect means that $\omega d\sigma/d\omega$, which for bremsstrahlung on a static potential should be roughly constant, increases discontinuously as the frequency ω rises from values considerably below the ionization potentials of any of the subshells I_i, to values considerably above them [20, 21] (see Chapter 7). The magnitude of the jump is proportional to the square of the number of electrons in a given subshell.

The scattering cross section is determined by the square of the sum of the amplitudes of Thomson and transition scattering [13, 14]. The scattering spectra for electrons and positrons moving at high velocities are, therefore, different when polarization (transition) scattering is taken into account [14]. Thus, the total bremsstrahlung is different for electrons and positrons when the polarization term is included, even in the first Born approximation [16].

It is evident from the preceding discussion that the term "bremsstrahlung" is fully justified with respect to the polarization effect. In fact, the correct physical picture of bremsstrahlung on atoms at high frequencies is obtained only if both effects (traditional and polarization) are taken into account together. Of course, it is not by chance that the authors who first examined polarization radiation in collisions referred to it as bremsstrahlung (transition bremsstrahlung [12], dynamic bremsstrahlung [22], atomic bremsstrahlung [17]).

At high frequencies, therefore, the shielding picture is replaced by a picture of scattering on the nucleus and on every atomic electron. The partial amplitudes of each scattering event combine into the total bremsstrahlung amplitude. This makes it possible to eliminate a particular difficulty in the theory of bremsstrahlung of relativistic particles on atoms (the Bethe–Heitler theory). This difficulty arises because the approximate expression for the Hamiltonian does not yield a limit corresponding to the case of a nuclear charge of zero in the bremsstrahlung problem [22] (of course, this case is not realized, since then there would be no bound electrons, but such a transition can be made formally). For concreteness we shall consider bremsstrahlung on a hydrogen atom. It is clear that when the nuclear field vanishes ($Z = 0$), scattering of the incident particle on the free electron still remains. Thus, the general formula for the bremsstrahlung cross section should formally transform to the formula for bremsstrahlung during collisions of incident particles with a free electron. This limit cannot be obtained in the Bethe–Heitler theory; this is prevented by the approximation used for the Hamiltonian (the shielding approximation). An exact approach allows us to obtain the limiting transition to the case $Z \to 0$. In order to do this, we have to include bremsstrahlung processes both without excitation and with simultaneous excitation of the atom. This yields a formula which transforms to the standard formula for bremsstrahlung during scattering of an electron on an electron when $Z = 0$ (when the incident particle is an electron).

Here we should emphasize that there are different kinds of electron collisions with atoms that cause emission of a photon: first, when the atom does not change its state as a result of a collision and second, when the atom changes its state (is excited or deexcited) as a result of a collision in which a photon is emitted. We shall refer to the latter process as inelastic bremsstrahlung. Up to now we have considered the elastic effect. In this consideration during an interaction, only a virtual change in the states of the electrons takes place. This actually reduces to having the atomic polarization induced by the incident particle. In its final state, the atom acquires momentum as a whole and remains in its initial electronic state.

The same thing happens in a plasma: an ion gains momentum, but, until it "readjusts" to the final velocity of the ion, the cloud has a variable polarization and radiates. It can be said that in these cases, the "classical atom" and a conventional atom do not change their electronic state during the radiation process. This is elastic bremsstrahlung. (Here we are considering elasticity with respect to the electronic state of the atom.)

Inelastic processes, however, can also occur during emission of bremsstrahlung photons. They cannot, of course, be described in the shielding approximation, but this can be done with a rigorous theory that includes the dynamic polarization (see Chapters 4–8 and 11). These inelastic processes, in turn, can be of two types: those in which an atomic electron is excited to a discrete level and those in which an atomic electron is removed by ionization. The total cross section for polarization bremsstrahlung is obtained by summing over all excitation processes. Elastic bremsstrahlung predominates at relatively low frequencies. At high frequencies a radiation process accompanied by ionization of the atom is more probable.

How does the redistribution of the energy and momentum of the particles take place during bremsstrahlung? Because of the difference in the dispersion laws for the incident particle and photon, only part of the change in the momentum of the incident particle is carried away by the photon. During an elastic process, all (if the atom is infinitely heavy) of the change in the energy of the incident particle is transferred to the photon, while practically all of the change in the momentum of the incident particle is transferred to the atom as a whole. At high frequencies, the situation is different: "inelastic" processes become more probable. Now the excess momentum is transferred to the electron that is ejected from the atom. Radiation, of course, does take place because of oscillations in the dipole moment of this electron, as before. This process should be distinguished from the radiation by so-called δ-electrons. In the case considered here, the ejected electron radiates in a collision with "its own" atom, but, of course, it can emit a bremsstrahlung photon during a collision with other atoms of the material. The latter process is emission by δ-electrons. It can also be said that these are direct and cascade processes. Here the difference between the processes is utterly obvious.

The conservation of energy also shows up differently in inelastic processes. A substantial part of the energy of the incident particle is transferred to the electron during ionization. When the incident particle is scattered by excited atoms, the maximum bremsstrahlung frequency is also changed. The following process becomes possible: all the kinetic energy of the incident particle T_i, and the deexcitation energy of the atom (the energy ΔE of the atomic transition from the excited state into the ground state) are transferred to the photon, that is,

$$\hbar\omega_{\max}=T_i+\Delta E.$$

The bremsstrahlung of electrons (and positrons) on the simplest atoms (positronium [23] and meso-hydrogen atom [24]) is of some interest. Positronium differs from ordinary hydrogen in lacking a static field associated with the nucleus, so that the bremsstrahlung that takes place during the scattering of fast particles on it is completely of the polarization type (atomic). As opposed to the case of hydrogen, when fast but nonrelativistic positrons or electrons are scattered on positronium, the inelastic bremsstrahlung processes are of the same order as the elastic processes, even at high frequencies.

Traditional bremsstrahlung, i.e., radiation in a static field, occurs only when charged particles collide. As for polarization bremsstrahlung, it can also occur during collisions of neutral atoms. The essence of this effect lies in the fact that the electronic structure of the atom is taken into account. Strictly speaking, since a neutral atom consists of charged particles, scattering of charged particles is involved when atoms collide. Bremsstrahlung occurs because the partial amplitudes of the bremsstrahlung from the individual particles do not always cancel out (they do cancel out, for example, during collisions of identical nonrelativistic atoms). An analogous situation arises in plasmas during collisions of two ions surrounded by clouds of electrons, i.e, of two seemingly classical atoms (see Chapter 2).

1.5. Polarization Bremsstrahlung on Atoms and Partially Ionized Ions in Plasmas

The case of partially ionized ions in plasmas is of importance for practical applications. This is of direct relevance to the emission from low-temperature plasmas. Here we encounter two simultaneous polarization effects: polarization bremsstrahlung of the shielding clouds and polarization bremsstrahlung of bound electrons in ions or atoms. The charge of the shielding cloud supplements the charge of the bound electrons so that, together, they completely shield the charge of the nucleus. One can say then that plasmas contains a unique kind of "atom" in which the bound electrons form a quantum system and the plasma electrons, a classical one. A positively charged polarization cloud, corresponding to a deficiency of electrons (or to the presence of seemingly positively charged holes), develops around negatively charged ions.

According to general principles, both polarization processes and the traditional radiation contribute to the total bremsstrahlung amplitude. Thus, corresponding interference terms show up in the total cross section. This sort of interference occurs in dense plasmas. In an ordinary, rarefied plasma, the polarization effects on bound and plasma electrons show up in different frequency bands. The general nature of these effects is distinctly clear from the fact that, under certain conditions, the formula for the cross section on bound electrons of an ion in a plasma transforms to the corresponding formula for emission on the plasma electrons of the cloud [25] [see Eqs. (6.51) and (6.46) of Chapter 6].

As for the different inelastic radiation processes involving the atoms and ions in a plasma, they can be significantly affected by polarization processes. Thus, recombination radiation is possible not only from an incident particle but also from an electron in the cloud. If one of the bound electrons is removed from an atom or ion as a result of an interaction with an incident particle, then polarization radiation from near the ejected electron is significant (it is "enveloped" by its own polarization cloud).

We conclude this section by emphasizing the important general properties of polarization bremsstrahlung in plasmas and on bound electrons (these will be discussed in more detail later).

1. The radiation from heavy nonrelativistic incident particles at low frequencies is primarily polarization radiation.

2. Polarization radiation from relativistic particles predominates at sufficiently low frequencies because ordinary bremsstrahlung is suppressed by the density effect (this applies to both light and heavy particles).

3. For relativistic incident particles, the emission at large angles to the direction of motion of the particles is mainly polarization bremsstrahlung, since traditional bremsstrahlung is concentrated in a narrow cone of angles around the direction of motion of an incident particle.

4. There is a close analogy between polarization bremsstrahlung during collisions of neutral atoms and ions and of electrons or ions in plasmas. Indeed, since it is surrounded by a shielding cloud, a charged particle in a plasma is a quasineutral structure. Thus, even in plasmas, collisions do take place among unique sorts of neutral atoms. The appearance of bremsstrahlung during collisions of neutral atoms at first seems paradoxical, but, as the above discussion made clear, it is the structure of these particles which is important here, so that the process actually involves scattering of the charged particles contained in these neutral particles. Then, as we have noted, the scattering amplitudes of the individual charged

particles contained in these quasineutral structures do not always cancel out (see [13, 21, 26–37] and Chapter 7).

1.6. Polarization Bremsstrahlung of Nuclear Particles

The theoretical study of polarization bremsstrahlung in collisions of nucleons with nuclei goes back to the work of Hubbard and Rose [38]. The polarization of a nucleus during collisions with nucleons and other nuclides plays an important role in the development of the bremsstrahlung spectrum of γ-rays with energies on the order of the binding energy between nucleons and nuclei.

As applied to nuclear collisions, polarization bremsstrahlung is understood to mean the emission of γ-rays as a result of the polarization of either the target or the incident particle. In this case, traditional bremsstrahlung is radiation by a proton or an undeformed nucleus in the static field of a target nucleus. Polarization bremsstrahlung generates peaks in the bremsstrahlung spectrum of nucleons on nuclei at energies corresponding to the so-called giant nuclear resonances (dipole and quadrupole). In terms of the terminology used here this is nothing other than resonance polarization bremsstrahlung. The specific feature of bremsstrahlung in nuclear collisions shows up through the large contribution of quadrupole photons to the total bremsstrahlung spectrum.

At frequencies close to the gigantic resonances (unlike in atoms, nuclear discrete excitations have a very large width, on the order of the energy of the resonance itself), polarization bremsstrahlung exceeds traditional bremsstrahlung by almost an order of magnitude. Since the interactions of incident protons and neutrons with nucleons in a nucleus are roughly the same, the polarization induced by them in a target nucleus and, therefore, the polarization bremsstrahlung of a proton and neutron are of the same order of magnitude. As in atomic collisions, the contribution of polarization bremsstrahlung increases as the mass of the incident nucleus is raised. Identical nuclei (or, more precisely, colliding pairs with equal charge to mass ratios, such as a deuteron and an α-particle) do not have a dipole moment, so that only quadrupole photons are emitted.

Theoretical studies of the contribution of polarization bremsstrahlung to the total spectrum of the radiation from collisions of nuclear particles are difficult because the internucleonic interaction is considerably more complicated that the interelectronic interaction, and there is no expression in closed analytic form for that interaction which is valid over a wide range of energies of the colliding particles.

Polarization bremsstrahlung is probably, in fact, the explanation for the wide maximum in the C^{12} spectrum in the energy range of the "gigantic" quadrupole resonance that is observed in collisions of both comparatively slow [39, 40] and fast, relativistic, C^{12} nuclei. The fact that the ordinary mechanism for bremsstrahlung, which neglects the polarization of the nucleon distribution within nuclei as they collide, cannot explain the experimental data has been pointed out by Nyman [41].

The contribution from polarization bremsstrahlung in collisions of μ-mesons with nuclei should be extremely large. Charged mesons can efficiently polarize a target nucleus, especially when the energy transfer lies within the region of the "gigantic" dipole resonance and the total spectrum then acquires a significant peak at that energy. The contribution of polarization bremsstrahlung should also be great when π-mesons, including π^0-mesons, are scattered on nuclei. Here the polarization of a π^0-meson by a nucleus must be taken into account, as well as the polarization of the nucleus by the meson.

As for mesic hydrogen, bremsstrahlung is the only reaction channel through which an incident electron can lose energy during scattering up to the threshold for excitation of the target atom.

Polarization bremsstrahlung should also be important in collisions of other structured particles (nucleons and mesons) with one another if we regard them as being made up of quarks and becoming polarized during the collision process.

The creation of gluons (the quanta of the chromodynamic field in the physics of strong interactions) is essentially similar to the polarization bremsstrahlung of neutral particles. The similarity is, in many respects, determined by the fact that, as with a photon in collisions of neutral atoms, a gluon is emitted only by a component of the elementary particle itself, i.e., a quark (or another gluon). Polarization bremsstrahlung should also be important for emission within a medium and not only during individual collisions of isolated pairs of particles (because of the polarization of the "neighbors" in the target, as in the case of transition bremsstrahlung). The process is more complicated within a medium, however, because of the superposition (coherent or incoherent, depending on the properties of the medium) of the radiation from individual collisions.

1.7. Experiments and the Theory of Polarization Bremsstrahlung

Polarization bremsstrahlung is, in part, a relatively new research topic. Thus, the number of experiments which have been interpreted as a manifestation of this effect is still small. We now discuss some of them.

Laser Breakdown in Alkali Metal Vapors. The basis of the avalanche theory of laser breakdown, which was first proposed by Zeldovich and Raizer [27] and is now generally accepted, is the concept of inverse bremsstrahlung. Inverse bremsstrahlung is the absorption of a photon by an electron when the electron is scattered in the field of an atom or ion. The absorption and bremsstrahlung are related by the Einstein relation. It seems that the traditional theory of bremsstrahlung yields a cross section which will not provide agreement between the avalanche theory of laser breakdown and experiments in alkali metal vapors (Rb, Cs). Calculations [18] show that the contribution of the polarization amplitude at the laser frequency to inverse bremsstrahlung is two orders of magnitude greater than that of ordinary inverse bremsstrahlung in a static field. Once this fact is included, the avalanche theory of breakdown is consistent with experiments on laser breakdown in these gases [18].

Scattering of Electrons on Atoms. The polarization mechanism has been used to explain a broad band in the x-ray emission spectrum in solid targets of barium, lanthanum, and cerium at frequencies corresponding to ≈100 eV [28, 29]. Since the photoabsorption corresponding to this band is known, one can obtain the polarizability corresponding to this transition (to do this one uses the optical theorem and the dispersion relation). The cross section for polarization bremsstrahlung has been calculated from the known polarizability. The theoretical and experimental dependences of the bremsstrahlung cross section on the photon frequency are in agreement. An analogous study has been made [30] for an inert gas (xenon).

We now consider this question in more detail. The point is that the contribution of polarization bremsstrahlung is large, even for intermediate energies of the incident electron (tens and hundreds of electron volts), when the energy of the photon is close to the ionization potential of one of the atomic shells. Polarization bremsstrahlung becomes especially important if the energy of the electron in the final state is low. This mechanism has been used [28] to explain a sharp increase in the photon yield at the edge of the spectrum in solid barium, lanthanum, and cerium bombarded by electrons with energies close to the threshold of the $3d^{10}$-shell of these elements.

The formula obtained in [18, 29] for the total bremsstrahlung cross section of fast electrons has been used to interpret experimental data on the emission of photons during electron bombardment of metallic lanthanum at photon energies close to the ionization potential of its $4d^{10}$-subshell [31]. Cross sections for bremsstrahlung (of course, including polarization bremsstrahlung) have been obtained for a number of atoms (argon, xenon, and lanthanum) [32, 32a] in the Born approximation for the incident electron. To very high accuracy, the total bremsstrahlung spectrum, $d\sigma/d\omega$, is simply the sum of the contributions from traditional and polarization bremsstrahlung. In this case, the contribution to the total spectrum from the interference of the amplitudes of these processes is small. This is not surprising, since the contributions from the two are determined by different impact parameters. The comparatively long-range (much greater than the radius of the atom) dipole interaction of the incident particle with the electrons in the target atom produces polarization bremsstrahlung, while interactions over distances on the order of or less than the radius of the atom produce ordinary bremsstrahlung, which originates in electron scattering in the shielded static field of the nucleus. If, however, the atom has a significant quadrupole moment, then the contribution of traditional bremsstrahlung will also be determined by distances in excess of the size of the atom. Then the interference of polarization and traditional bremsstrahlung will be important in the total radiation spectrum. It turns out that the contribution of polarization bremsstrahlung to the total bremsstrahlung spectrum is generally no less than that of traditional bremsstrahlung [29, 32, 32a]. These calculations have been used to explain the experimentally observed emission peak at energies close to the ionization potential of the $4d^{10}$-subshell of xenon when it is bombarded by electrons with energies of about 1 keV [30].

Proton Bremsstrahlung. During proton bombardment of solid targets, a low-frequency background develops which has long attracted the attention of investigators [33–36]. It is usually explained as bremsstrahlung of secondary electrons that have been knocked out by protons in the target [33] or in terms of a number of other effects. Ishii and Morita [37] have calculated the cross section for polarization bremsstrahlung during scattering of protons on aluminum atoms (the calculation was done in the Born approximation, as in [19, 21]). They showed that for protons with energies of 1 MeV, agreement between theory and experiment can be attained at photon energies of roughly 2–5 keV by including polarization bremsstrahlung. In this energy interval, the polarization radiation is roughly an order of magnitude greater than the other forms of x-ray background emission.

Although these examples show that interesting manifestations of polarization bremsstrahlung have already been observed, further experimental studies are obviously needed.

Here, of course, we have omitted polarization bremsstrahlung during nuclear collisions, which was discussed in the previous section.

1.8. Some Historical Comments

We conclude with a discussion of how the concept of polarization bremsstrahlung arose. The first papers, which provided an impetus for further work, appeared in the beginning and middle of the 1970s [10–12, 15–18]. In this chapter we have conducted a parallel discussion for atoms and plasmas, so as to emphasize the physical generality of polarization bremsstrahlung on atoms and on ions in plasmas. In reality these effects were studied independently of one another for a long time. The initial considerations which led to the concepts of polarization transition bremsstrahlung and polarization (atomic, dynamic) bremsstrahlung on atoms were also different.

We shall first discuss radiation produced by scattering on atoms. Here the starting point was a more exact formulation of the bremsstrahlung problem. The shielding approximation, which was originally used to solve this problem, is based on a certain physical model in which a nucleus is shielded by the electrostatic charge of the atomic electron cloud. This model, of course, does not correspond to the exact Hamiltonian for the problem, but to an approximation. Everyone understood that this was an approximation, but somehow it was implicitly assumed that this approximation would not result in large errors. An exception to the general misunderstanding of the situation was the the early work of Landau and Rumer [42], who derived the proper correction for a special case almost intuitively. The shielding approximation is so popular because of its clarity, as well as because of the fact that it is exact for elastic scattering. As far as we know, the need to use the exact Hamiltonian was first pointed out by Percival and Seaton [43]. Their specific calculations, however, were done only for the resonance region. The most characteristic range of x-ray frequencies for bremsstrahlung corresponds to the condition $\hbar\omega \gg I_i$ and is nonresonant, at least for the outer shells of an atom. Thus, despite an obvious inaccuracy in the statement of the problem, it is clear that no one wanted to solve it exactly for high frequencies. In all the textbooks and monographs known to us, the bremsstrahlung problem is solved in the shielding approximation. Just how deeply ingrained the habit of accepting the shielding approximation is, can be seen, among other places, in the classic book by Mott and Massey [44]. They discuss Percival and Seaton's article in detail in connection with the polarization of the resonance radiation produced during electron–atom collisions. That the article has any connection at all with the theory of bremsstrahlung is completely unmentioned.

The situation has changed significantly only since the middle of the 1970s, when it was demonstrated by numerical calculations for hydrogen [16] and more complex atoms [17] that the new formulation of the bremsstrahlung problem also changes the qualitative results outside the region of the exact resonance [43].

Subsequent work on polarization bremsstrahlung on atoms has also revealed another important fact: without an exact statement of the problem, at high frequencies ($\hbar\omega \gg I_i$) one cannot even obtain a correct qualitative picture of bremsstrahlung, much less correct quantitative results. This has shown up especially clearly in two effects, bremsstrahlung by electrons in distant collisions with atoms [16a] and bremsstrahlung by protons [19]. In both cases, the physical picture changes radically. The correct picture can be obtained only when the total amplitude of the radiation includes the contributions from scattering on the static field of the shielded nucleus and from polarization.

Clarification of the physical mechanism has also been aided by some papers [17, 18, 18a] where bremsstrahlung is interpreted in terms of the polarization mechanism. Spontaneous bremsstrahlung can be represented as the result of the polarization of an atom by the field of an incident particle (see Chapter 7). Stimulated bremsstrahlung develops as a result of the dynamic polarization of an atom by an external electromagnetic field and of the scattering of the incident particle by the induced atomic dipole. This graphic interpretation was offered by Zon [18]. Ultimately, any of these mechanisms is suitable for explaining either spontaneous or stimulated polarization radiation, since they are coupled by the Einstein relations.

It is evident from the above discussion that polarization bremsstrahlung on the shielding clouds of the particles in a plasma and polarization bremsstrahlung on atoms have fundamentally the same physical mechanism. This question has been examined in more detail by Astapenko [25] and in Chapter 6 of this book. It is of independent interest, since both effects (polarization bremsstrahlung on the electrons in the cloud and polarization bremsstrahlung on the bound electrons of atoms and ions) occur during scattering on ions or atoms with electronic structure situated in plasmas. The common nature of plasma and atomic bremsstrahlung manifests itself especially clearly when, under certain conditions, the bound electrons of a plasma ion and the electrons in the Debye shield contribute equally to the amplitude of the radiation. In both cases, radiation is caused by the dynamic polarization induced by the field of an incident particle.

Studies of polarization radiation on the "clouds" surrounding particles have undergone a somewhat different historical development. As noted above, any particles in an arbitrary medium will have polarization "clouds," and for real targets, the "cloud" produces additional polarization radiation. The polarization radiation from clouds of particles in plasmas also attracted attention only at the beginning of the 1970s [10–12]. The phenomenon of Debye shielding of charges situated in a plasma has essentially been known for a very long time, but this is static shielding, which produces no radiation. In addition, as we shall see below, it is clear from the example of plasma physics that polarization radiation from clouds is a more complicated physical effect than polarization radiation on isolated atoms, as it is a collective effect. This sort of collective effect must, in general, be taken into account in interpreting experiments on the radiation from particles incident on solids made up of neutral atoms as well as on plasma targets.

Let us clarify this point using plasmas as an example. If we treat a plasma as a continuous medium with average characteristics such as density, temperature, etc., that are constant in space and time, then an incident particle will not meet any electrodynamic inhomogeneities and will not radiate at all. Thus, radiation is related to fluctuations. This has also long been known, especially for scattering processes. Rayleigh scattering on atoms has long been treated as scattering on fluctuations. In a certain approximation, scattering on fluctuations in an ideal gas reduces to the sum of the scattering on the individual atoms [45]. Examples exist, however, in which this additivity does not occur. The polarization of the surrounding particles has an effect. In a plasma this polarization of the other particles is decisive, rather than of secondary importance.

Dynamic polarization phenomena during scattering in plasmas were first reported by Dougherty and Farley and by Rosenbluth and Rostoker [46], although these results have only recently been interpreted [47]. At the same time, results from the theory of scattering on plasma fluctuations have been used in an enormous number of experiments on laser diagnostics of the temperature of plasmas.

The nontriviality of the result obtained from the theory of fluctuations for the scattering of electromagnetic waves lies in the following: suppose that an additional charge e_i is inserted in a quasineutral plasma. It will be shielded at the Debye radius by the static potential created by other plasma particles (of charge $-e_i$), while a charge e_i of plasma particles will go to infinity. During an interaction with electromagnetic waves, the shielding charge will begin to oscillate and create a dynamic polarization which will radiate. This effect subsequently became known as transition scattering [14, 13] and the term "nonlinear scattering" [48] has also been used, since the oscillations of the cloud are the result of a nonlinear interaction of the plasma electrons with two fields, that of the charge e_i and that of incident wave.

The situation would seem to be fairly simple for an external charge (inserted into a plasma).

The situation is more complicated for fluctuations. Now, all charges are equivalent and all participate in the scattering. Every particle participates in both the averaged and the fluctuating motion, so that each particle is a scattering center and simultaneously participates in the creation of a dynamic polarization cloud around a scattering center. What the result will be in this case is not obvious at first glance. According to the interpretation given by Ginzburg and Tsytovich [47], the result obtained first by Rosenbluth and Rostoker indicates that scattering of electromagnetic waves on the fluctuations in a uniform plasma proceeds as if the plasma consisted of a set of electrons surrounded by their own polarization clouds and ions surrounded by their clouds. The number of such "dressed" electrons and ions is equal to the number of "free" electrons and ions in the plasma. All of the particles are effectively combined into complexes, but the number of these complexes is equal to the original number of particles. This can happen only if each individual particle participates in two types of microscopic motion during the fluctuations: i.e., it acts as a scattering center and as a particle in a polarization cloud.

This complicated organization of the motion plays a fundamental role, especially during scattering with a small frequency change (the latter because of Doppler shifting as the particles move).

Scattering on a polarization cloud has, as noted above, come to be referred to as transition scattering [14]. In a number of cases (during scattering on heavy particles) transition scattering predominates, since heavy particles and ions move little in the field of an incident wave [14, 48]. For light particles (electrons), the amplitude of transition scattering is almost equal and opposite in sign to the amplitude of Thomson scattering. A paradoxical situation arises in which the cross section for scattering on ions in a plasma becomes equal to the cross section for Thomson scattering on electrons, while the cross section for scattering on electrons is much smaller than the Thomson scattering cross section (equal to zero in the dipole approximation).

This was all clarified considerably later, roughly in the mid 1960s, when a nonlinear plasma theory was developed for the scattering of (Langmuir) plasma waves. The nonlinear interaction of the waves is related to stimulated scattering on particles [48]. Equations were first derived for the nonlinear interaction of random plasma waves [49] (from the theory of fluctuations). Later it was found that these equations could be derived from the concepts of stimulated scattering on particles if both Thomson and transition scattering were included in the scattering amplitude [50]. Still later, methods were proposed for calculating spontaneous scattering in the theory of fluctuations [51] and it was found that scattering on fluctuations yields a result that corresponds to the sum (in the scattering amplitude) of transition and Thomson scattering. And only comparatively recently has attention been paid to the fact that the results of Rosenbluth and Rostoker for scattering of laser light on fluctuations correspond exactly to the sum of transition and Thomson scattering.

The fact that in all cases when scattering on ions is determined by the dynamic polarization of the cloud (transition scattering), the ion receives the recoil momentum during scattering, is of extreme importance for applications in plasma physics (in particular, for the nonlinear interactions corresponding to stimulated scattering). The cloud, however, is motionless and it seems somewhat surprising that the light particles (the electrons in the cloud) on which the scattering takes place can transfer momentum (and energy) to the ions. The answer is simple: until the cloud has readjusted to the final state of the ion, it oscillates and the radiation process cannot be regarded as finished. In its initial and final states, the ion has a fully determined momentum (velocity) and its cloud is uniquely determined by its velocity. Thus, the correct results for transition and conventional Thomson scattering for individual external particles inserted in a plasma add up. Scattering on fluctuations corresponds to the sum of the scattering on such "dressed" particles. They can be regarded as an arbitrary test particle in the plasma. These ideas are the foundation of modern plasma physics, which was laid toward the end of the 1960s.

We now return to bremsstrahlung. In order to take the first step toward observing the new mechanism for bremsstrahlung known as transition bremsstrahlung, it is sufficient to rely on two facts. First, the probability of bremsstrahlung on a particle in vacuum can be written as the product of two factors: the square of the probability amplitude of the spatial Fourier component of virtual Coulomb field of the colliding particles and the probability of Thomson scattering of the virtual wave into a propagating electromagnetic wave (i.e., the wave on the light cone [52]). The second fact is that the probability of scattering in a plasma is determined by the sum of the amplitudes of Thomson and transition scattering. From this we can conclude at once that a new mechanism for bremsstrahlung must exist, which is associated with the transition scattering of a virtual wave into a real wave. The amplitudes of the ordinary effect and this new effect will interfere and form a total bremsstrahlung amplitude. This physical picture had become clear by the end of the 1960s, when the physics of transition scattering in plasmas described above was discovered.

In order to prove the validity of this picture, however, it was necessary to examine the entire question from the standpoint of fluctuations, since it is fluctuations which determine the radiation originating in the interior of a plasma. The first proof was given by Tsytovich [10] for the case of bremsstrahlung of longitudinal waves. He showed that the theory of fluctuations contributes two terms to the probability amplitude for bremsstrahlung of longitudinal waves owing to Thomson and transition scattering of virtual waves. Rather more effort was required to generalize this result to waves with arbitrary polarization in a plasma and to nonequilibrium particle distributions [11]. After this, real confidence developed in the physical correctness of this picture and in the possibility of using it for calculations of the actual bremsstrahlung per unit volume of plasma. The case of relativistic particles, where ordinary bremsstrahlung is suppressed by the density effect [53, 54], was of particular interest.

For a long time no unified method of calculating transition bremsstrahlung without fluctuation theory was available. Such a method was proposed only by Ginzburg and Tsytovich [13]. In this way a cycle of research seemed to have been completed: the bremsstrahlung of individual particles and its correspondence with the theory of fluctuations had been examined (the same cycle of problems which had been the object of intensive study in the theory of scattering in plasmas).

We now return to the relationship between transition bremsstrahlung and polarization bremsstrahlung on atoms. A quantum mechanical approach has been used in the theory of radiation on atoms. A comparison of transition and polarization bremsstrahlung on atoms was difficult because different mathematical apparatus had been used to study them. The results [25] of a quantum mechanical description of transition bremsstrahlung in plasmas by the methods used to calculate radiation on atoms were in exact agreement with [12]. Those methods [25] also yielded something new – it became possible to examine bremsstrahlung on an ion with electronic structure. In this case, both polarization transition and polarization atomic bremsstrahlung (that is, bremsstrahlung on free and bound electrons of the ion) play a role.

In the above remarks we have discussed the considerations that led the authors of this book to the concept of polarization bremsstrahlung. In concluding, it seems appropriate, once again, to recall the work that preceded the creation of the concepts to be discussed here and which, to a greater or lesser degree, is related to this area of research.

With regard to radiation on atoms and nuclei, we would first like, once again, to acknowledge a debt to the work of Percival and Seaton [43] and Hubbard and Rose [38]. Percival and Seaton treated an atom and incident electron as a single system that emits a photon. They made the very penetrating comment that their work is a generalization of the theory of bremsstrahlung. Now it is difficult to say why the work of Percival and Seaton was not developed further, as it undoubtedly deserved. This is all the more so, because the inadequacy of the shielding approximation was clear even in the work of Landau and Rumer [42] and, as applied to the nucleus, in that of Hubbard and Rose [38].

The following circumstance may have played a role in this course of events: a study of resonant bremsstrahlung outside the region of the exact resonance would have required well-developed methods for calculating the sums over intermediate atomic states, as well as the associated methods for calculating the atomic polarizabilities. A rigorous justification of the formulas for the cross section for polarization bremsstrahlung on atoms at high frequencies also requires methods of this type. For the most part, substantial experience in applying these methods was only gained later [55–57].

As for transition bremsstrahlung, we must first acknowledge the work of Ginzburg and Frank [58] on transition radiation. This type of radiation occurs because the polarization surrounding a charge varies as the charge passes through inhomogeneities in a medium. These inhomogeneities might be the "clouds" surrounding particles or isolated atoms. Polarization bremsstrahlung and polarization radiation on atoms fit into a general scheme of transition scattering of virtual waves associated with particles [1, 13]. Naturally, we must acknowledge the fundamental work of Tamm and Frank [59] on the theory of Cerenkov radiation. It might seem that polarization bremsstrahlung, which takes place as an incident particle with a constant velocity passes near an isolated polarized atom, is an elementary event (on a microscopic level) that ultimately yields Cerenkov radiation.

In this case, naturally, the atoms must be polarized (here we avoid the fact that the polarizability of a medium is not strictly equal to the sum of the polarizabilities of the atoms), but Cerenkov radiation, as noted clearly in the paper by Tamm and Frank, is coherent. If we leave out the coherent component, then the fluctuating component will be transition bremsstrahlung, as shown by Kapitsa [60] and, especially clearly, by Ginzburg and Tsytovich [13]. (We note, by the way, that radiation during the scattering of fast charged particles on fluctuations was first examined by Kapitsa [60] and Ter-Mikaélyan [61]); refined versions of Kapitsa's formulas [60] are given in the book by Ginzburg and Tsytovich [13].) Naturally, when the conditions for Cerenkov radiation are not satisfied, only transition bremsstrahlung remains.

It might be appropriate to end this review with Dirac's comment to the effect that science develops in stages but no single stage is the last.

Chapter 2

POLARIZATION BREMSSTRAHLUNG IN DISTANT COLLISIONS

V. N. Tsytovich

2.1. Statement of the Problem

In ordinary bremsstrahlung, near collisions lead to the appearance of rather high frequencies, whereas distant collisions yield rather low frequencies. Since, as we have noted, polarization bremsstrahlung is most efficient at relatively low frequencies, it becomes necessary to examine the effects of distant collisions. Perturbation theory can be used to study them under the assumption that an incident particle produces a small perturbation in the polarization cloud of a target particle (see [1]). For the time being, we shall not specify whether this cloud is produced by bound electrons, as in a neutral atom, or by free electrons, as in a plasma.

We shall assume that the dynamic polarizability of the target particle is known and that the dynamic polarizability of the incident particle is zero or negligibly small (a medium always has a dynamic polarizability, but it can be neglected for incident particles with high enough velocities). In addition, having assumed that the frequency is rather low, we shall use the classical theory of radiation (we shall neglect quantum mechanical effects in radiation processes, but not in the polarizability of bound electrons).

The problem is then to find the alternating current created by the polarization cloud of the particle (atom) and to calculate the intensity of the radiation from this current. We shall use Landau's method to evaluate the intensity of this radiation. In general, that method can be used to find the radiated power, including when the radiation is not pure dipole radiation. In many cases it is enough to limit ourselves to the dipole approximation, but this is not always justified for ultrarelativistic particles. Landau's method was first used to calculate the Cerenkov radiation emitted by a charge moving faster than the speed of light in a medium. It involves calculating the the work done by the radiation field on the current that creates it using the equation

$$Q = -\int \mathbf{Ej}\, d\mathbf{r}, \tag{2.1}$$

where Q is the total radiated power and $\mathbf{E}$ is the radiation field of the current $\mathbf{j}$.

In the absence of dissipation, the radiated power is equal (with a negative sign) to the work done by the radiation field on the current that produces it. This fact was used by Landau in studying Cerenkov radiation and is the simplest way of deriving the Tamm–Frank formula [2]. Equation (2.1) is purely classical. Thus, we shall not use any concepts from the quantum theory of radiation here. As for the current $\mathbf{j}$, it may also be related to perturbations in the quantum mechanical motion of the bound electrons. In fact, there is no inconsistency here if we speak in terms of a radiation effect averaged over many sources (or many independent sources) and neglect the quantum mechanical fluctuations in the radiation.

Of course, the validity of Eq. (2.1) is restricted by the condition

$$\hbar\omega \ll T_i, \tag{2.2}$$

where ω is the frequency of the emitted light and T_i is the kinetic energy of the incident particle. Questions relating to the quantum theory and a number of related details will be discussed separately below, but here we limit ourselves to the method just described.

The Landau method has been used successfully to calculate transition radiation [3] and other classical particle emission processes, such as polarization (transition) bremsstrahlung [1]. In the latter case [1], expressions for the nonlinear currents were used. In the following analysis we shall use a somewhat simplified treatment without introducing nonlinear currents. In doing this we lose generality, but there is a gain in clarity. In Chapter 3 we shall show how polarization (transition) bremsstrahlung can be related to the nonlinear polarizability.

Before discussing these questions, we shall show how Eq. (2.1) can be used to obtain the ordinary bremsstrahlung power. First, we examine the radiation from a light incident particle in the field of randomly positioned heavy ions in a way that differs slightly from the method demonstrated in Chapter 1. Let ρ_0, z_0 be the cylindrical coordinates of an ion relative to the trajectory of the incident particle (with velocity v_0 along the z axis) up to time $t = 0$. In the approximation where deviations from the initial trajectory owing to the accelerating force can be neglected, the acceleration experienced by the incident particle on this ion has the form ($v_0 \ll c$)

$$\ddot{\mathbf{r}} = -\frac{e_0 e_i}{2\pi^2 m_0}\int \frac{i\mathbf{k}}{(\varkappa^2 + k_z^2)}\, dk_z\, d\varkappa \exp\left(-i\varkappa\rho_0 - ik_z z_0 + ik_z v_0 t\right). \tag{2.3}$$

Here we use the fact that the field of the ion has the form ($k = \{k_z, \kappa\}$, $\kappa = \mathbf{k}_\perp$, and the z axis is directed along v_0)

$$\mathbf{E} = \frac{e_i(\mathbf{r}-\mathbf{r}_0)}{|\mathbf{r}-\mathbf{r}_0|^3} = -\frac{e_i}{2\pi^2}\int \frac{i\,\mathbf{k}}{k^2}\,d\mathbf{k}\exp[i\mathbf{k}(\mathbf{r}-\mathbf{r}_0)]. \tag{2.4}$$

The total radiation is given by the sum over all the ions or by the integral over $n_i d\mathbf{r}_0$ (where n_i is the density of ions with charge $e_i = Ze$):

$$Q = (2e_0^2/3c^3)\int (\ddot{\mathbf{r}})^2 n_i\, d\mathbf{r}_0. \tag{2.5}$$

Given that

$$\int \exp[i(\mathbf{k}+\mathbf{k}')\mathbf{r}_0]\,d\mathbf{r}_0 = (2\pi)^3\delta(\mathbf{k}+\mathbf{k}'), \tag{2.6}$$

and that $k_z = \omega/v_0$, we obtain

$$Q = \int \frac{8e_i^2e_0^4 n_i\, d\omega\, d\varkappa}{3\pi c^3 m_0^2 v_0(\varkappa^2+\omega^2/v_0^2)} = \int_0^\infty Q_\omega\, d\omega = \int_0^\infty \frac{16e_i^2e_0^4 n_i}{3m_0^2c^3v_0}\ln\frac{\varkappa_{\max}v_0}{\omega}\,d\omega. \tag{2.7}$$

It is evident that $\kappa_{\max} = 1/\rho_{\min}$ and we obtain the earlier result (1.2) for Q_ω.

This result can also be obtained by Landau's method from Eq. (2.1). In fact, Eq. (2.3) can be used to find the additional velocity $\delta\mathbf{v}(t)$ and the additional displacement $\delta\mathbf{r}(t)$ gained by the incident particle because of the interaction with a motionless ion. The additional current $\delta\mathbf{j}$ owing to these changes in the motion of the incident particle is given by

$$\delta\mathbf{j}_{\mathbf{k},\omega} = \frac{e_0}{(2\pi)^4}\int dt\exp[i(\omega-k_zv_0)t]\,[\delta\mathbf{v}(t) - i\,\mathbf{v}_0(\mathbf{k}\delta\mathbf{r}(t))]. \tag{2.8}$$

Simply substituting Eq. (2.8) in Eq. (2.1) and integrating this expression over the factor $n_i d\mathbf{r}_0$ immediately gives Eq. (2.7). The approach of integrating over the positions of the scattering centers, which has been demonstrated using both Eqs. (2.5) and (2.1), is extremely convenient for simple derivations of the polarization bremsstrahlung formula.

2.2. Polarization Bremsstrahlung of Electrons on a Charge in a Plasma

In speaking of a charge in a plasma, we shall mean either a fully ionized ion or another structureless charge of magnitude $e_i = Ze$. According to this hypothesis, the polarization is created by the plasma electrons, whose distribution $f_\mathbf{p}$ with respect to momentum $\mathbf{p}$ is determined by the kinetic equation

$$\partial f_{\mathbf{p}}/\partial t + \mathbf{v}\partial f_{\mathbf{p}}/\partial \mathbf{r} - e\,\mathbf{E}\partial f_{\mathbf{p}}/\partial \mathbf{p} = 0, \qquad n_e = \int f_{\mathbf{p}}\, d\mathbf{p}/(2\pi)^3. \tag{2.9}$$

The steady-state distribution of electrons in the field of a charge e_i is determined by the equation

$$\mathbf{v}\,\partial \delta f_{\mathbf{p}}^{(s)}/\partial \mathbf{r} - e\mathbf{E}^{(i)}\partial \Phi_{\mathbf{p}}/\partial \mathbf{p} = 0, \tag{2.10}$$

where $\Phi_{\mathbf{p}}$ is the uniform, steady-state distribution of the electrons far from the charge and $\delta f_{\mathbf{p}}{}^{(s)}$ is the perturbation of this distribution near the charge, that causes Debye shielding.

We now write the the static field of a charge e_i in the form of an expansion in terms of spatial harmonics,

$$\mathbf{E}^{(i)} = \int \mathbf{E}_{\mathbf{k}}^{(i)} e^{i\mathbf{kr}} d\,\mathbf{k}. \tag{2.11}$$

The perturbation $\delta f_{\mathbf{p}}{}^{(s)}$ also must have a form similar to the expansion (2.11), i.e.,

$$\delta f_{\mathbf{p}}^{(s)} = \int \delta f_{\mathbf{p},\mathbf{k}}^{(s)} e^{i\mathbf{kr}}\, d\mathbf{k}. \tag{2.12}$$

Equating the coefficients in front of $e^{i\mathbf{kr}}$ after substituting Eqs. (2.11) and (2.12) in Eq. (2.10), we obtain

$$i\,(\mathbf{kv})\,\delta f_{\mathbf{p},\mathbf{k}}^{(s)} - e\mathbf{E}_{\mathbf{k}}^{(i)}\partial \Phi_{\mathbf{p}}/\partial \mathbf{p} = 0. \tag{2.13}$$

The static field should satisfy the Poisson equation

$$\operatorname{div} \mathbf{E}^{(i)} = 4\pi\rho, \tag{2.14}$$

with

$$\rho = \int \rho_{\mathbf{k}} e^{i\mathbf{kr}}\, d\mathbf{k}, \qquad i\,(\mathbf{k}\mathbf{E}_{\mathbf{k}}^{(i)}) = 4\pi\rho_{\mathbf{k}}^{(e_i)} + 4\pi\rho_{\mathbf{k}}^{(s)},$$

where $\rho_{\mathbf{k}}{}^{(s)}$ is the charge density of the plasma electrons, $\rho_{\mathbf{k}}{}^{(e_i)}$ is the density of the stationary charge e_i, and

$$\rho_{\mathbf{k}}^{(s)} = -e\int \delta f_{\mathbf{p},\mathbf{k}}^{(s)} \frac{d\mathbf{p}}{(2\pi)^3} = ie^2 \int \frac{\mathbf{E}_{\mathbf{k}}^{(i)}}{(\mathbf{kv})} \frac{\partial \Phi_{\mathbf{p}}}{\partial \mathbf{p}} \frac{d\mathbf{p}}{(2\pi)^3}\,. \tag{2.15}$$

For an electrostatic field ($\mathbf{E}_{\mathbf{k}}{}^{(i)}$ directed along $\mathbf{k}$) we have

$$\mathbf{E}_{\mathbf{k}}^{(i)} = (\mathbf{k}/k^2)\,(\mathbf{k}\mathbf{E}_{\mathbf{k}}^{(i)}) \tag{2.16}$$

and Eq. (2.14) can be rewritten in the form

$$i\,(\mathbf{k}\mathbf{E}_{\mathbf{k}}^{(i)})\left[1-\frac{4\pi e^2}{k^2}\int\frac{1}{(\mathbf{k}\mathbf{v})}\left(\mathbf{k}\,\frac{\partial\Phi_{\mathbf{p}}}{\partial\mathbf{p}}\right)\frac{d\,\mathbf{p}}{(2\pi)^3}\right]=4\pi\rho_{\mathbf{k}}^{(e_i)}. \tag{2.17}$$

Let $\Phi_{\mathbf{p}}$ correspond to a thermal distribution of electrons with temperature T_e. Then we have

$$(\mathbf{k}\,\partial\Phi_{\mathbf{p}}/\partial\mathbf{p})=-(\mathbf{k}\mathbf{v})\,\Phi_p/T_e \tag{2.18}$$

and, therefore,

$$i\,(\mathbf{k}\mathbf{E}_{\mathbf{k}}^{(i)})\,(1+4\pi n_e e^2/k^2T_e)=i\,(\mathbf{k}\mathbf{E}_{\mathbf{k}}^{(i)})\,\varepsilon_{\mathbf{k}}=4\pi\rho_{\mathbf{k}}^{(e_i)}, \tag{2.19}$$

where

$$\varepsilon_{\mathbf{k}}=1+1/k^2r_d^2=1+4\pi n_e e^2/k^2T_e\ \ (r_d^2=v_{Te}^2/\omega_{pe}^2=T_e/4\pi n_e e^2) \tag{2.20}$$

is the static dielectric constant of the plasma.

Now, from Eq. (2.19) we have

$$\mathbf{E}_{\mathbf{k}}^{(i)}=\frac{4\pi\mathbf{k}\rho_{\mathbf{k}}^{(e_i)}}{ik^2\varepsilon_{\mathbf{k}}}=-\frac{4\pi i\mathbf{k}\rho_{\mathbf{k}}^{(e_i)}}{k^2+1/r_d^2}. \tag{2.21}$$

For a point charge

$$\rho_{\mathbf{k}}^{(e_i)}=e_i/(2\pi)^3,$$

that is,

$$\mathbf{E}_{\mathbf{k}}^{(i)}=-\frac{1}{2\pi^2}\,\frac{i\mathbf{k}e_i}{(k^2+1/r_d^2)}. \tag{2.22}$$

Here we have introduced the standard classical theory of Debye shielding of the field of a charge only for the purpose of completeness in the discussion and in order to point out clearly the sorts of assumptions that lie at the basis of the theory of Debye shielding. We also meant to show that the dynamic and static polarizabilities can be obtained in a unified way from Eq. (2.9).

As for the limits of applicability of Eq. (2.22), we have assumed that the plasma electrons are only weakly perturbed by the static charge e_i. This means, of course, that the charge e_i is fairly small and the thermal velocity of the electrons is fairly high (i.e., the fraction of the electrons which have low velocities comparable to those of the electrons in the bound levels of the charge e_i must be small). Shielding is caused by the free passing electrons.

The shielding cloud is described by $\delta f_{\mathbf{p},\mathbf{k}}^{(s)}$ and from Eq. (2.13) we have

$$\delta f_{\mathbf{p},\mathbf{k}}^{(s)}=\frac{e}{ik^2T_e}\,(\mathbf{k}\mathbf{E}_{\mathbf{k}}^{(i)})\,\Phi_{\mathbf{p}}=-\frac{ee_i\Phi_{\mathbf{p}}}{2\pi^2T_e\,(k^2+1/r_d^2)}. \tag{2.23}$$

We now consider the effect of the field of the incident particle on the distribution of the shielding charge. In the general case we shall regard the incident particle of charge e_0 ($e_0 = -e$ for an electron) as relativistic. Then the magnetic and electric fields of the incident particle may be of the same order of magnitude. We shall assume, however, that the plasma particles are nonrelativistic. Thus, in Eq. (2.9) the term with the Lorentz force $[\mathbf{vH}]/c$ ($\mathbf{E}$ must be replaced by $\mathbf{E} + [\mathbf{vH}]/c$) can be neglected. We denote the field of the incident particle by $\mathbf{E}^{(0)}$. We shall assume that this particle is fast and neglect the polarization charge near it; thus, it creates the same field as if it were in a vacuum. We shall also neglect the ordinary bremsstrahlung produced by the change in the particle's velocity during the collision (i.e., as the particle is slowed down or accelerated). Thus, the velocity can be treated as constant throughout the entire collision process. We denote the velocity of the incident particle by $\mathbf{v}_0$. Then its charge density depends on time as

$$\begin{gathered}\rho^{(0)} = [e_0/(2\pi)^3] \int d\mathbf{k} \exp [i\mathbf{k}(\mathbf{r} - \mathbf{v}_0 t) + i\mathbf{k}\mathbf{r}_0], \\ \mathbf{r}_{0,\perp} = \boldsymbol{\rho}_0, \quad r_{0,\parallel} = z_0. \end{gathered} \tag{2.24}$$

Here $\boldsymbol{\rho}_0$ is a vector representing the distance perpendicular to the velocity at which the incident particle passes by the ion (recall that the velocity is regarded as constant) and z_0 is the distance of the incident particle along its trajectory from the point of closest approach to the ion at time $t = 0$. The charge density (2.24) also depends on the time and should create a time-varying field $\mathbf{E}^{(0)}$ which, in turn, should produce a time-varying polarization. For the fields of the particle we write

$$\begin{gathered}\mathbf{E}^{(0)} = \int \mathbf{E}_{\mathbf{k}}^{(0)} \exp [i\mathbf{k}(\mathbf{r} - \mathbf{v}_0 t) + i\mathbf{k}\mathbf{r}_0]\, d\mathbf{k}, \\ \mathbf{H}^{(0)} = \int \mathbf{H}_{\mathbf{k}}^{(0)} \exp [i\mathbf{k}(\mathbf{r} - \mathbf{v}_0 t) + i\mathbf{k}\mathbf{r}_0]\, d\mathbf{k}. \end{gathered} \tag{2.25}$$

Maxwell's equations

$$\begin{gathered}\operatorname{div} \mathbf{E}^{(0)} = 4\pi\rho^{(0)}, \quad \operatorname{curl} \mathbf{H}^{(0)} = \frac{4\pi}{c} \mathbf{j}^{(0)} + \frac{1}{c} \frac{\partial \mathbf{E}^{(0)}}{\partial t}, \\ \operatorname{curl} \mathbf{E}^{(0)} = -\frac{1}{c} \frac{\partial \mathbf{H}^{(0)}}{\partial t}, \quad \mathbf{j}^{(0)} = \mathbf{v}_0 \rho^{(0)} \end{gathered} \tag{2.26}$$

yield

$$[\mathbf{k}\mathbf{E}_{\mathbf{k}}^{(0)}] = \frac{(\mathbf{k}\mathbf{v}_0)}{c} \mathbf{H}_{\mathbf{k}}^{(0)}, \quad i\,[\mathbf{k}\mathbf{H}_{\mathbf{k}}^{(0)}] = \frac{e_0 \mathbf{v}_0}{2\pi^2 c} - i \frac{(\mathbf{k}\mathbf{v}_0)}{c} \mathbf{E}^{(0)}, \tag{2.27}$$

or

$$\mathbf{E}^{(0)} = \frac{e_0 [\mathbf{k} - \mathbf{v}_0 (\mathbf{k}\mathbf{v}_0)/c^2]}{i 2\pi^2 [k^2 - (\mathbf{k}\mathbf{v}_0)^2/c^2]}. \tag{2.28}$$

We can find when the polarization near the incident particle is negligible. First, it is clear from Eq. (2.25) that the characteristic frequencies of the field of the incident particle are $\mathbf{kv}_0$. When $v \gg v_{Te}$ (i.e., if the velocity of the incident particle is much greater than the average thermal speed of the plasma particles), the dielectric constant will not be of the static type, but of the ordinary type, $\varepsilon = 1 - \omega_{pe}^2/\omega^2$. Thus, the required criterion reduces to $\omega \gg \omega_{pe}$.

Actually, as we shall see later, $\mathbf{kv}_0 = \omega$ [below we have $\mathbf{k} \to \mathbf{k} - \mathbf{q}$ and $(\mathbf{k} - \mathbf{q})\mathbf{v}_0 = \omega$; see Eqs. (2.33) and (2.46)]. In other words, the only criteria would be

$$v_0 \gg v_{Te}, \qquad \omega \gg \omega_{pe}. \tag{2.29}$$

We now consider the type of dynamic polarization that develops because of the effect of the field of the incident particle on the polarization cloud of the charge e_i. To do this we use Eq. (2.9) in the form

$$\begin{gathered} f_{\mathbf{p}} = \Phi_{\mathbf{p}} + \delta f_{\mathbf{p}}^{(s)} + \delta f_{\mathbf{p}}^{(\mathrm{din})}, \\ \frac{\partial \delta f_{\mathbf{p}}^{(\mathrm{din})}}{\partial t} + \frac{\mathbf{v}\partial \delta f_{\mathbf{p}}^{(\mathrm{din})}}{\partial \mathbf{r}} = e\left(\mathbf{E}^{(0)} \frac{\partial}{\partial \mathbf{p}} \delta f_{\mathbf{p}}^{(s)}\right) = \\ = e\int \mathbf{E}_{\mathbf{k}}^{(0)}\, d\mathbf{k} \exp\left[i(\mathbf{kr} - \mathbf{kv}_0 t) + i\mathbf{kr}_0\right] \frac{\partial}{\partial \mathbf{p}} \delta f_{\mathbf{p},\mathbf{q}}^{(s)} e^{i\mathbf{qr}}\, d\mathbf{q} = \\ = e\int \left(\mathbf{E}_{\mathbf{k}-\mathbf{q}}^{(0)} \frac{\partial}{\partial \mathbf{p}}\right) \delta f_{\mathbf{p},\mathbf{q}}^{(s)}\, d\mathbf{k}\, d\mathbf{q} \exp\left[i\,\mathbf{kr} - i(\mathbf{k} - \mathbf{q})\mathbf{v}_0 t + i(\mathbf{k} - \mathbf{q})\mathbf{r}_0\right]. \end{gathered} \tag{2.30}$$

Here we have made the change of variables $\mathbf{k} \to \mathbf{k} - \mathbf{q}$. The right-hand side of Eq. (2.30) describes a given, time-dependent source of perturbations.

Thus, $\delta f_{\mathbf{p}}^{(\mathrm{din})}$ will also depend on the time:

$$\delta f_{\mathbf{p}}^{(\mathrm{din})} = \int \delta f_{\mathbf{p},\mathbf{k},\omega}^{(\mathrm{din})} \exp\left[i(\mathbf{kr} - \omega t)\right] d\mathbf{k}\, d\omega. \tag{2.31}$$

Here we have written down an expansion of $\delta f_{\mathbf{p}}^{\mathrm{din}}$ over the possible frequencies. Each frequency is excited in accordance with the right-hand side of Eq. (2.30); that is, we can write

$$\delta f_{\mathbf{p},\mathbf{k},\omega}^{(\mathrm{din})} = \int \delta f_{\mathbf{p},\mathbf{k},\mathbf{q}}^{(\mathrm{din})} \delta\left[\omega - (\mathbf{k} - \mathbf{q})\mathbf{v}_0\right] \exp\left[i(\mathbf{k} - \mathbf{q})\mathbf{r}_0\right] d\mathbf{q}. \tag{2.32}$$

We have written the perturbations in a form that isolates the conservation of energy in an elementary emission event,

$$\omega = (\mathbf{k} - \mathbf{q})\mathbf{v}_0. \tag{2.33}$$

It turns out that this is the same conservation law as in ordinary bremsstrahlung for distant collisions. This can be confirmed from the following (still quantum mechanical) equations: let us consider an emission event in which the incident particle has energy $\varepsilon_i = (p_i^2c^2 + m_0^2c^4)^{1/2}$ and the energy of its final state is $\varepsilon_f = (p_f^2c^2 + m_0^2c^4)^{1/2}$. This process is radiation in which a momentum of $\hbar\mathbf{q}$ is transferred to the nucleus.

The conservation of energy and momentum is described by

$$\varepsilon_f + \hbar\omega = \varepsilon_i,$$
$$\mathbf{p}_f + \hbar\mathbf{k} - \hbar\mathbf{q} = \mathbf{p}_i. \tag{2.34}$$

This gives

$$\varepsilon_i = (p_i^2c^2 + m_0^2c^4)^{1/2} = [(\mathbf{p}_i - \hbar\mathbf{k} + \hbar\mathbf{q})^2 c^2 + m_0^2c^4]^{1/2} + \hbar\omega. \tag{2.35}$$

Distant collisions correspond to small momentum transfers and to a small momentum of the emitted photon compared to the initial momentum of the incident particle:

$$\varepsilon_f \approx \varepsilon_i - \hbar(\mathbf{k} - \mathbf{q})\mathbf{v}_0, \quad \mathbf{v}_0 = \mathbf{p}_0/\varepsilon_0 \approx \mathbf{p}_i/\varepsilon_i \approx \mathbf{p}_f/\varepsilon_f. \tag{2.36}$$

When Eq. (2.35) is substituted in Eq. (2.36), Planck's constant $\hbar$ cancels out and we obtain the classical result, which is the same as Eq. (2.33).

Even before any specific calculations of the radiation intensity have been carried out, therefore, we have found an intimate relationship between traditional bremsstrahlung and polarization bremsstrahlung.

If we substitute Eq. (2.32) in Eq. (2.31), then it is possible to write the perturbation of the polarization charge in another form:

$$\delta f_{\mathbf{p}}^{(\mathrm{din})} = \int \delta f_{\mathbf{p},\mathbf{k},\mathbf{q}}^{(\mathrm{din})} \exp[i\mathbf{k}\mathbf{r} - i(\mathbf{k} - \mathbf{q})\mathbf{v}_0 t + i(\mathbf{k} - \mathbf{q})\mathbf{r}_0]\, d\mathbf{k}\, d\mathbf{q}. \tag{2.37}$$

Substituting Eq. (2.37) in Eq. (2.30) and equating the coefficients of the corresponding exponents, we obtain

$$[i(\mathbf{k} - \mathbf{q})\mathbf{v}_0 + i(\mathbf{k}\mathbf{v})]\,\delta f_{\mathbf{p},\mathbf{k},\mathbf{q}}^{(\mathrm{din})} = e\mathbf{E}_{\mathbf{k}-\mathbf{q}}^{(0)} \frac{\partial}{\partial \mathbf{p}} \delta f_{\mathbf{p},\mathbf{q}}^{(s)}. \tag{2.38}$$

We have already assumed that the velocity $\mathbf{v}_0$ of the charge is much greater than that of the plasma electrons [see Eq. (2.29)], so that the term with $(\mathbf{kv})$ on the left-hand side of Eq. (2.38) can be neglected. Thus, we have

$$\delta f_{\mathbf{p},\mathbf{k},\mathbf{q}}^{(\mathrm{din})} = \frac{e\,\mathbf{E}_{\mathbf{k}-\mathbf{q}}^{(0)}}{i(\mathbf{k} - \mathbf{q})\mathbf{v}_0} \frac{\partial}{\partial \mathbf{p}} \delta f_{\mathbf{p},\mathbf{q}}^{(s)}. \tag{2.39}$$

In order to write down the final result for the perturbation in the distribution of the shielding electrons by the incident particle, it is sufficient to substitute the earlier expressions for $\mathbf{E}_{\mathbf{k}-\mathbf{q}}^{(0)}$ [Eq. (2.28)] and $\delta f_{\mathbf{p},\mathbf{q}}^{(s)}$ [Eq. (2.23)] in Eq. (2.39). We shall not do this yet, but first find the polarization current produced by this perturbation:

$$\delta\mathbf{j} = -e\int \mathbf{v}\delta f_{\mathbf{p}}^{(\mathrm{din})}\frac{d\mathbf{p}}{(2\pi)^3} = -e\int \mathbf{v}\delta f_{\mathbf{p},\mathbf{k},\mathbf{q}}^{(\mathrm{din})}\frac{d\mathbf{p}\,d\mathbf{k}\,d\mathbf{q}}{(2\pi)^3}\exp\left[i\mathbf{k}\mathbf{r} - i(\mathbf{k}-\mathbf{q})\mathbf{v}_0 t + i(\mathbf{k}-\mathbf{q})\mathbf{r}_0\right]. \tag{2.40}$$

Substituting Eq. (2.39) into Eq. (2.40) and integrating by parts with respect to the momenta $\mathbf{p}$, we find that $-\partial/\partial\mathbf{p}$ will act on $\mathbf{v}$ and (since the plasma particles are nonrelativistic) to yield $-1/m$. Thus,

$$\delta\mathbf{j} = \frac{ie^2}{m}\int \frac{\mathbf{E}_{\mathbf{k}-\mathbf{q}}^{(0)}}{(\mathbf{k}-\mathbf{q})\,\mathbf{v}_0}\,\delta f_{\mathbf{p},\mathbf{q}}^{(s)}\frac{d\mathbf{p}\,d\mathbf{k}\,d\mathbf{q}}{(2\pi)^3}\exp\left[i\mathbf{k}\mathbf{r} - i(\mathbf{k}-\mathbf{q})\mathbf{v}_0 t + i(\mathbf{k}-\mathbf{q})\mathbf{r}_0\right]. \tag{2.41}$$

If we substitute Eq. (2.23) here, then the integral with respect to the momenta of $\Phi_{\mathbf{p}}$ yields the total number of electrons n_e which can conveniently be expressed in terms of $\omega_{pe}{}^2 = 4\pi n_e e^2/m$. Using $\omega_{pe}/v_{Te} = 1/r_d$ and $v_{Te} = (T_e/m)^{1/2}$ we obtain

$$\frac{\partial}{\partial t}\,\delta\mathbf{j}_i = -\frac{ee_i}{m(2\pi)^3}\int \frac{\mathbf{E}_{\mathbf{k}-\mathbf{q}}^{(0)}}{(q^2r_d^2+1)}\exp\left[i\mathbf{k}\mathbf{r} - i(\mathbf{k}-\mathbf{q})\mathbf{v}_0 t + i(\mathbf{k}-\mathbf{q})\mathbf{r}_0\right]d\mathbf{k}\,d\mathbf{q},$$

$$\delta\mathbf{j}_i = \int \delta j_{\mathbf{k},\mathbf{q}}\exp\left[i\,\mathbf{k}\mathbf{r} - i(\mathbf{k}-\mathbf{q})\mathbf{v}_0 t + i(\mathbf{k}-\mathbf{q})\mathbf{r}_0\right]d\mathbf{k}\,d\mathbf{q}, \tag{2.42}$$

where

$$\delta\mathbf{j}_{\mathbf{k},\mathbf{q}} = +\frac{ee_i}{(2\pi)^3 m}\,\frac{\mathbf{E}_{\mathbf{k}-\mathbf{q}}^{(0)}}{(q^2r_d^2+1)\,i(\mathbf{k}-\mathbf{q})\,\mathbf{v}_0}.$$

This result is already sufficient for use in the Landau method and Eq. (2.1). To do this we must find the radiation field in terms of the current (2.42) using Maxwell's equations:

$$\Delta\mathbf{E} - \operatorname{grad}\operatorname{div}\mathbf{E} - \frac{1}{c^2}\frac{\partial^2}{\partial t^2}\mathbf{E} = \frac{4\pi}{c^2}\frac{\partial}{\partial t}\delta\mathbf{j}. \tag{2.43}$$

Clearly, the field $\mathbf{E}$ must be written as an expansion of the same type used for the perturbation in the current $\delta\mathbf{j}$ (2.42) which generates this field:

$$\mathbf{E} = \int \mathbf{E}_{\mathbf{k},\mathbf{q}}\,d\mathbf{k}\,d\mathbf{q}\exp\left[i\mathbf{k}\mathbf{r} - i(\mathbf{k}-\mathbf{q})\mathbf{v}_0 t + i(\mathbf{k}-\mathbf{q})\mathbf{r}_0\right], \tag{2.44}$$

with

$$k^2\mathbf{E}_{\mathbf{k},\mathbf{q}} - \mathbf{k}(\mathbf{k}E_{\mathbf{k},\mathbf{q}}) - \frac{[(\mathbf{k}-\mathbf{q})\,\mathbf{v}_0]^2}{c^2}\mathbf{E}_{\mathbf{k},\mathbf{q}} = \frac{4\pi i}{c^2}(\mathbf{k}-\mathbf{q})\,\mathbf{v}_0\delta\mathbf{j}_{\mathbf{k},\mathbf{q}}. \tag{2.45}$$

This yields

$$\mathbf{E}_{\mathbf{k},\mathbf{q}} = \frac{4\pi ee_i\left[\mathbf{E}_{\mathbf{k}-\mathbf{q}}^{(0)} - \mathbf{k}(\mathbf{k}\mathbf{E}_{\mathbf{k}-\mathbf{q}}^{(0)})/k^2\right]}{mc^2(2\pi)^3\left[k^2 - ((\mathbf{k}-\mathbf{q})\,\mathbf{v}_0)^2/c^2\right](1+q^2r_d^2)}. \tag{2.46}$$

Equation (2.45) is solved very simply [as is Eq. (2.27)]: Eq. (2.45) is multiplied by **k**, the scalar product $(\mathbf{k}\mathbf{E}_{\mathbf{k},\mathbf{q}})$ is found, and this term is transferred to the right hand side to obtain Eq. (2.46).

We now use Eq. (2.1) to calculate the radiation intensity. To do this, we must substitute the expansions (2.46) and (2.44) in Eq. (2.1) to obtain

$$Q_i = -\int \delta\mathbf{j}\mathbf{E}\,dr = -\int \delta\mathbf{j}_{\mathbf{k},\mathbf{q}}\mathbf{E}_{\mathbf{k}',\mathbf{q}'}\exp[i(\mathbf{k}+\mathbf{k}')\mathbf{r} - i(\mathbf{k}-\mathbf{q}+ + \mathbf{k}'-\mathbf{q}')\mathbf{v}_0 t + i(\mathbf{k}-\mathbf{q}+\mathbf{k}'-\mathbf{q}')\mathbf{r}_0]\,dr\,d\mathbf{k}d\mathbf{k}'\,d\mathbf{q}d\mathbf{q}' = -(2\pi)^3\int \delta\mathbf{j}_{\mathbf{k},\mathbf{q}}\mathbf{E}_{-\mathbf{k},\mathbf{q}'}\exp[i(\mathbf{q}+\mathbf{q})(\mathbf{v}_0 t - \mathbf{r}_0)]\,d\mathbf{k}\,d\mathbf{q}d\mathbf{q}'. \quad (2.47)$$

Here we have used the fact that

$$\int \exp[i(\mathbf{k}+\mathbf{k}')\mathbf{r}]\,dr = (2\pi)^3\,\delta(\mathbf{k}+\mathbf{k}').$$

As in the elementary theory of ordinary bremsstrahlung [see Eq. (2.7)], we are considering a situation in which the ions are positioned randomly but, on the average, uniformly with a density of n_i. Then $\mathbf{r}_0$ is the coordinate of an ion relative to the position of the incident particle at time $t = 0$.

The total intensity of the radiation created by all the ions will be

$$Q = \int Q_i n_i\,dr_0 = -(2\pi)^6 n_i \int \delta\mathbf{j}_{\mathbf{k},\mathbf{q}}\mathbf{E}_{-\mathbf{k},-\mathbf{q}}\,d\mathbf{k}\,d\mathbf{q}. \quad (2.48)$$

Substituting Eqs. (2.42) and (2.46), we obtain

$$Q = \frac{4\pi e^2 e_i^2 n_i}{m^2c^2}\int \frac{d\mathbf{k}\,d\mathbf{q}\,|[\mathbf{k}\mathbf{E}^{(0)}_{\mathbf{k}-\mathbf{q}}]|^2}{i(\mathbf{k}-\mathbf{q})\mathbf{v}_0(1+q^2r_d^2)^2[k^2-((\mathbf{k}-\mathbf{q})\mathbf{v}_0)^2/c^2]}. \quad (2.49)$$

In view of the fact that all quantities under the integral of Eq. (2.49) are odd, when the substitutions $\mathbf{k} \to -\mathbf{k}$ and $\mathbf{q} \to -\mathbf{q}$ are made, only the imaginary part remains:

$$\operatorname{Im}\frac{1}{k^2-((\mathbf{k}-\mathbf{q})\mathbf{v}_0)^2/c^2} = \pi\frac{(\mathbf{k}-\mathbf{q})\mathbf{v}_0}{|(\mathbf{k}-\mathbf{q})\mathbf{v}_0|}\delta\left[k^2-\frac{((\mathbf{k}-\mathbf{q})\mathbf{v}_0)^2}{c^2}\right],$$

This finally yields

$$Q = \frac{4\pi^2 e^2 e_i^2 n_i}{m^2c^3}\int \frac{d\mathbf{k}\,d\mathbf{q}\,|[\mathbf{k}\mathbf{E}^{(0)}_{\mathbf{k}-\mathbf{q}}]|^2}{k^3(1+q^2r_d^2)^2}\delta\left[k^2-\frac{((\mathbf{k}-\mathbf{q})\mathbf{v}_0)^2}{c^2}\right], \quad (2.50)$$

where $\mathbf{E}_{\mathbf{k}-\mathbf{q}}^{(0)}$ is the field of the incident particle given by Eq. (2.28). Given that $(\mathbf{k}-\mathbf{q})\mathbf{v}_0 = kc$ because of the δ-function in Eq. (2.50), this field can be transformed to

$$\mathbf{E}^{(0)}_{\mathbf{k}-\mathbf{q}} = -\frac{e_0}{i2\pi^2}\frac{\mathbf{q}+\mathbf{v}_0 k/c}{(\mathbf{k}-\mathbf{q})^2-k^2}. \quad (2.51)$$

Here we have dropped the term proportional to **k**, since this term does not appear in the vector product in Eq. (2.50).

We, therefore, have [1, 5–7]

$$Q = \frac{e^2 e_i^2 e_0^2 n_i}{\pi^2 m^2 c^3} \int \frac{d\mathbf{k}\, d\mathbf{q}\, |\, [\mathbf{kq}] + k\, [\mathbf{kv}_0]/c\, |^2}{k^3 (1 + q^2 r_d^2)^2\, [(\mathbf{k} - \mathbf{q})^2 - k^2]^2}\, \delta\, [k^2 - ((\mathbf{k} - \mathbf{q})\, \mathbf{v}_0)^2/c^2]. \tag{2.52}$$

2.3. Radiation from Nonrelativistic Incident Particles

We now consider nonrelativistic incident particles [7], i.e.,

$$v_0 \ll c. \tag{2.53}$$

Then, $(\mathbf{kv}_0) \ll kc$ and, in view of the assumption that $\omega \gg \omega_{pe}$ $[\varepsilon(\omega) \simeq 1]$, we have

$$|\,\omega\,| = kc = |\,(\mathbf{qv}_0)\,|.$$

This means that the projection of the momentum transfer on the particle velocity obeys $(\mathbf{qv}_0)/v_0 = q_{\parallel} = kc/v_0$, i.e., $q \geq q_{\parallel} \gg k$ (the projection of the vector is less than or equal to its length, $q_{\parallel} \leq q$). This inequality can be written differently as

$$\lambda = 2\pi/k \gg 2\pi/q \approx d, \tag{2.54}$$

where d is the size of the inhomogeneity in the distribution of the polarization charges near the ion (in general, d should be on the order of r_d). In this sense the condition (2.54) corresponds to the dipole approximation. Here, in the first approximation with respect to the parameter of Eq. (2.54) and, therefore, to that of Eq. (2.53), we have, on noting that $k/q \sim v_0/c$,

$$Q \approx \frac{e^2 e_i^2 e_0^2 n_i}{\pi^2 m^2 c^3} \int \frac{d\mathbf{k}\, d\mathbf{q}\, [\mathbf{kq}]^2}{k^3 q^4 (1 + q^2 r_d^2)^2}\, \delta \left(k^2 - \frac{(\mathbf{qv}_0)^2}{c^2} \right). \tag{2.55}$$

Note that this corresponds to neglecting terms in v_0^2/c^2 in the field of the incident particle (2.28), i.e., to using the expression

$$\mathbf{E}_{\mathbf{q}}^{(0)} = e_0 \mathbf{q}/i 2\pi^2 q^2$$

for it, which corresponds to the purely electrostatic field from a charge e_0 in a vacuum (shielding can be neglected when $v_0 \gg v_{Te}$).

We denote the component of the vector perpendicular to the velocity of the incident particle by $\mathbf{q}_{\perp}$ and the parallel component by $\mathbf{q}_{\parallel}$, and introduce the solid angle $d\Omega_{\mathbf{k}}$ over which waves within $d\mathbf{k} = d\Omega_{\mathbf{k}} k^2 dk$ $(\omega \simeq kc)$ are emitted. We then obtain

$$Q = \frac{e^2 e_i^2 e_0^2 n_i}{\pi^2 m^2 c^3 v_0} \int d\Omega_{\mathbf{k}} \int_0^{\infty} d\omega \int d\mathbf{q}_{\perp} \frac{\{[\mathbf{nq}_{\perp}]^2 + [\mathbf{nv}_0]^2\, \omega^2/v_0^4\}}{\left(q_{\perp}^2 + \dfrac{\omega^2}{v_0^2}\right)^2 \left[1 + r_d^2 \left(q_{\perp}^2 + \dfrac{\omega^2}{v_0^2}\right)\right]^2}. \tag{2.56}$$

Here $\mathbf{q}$ is written in the form of two components as

$$\mathbf{q} = \mathbf{q}_{\perp} + \mathbf{v}_0 (\mathbf{q}\mathbf{v}_0)/v_0^2 = \mathbf{q}_{\perp} + \omega \mathbf{v}_0/v_0^2,$$

and $\mathbf{n} = \mathbf{k}/k$ is the unit vector in the direction of the emitted wave.

The first term in the curly brackets of Eq. (2.56) corresponds effectively to a dipole oscillating perpendicular to the velocity of the incident particle and the second, to a dipole oscillating parallel to it. In reality, the dipole oscillates at an angle to that velocity. The crossover term vanishes when it is noted that the momentum transfers are symmetric in the plane perpendicular to the velocity of the incident particle. It is convenient to take the momentum transfer in units of ω/v_0 and introduce the variable

$$\xi = q_{\perp}^2 v_0^2/\omega^2. \tag{2.57}$$

The result (2.56) can then be written in the form

$$Q = \frac{2e^2 e_i^2 e_0^2 n_i}{m^2 c^3 v_0} \int\limits_0^{\pi} \sin\theta \, d\theta \int\limits_0^{\infty} d\omega \int\limits_0^{\infty} \frac{d\xi \, [\xi (1 - {}^1\!/_2 \sin^2\theta) + \sin^2\theta]}{(\xi + 1)^2 \, [(\xi + 1)\, \omega^2 r_d^2/v_0^2 + 1]^2}, \tag{2.58}$$

where

$$d\Omega_{\mathbf{k}} = 2\pi \sin\theta \, d\theta, \quad \overline{[\mathbf{n}\mathbf{q}_{\perp}]^2} = q_{\perp}^2 (1 - {}^1\!/_2 \sin^2\theta).$$

After integrating with respect to ξ, we obtain the angular and frequency distribution of the radiation.

Equation (2.58) shows clearly that the result for this case depends only on a single parameter, $\omega^2 r_d^2/v_0^2 = \omega^2 v_{Te}^2/\omega_{pe}^2 v_0^2$.

Let us consider the case

$$\omega_{pe} \ll \omega \ll \omega_{pe} v_0/v_{Te}. \tag{2.59}$$

Then in the second term of Eq. (2.58) (proportional to $\sin^2\theta$) we can drop $\omega^2 r_d^2/v_0^2$ in the denominator compared to unity, while in the first term, this can only be done when $\xi \ll \xi_*$, where $1 \ll \xi_* \ll v_0^2 r_d^2/\omega^2$, so that

$$\int\limits_0^{\infty} \frac{d\xi}{(\xi + 1)^2} = 1, \quad \int\limits_0^{\xi_*} \frac{\xi \, d\xi}{(\xi + 1)^2} \approx \ln \xi_* - 1. \tag{2.60}$$

In the integral from ξ to infinity, however, we can use the fact that

$$\int\limits_{\xi_*}^{\infty} \frac{d\xi}{\xi \, (\xi \omega^2 r_d^2/v_0^2 + 1)^2} = \int\limits_{\xi_* \omega^2 r_d^2/v_0^2}^{\infty} \frac{d\xi}{\xi \, (\xi + 1)^2} = -\ln \frac{\omega^2 r_d^2}{v_0^2} \xi_* - 1. \tag{2.61}$$

Thus, we have

$$Q=\frac{2e^2e_i^2e_0^2n_i}{m^2c^3v_0}\int_0^{\pi}\sin\theta\,d\theta\int_0^{\infty}d\omega\left\{(2-\sin^2\theta)\left(\ln\frac{v_0\omega_{pe}}{v_{Te}\omega}-1\right)+\sin^2\theta\right\}. \quad (2.62)$$

Integrating over the angle yields the frequency spectrum

$$Q=\int_0^{\infty}Q_\omega\,d\omega. \quad (2.63)$$

For $\omega \ll \omega_{pe}v_0/v_{Te}$, we have

$$Q_\omega=\frac{16e^2e_i^2e_0^2n_i}{3m^2c^3v_0}\left(\ln\frac{v_0\omega_{pe}}{v_{Te}\omega}-\frac{1}{2}\right). \quad (2.64)$$

As mentioned above, polarization bremsstrahlung contains the square of the charge of the incident particle, e_0, and the square of the charge of the polarized particle e (but not e_0^4, as in the case of ordinary bremsstrahlung), as well as the mass m of the polarized particle, rather than the mass m_0 of the incident particle.

For $\omega \gg \omega_{pe}v_0/v_{Te}$, Eq. (2.58) gives

$$Q=\frac{e^2e_i^2e_0^2n_i}{6m^2c^3v_0}\int_0^{\pi}\sin\theta\,d\theta\int_0^{\infty}\frac{d\omega v_0^4}{\omega^4r_d^4}(2+3\sin^2\theta). \quad (2.65)$$

Integrating over the angle gives

$$Q_\omega=\frac{4e^2e_i^2e_0^2n_i}{3m^2c^3v_0}\left(\frac{v_0\omega_{pe}}{\omega v_{Te}}\right)^4. \quad (2.66)$$

Therefore, the spectrum is flat out to $\omega \sim \omega_{pe}v_0/v_{Te}$ and then falls off with frequency as $1/\omega^4$.

The integral over all frequencies converges, so that the contribution of quantum mechanical effects must be extremely small. The integral with respect to ω can be taken immediately in Eq. (2.58) if we are concerned with the total radiated power, rather than the spectral density, to give

$$Q=\frac{e^2e_i^2e_0^2n_i\omega_{pe}\pi}{2\,m^2c^3v_{Te}}\int_0^{\pi}\sin\theta\,d\theta\int_0^{\infty}\frac{d\xi}{(\xi+1)^{5/2}}\left[\xi\left(1-\frac{1}{2}\sin^2\theta\right)+\sin^2\theta\right]=$$

$$=\frac{2e^2e_i^2e_0^2n_i\omega_{pe}\pi}{3m^2c^3v_{Te}}\int_0^{\pi}\sin\theta\,d\theta=\frac{4e^2e_i^2e_0^2n_i\omega_{pe}\pi}{3m^2c^3v_{Te}}. \quad (2.67)$$

2.4. Radiation from Ultrarelativistic Incident Particles

Since the radiation comes from nonrelativistic polarization electrons, we should not expect the radiation to be very highly directed in the case of an ultrarelativistic incident particle with

$$v_0\to c,\ \varepsilon_0=\sqrt{p^2c^2+m_0^2c^4}\gg m_0c^2 \quad (2.68)$$

For ultrarelativistic particles it is convenient to replace the momentum **q** transferred from the nucleus to the incident particle by **q**', the total momentum lost by the incident particle (**q**' = **q** − **k**), and integrate with respect to **q**' instead of **q**. We shall denote $(\mathbf{q}'\mathbf{v}_0)v_0$ by $q_\parallel'$ and $\mathbf{q}' - \mathbf{q}_\parallel'$ by $\mathbf{q}_\perp'$. Then, we have

$$q_\parallel' v_0 = -\omega = -kc,$$

$$(\mathbf{k}-\mathbf{q})^2 - k^2 = q'^2 - k^2 = (\mathbf{q}_\perp')^2 + \omega^2/v_0^2 - \omega^2/c^2 = (\mathbf{q}_\perp')^2 + (\omega^2/c^2)(m_0c^2/\varepsilon_0)^2,$$

and (2.69)

$$[\mathbf{k}\mathbf{v}_0]\,k/c + [\mathbf{k}\mathbf{q}] = [\mathbf{k}\mathbf{v}_0]\,(k/c)(1 - c^2/v_0^2) + [\mathbf{k}\mathbf{q}_\perp'] \approx [\mathbf{k}\mathbf{q}_\perp']. \tag{2.70}$$

The last equality is a consequence of Eq. (2.68).

In addition, we have

$$1 + q^2 r_d^2 = 1 + (\mathbf{q}' - \mathbf{k})^2 r_d^2. \tag{2.71}$$

Under conditions such that

$$r_d\omega/c \ll 1, \tag{2.72}$$

the terms with $k^2 r_d^2$ and $(q_\parallel')^2 r_d^2$ are small in Eq. (2.71). The only case in which Eq. (2.71) differs from unity is when $\mathbf{q}_\perp'$ is large, i.e.,

$$1 + q^2 r_d^2 \approx 1 + (q_\perp')^2 r_d^2. \tag{2.73}$$

With these simplifications, Eq. (2.52) then yields

$$Q = \frac{e^2 e_i^2 e_0^2 n_i}{\pi^2 m^2 c^3 v_0} \int \frac{dk\, d\mathbf{q}_\perp'}{k^4} \frac{[\mathbf{k}\mathbf{q}_\perp']^2}{[(\mathbf{q}_\perp')^2 + (\omega^2/c^2)(m_0c^2/\varepsilon_0)^2]^2 (1 + (\mathbf{q}_\perp')^2 r_d^2)^2}. \tag{2.74}$$

Integrating with respect to the angular part of the vector $\mathbf{q}_\perp'$, we obtain

$$Q = \frac{8e^2 e_i^2 e_0^2 n_i}{3m^2 c^3 v_0} \int_0^\infty d\omega \int_0^\infty \frac{dq_\perp'^2 q_\perp'^2}{[q_\perp'^2 + (\omega^2/c^2)(m_0c^2/\varepsilon_0)^2]^2 (1 + q_\perp'^2 r_d^2)^2}. \tag{2.75}$$

This integral is calculated in the same way as for the nonrelativistic case with the result

$$Q_\omega = \frac{16 e^2 e_0^2 e_i^2 n_i}{3m^2 c^4} \left(\ln \frac{\omega_{pe} c \varepsilon_0}{\omega v_{Te} m_0 c^2} - 1 \right). \tag{2.76}$$

The angular distribution of the radiation corresponds to a dipole oriented perpendicular to the velocity of the incident particle (i.e., along $\mathbf{q}_\perp'$). If, however, we introduce the angle θ between the velocity of the incident particle and the wave vector of the emitted wave, then we have

$$\overline{[\mathbf{k}\mathbf{q}'_\perp]} = k^2 q'^2_\perp (1 - {}^1/_2 \sin^2\theta) \tag{2.77}$$

and the angular dependence is of the same type as in the nonrelativistic formula (2.58) [the term with $\sin^2\theta$ has a small factor of order $(m_0c^2/\varepsilon_0)^2$ in the ultrarelativistic limit].

Ultrarelativistic particles can radiate within the frequency range

$$\frac{c}{r_d} \ll \omega \ll \frac{c}{r_d}\left(\frac{\varepsilon_0}{m_0c^2}\right)^2 \tag{2.78}$$

where nonrelativistic particles do not radiate. In this case, the intensity falls off as $1/\omega^2$, i.e.,

$$Q_\omega = \frac{2e^2e_i^2e_0^2n_i}{m^2c^4}\left(\frac{c}{\omega r_d}\right)^2\left(\ln\frac{\varepsilon_0}{\sqrt{2}m_0c^2} - \frac{1}{2}\right), \tag{2.79}$$

and, finally, for

$$\omega \gg \frac{c}{r_d}\left(\frac{\varepsilon_0}{m_0c^2}\right)^2 \tag{2.80}$$

we obtain a dependence that is proportional to $1/\omega^4$, i.e.,

$$Q_\omega = \frac{2e^2e_i^2e_0^2n_i}{3m^2c^4}\left(\frac{c}{\omega r_d}\right)^2\left[\frac{c}{\omega r_d}\left(\frac{\varepsilon_0}{m_0c^2}\right)^2\right]^2. \tag{2.81}$$

Of course, the total intensity (the integral over frequency) converges (unlike in the case of ordinary bremsstrahlung) and a quantum mechanical calculation is not required. It is also possible to find the total intensity by starting with Eq. (2.48) in the general case. Introducing $q_\perp'$ and $q_\parallel'$, we obtain

$$Q = \frac{e^2e_i^2e_0^2n_i}{m^2c^3v_0\pi^2}\int_0^\infty d\omega \int d\Omega_k \int_0^\infty q'_\perp dq'_\perp \frac{1}{[q'^2_\perp + (\omega^2/v_0^2)(m_0c^2/\varepsilon_0)^2]^2} \times$$
$$\times \frac{\{[\mathbf{n}\mathbf{q}'_\perp] + (\omega/v_0^2)[\mathbf{n}\mathbf{v}_0](m_0c^2/\varepsilon_0)^2\}\, d\varphi}{\{r_d^2[q'^2_\perp + (\omega^2/v_0^2)(1-(v_0/c)\cos\theta)^2 + (\omega^2/c^2)\sin^2\theta - 2(\omega/c)q'_\perp \sin\theta\cos\varphi] + 1\}^2}. \tag{2.82}$$

We introduce

$$\lambda = -q'_\perp v_0/\omega. \tag{2.83}$$

Then, we have

$$Q=\frac{e^2e_0^2e_i^2n_i}{\pi^2m^2c^3v_0}\int_0^\infty d\omega\int d\Omega_k\int_0^{2\pi}d\varphi\int_0^\infty\lambda d\lambda\left\{[\mathbf{n}\boldsymbol{\lambda}]+\left(\frac{m_0c^2}{\varepsilon_0}\right)^2[\mathbf{n}\mathbf{v}_0]/v_0\right\}^2\times$$
$$\times\left[\lambda^2+\left(\frac{m_0c^2}{\varepsilon_0}\right)^2\right]^{-2}\left\{r_d^2\frac{\omega^2}{v_0^2}\left[\lambda^2+\left(1-\frac{v_0}{c}\cos\theta\right)^2+\right.\right.$$
$$\left.\left.+\frac{v_0^2}{c^2}\sin^2\theta-\frac{2v_0}{c}\lambda\sin\theta\cos\varphi\right]+1\right\}^{-2}. \tag{2.84}$$

The integral with respect to ω can be evaluated to yield

$$Q=\frac{e^2e_i^2e_0^2n_i}{2m^2c^3r_d}\int_0^\pi\sin\theta\,d\theta\int_0^{2\pi}d\varphi\int_0^\infty\frac{\lambda\,d\lambda\,\{[\mathbf{n}\boldsymbol{\lambda}]+[\mathbf{n}\mathbf{v}_0]\,m_0^2c^4/v_0\varepsilon_0^2\}^2}{[\lambda^2+(m_0c^2)^2/\varepsilon_0^2]^2}\times$$
$$\times\left[\lambda^2-2\frac{v_0}{c}\lambda\sin\theta\cos\varphi+\left(1-\frac{v_0}{c}\cos\theta\right)^2+\frac{v_0^2}{c^2}\sin^2\theta\right]^{-1/2}. \tag{2.85}$$

The last integral is a function of v_0/c alone. The final result of the integration has the form

$$Q=\frac{4\pi e^2e_0^2e_i^2n_i}{3m^2c^3r_d}\begin{cases}1, & v_0\ll c,\\ \frac{8}{5}\ln\frac{\varepsilon_0}{m_0c^2}, & v\to c,\ \varepsilon_0\gg m_0c^2.\end{cases} \tag{2.86}$$

This analysis shows that the overall intensity depends strongly on the Debye (shielding) radius, while in the region where the spectral intensity is greatest, it depends only logarithmically on the Debye radius [see Eqs. (2.64) and (2.76)]. Thus, to within logarithmic accuracy, we can neglect $k^2r_d^2$ compared to unity in the spectral intensity. This means that in Eq. (2.84) the factor in the curly brackets can be set equal to unity when integrating over λ to $\lambda_{max}\sim\xi v_0/\omega r_d$ [see Eq. (2.83)]. We then have

$$Q_\omega=\frac{8e^2e_i^2e_0^2n_i}{3m^2c^3v_0}\int_0^{\lambda_{max}^2}d\lambda^2\,\frac{\lambda^2+(m_0c^2/\varepsilon_0)^4}{[\lambda^2+(m_0c^2/\varepsilon_0)^2]^2}=\frac{16e^2e_i^2e_0^2n_i}{3m^2c^3v_0}\left(\ln\xi\frac{v_0\varepsilon_0}{\omega r_dm_0c^2}-\frac{v_0^2}{c^2}\right). \tag{2.87}$$

In the nonrelativistic limit ($v_0\ll c$), we obtain Eq. (2.64) with $\xi=1/(e)^{1/2}=0.7$ and in the ultrarelativistic limit, we obtain Eq. (2.76) with $\xi=1$.

2.5. Resonant Polarization Bremsstrahlung and the Role of the Density Effect

The dependence of the spectral density of the radiation on the electron density is referred to as the density effect. It originates in the density dependence of the refractive index for the emitted electromagnetic waves. The electron density also appears in the Debye shielding radius $r_d = \omega_{pe}/v_{Te}$ and, because of the quasineutrality of the plasma, n_i and n_e are coupled. This coupling, however, is not direct (since bremsstrahlung on a small number of impurity ions with a high Z may be of interest). The quasineutrality condition has the form $n_e = \Sigma_i n_i Z_i$.

The difference between the refractive index and unity can have an effect only for $\omega \to \omega_{pe}$ or for ultrarelativistic particles.

Indeed, for electromagnetic waves in a plasma we have

$$\varepsilon(\omega) = 1 - \omega_{pe}^2/\omega^2; \qquad k^2c^2 = \omega^2\varepsilon(\omega) = \omega^2 - \omega_{pe}^2 \tag{2.88}$$

so that when $\omega \to \omega_{pe}$, we have $\varepsilon(\omega) \to 0$ and $k \to 0$ and the intensity of the radiation should go to zero. On the other hand, for relativistic particles, even when $\omega \gg \omega_{pe}$, the quantity $(\omega_{pe}/\omega)^2$ may be of the same order of magnitude as $(m_0c^2/\varepsilon_0)^2$ and this will be reflected in the radiated power.

The generalization of Eq. (2.52) to the case $\varepsilon(\omega) \neq 1$ is fairly simple, so we present the result here. First, however, it is convenient to rewrite Eq. (2.52) in the form

$$Q = \frac{e^2 e_0^2 e_i^2 n_i}{\pi^2 m^2 c^4} \int \frac{dkdq\,([kv_0] + c\,[kq]/k)^2}{k^2(1+q^2r_d^2)^2((k-q)^2 - k^2)^2}\,\delta(kc - (k-q)\,v_0). \tag{2.89}$$

This generalization reduces to the following formula:

$$Q = \frac{e^2 e_0^2 e_i^2 n_i}{\pi^2 m^2 c^4} \int \frac{dkdq\,([kv_0] + [kq]\sqrt{k^2c^2 + \omega_{pe}^2}/k^2)^2}{k^2(1+q^2r_d^2)^2((k-q)^2 - k^2)^2}\,\delta(\sqrt{k^2c^2 + \omega_{pe}^2} - (k-q)\,v_0). \tag{2.90}$$

For frequencies $\omega \to \omega_{pe}$, $k \to 0$ and the second term in the numerator of Eq. (2.90) predominates, so that we obtain

$$Q \approx \frac{e^2 e_0^2 e_i^2 n_i}{\pi^2 m^2 c^4} \int \frac{dkdq\,\omega_{pe}^2\,[kq]^2}{k^6 q^4 (1+q^2r_d^2)^2}\,\delta(\omega_{pe} + qv_0). \tag{2.91}$$

We shall be interested in the integrated (over all angles) radiation. Then we have

$$\int d\Omega_k \frac{[kq]^2}{k^2} = \frac{8\pi}{3} q^2, \tag{2.92}$$

that is,

$$Q = \frac{8e^2 e_0^2 e_i^2 n_i \omega_{pe}^2}{3m^2c^3 v_0} \int \frac{dk}{k^2 c} \int_0^\infty \frac{dq_\perp^2}{(q_\perp^2 + \omega_{pe}^2/v_0^2)\,[1 + r_d^2(q_\perp^2 + \omega_{pe}^2/v_0^2)]^2}, \tag{2.93}$$

where $\mathbf{q}_\perp = \mathbf{q} - \mathbf{v}_0(\mathbf{q}\mathbf{v}_0)/v_0^2$.

Since $c^2kdk = \omega d\omega \sim \omega_{pe}d\omega$, we can write

$$\frac{dk}{k^2c} = \frac{\omega_{pe}d\omega}{k^3c^3} = \frac{\omega_{pe}d\omega}{(\omega^2-\omega_{pe}^2)^{3/2}}. \qquad (2.94)$$

For $v_0 \gg v_{Te}$, we have $\omega_{pe}^2 r_d^2/v_0^2 = v_{Te}^2/v_0^2 \ll 1$.

The integral with respect to $q_\perp^2$ can be calculated exactly, but here we shall calculate it approximately by introducing the quantity q_*, where

$$\omega_{pe}^2/v_0^2 \ll q_*^2 \ll 1/r_d^2. \qquad (2.95)$$

Then, in the region $q_\perp^2 \ll q_*^2$ the integral is given approximately by

$$\int_0^{q_*^2} \frac{dq_\perp^2}{q_\perp^2 + \omega_{pe}^2/v_0^2} = \ln\frac{q_*^2 v_0^2}{\omega_{pe}^2}, \qquad (2.96)$$

and for $q_\perp^2 \gg q_*^2$, it is given by

$$\int_{q_*^2}^{\infty} \frac{dq_\perp^2}{q_\perp^2(1+q_\perp^2 r_d^2)^2} = \ln\frac{1}{q_*^2 r_d^2} - 1. \qquad (2.97)$$

Thus, given Eqs. (2.94), (2.96), and (2.97), we obtain

$$Q_\omega = \frac{16e^2e_0^2e_i^2n_i}{3m^2c^3v_0}\frac{\omega_{pe}^3}{(\omega^2-\omega_{pe}^2)^{3/2}}\left(\ln\frac{v_0}{\omega_{pe}r_d} - \frac{1}{2}\right) \qquad (2.98)$$

It is noteworthy that the radiated power approaches infinity as $\omega \to \omega_{pe}$. This has a clear explanation in that the frequency ω_{pe} is the resonant frequency of the shield cloud. Thus, it is a manifestation of a typical polarization radiation effect, namely resonant polarization bremsstrahlung.

A comparison with the formulas for ordinary bremsstrahlung shows that polarization (transition) bremsstrahlung becomes predominant when $\omega \to \omega_{pe}$. Here, it is true, we must make one caveat about the validity of Eq. (2.98). In deriving Eq. (2.98), we have used the condition

$$v_0 \gg v_{Te}. \qquad (2.99)$$

Thus, the polarization bremsstrahlung is dominant for sufficiently fast particles. Even for them, however, the radiated power cannot go to infinity as $\omega \to \omega_{pe}$.

An analysis (on which we shall dwell in detail here, although it can be carried out following the same computational scheme used above) shows that there is another criterion which indicates that Eq. (2.98) is not applicable near the resonance $\omega = \omega_{pe}$, i.e.,

$$|\omega - \omega_{pe}|/\omega_{pe} \gg v_{Te}^2/v_0^2. \qquad (2.100)$$

The inequality (2.99) shows precisely that the right-hand side of Eq. (2.100) is small, i.e., the resonance is clearly distinct.

When the inequality opposite to Eq. (2.100), i.e.,

$$|\omega-\omega_{pe}|/\omega_{pe} \ll v_{Te}^2/v_0^2, \tag{2.101}$$

is satisfied, a calculation yields [1, 7]

$$Q_\omega = \frac{2e^2e_0^2e_i^2n_i}{m^2c^3v_0}\left(\frac{v_0}{v_{Te}}\right)^4 \frac{\sqrt{\omega^2-\omega_{pe}^2}}{\omega_{pe}}. \tag{2.102}$$

This formula shows that the spectral intensity of the polarization bremsstrahlung goes to zero when $\omega \to \omega_{pe}$. However, the resonance is clearly distinct and the spectral intensity reaches a maximum at

$$\omega-\omega_{pe} \approx \omega_{pe}v_{Te}^2/v_0^2, \tag{2.103}$$

when Eqs. (2.98) and (2.102) both yield estimates of the same order of magnitude [we assume that the logarithm in Eq. (2.98) is of order unity]:

$$(Q_\omega)_{\max} \approx \xi e^2e_0^2e_i^2n_iv_0^2/m^2c^3v_{Te}^3, \tag{2.104}$$

where ξ is of order unity.

This value is $(v_0/v_{Te})^3$ times greater than the spectral density of ordinary bremsstrahlung.

It is worth noting that that for thermal particles $v_0 \sim v_{Te}$ and it follows from Eq. (2.104) that here the polarization bremsstrahlung is not especially stronger than ordinary bremsstrahlung. Perhaps this is the reason why such a "clear" effect was overlooked and no attention was paid to it until recently. Now the effect is undoubtedly of interest because nonthermal distributions with "tails" extending far into the high energy region have been observed in plasmas. The number of these fast particles can be quite large (a percent or more). The spectral power of ordinary bremsstrahlung is proportional to $1/v_0$; that is, for the thermal particles, $1/v_{Te} \propto 1(T_e)^{1/2}$. If the fast particles in the "tail" can be characterized by a temperature T_e', then we have $Q_{\omega,\max} \propto T_e'/T_e^{3/2}$ according to Eq. (2.104).

We now consider the effect of having $\varepsilon(\omega) \neq 1$ (a refractive index unequal to unity for the emitted waves) for ultrarelativistic particles. It turns out that this inequality is important if

$$(m_0c^2/\varepsilon_0)^2 \lesssim \omega_{pe}^2/\omega^2. \tag{2.105}$$

Instead of Eq. (2.76), from Eq. (2.90) we obtain [1, 6]

$$Q_\omega = \frac{16e^2e_0^2e_i^2n_i}{3m^2c^4}\left[\ln\frac{c\omega_{pe}}{\omega v_{Te}\sqrt{(m_0c^2/\varepsilon_0)^2+(\omega_{pe}/\omega)^2}}-1\right]. \tag{2.106}$$

Thus, when Eq. (2.105) is satisfied, we have

$$Q_\omega = \frac{16e^2e_0^2e_i^2n_i}{3m^2c^4}\left(\ln\frac{c}{v_{Te}} - 1\right), \tag{2.107}$$

where the logarithm depends on ω only in the high-frequency region.

Equation (2.76), however, is applicable only in a limited region, since the inequality (2.72) still applies, or

$$\omega_{pe}\varepsilon_0/m_0c^2 \ll \omega \ll \omega_{pe}c/v_{Te}, \tag{2.108}$$

that is,

$$\varepsilon_0/m_0c^2 \ll c/v_{Te}. \tag{2.109}$$

If, on the other hand,

$$\varepsilon_0/m_0c^2 \gg c/v_{Te}, \tag{2.110}$$

then for frequencies $\omega \ll \omega_{pe}c/v_{Te}$, we clearly have $\omega \ll \omega_{pe}\varepsilon_0/m_0c^2$ and Eq. (2.107) is valid.

Meanwhile, for $\omega \gg \omega_{pe}c/v_{Te}$, instead of Eq. (2.79) we obtain

$$Q_\omega = \frac{2e^2e_0^2e_i^2n_i}{m^2c^4}\left(\frac{c}{\omega r_d}\right)^2\left\{\ln\frac{1}{\sqrt{2}\sqrt{(m_0c^2/\varepsilon_0)^2 + \omega_{pe}^2/\omega^2}} - \frac{1}{2}\right\}. \tag{2.111}$$

When $\omega_{pe}\varepsilon_0/m_0c^2 \gg \omega \gg \omega_{pe}c/v_{Te}$, the expression under the logarithm is frequency dependent with

$$Q_\omega = \frac{2e^2e_0^2e_i^2n_i}{m^2c^4}\left(\frac{c}{\omega r_d}\right)^2\left(\ln\frac{\omega}{\omega_{pe}\sqrt{2}} - \frac{1}{2}\right) \tag{2.112}$$

and Eq. (2.79) is valid only for $\omega \gg \omega_{pe}\varepsilon_0/m_0c^2$. If, on the other hand, the energies of the incident particles are small and the inequality (2.109), rather than (2.110), is satisfied, then the condition $\omega \gg \omega_{pe}c/v_{Te}$ automatically implies $\omega \gg \omega_{pe}\varepsilon_0/m_0c^2$ and Eq. (2.79) is valid. Finally, when

$$\omega \gg \omega_{pe}(c/v_{Te})(\varepsilon_0/m_0c^2)^2 \tag{2.113}$$

we automatically have $\omega \gg \omega_{pe}\varepsilon_0/m_0c^2$; that is, Eq. (2.81), in which the spectral density of the radiation intensity, Q_ω, is proportional to the fourth power of the particle energy, always holds.

Therefore, this analysis of the role of the density effect in the polarization radiation of relativistic particles shows that it only affects the corresponding argument in the logarithm.

The role of the density effect is entirely different in the ordinary bremsstrahlung of relativistic particles [1, 6]. First, shielding is important over a wider range of frequencies. Whereas the shielding criterion for nonrelativistic particles (the appearance of $1/r_d$ instead of ω/v_0 in the logarithm) is $\omega \ll \omega_{pe} v_0/v_{Te}$, in the ultrarelativistic limit it reduces to

$$\omega \ll \omega_{pe} (c/v_{Te}) (\varepsilon_0/m_0c^2)^2 \tag{2.114}$$

and the frequency range $\omega \ll \omega_{pe}\varepsilon_0/m_0c^2$ where the density effect may be important is clearly within a range that obeys the inequality (2.114). The calculations are extremely simple and can be carried out in the same way as those used to derive Eq. (1.2). The generalization of Eq. (1.2) to the case where the density effect must be taken into account has the form [6, 8, 9]

$$Q_\omega = \frac{16 e_0^4 e_i^2 n_i}{3 m_0^2 c^4} \frac{1}{[1 + (\omega_{pe}/\omega)^2 (\varepsilon_0/m_0c^2)^2]} \ln \frac{\rho_{max}}{r_d}. \tag{2.115}$$

Thus, even for electrons (for which $e_0 = -e$ and $m_0 = m$), when $\omega \ll \omega_{pe}\varepsilon_0/m_0c^2$, ordinary bremsstrahlung is strongly suppressed by the density effect and transition polarization bremsstrahlung becomes predominant. In this case it predominates at frequencies near ω_{pe}, but also over a wide range of frequencies up to $\omega \sim \omega_{pe}\varepsilon_0/m_0c^2$ [3, 6].

Until recently, this circumstance had also been ignored. From an experimental point of view, this is an important effect which is fully verifiable, especially for

$$\varepsilon_0/m_0c^2 \gg c/v_{Te}, \tag{2.116}$$

when a "dip" should be observed in the total emission intensity.

2.6. Polarization Radiation of Charged Particles on Dipoles. Model for Polarization Bremsstrahlung on Bound Electrons

We now consider a point polarizable dipole **p**, next to which a charge e_0 passes with a constant velocity at a distance of $\mathbf{r}_0$ (ρ_0, z_0 denote the coordinates at $t = 0$ and the dipole is located at $r = 0$). The magnitude of the dipole is zero in the absence of a perturbing electric field. A field with frequency ω polarizes the dipole and creates a dipole moment

$$\mathbf{p}_\omega = \alpha(\omega) \mathbf{E}_\omega \delta(\mathbf{r}), \tag{2.117}$$

where $\alpha(\omega)$ is the polarizability and $\mathbf{p}_\omega$ is the dipole moment per unit volume.

We shall treat the field of an incident charge e_0 as the acting field [see Eqs. (2.25) and (2.28)]. Then we have

$$\mathbf{p} = \int \alpha((\mathbf{k}\mathbf{v}_0)) \mathbf{E}_\mathbf{k}^{(0)} \frac{d\mathbf{k}\, d\mathbf{q}}{(2\pi)^3} \exp\{i\mathbf{k}(\mathbf{r} - \mathbf{v}_0 t) + i\mathbf{k}\mathbf{r}_0 + i\mathbf{q}\mathbf{r}\}. \tag{2.118}$$

Here we have used the fact that $\omega = \mathbf{k}\mathbf{v}_0$.

As before, we make the substitution $\mathbf{k} \to \mathbf{k} - \mathbf{q}$ in order to obtain the spatial component of $\mathbf{p}$ in the form $\exp(i\mathbf{kr})$:

$$\mathbf{p} = \int \alpha((\mathbf{k}-\mathbf{q})\mathbf{v}_0)\,\mathbf{E}^{(0)}_{\mathbf{k}-\mathbf{q}} \exp\{i\mathbf{kr} - i(\mathbf{k}-\mathbf{q})\mathbf{v}_0 t + i(\mathbf{k}-\mathbf{q})\mathbf{r}_0\}\, d\mathbf{k}\, d\mathbf{q}\,(2\pi)^{-3}. \quad (2.119)$$

The current density $\delta\mathbf{j}_p$ will be equal to

$$\delta\mathbf{j}_p = -\int i(\mathbf{k}-\mathbf{q})\mathbf{v}_0 \alpha((\mathbf{k}-\mathbf{q})\mathbf{v}_0)\,\mathbf{E}^{(0)}_{\mathbf{k}-\mathbf{q}} \exp[i\mathbf{kr} - i(\mathbf{k}-\mathbf{q})\mathbf{v}_0 t + i(\mathbf{k}-\mathbf{q})\mathbf{r}_0]\, d\mathbf{k}\, d\mathbf{q}\,(2\pi)^{-3}. \quad (2.120)$$

Let us compare this expression with Eq. (2.42) for a charge shielded by a "cloud" in a plasma. Equation (2.42) can be rewritten in the form

$$\delta\mathbf{j}_i = \int \frac{d\mathbf{k}\, d\mathbf{q}}{i(\mathbf{k}-\mathbf{q})\mathbf{v}_0} \frac{ee_i \mathbf{E}^{(0)}_{\mathbf{k}-\mathbf{q}}}{m(q^2 r_d^2 + 1)} \exp[i\mathbf{kr} - i(\mathbf{k}-\mathbf{q})\mathbf{v}_0 t + i(\mathbf{k}-\mathbf{q})\mathbf{r}_0]\,(2\pi)^{-3}. \quad (2.121)$$

A comparison shows that the result for a dipole can be obtained by making the substitution

$$\frac{ee_i}{m(1+q^2 r_d^2)} \to \omega^2 \alpha(\omega), \quad \omega = (\mathbf{k}-\mathbf{q})\mathbf{v}_0. \quad (2.122)$$

This substitution can also be made in Eq. (2.90) on noting that $(k^2c^2 + \omega_{pe}{}^2)^{1/2}$ must be replaced by the frequency $\omega_\mathbf{k}$ obtained by solving the propagation equation for electromagnetic waves,

$$k^2 = (\omega_\mathbf{k}^2/c^2)\,\varepsilon(\omega_\mathbf{k}) = (\omega_\mathbf{k}^2/c^2)\, n^2(\omega_\mathbf{k}), \quad (2.123)$$

where $n(\omega)$ is the refractive index of a medium consisting of dipoles:

$$Q = \frac{e_0^2 n_p}{\pi^2 c^4} \int \frac{d\mathbf{k}\, d\mathbf{q}}{k^2} \frac{([\mathbf{k}\mathbf{v}_0] + \omega_\mathbf{k}[\mathbf{kq}]/k^2)^2}{((\mathbf{k}-\mathbf{q})^2 - k^2)^2} (\omega_\mathbf{k}^2 \alpha(\omega_\mathbf{k}))^2\, \delta(\omega_\mathbf{k} - (\mathbf{k}-\mathbf{q})\mathbf{v}_0) \quad (2.124)$$

(here n_p is the density of dipoles).

For a gaseous dipole medium, we have

$$\varepsilon(\omega) = 1 + 4\pi\alpha(\omega) n_p. \quad (2.125)$$

The assumption of point dipoles is valid only when the components of the transverse momentum transfer are less than $1/R_a$, where R_a is the size of the dipole (atom). Thus, in the final formulas the quantity r_d that appeared previously under the logarithm must be replaced by R_a. For nonrelativistic particles, when

$$\omega \ll v_0/R_a \quad (2.126)$$

Eq. (2.64) is replaced by the formula that results from substituting Eq. (2.122) and $r_d \to R_a$ [10]:

$$Q_\omega = \frac{16}{3} \frac{|\omega^2 \alpha(\omega)|^2 e_0^2 n_p}{c^3 v_0} \ln \frac{v_0}{\omega R_a}. \tag{2.127}$$

Here we have omitted the term 1/2 under the logarithm, since the accuracy of the argument of the logarithm is inadequate in any case.

A rigorous quantum mechanical theory should yield an expression for R_a in the logarithm and a value on the order of unity for the expression under the logarithm, as well as an expression for $\alpha(\omega)$. It is clear from the very beginning, however, that R_a is on the order of the size of an atom or molecule (i.e., on the order of the size of the Bohr orbit for the hydrogen atom, so it contains $\hbar$) and, since R_a appears logarithmically, there is little value in obtaining a more precise value for use in rough calculations. As for the polarizability $\alpha(\omega)$, it is also given by expressions containing $\hbar$ in the quantum theory. However, experimental values of $\alpha(\omega)$ can be substituted directly in Eq. (2.125). The formulas for the intensity of the polarization radiation at frequencies $\omega > v_0/R_a$ already contain R_a explicitly in front of the logarithm, so that the correct formulas will depend on the specific value of R_a and will require a more exact quantum mechanical calculation.

For ultrarelativistic particles without the density effect [$\varepsilon(\omega)$ is given by the plasma formula at high frequencies] when

$$\omega \ll c/R_a \tag{2.128}$$

we obtain

$$Q_\omega = \frac{16}{3} \frac{|\omega^2 \alpha(\omega)|^2}{c^4} e_0^2 n_p \ln \left(\frac{c}{\omega R_a} \frac{\varepsilon_0}{m_0 c^2} \right). \tag{2.129}$$

For an ion with charge e_i which is not completely ionized and has bound electrons which are described by a polarizability $\alpha(\omega)$, we must note that it represents both a charge e_i and a variable dipole $\mathbf{p}(\omega)$. The static charge e_i is shielded by a Debye cloud if the ion is in a plasma. Then the following substitution must be made when writing down the equations:

$$\frac{ee_i}{m(1+q^2 r_d^2)} \to \frac{ee_i}{m(1+q^2 r_d^2)} + \omega^2 \alpha(\omega). \tag{2.130}$$

If $R_a \ll r_d$, then when $\omega \ll v_0/r_d$, both the bound and free electrons will contribute. When $v_0/r_d \ll \omega \ll v_0/R_a$, the free electrons are "switched off" and their contribution falls off as $1/\omega^4$. The radiation on the bound electrons is still important. If the atom has different shells with $R_1 > R_2 > R_3$..., then the different shells will gradually be "switched off" as the frequency increases. We shall examine these effects quantitatively in a later theoretical analysis and compare the theoretical results with experimental data. We must still say a few words about Eq. (2.130).

For wavelengths much greater than the Debye radius ($q^2r_d^2 \ll 1$), one can write the polarizability of the Debye shielding cloud in the form

$$\alpha_d = ee_i/m\omega^2. \tag{2.131}$$

Then Eq. (2.130) will have the form

$$ee_i/m \to \omega^2(\alpha(\omega) + \alpha_d(\omega)) \tag{2.132}$$

or

$$\alpha_d(\omega) \to \alpha(\omega) + \alpha_d(\omega), \tag{2.133}$$

that is, instead of the polarizability of the cloud, we have the sum of the polarizabilities of the cloud and the bound electrons.

The physical meaning of Eq. (2.131) is extremely simple. For coherent radiation, when $k \ll 1/r_d$, the cloud oscillates as a whole, i.e.,

$$\mathbf{p} = e_i\delta\mathbf{r}, \tag{2.134}$$

where $\delta\mathbf{r}$ is the displacement of the electrons in the field $\mathbf{E}$ and is given by

$$\delta\mathbf{r} = e\mathbf{E}/m\omega^2, \tag{2.135}$$

and this displacement is determined by the charge to mass ratio of the electron, since the cloud is made up of electrons. Substituting Eq. (2.135) in Eq. (2.134) immediately yields (2.131).

2.7. Polarization Bremsstrahlung on Dipole Molecules in Plasmas

A molecule can be in a state such that it has a dipole moment even when there is no external field. This static dipole moment must also be included in Eq. (2.117). We now consider the radiation from a static dipole in a plasma:

$$\mathbf{p} = \mathbf{p}_0\delta(\mathbf{r}). \tag{2.136}$$

A shielding cloud develops around the dipole in the plasma.

Instead of Eq. (2.22), the static field will obey the equation

$$\mathbf{E}_{\mathbf{k}}^{\mathbf{p}_0} = \frac{\mathbf{k}\,(\mathbf{k}\mathbf{p}_0)}{2\pi^2\,(k^2 + 1/r_d^2)}, \tag{2.137}$$

while the perturbation in the particle distribution will be given by [instead of Eq. (2.23)]

$$\delta f^{(p_0)}_{\mathbf{p},\mathbf{k}} = \frac{ie\,(\mathbf{k}\mathbf{p}_0)\,\Phi_{\mathbf{p}}}{2\pi^2 T_e\,(k^2 + 1/r_d^2)} . \tag{2.138}$$

This means that in all formulas we must make the substitution

$$e_0 \rightarrow -i\,(\mathbf{q}\mathbf{p}_0). \tag{2.139}$$

Thus, for example, instead of Eq. (2.48) we obtain (in the same approximations for which Eq. (2.89) is valid)

$$Q = \frac{e^2 e_0^2 p_0^2 n_{\mathbf{p}_0}}{3\pi^2 m^2 c^4} \int \frac{d\mathbf{k}\, d\mathbf{q} \cdot q^2\,([\mathbf{k}\mathbf{v}_0] + c\,[\mathbf{k}\mathbf{q}]/k)^2}{k^2\,(q^2 r_d^2 + 1)^2\,((\mathbf{k}-\mathbf{q})^2 - k^2)^2}\, \delta\,(kc - (\mathbf{k}-\mathbf{q})\,\mathbf{v}_0), \tag{2.140}$$

where $n_{\mathbf{p}0}$ is the number of dipoles per unit volume.

Here, for the sake of simplicity it was assumed that the dipoles are oriented randomly, i.e.,

$$\overline{(\mathbf{q}\mathbf{p}_0)^2} = {}^1\!/_3\, q^2 p_0^2. \tag{2.141}$$

On integrating over the exit angles of the emitted photons, for nonrelativistic incident particle velocities we obtain

$$Q_\omega = \frac{8e^2 e_0^2 p_0^2 n_{\mathbf{p}_0}}{9m^2 c^3 v_0} \int_0^\infty d\omega \int_0^\infty \frac{dq_\perp^2}{[1 + r_d^2\,(q_\perp^2 + \omega^2/v_0^2)]^2} . \tag{2.142}$$

Then

$$Q_\omega = \frac{8e^2 e_0^2 p_0^2 n_{\mathbf{p}_0}}{9m^2 c^3 v_0 r_d^2} \frac{1}{[1 + \omega^2 r_d^2/v_0^2]} . \tag{2.143}$$

The spectrum will be flat for $\omega \le v_0/r_d$, and for $\omega \gg v_0/r_d$ the spectrum falls off as $1/\omega^2$. The explanation of this result is the same as above for a charge. The polarization cloud oscillates as a whole (coherently) only for wavelengths greater than the Debye radius.

The total radiation intensity is equal to

$$Q = 4\pi e^2 e_0^2 p_0^2 n_{\mathbf{p}_0}/9m^2 c^3 r_d^3. \tag{2.144}$$

If we write

$$\mathbf{p}_0 = e_i \mathbf{r}_0, \tag{2.145}$$

where $\mathbf{r}_0$ is the characteristic size of the dipole, then the radiation from a dipole is a factor of

$$(r_0/r_d)^2 \ll 1 \tag{2.146}$$

times smaller than the emission from a charge in the region where the spectrum is flat. This inequality is necessary because the dipole was assumed to be a point, which means that its dimensions are small compared to the Debye radius.

For ultrarelativistic incident particles with $\omega \ll c/r_d$, the additional factor q^2 in the numerator of the spectral density can be replaced in practice by $q_\perp^2$, so that instead of Eq. (2.75), we obtain

$$Q_\omega = \frac{8e^2 e_0^2 p_0^2 n_{p_0}}{3m^2 c^4} \int_0^\infty dq_\perp^2 q_\perp^4 (1 + q_\perp^2 r_d^2)^{-2} [q_\perp^2 + (\omega/c)^2 (mc^2/\varepsilon_0)^2]^{-2}. \tag{2.147}$$

Introducing the variable $x = q_\perp^2 r_d^2$, we have

$$Q_\omega = \frac{8e^2 q_0^2 p_0^2 n_{p_0}}{m^2 c^4 r_d^2} \int_0^\infty \frac{x^4\, dx}{(1+x)^2 \left[x^2 + \frac{\omega^2}{c^2} r_d^2 \left(\frac{m_0 c^2}{\varepsilon_0}\right)^2\right]^2}, \tag{2.148}$$

and, since $m_0 c^2 \ll \varepsilon_0$ and $r_d^2 \omega^2 \ll c^2$, this gives

$$Q_\omega = 8e^2 e_0^2 p_0^2 n_{p_0} / m^2 c^4 r_d^2. \tag{2.149}$$

2.8. Total Bremsstrahlung for Distant Collisions

We have shown above that when the scatterers have different moments (charge, dipole, etc.) and different types of bound and free electrons, one must add the polarizabilities (amplitudes) of the radiation source in order to calculate the total radiated power. In other words, the different mechanisms interfere. As already noted, however, it is also impossible to separate the resulting radiation from ordinary bremsstrahlung.

In Chapter 1 the ordinary bremsstrahlung was computed from the acceleration of the incident particle by the field of a shielded ion acting on it from the side, under the assumption that the acceleration itself is determined by the unperturbed motion.

This, of course, is permissible for distant collisions. In this approximation it is easy to find the acceleration, as well as the perturbation in the velocity, $\delta\mathbf{v}(t)$, and the perturbation in the trajectory, $\delta\mathbf{r}(t)$. From these it is not difficult to find the general form for the current created by the particle and then, using Eq. (2.1), to find the radiated power Q. Naturally, for nonrelativistic particles (see Sec. 2.1), we obtain the formulas which appeared earlier [see Eq. (1.5)]. But it is also easy to write down the result for the general case of a relativistic incident particle [1]:

$$\delta \mathbf{j}_T = \frac{4\pi e_0^2 e_i}{m_0 (2\pi)^3} \int \frac{dk\, dq r_d^2 [\mathbf{v}_0 (\mathbf{kq}) + (\mathbf{qv}_0)(\mathbf{q} + \mathbf{v}_0 ((\mathbf{k}-\mathbf{q})\mathbf{v}_0)/c^2)]}{(\mathbf{qv}_0)^2 (1 + q^2 r_d^2)} \times$$
$$\times \exp[i\mathbf{kr} + i(\mathbf{k}-\mathbf{q})\mathbf{r}_0 - i(\mathbf{k}-\mathbf{q})\mathbf{v}_0 t]. \qquad (2.150)$$

A comparison with the expression for the polarization radiation current (2.42) shows that the total current, corresponding to ordinary as well as polarization bremsstrahlung (i.e., general bremsstrahlung), is obtained from Eq. (2.42) by making the following substitution:

$$\Gamma^{(0)}_{\mathbf{k}-\mathbf{q}} \rightarrow \mathbf{E}^{(0)}_{\mathbf{k}-\mathbf{q}} + \mathbf{E}^{(T)}_{\mathbf{k}-\mathbf{q}}, \qquad (2.151)$$

with

$$\mathbf{E}^{(T)}_{\mathbf{k}-\mathbf{q}} = -\frac{e_0^2 m r_d^2 (\mathbf{k}-\mathbf{q})\, \mathbf{v}_0}{2\pi^2 e m_0 (\mathbf{qv}_0)} \{\mathbf{v}_0 (\mathbf{kq}) + [\mathbf{q} + \mathbf{v}_0 (\mathbf{v}_0 (\mathbf{k}-\mathbf{q}))/c^2] (\mathbf{qv}_0)\}. \qquad (2.152)$$

Of course, r_d appears here only because the factor $(1 + q^2 r_d^2)$ has been isolated in Eq. (2.150). When $r_d \rightarrow \infty$ (no shielding), the current (2.150) is independent of r_d (as it should be). In the nonrelativistic limit $v_0 \ll c$, as we can see, $k \ll q$ and Eq. (2.152) becomes extremely simple:

$$\mathbf{E}^{(T)}_{\mathbf{k}-\mathbf{q}} = e_0^2 m r_d^2 \mathbf{q} / 2\pi^2 e m_0. \qquad (2.153)$$

From this and Eq. (2.50), we immediately obtain Eq. (1.5).

Equation (2.151) makes it possible to take the interference of polarization bremsstrahlung and the usual bremsstrahlung into account. This interference is very important for bremsstrahlung of nonrelativistic electrons. At sufficiently high frequencies, the amplitude associated with the shielding and the amplitude of the polarization bremsstrahlung are compensated.

If it is still necessary to include bound electrons, then in the first term of Eq. (2.151), we replace the coefficient in accordance with Eq. (2.122). When a constant dipole moment is present, one must also add the terms corresponding to the substitution (2.139) in both terms of Eq. (2.151). This yields the total bremsstrahlung intensity for distant collisions in the most general form.

We conclude with a number of general comments on polarization bremsstrahlung in distant collisions.

The examples given here do not, of course, exhaust all the possible manifestations of polarization radiation in distant collisions. Polarization radiation can originate in the polarization of the surrounding medium, as well as of the particles themselves. Thus, when interpreting experiments on real targets, we must take both effects into account. Polarization radiation can even occur in a gas of neutral molecules with constant dipole moments. Each dipole surrounds itself with oppositely oriented dipoles. The dipole itself and the cloud oscillate because the field of the incident particle is nonuniform. For wavelengths greater than the dimensions of the cloud, these oscillations are in opposite phase and quench one another. If, however, the medium has dipole particles of different types, then complete quenching does not take place. Broad atmospheric showers of relativistic particles, especially, can generate electromagnetic waves (including radio waves) in the atmosphere.

The character of the polarized dipole moments can be utterly different. Dipoles are not necessarily molecules; they can be polarized dielectric or other dust particles. Unlike transition radiation, the generation of which requires that the size of the dust particles be greater than the zone in which the radiation develops, here it is only necessary that the wavelength be greater than the size of a dust particle. Cosmic rays may generate radiation on dust particles, especially since this kind of radiation does not at all require a change in the velocity of the incident particles, which might be heavy particles of cosmic rays (such as protons).

These calculations have shown that the basic features of polarization bremsstrahlung can be described in classical terms. This does not mean that a quantum mechanical approach is not needed, especially for polarization radiation by bound electrons. The limits of applicability of the classical description will be clarified below.

As we have pointed out, ordinary bremsstrahlung is inseparable from polarization bremsstrahlung. Interference effects may be important. The bremsstrahlung mechanism also exists for close collisions, for which polarization radiation does not vanish, by any means. And finally, bremsstrahlung requires a quantum mechanical analysis, if only because the maximum emitted frequency is determined by quantum mechanical conditions.

Chapter 3

FLUCTUATIONS IN IDEAL GASES AND PLASMAS AND BREMSSTRAHLUNG

V. N. Tsytovich

3.1. Fluctuations in Nonequilibrium Media and Bremsstrahlung

The role of fluctuations has already been mentioned in Chapter 1. Here we can distinguish two questions: one concerning the polarization radiation of fast particles in a fluctuating medium and the other, the radiation by the medium as such because of the fluctuations. The first question is closer to the problem discussed in Chapters 1 and 2. If the fluctuations are similar in origin to those in a gas of free particles, then the final result should correspond to the sum of the polarization radiation on the individual ions, atoms, and molecules. In the above discussion, we found the radiation from a fast particle by summing the radiation on the individual ions, atoms, and molecules. A proof of the statement that the emission from a fast particle on plasma fluctuations (in the approximation of an ideal plasma) is the same as the sum of the emission from the individual ions has been given in [1, 2].

Radiation on medium fluctuations was first examined by Kapitsa [3] and Ter-Mikaélyan [3a] (in order to make the results agree with the formulas for polarization bremsstrahlung, one must [4] eliminate a number of errors such as the coefficient $1/2\pi^2$ left out of [3]). Formulas are also given in [3] for arbitrary polarization $\alpha(\omega)$ of the polarized particles and these can be used for the case of a gas with polarizable molecules.

It is then easy to confirm that the above statement that the emission from fluctuations is the same as the sum of the polarization emissions is also correct for an ideal gas of polarizable molecules or atoms. Thus, this result applies to any gaseous medium or plasma containing ions, atoms, or molecules.

We now consider the fluctuation radiation from the medium itself. In this case, we assume that there are thermal free particles in the medium; that is, the medium must be a plasma, perhaps partially ionized, that contains both ions and neutral molecules (and, of course, electrons).

A proof that the bremsstrahlung from a fully ionized plasma is strictly determined by the square of the sum of the amplitudes of the polarization and ordinary bremsstrahlung is given in [1, 2]. This proof applies to an arbitrary nonequilibrium particle distribution. This last point is very important, since, in principle, a fast nonthermal particle may emit in a different way from a thermal one (quantitatively, this is indeed so, but here we are speaking of the mechanism, which in this case is the same).

Since the proof is given for arbitrary nonequilibrium distributions, the nonequilibrium distribution might consist of a single fast particle against a background formed by the remaining thermal particles. This, however, is the previous case. Thus, the results of [1, 2] are considerably more general that those which might be obtained by comparing the emission from fast particles in a fluctuating medium with the emission from individual particles. They do apply to any nonequilibrium distribution.

Of course, there is as yet no such proof for an incompletely ionized plasma containing partially ionized atoms and polarizable molecules, but we can scarcely doubt, on the basis of what is now known, that such a proof could be provided (for an ideal gas).

This (if we assume that a proof exists) leads to the important result that the emission from a partially ionized plasma is always determined by polarization bremsstrahlung and ordinary bremsstrahlung and not just by ordinary bremsstrahlung.

We have previously considered an elementary event of polarization bremsstrahlung involving three particles: a photon, ion (atom), and electron (or other incident particle). For simplicity the ion was assumed to be at rest. Of course, this simplification can be dropped for an individual event, while in a medium consisting of many particles the ions (atoms) must be regarded as moving (it is impossible to transform to a coordinate system that is at rest relative to a scattering center, since we have singled out a reference system in which the medium is at rest). Here we come upon the question of the kinetics of all the particles that participate in the process. We are speaking of changes in the distributions of the incident particles (electrons), scattering particles (ions, atoms), and radiation. The process leading to changes in the distribution of the radiation, known as inverse bremsstrahlung, should include the inverse of polarization bremsstrahlung.

In a number of cases, two of these three components can be regarded as given. Then it is possible to study the behavior of the third with a given distribution of the other two. This involves the linear absorption of electromagnetic waves radiation field and a given ion (atom) distribution. These problems have a limited applicability. When the energy input is high (as in laser irradiation), we must examine the kinetics of all three components. In a number of cases, special consideration of the kinetics of any one of the components is dictated by the physics of the process. Thus, the absorption of radiation, as an integral effect, may not be determined by fast electrons, whose kinetics is of interest in the ionization of plasmas, breakdown, etc.

We have mentioned the absorption of electromagnetic waves as the inverse of polarization bremsstrahlung. In a fully ionized plasma, polarization bremsstrahlung originates in the clouds surrounding the particles. Usually, the absorption of electromagnetic waves is calculated in a fully ionized plasma using the collision integral of Landau [5]. Why, then, has the absorption associated with polarization bremsstrahlung been ignored? The answer to this question lies in the fact that the additional terms in the absorption corresponding to inverse polarization radiation can actually be obtained by using the Landau–Balescu collision integral [6] in place of the Landau collision integral. The former differs from the latter in that Debye shielding of the field of the colliding particles is taken into account by a factor $1/|\varepsilon|^2$ (where ε is the dielectric constant) in the collision integral. Usually, the damping of electromagnetic waves is calculated as a perturbation by the electromagnetic wave of the distributions of the colliding particles, $\delta\Phi_{\mathbf{p}}$ and $\delta\Phi_{\mathbf{p}'}$, which appear in the collision integral, but the perturbation in the factor $1/|\varepsilon|^2$, which also depends on $\delta\Phi_{\mathbf{p}}$ and $\delta\Phi_{\mathbf{p}'}$, is neglected. These perturbations correspond to a change in the polarization cloud of the particles and yield inverse polarization bremsstrahlung. Of course, a large contribution to absorption may come from terms associated with the direct change in $\Phi_{\mathbf{p}}$ and $\Phi_{\mathbf{p}'}$, but not always. It must also be noted that the absorption is determined by an integral over all the particles. Ion–ion collisions are an example where the perturbations of $1/|\varepsilon|^2$ may be significant. In an analysis of absorption on polarized atoms, terms describing inverse polarization bremsstrahlung naturally appear in the collision integral.

3.2. The Balance Equation and the Probabilities of Bremsstrahlung

We now examine the kinetics of bremsstrahlung for particles of type α on particles of type β, with distribution functions Φ^α and Φ^β respectively . That is, we are considering bremsstrahlung in the case where a large number of both incident particles and centers are present (Φ^α and Φ^β, for example) and, in general, may be moving, rather than the bremsstrahlung of individual particles on randomly positioned centers (in general, including polarization effects). In this sense, the division into incident particles and scattering centers is arbitrary. In particular, for a beam of incident particles with momentum $\mathbf{p}_0$, we have $\Phi_{\mathbf{p}}^\alpha = (2\pi)^3 n_\alpha \delta(\mathbf{p} - \mathbf{p}_0)$, while for motionless scattering centers, $\Phi^\beta = (2\pi)^3 n_\beta \delta(\mathbf{p})$. The coefficient $(2\pi)^3$ is chosen for normalization:

$$n_\alpha = \int \Phi_{\mathbf{p}}^{\alpha} d\mathbf{p}/(2\pi)^3, \qquad n_\beta = \int \Phi_{\mathbf{p}}^{\beta} d\mathbf{p}/(2\pi)^3, \tag{3.1}$$

where n_α and n_β are the densities of particles of type α and β per cm^3.

Let us consider the case in which the bremsstrahlung photons have random phases, e.g., when they originate from the fluctuation radiation of the medium. They can then be characterized by their distribution functions, specifically by the number $N_{\mathbf{k}}^{\sigma}$ of photons, which is defined so that the energy of the photons is given by ($\hbar$ is Planck's constant and σ is the type of polarization of the photons)

$$E^{\sigma} = \int \hbar\omega_{\mathbf{k}} N_{\mathbf{k}}^{\sigma}\, d\mathbf{k}/(2\pi)^3, \tag{3.2}$$

and their number per cubic centimeter is

$$N^{\sigma} = \int N_{\mathbf{k}}^{\sigma}\, d\mathbf{k}/(2\pi)^3. \tag{3.3}$$

Thus, the three components involved in bremsstrahlung have the distribution functions $\Phi_{\mathbf{p}}^{\alpha}$, $\Phi_{\mathbf{p}}^{\beta}$, and $N_{\mathbf{k}}^{\sigma}$.

The problem consists of describing the kinetics of the variation in $\Phi_{\mathbf{p}}^{\alpha}$, $\Phi_{\mathbf{p}}^{\beta}$, and $N_{\mathbf{k}}^{\sigma}$ owing to bremsstrahlung. To do this, we can introduce two probabilities: the probability of particle collisions and the probability of bremsstrahlung in such collisions. It will be shown that the bremsstrahlung probability automatically includes polarization bremsstrahlung, provided the theory of fluctuation radiation is developed rigorously. For now, we shall derive the kinetic equation phenomenologically by introducing the concept of probabilities. During the collisions a momentum $\mathbf{q}$ is transferred from one particle to another ($\hbar = 1$); i.e.,

$$\mathbf{p}_\alpha = \mathbf{p}_\alpha + \mathbf{q}; \qquad \mathbf{p}_\beta = \mathbf{p}_\beta' - \mathbf{q}; \qquad \varepsilon_{\mathbf{p}_\alpha} + \varepsilon_{\mathbf{p}_\beta} = \varepsilon_{\mathbf{p}_\alpha'} + \varepsilon_{\mathbf{p}_\beta'}. \tag{3.4}$$

Here the primes denote the final states.

We denote the corresponding probability per unit time within the interval of momentum transfers $d\mathbf{q}/(2\pi)^3$ by $w_{\mathbf{p}_\alpha,\mathbf{p}_\beta}(\mathbf{q})$.

The balance equation for Φ^α owing to collisions has the obvious form

$$\frac{d\Phi_{\mathbf{p}_\alpha}^{\alpha}}{dt} = -\int \frac{d\mathbf{p}_\beta\, d\mathbf{q}}{(2\pi)^6} \left[w_{\mathbf{p}_\alpha,\mathbf{p}_\beta}(q)\, \Phi_{\mathbf{p}_\alpha}^{\alpha} \Phi_{\mathbf{p}_\beta}^{\beta} - w_{\mathbf{p}_\alpha+\mathbf{q},\mathbf{p}_\beta-\mathbf{q}}(\mathbf{q})\, \Phi_{\mathbf{p}_\alpha+q}^{\alpha} \Phi_{\mathbf{p}_\beta-\mathbf{q}}^{\beta}\right]. \tag{3.5}$$

For $q \ll |\mathbf{p}_\alpha|, |\mathbf{p}_\beta|$, an expansion in $\mathbf{q}$ yields [1]

$$\frac{d\Phi_{\mathbf{p}_\alpha}^{\alpha}}{dt} = \hat{I}_{\mathbf{p}_\alpha}^{\mathrm{col}} \Phi_{\mathbf{p}_\alpha}^{\alpha} = \frac{1}{2} \int \frac{\partial}{\partial p_{\alpha,i}} w_{\mathbf{p}_\alpha,\mathbf{p}_\beta}^{(0)}(\mathbf{q}) \left(\Phi_{\mathbf{p}_\beta}^{\beta} \frac{\partial \Phi_{\mathbf{p}_\alpha}^{\alpha}}{\partial p_{\alpha,j}} - \Phi_{\mathbf{p}_\alpha}^{\alpha} \frac{\partial \Phi_{\mathbf{p}_\beta}^{\beta}}{\partial p_{\beta,j}} \right) q_i q_j \frac{d\mathbf{q}\, d\mathbf{p}_\beta}{(2\pi)^6}, \tag{3.6}$$

where $w_{\mathbf{p}_\alpha,\mathbf{p}_\beta}^{(0)}(\mathbf{q})$ is the zeroth (in terms of the parameter $q/|\mathbf{p}|$) approximation for the collision probability.

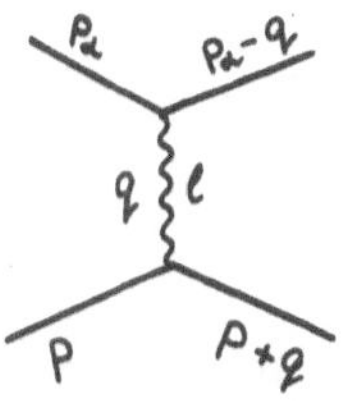

Fig. 3.1. Diagram of the scattering of two particles.

This equation can be derived simply from the theory of fluctuations. A collision can occur with momentum transfer to a virtual longitudinal wave or to a virtual transverse wave (Fig. 3.1). For nonrelativistic particles we can restrict ourselves to a virtual longitudinal wave. Then, we obtain [1, 6]

$$w^{(0)}_{\mathbf{p}_\alpha,\mathbf{p}_\beta}(\mathbf{q}) = \frac{4e_\alpha^2 e_\beta^2 \delta(\mathbf{q}\mathbf{v}_\alpha - \mathbf{q}\mathbf{v}_\beta)}{q^4 \,|\, \varepsilon^l(\mathbf{q}\mathbf{v}_\alpha, \mathbf{q})\,|^2}\,. \tag{3.7}$$

This result corresponds to the Landau–Balescu collision integral [6] which takes shielding of the colliding particles into account. The result was obtained in the form (3.7) by Balescu [6] in a complicated system of Bogolyubov correlation chains [8] and Tsytovich [1] using the so-called correspondence principle, and by Sitenko [7] from fluctuations. There is an analogous result for partially ionized plasmas [8] in which shielding can be partially determined by resonant molecules. The quantity ε^l in Eq. (3.7) depends on Φ^α and Φ^β, while their perturbation by a regular external wave also gives the inverse polarization bremsstrahlung effect mentioned above. This type of absorption is especially important when strongly polarizable molecules exist in the medium (see Chapter 11).

Collisions are a necessary element of the kinetics of bremsstrahlung and have an effect on the particle distributions. In addition, these distributions change as a result of the bremsstrahlung process itself.

Let us denote the momentum of a bremsstrahlung photon by $\mathbf{k}$ ($\hbar = 1$) and the change in the momentum of particle β during bremsstrahlung by $\mathbf{q}$. Then we have

$$\mathbf{p}_\beta = \mathbf{p}_\beta' + \mathbf{q} \tag{3.8}$$

and

$$\mathbf{p}_\alpha' = \mathbf{p}_\alpha + \mathbf{q} - \mathbf{k}. \tag{3.9}$$

We denote the probability of bremsstrahlung per second within $d\mathbf{k}/(2\pi)^3$ and $d\mathbf{q}/(2\pi)^3$ by $w_{\mathbf{p}_\alpha,\mathbf{p}_\beta}(\mathbf{q})$.

Including spontaneous and stimulated bremsstrahlung, we write the equation for particles of type α in the form [1]

$$\frac{d\Phi^{\alpha}_{\mathbf{p}_\alpha}}{dt} = \hat{I}^{\text{col}}_{\mathbf{p}_\alpha}\Phi_{\mathbf{p}_\alpha} - \int \{w_{\mathbf{p}_\alpha,\mathbf{p}_\beta}(\mathbf{q},\mathbf{k})\,[\Phi^{\alpha}_{\mathbf{p}_\alpha}\Phi^{\beta}_{\mathbf{p}_\beta}(N^{\sigma}_{\mathbf{k}}+1) -$$
$$- N^{\sigma}_{\mathbf{k}}\Phi^{\alpha}_{\mathbf{p}_\alpha+\mathbf{q}-\mathbf{k}}\Phi^{\beta}_{\mathbf{p}_\beta-\mathbf{q}}] + w_{\mathbf{p}_\alpha-\mathbf{q}+\mathbf{k},\mathbf{p}_\beta+\mathbf{q}}(\mathbf{q},\mathbf{k})\,[\Phi^{\alpha}_{\mathbf{p}_\alpha}\Phi^{\beta}_{\mathbf{p}_\beta}N^{\sigma}_{\mathbf{k}} -$$
$$- \Phi^{\alpha}_{\mathbf{p}_\alpha-\mathbf{q}+\mathbf{k}}\Phi^{\beta}_{\mathbf{p}_\beta+\mathbf{q}}(N^{\sigma}_{\mathbf{k}}+1)]\}\,d\mathbf{k}\,d\mathbf{q}\,d\mathbf{p}_\beta/(2\pi)^9. \tag{3.10}$$

Here we isolate the terms for spontaneous (without $N_{\mathbf{k}}^{\sigma}$) and stimulated (with $N_{\mathbf{k}}^{\sigma}$) emission. Assuming that the momentum transfer $\mathbf{q}$ and the momentum $\mathbf{k}$ of the emitted photon are small compared to the momenta of the particles, $\mathbf{p}_\alpha$ and $\mathbf{p}_\beta$, we obtain the following formulas for the spontaneous processes:

$$\left(\frac{d\Phi^{\alpha}_{\mathbf{p}_\alpha}}{dt}\right)^{\text{rad}}_{\text{sp}} = \frac{\partial}{\partial \mathbf{p}_\alpha}\mathbf{F}^{\text{sp,rad}}_{\mathbf{p}_\alpha}\Phi^{\alpha}_{\mathbf{p}_\alpha} \tag{3.11}$$

and

$$\mathbf{F}^{\text{sp,rad}}_{\mathbf{p}_\alpha} = \int (\mathbf{k}-\mathbf{q})\,w_{\mathbf{p}_\alpha,\mathbf{p}_\beta}(\mathbf{q},\mathbf{k})\,\Phi_{\mathbf{p}_\beta}\frac{d\mathbf{q}\,d\mathbf{k}\,d\mathbf{p}_\beta}{(2\pi)^9}. \tag{3.12}$$

It should be noted that for spontaneous processes the integral in Eq. (3.12) is taken over all $\mathbf{k}$ except those for which quantum mechanical effects are important. Thus, Eq. (3.12) can be used only for relatively small $\mathbf{k}$ (the classical region). In the quantum mechanical region, we must use a more exact expression derived from Eq. (3.10) that does not contain the derivative $\partial/\partial p_\alpha$, as does Eq. (3.11).

Equation (3.12) evidently describes the frictional force owing to spontaneous bremsstrahlung. For polarization transition bremsstrahlung, as we have shown, a drop in the intensity often occurs before quantum mechanical effects become important. Thus, Eq. (3.12) is applicable in those cases where polarization bremsstrahlung dominates. Since the emission at high frequencies is usually through ordinary bremsstrahlung, however, the use of Eqs. (3.11) and (3.12) is unjustified there.

These problems do not arise with stimulated bremsstrahlung effects, if $N_{\mathbf{k}}$ has a limited and fairly narrow frequency interval. When $|\mathbf{q}|, |\mathbf{k}| \ll p_\alpha, p_\beta$, we have

$$\left(\frac{d\Phi^{\alpha}_{\mathbf{p}_\alpha}}{dt}\right)^{\text{rad}}_{\text{ind}} = \frac{\partial}{\partial p_{\alpha,i}}D^{\alpha,\text{ind,rad}}_{ij}\frac{\partial\Phi_{\mathbf{p}_\alpha}}{\partial p_{\alpha,j}} + \frac{\partial}{\partial p_{\alpha,i}}F^{\text{ind,rad}}_{\mathbf{p}_\alpha,i}\Phi^{\alpha}_{\mathbf{p}_\alpha}, \tag{3.13}$$

where

$$D^{\alpha\,\text{ind,rad}}_{ij} = \int (\mathbf{k}-\mathbf{q})_i(\mathbf{k}-\mathbf{q})_j\,w_{\mathbf{p}_\alpha,\mathbf{p}_\beta}(\mathbf{q},\mathbf{k})\,N^{\sigma}_{\mathbf{k}}\Phi_{\mathbf{p}_\beta}\frac{d\mathbf{k}\,d\mathbf{q}\,d\mathbf{p}_\beta}{(2\pi)^9} \tag{3.14}$$

and

$$F^{\text{ind,rad}}_{\mathbf{p}_\alpha,i} = +\int (q_i - k_i)\left(\mathbf{q}\frac{\partial\Phi^{\beta}_{\mathbf{p}_\beta}}{\partial\mathbf{p}_\beta}\right)w_{\mathbf{p}_\alpha,\mathbf{p}_\beta}(\mathbf{q},\mathbf{k})\,N^{\sigma}_{\mathbf{k}}\frac{d\mathbf{k}\,d\mathbf{q}\,d\mathbf{p}_\beta}{(2\pi)^9}. \tag{3.15}$$

Since the momentum transfer **q** and the momentum **k** of the emitted wave are introduced asymmetrically for the particles α and β through Eq. (3.8) (if desired, they can be symmetrized by using **q** to denote the quantity **q** − **k**, on which the final result, naturally, does not depend), the expressions for the particles β are somewhat different. Thus, in the spontaneous force, the equation for $\Phi_{\mathbf{p}}^{\beta}$ includes $\mathbf{q}\Phi_{\mathbf{p}_\alpha}^{\alpha}$, rather than $(\mathbf{k}-\mathbf{q})\Phi_{\mathbf{p}}$; the D_{ij}^{ind} include $q_i q_j \Phi_{\mathbf{p}_\alpha}^{\alpha}$ instead of $(\mathbf{k}-\mathbf{q})_i(\mathbf{k}-\mathbf{q})_j\Phi_{\mathbf{p}}$; and, the F_i^{ind} include $q_i(\mathbf{k}-\mathbf{q})\partial\Phi_{\mathbf{p}_\alpha}^{\alpha}/\partial\mathbf{p}_\alpha$ instead of $(k_i - q_i)\,(\mathbf{q}\partial\Phi_{\mathbf{p}}/\partial\mathbf{p})$.

The balance equation for the photons has the form [1]

$$\frac{dN_{\mathbf{k}}^{\sigma}}{dt} = \int w_{\mathbf{p}_\alpha,\mathbf{p}_\beta}(\mathbf{q},\mathbf{k})\,[\Phi_{\mathbf{p}_\alpha}^{\alpha}\Phi_{\mathbf{p}_\beta}^{\beta}(N_k^{\sigma}+1) - N_{\mathbf{k}}^{\sigma}\Phi_{\mathbf{p}_\alpha+\mathbf{q}-\mathbf{k}}^{\alpha}\Phi_{\mathbf{p}_\beta-\mathbf{q}}^{\beta}]\,\frac{d\mathbf{p}_\alpha d\mathbf{p}_\beta\,d\mathbf{q}}{(2\pi)^9} = \left(\frac{dN_{\mathbf{k}}^{\sigma}}{dt}\right)_{\text{sp}} + \left(\frac{dN_{\mathbf{k}}^{\sigma}}{dt}\right)_{\text{ind}}. \tag{3.16}$$

For spontaneous emission we have

$$\left(\frac{dN_{\mathbf{k}}^{\sigma}}{dt}\right)_{\text{sp}} = \int w_{\mathbf{p}_\alpha,\mathbf{p}_\beta}(\mathbf{q},\mathbf{k})\,\Phi_{\mathbf{p}_\alpha}^{\alpha}\Phi_{\mathbf{p}_\beta}^{\beta}\,\frac{d\mathbf{q}\,d\mathbf{p}_\alpha\,d\mathbf{p}_\beta}{(2\pi)^9}. \tag{3.17}$$

In particular, the intensity of spontaneous emission per unit volume is given by

$$Q = \int \omega_{\mathbf{k}}^{\sigma}\left(\frac{dN_{\mathbf{k}}^{\sigma}}{dt}\right)_{\text{sp}}\frac{d\mathbf{k}}{(2\pi)^3} = \int Q_{\mathbf{p}_\alpha}^{\alpha}\Phi_{\mathbf{p}_\alpha}\frac{d\mathbf{p}_\alpha}{(2\pi)^3} = \int Q_{\mathbf{p}_\alpha,\mathbf{p}_\beta}^{\alpha,\beta}\Phi_{\mathbf{p}_\alpha}^{\alpha}\Phi_{\mathbf{p}_\beta}^{\beta}\,\frac{d\mathbf{p}_\alpha\,d\mathbf{p}_\beta}{(2\pi)^6}, \tag{3.18}$$

where

$$Q_{\mathbf{p}_\alpha}^{\alpha} = \int Q_{\mathbf{p}_\alpha,\mathbf{p}_\beta}^{\alpha,\beta}\Phi_{\mathbf{p}_\beta}^{\beta}\,\frac{d\mathbf{p}_\beta}{(2\pi)^3}$$
$$Q_{\mathbf{p}_\alpha,\mathbf{p}_\beta}^{\alpha,\beta} = \int \omega_{\mathbf{k}}^{\sigma}w_{\mathbf{p}_\alpha,\mathbf{p}_\beta}(\mathbf{q},\mathbf{k})\,\frac{d\mathbf{k}\,d\mathbf{q}}{(2\pi)^6}. \tag{3.19}$$

If the particles α are fast incident particles and β denotes heavy particles at rest, then the radiation from an isolated incident particle is given by a formula that follows from Eq. (3.18):

$$Q_{\mathbf{p}_\alpha}^{\alpha} = Q_{\mathbf{p}_\alpha,0}^{\alpha,\beta}n_\beta. \tag{3.20}$$

The stimulated effect can be written in the form of a growth rate (or damping rate) in the number of photons:

$$\left(\frac{dN_{\mathbf{k}}^{\sigma}}{dt}\right)_{\text{ind}} = 2\gamma_{\mathbf{k}}^{\sigma}N_{\mathbf{k}}^{\sigma}. \tag{3.21}$$

When $k,\, q \ll |\mathbf{p}_\alpha|,\, |\mathbf{p}_\beta|$, we obtain

$$2\gamma_{\mathbf{k}}^{\sigma} = \int w_{\mathbf{p}_\alpha,\mathbf{p}_\beta}(\mathbf{q},\mathbf{k})\left[\left(\mathbf{k}-\mathbf{q}\right)\frac{\partial\Phi_{\mathbf{p}_\alpha}^{\alpha}}{\partial\mathbf{p}_\alpha}\Phi_{\mathbf{p}_\beta} + \mathbf{q}\frac{\partial\Phi_{\mathbf{p}_\beta}^{\beta}}{\partial\mathbf{p}_\beta}\Phi_{\mathbf{p}_\alpha}\right]\frac{d\mathbf{p}_\alpha d\mathbf{p}_\beta d\mathbf{q}}{(2\pi)^9}. \tag{3.22}$$

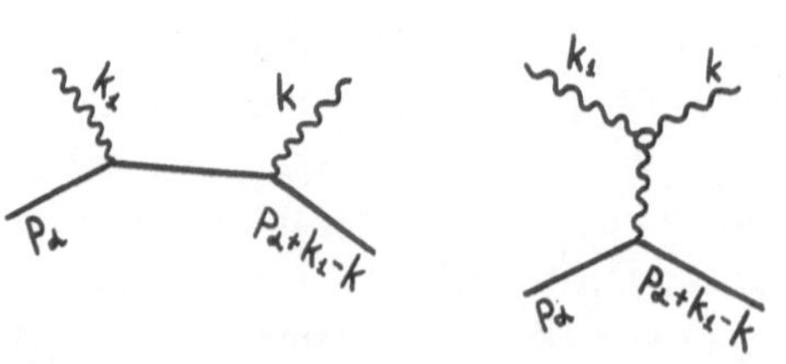

Fig. 3.2. Diagrams of the scattering of waves on particles: (a) Thomson scattering; (b) transition scattering.

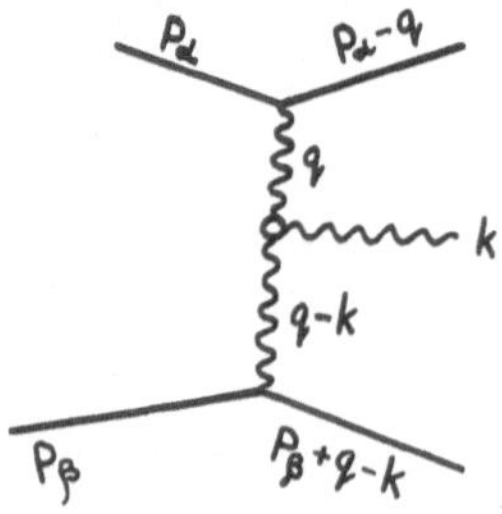

Fig. 3.3. Diagram of transition bremsstrahlung.

The quantity $\gamma_{\mathbf{k}}^{\sigma}$ can be either positive or negative for nonequilibrium particle distributions. In the first case, one speaks of a bremsstrahlung instability and in the second, of stimulated bremsstrahlung absorption or the inverse bremsstrahlung effect.

3.3. Nonlinear Currents and the Probabilities of Polarization Radiation

The conservation of energy in an elementary bremsstrahlung event has the form ($\hbar = 1$)

$$\varepsilon_{\mathbf{p}_\alpha} + \varepsilon_{\mathbf{p}_\beta} = \varepsilon_{\mathbf{p}_\beta - \mathbf{q}} + \varepsilon_{\mathbf{p}_\alpha + \mathbf{q} - \mathbf{k}} + \omega_{\mathbf{k}}^{\sigma} \tag{3.23}$$

and, when $q, k \ll |p_\alpha|, |p_\beta|$, yields

$$\omega_{\mathbf{k}}^{\sigma} = \mathbf{q}\mathbf{v}_\beta + (\mathbf{k} - \mathbf{q})\,\mathbf{v}_\alpha. \tag{3.24}$$

The bremsstrahlung probability must contain a δ-function which describes the conservation of energy (3.24) in an elementary event:

$$w_{\mathbf{p}_\alpha,\mathbf{p}_\beta}(\mathbf{q},\mathbf{k}) = v_{\mathbf{p}_\alpha,\mathbf{p}_\beta}(\mathbf{q},\mathbf{k})\,\delta(\omega_{\mathbf{k}}^{\sigma} - \mathbf{q}\mathbf{v}_\beta - (\mathbf{k}-\mathbf{q})\mathbf{v}_\alpha). \tag{3.25}$$

$v_{\mathbf{p}_\alpha,\mathbf{p}_\beta}$ is determined by the square of the matrix element for the process, which in the framework of the quasiclassical limit ($q, k \ll |\mathbf{p}_\alpha|, |\mathbf{p}_\beta|$) will be equal to the product of the matrix element for formation of a virtual field q and for scattering of the virtual field into an electromagnetic wave. It is known that two types of scattering can occur: ordinary Thomson scattering (Fig. 3.2a) and transition scattering (Fig. 3.2b).

Transition scattering takes place on the polarization "cloud" surrounding a charge, and is nonlinear in this sense (see the circle in Fig. 3.2b). Thus, polarization bremsstrahlung should be described by the diagram of Fig. 3.3.

Polarization bremsstrahlung, therefore, must be determined by nonlinear currents in the system. The polarization bremsstrahlung of bound electrons is obviously obtained in a similar fashion (see below, where the field of the nucleus is taken into exact account for the bound electrons, rather than through a perturbation theory approximation), but here we limit ourselves to the simplest case of free colliding particles, i.e., a plasma. More precisely, we shall consider arbitrary nonlinear media, but use perturbation theory to obtain the fields of the colliding particles (for bound electrons, the field of the nucleus cannot be treated by perturbation theory). In the following discussion of the coupling of polarization radiation with nonlinear currents in the system, we shall follow Ginzburg and Tsytovich [4].

Because the field of the colliding particles can be treated by perturbation theory, we shall use a general expression for the nonlinear current,

$$j_i(k) = \int S_{i,j,l}(k;q)\,E_{j,q}E_{l,k-q}dq, \qquad q = \{q_0, \mathbf{q}\},\ dq = dq_0 d\mathbf{q}, \tag{3.26}$$

where $S_{ijl}(k;q)$ is a tensor which is expressed in terms of the known nonlinear polarization tensor.

For the fields $\mathbf{E}$ in Eq. (3.26) it is enough to substitute the fields of two colliding particles, assuming, as usual, that the tensor $S_{i,j,l}$ is symmetrized with respect to j and l:

$$j_{k,i} = 2\int S_{i,j,l}(k;q)\,E^{\beta}_{j,q}E^{\alpha}_{l,k-q}dq, \tag{3.27}$$

where the field of the particle moving freely in the medium is determined from Maxwell's equations by its current as

$$E^{\alpha;\beta}_{k,i} = G_{il,k}\,j^{\alpha;\beta}_{k,l}, \tag{3.28}$$

with $G_{ij,k}$ for an isotropic medium given by

$$G_{i,j}(k) = -\frac{4\pi i k_i k_j}{\omega k^2 \varepsilon_k^l} + \left(\delta_{ij} - \frac{k_i k_j}{k^2}\right)\frac{4\pi i\omega}{c^2\left(k^2 - \dfrac{\omega^2}{c^2}\varepsilon_k^t\right)}. \tag{3.29}$$

The Green function (3.29) describes two virtual lines in the diagram (Fig. 3.3) for transition polarization bremsstrahlung. The first term of Eq. (3.29) corresponds to a virtual longitudinal wave and the second, to a virtual transverse wave.

The particle currents $j_{k,i}{}^{\alpha,\beta}$ in Eq. (3.28) are given by the ordinary expressions for the classical currents of charged particles moving uniformly and rectilinearly at velocities $\mathbf{v}_\alpha$ and $\mathbf{v}_\beta$:

$$\mathbf{j}_k^{\alpha,\beta} = \frac{e_{\alpha,\beta}\mathbf{v}_{\alpha,\beta}}{(2\pi)^3}\,\delta(\omega - \mathbf{k}\mathbf{v}_{\alpha,\beta})\exp(i\mathbf{k}\mathbf{r}_{0,\alpha,\beta}), \tag{3.30}$$

where e_α and e_β are the charges of the particles; $\mathbf{r}_{0,\alpha}$ and $\mathbf{r}_{0,\beta}$ are the vectors of the impact parameter (for simplicity, the distance between the particles at $t = 0$ can be specified by $\mathbf{r}_{0,\beta} = 0$ with $\mathbf{r}_0 = \mathbf{r}_{0,\alpha} - \mathbf{r}_{0,\beta}$).

Then the current which produces the bremsstrahlung photon can be written in the form

$$\begin{aligned} j_{k,t} = {} & \frac{2e_\alpha e_\beta}{(2\pi)^3}\int S_{i,j,l}(\omega, \mathbf{k}; \mathbf{q}\mathbf{v}_\beta, \mathbf{q})\, G_{l,s}(\mathbf{q}\mathbf{v}_\beta, \mathbf{q}) \times \\ & \times G_{j,n}(\omega - \mathbf{q}\mathbf{v}_\alpha, \mathbf{k} - \mathbf{q})\, v_{\beta,s} v_{\alpha,n}\delta(\omega - \mathbf{q}\mathbf{v}_\beta - \\ & - (\mathbf{k} - \mathbf{q})\mathbf{v}_\alpha)\exp(i\mathbf{r}_0(\mathbf{k} - \mathbf{q}))\, d\mathbf{q}. \end{aligned} \tag{3.31}$$

The δ-function in Eq. (3.31) includes the conservation law (3.24) for bremsstrahlung emission of a photon with $\omega_{\mathbf{k}}{}^\sigma$ if we just set $\omega = \omega^\sigma$.

The power radiated by the current $\mathbf{j}$ can be found by Landau's method as in the previous chapter. After this, one can find the power radiated in a medium with randomly distributed particles by averaging over the impact parameters r_0. Comparing this power with Eqs. (3.18) and (3.19), we obtain [1, 4]

$$v_{p_\alpha, p_\beta}(\mathbf{q}, \mathbf{k}) = \frac{2(2\pi)^2[\mathbf{k}\mathbf{M}^{\alpha,\beta}]^2}{k^2\left|\dfrac{\partial}{\partial\omega}\omega^2\varepsilon_k^t\right|_{\omega=\omega_{\mathbf{k}}^\sigma}}, \tag{3.32}$$

where

$$M_i^{\alpha\beta} = 2e_\alpha e_\beta S_{ijl}(\omega_{\mathbf{k}}^\sigma, \mathbf{k}; \mathbf{q}\mathbf{v}_\beta, \mathbf{q})\, v_{\beta,s} v_{\alpha,n} G_{l,s}(\mathbf{q}\mathbf{v}_\beta, \mathbf{q}) \times G_{j,n}(\omega_{\mathbf{k}}^\sigma - \mathbf{q}\mathbf{v}_\beta, \mathbf{k} - \mathbf{q}). \tag{3.33}$$

In the general case, including ordinary bremsstrahlung reduces to making the substitution

$$\mathbf{M}^{\alpha\beta} \to \mathbf{M}^{\alpha\beta} + \mathbf{M}^\alpha + \mathbf{M}^\beta, \tag{3.34}$$

where $\mathbf{M}^\alpha$ and $\mathbf{M}^\beta$ are the effects owing to Thomson scattering of virtual photons on particle α and particle β (naturally, if α is a light particle and β is a heavy ion, then it is only necessary to include $\mathbf{M}^\alpha$).

Expressions for $\mathbf{M}^\alpha$ and $\mathbf{M}^\beta$ follow from Eq. (2.152)

$$M_i^\alpha = e_\beta (2\pi)^3 \Lambda_{ij}^\alpha (\omega_{\mathbf{k}}^\sigma, \mathbf{k}; \mathbf{q}\mathbf{v}_\beta, \mathbf{q}) G_{j,l} (\mathbf{q}\mathbf{v}_\beta, \mathbf{q}) v_{\beta,l}, \tag{3.35}$$

and

$$M_i^\beta = e_\alpha (2\pi)^3 \Lambda_{ij}^\beta (\omega_{\mathbf{k}}^\sigma, \mathbf{k}; \omega - \mathbf{q}\mathbf{v}_\beta, \mathbf{k} - \mathbf{q}) G_{j,l} (\omega_{\mathbf{k}}^\sigma - \mathbf{q}\mathbf{v}_\beta, \mathbf{k} - \mathbf{q}) v_{\alpha,l} \tag{3.36}$$

where

$$\Lambda_{ij}^\alpha (\omega, \mathbf{k}; \omega_1, \mathbf{k}_1) = \frac{e_\alpha^2 \sqrt{1 - v_\alpha^2/c^2 i c^3}}{m_\alpha \omega_1 (2\pi)^3 (\omega - \mathbf{k}\mathbf{v}_\alpha)^2} \{\delta_{ij} (\omega - \mathbf{k}\mathbf{v}_\alpha)^2 +$$
$$+ (\omega - \mathbf{k}\mathbf{v}_\alpha)(v_{\alpha,i} k_j + v_{\alpha,j} k_{1,i}) + v_{\alpha,i} v_{\alpha,j} (\mathbf{k}\mathbf{k}_1 - \omega\omega_1/c^2)\}. \tag{3.37}$$

In this way, we have written down the probability of bremsstrahlung (in the classical limit) in a general form that also includes transition polarization bremsstrahlung.

3.4. Fluctuations and Polarization Bremsstrahlung

It is well known that homogeneous media and, in particular, uniform plasmas do not radiate. Here, of course, we are speaking of homogeneity on the average. If the fluctuations are present, the distribution function must be broken up into fluctuating and regular parts, i.e.,

$$f_{\mathbf{p}_\alpha}^\alpha = \Phi_{\mathbf{p}_\alpha}^\alpha + \delta f_{\mathbf{p}_\alpha}^\alpha. \tag{3.38}$$

The assertion of homogeneity and stationarity applies to $\Phi_{\mathbf{p}\alpha}{}^\alpha$, for which

$$\langle f_{\mathbf{p}_\alpha}^\alpha \rangle = \Phi_{\mathbf{p}_\alpha}^\alpha, \quad \langle \delta f_{\mathbf{p}_\alpha}^\alpha \rangle = 0. \tag{3.39}$$

In order to describe a system of particles using $\Phi_{\mathbf{p}\alpha}{}^\alpha$, we must average the initial kinetic equation for the plasma,

$$\frac{\partial f_{\mathbf{p}_\alpha}^\alpha}{\partial t} + \mathbf{v}_\alpha \frac{\partial f_{\mathbf{p}_\alpha}^\alpha}{\partial \mathbf{r}} + e_\alpha \left(\mathbf{E} + \left[\frac{\mathbf{v}_\alpha}{c} \mathbf{H} \right] \right) \frac{\partial f_{\mathbf{p}_\alpha}^\alpha}{\partial \mathbf{p}_\alpha} = 0 \tag{3.40}$$

over the statistical ensemble. We may assume that there are no regular fields (although one can include both $\langle \mathbf{E} \rangle$ and $\langle \mathbf{H} \rangle$). Then $\delta\mathbf{E}$ and $\delta\mathbf{H}$ are expressed in terms of δf with the aid of Maxwell's equations. The equation for δf is obtained from Eq. (3.40) by subtracting the averaged equation.

In the linear approximation we have

$$i(\omega - \mathbf{k}\mathbf{v}_\alpha)\,\delta f^\alpha_{\mathbf{p},k} = e_\alpha\left(\delta\mathbf{E}_k + \left[\frac{\mathbf{v}_\alpha}{c}\,\mathbf{H}_k\right]\right)\partial\Phi^\alpha_{\mathbf{p}_\alpha}/\partial\mathbf{p}_\alpha, \tag{3.41}$$

while the averaged equation will have the form

$$\frac{\partial\Phi^\alpha_{\mathbf{p}_\alpha}}{\partial t} + \mathbf{v}_\alpha\frac{\partial\Phi^\alpha_{\mathbf{p}_\alpha}}{\partial\mathbf{r}} = -e_\alpha\left\langle\left(\delta\mathbf{E} + \left[\frac{\mathbf{v}_\alpha}{c}\,\delta\mathbf{H}\right]\right)\frac{\partial\delta f^\alpha_{\mathbf{p}_\alpha}}{\partial\mathbf{p}_\alpha}\right\rangle. \tag{3.42}$$

We can derive the Landau–Balescu collision integral (3.7) and (3.8) from Eqs. (3.41) and (3.42) in two short lines. For this it is sufficient to do the following: a) drop the term with the magnetic field in Eqs. (3.41) and (3.42) for the case of nonrelativistic particles; b) express δE in terms of δf through Poisson's equation

$$i(\mathbf{k}\delta\mathbf{E}_k) = 4\pi\sum_\beta e_\beta\int\delta f^\beta_{\mathbf{p}_\beta}\,d\mathbf{p}_\beta/(2\pi)^3 \tag{3.43}$$

and, most importantly, c) when solving Eq. (3.41) include the fluctuations which exist in the system when no fields are present [the solution of the homogeneous equation given by the left-hand side of Eq. (3.41)], as well as the fluctuations induced by the fluctuating fields δE. It is clear that these δf will be proportional to $\delta(\omega - \mathbf{k}\mathbf{v})$. The proportionality coefficient is easily found if we use the standard theorem of statistical mechanics which states that the mean square fluctuation in a gas of free particles within a volume V is equal to the average number of particles in that volume. This yields

$$\langle\delta f^{\alpha(0)}_{\mathbf{p},k}\delta f^{\alpha(0)}_{\mathbf{p}',k'}\rangle = \Phi^\alpha_{\mathbf{p}}\delta(\mathbf{p}-\mathbf{p}')\,\delta(k+k')\,\delta(\omega - \mathbf{k}\mathbf{v}). \tag{3.44}$$

Thus, for electrostatic fluctuating fields we have

$$\delta f^\alpha_{\mathbf{p},k} = \delta f^{\alpha(0)}_{\mathbf{p},k} + \frac{e_\alpha\delta\mathbf{E}^{(0)}_k}{i(\omega - \mathbf{k}\mathbf{v}_\alpha)}\frac{\partial\Phi^\alpha_{\mathbf{p}_\alpha}}{\partial\mathbf{p}_\alpha} \tag{3.45}$$

in Eq. (3.41). Substituting this equation together with Eq. (3.43) in Eq. (3.42) immediately leads to the collision integral given by Eqs. (3.7) and (3.8)

Bremsstrahlung effects are obtained by doing the calculations for higher-order terms in the charges of the particles and, in particular, for terms that are nonlinear in δf. It is then easy to obtain several terms in these balance equations for $\Phi_{\mathbf{p}_\alpha}$, $\Phi_{\mathbf{p}_\beta}$, and $N_\mathbf{k}{}^\sigma$. Previously the frictional force F owing to bremsstrahlung has been used [1, 2]. However, it is simpler to illustrate the result with the example of spontaneous polarization bremsstrahlung. To do this, we must obtain an equation for N^σ. $N_\mathbf{k}{}^\sigma$ itself is given in terms of the free transverse electromagnetic field $\delta E_k{}^{\sigma(0)}$ by taking the average $\langle\delta E_{k,i}{}^{\sigma(0)}\delta E_{k',j}{}^{\sigma(0)}\rangle$. $\delta E_k{}^\sigma$ is determined by solving Maxwell's equations with right-hand terms caused by the fluctuations in the particle distribution. When solving this equation, we must take the solution of the free equation into account (as when solving the equation for δf) along with the forced term. This yields $\delta E_{k,i}^{\sigma(0)}$, whose mean square gives $N_{\mathbf{k}}^{\sigma}$.

In setting up the equation for $N_{\mathbf{k}}^{\sigma}$ it is easiest to multiply the Maxwell equation by $\delta\mathbf{E}$. Then $\delta\mathbf{E}\delta\mathbf{j}$, that is, the work performed by the field on the current, appears on the right (compare with Landau's method in Sec. 2.1). This is actually the standard energy balance equation. After averaging, $dN_{\mathbf{k}}^{\sigma}/dt$ appears on the left and $\delta\mathbf{E}\delta\mathbf{j}$ appears on the right. This is the same expression used to calculate the radiated power in the Landau method, but here it is for fluctuating currents and fields.

Later, to obtain the spontaneous emission we can go to the limit $N_{\mathbf{k}}^{\sigma} \to 0$ in $\delta\mathbf{E}\delta\mathbf{j}$ (the stimulated term drops out). This means, however, that $\delta\mathbf{E}$ can be expressed in terms of $\delta\mathbf{j}$ by using the forced solution of the Maxwell equation. Then we obtain an expression for the radiated power that contains $\langle[\mathbf{k}\delta\mathbf{j}]^2\rangle$. Here we can substitute the different $\delta\mathbf{j}$. Three contributions appear. Two of them are associated with bremsstrahlung and we shall discuss the third somewhat later.

Polarization bremsstrahlung is described by the following term. In the equation for δf we include the nonlinear term that was dropped in Eq. (3.41),

$$i(\omega - \mathbf{k}\mathbf{v}_\alpha)\,\delta f^{\alpha}_{\mathbf{p}_\alpha,k} = e_\alpha \int \left(\delta\mathbf{E}_q + \left[\frac{\mathbf{v}_\alpha}{c}\,\delta\mathbf{H}_q\right]\right)\frac{\partial}{\partial\mathbf{p}_\alpha}\,\delta f^{\alpha}_{\mathbf{p}_\alpha,k-q}\,dq, \tag{3.46}$$

and (in the first approximation) we substitute Eq. (3.41) in place of δf on the right.

This yields

$$\delta\mathbf{j} = \int e_\alpha \mathbf{v}_\alpha \frac{\delta f^{\alpha}_{\mathbf{p}_\alpha,k}}{(2\pi)^3}\,d\mathbf{p}_\alpha = \int S_{ijl}(k;q)\,(\delta E_{j,q}\delta E_{l,k-q} - \langle\delta E_{j,q}\delta E_{l,k-q}\rangle)\,dq, \tag{3.47}$$

where

$$S_{ijl} = -\frac{1}{2}\sum_{\beta} e_\beta^3 \int \frac{1}{\omega - \mathbf{k}\mathbf{v}_\beta}\left\{\left(\frac{\mathbf{q}}{q}\frac{\partial}{\partial\mathbf{p}_\beta}\right)\frac{1}{\omega - q_0 - (\mathbf{k}-\mathbf{q})\,\mathbf{v}_\beta}\,\times\right.$$

$$\left.\times\,\frac{(\mathbf{k}-\mathbf{q})}{|\mathbf{k}-\mathbf{q}|}\frac{\partial}{\partial\mathbf{p}_\beta} + \left(\frac{(\mathbf{k}-\mathbf{q})}{|\mathbf{k}-\mathbf{q}|}\frac{\partial}{\partial\mathbf{p}_\beta}\right)\frac{1}{q_0 - (\mathbf{q}\mathbf{v}_\beta)}\left(\frac{\mathbf{q}}{q}\frac{\partial}{\partial\mathbf{p}_\beta}\right)\right\}\Phi^{\beta}_{\mathbf{p}_\beta}\frac{d\mathbf{p}_\beta}{(2\pi)^3}\,. \tag{3.48}$$

We have obtained the current (3.26) for a specific form of the nonlinear polarizability coefficients S_{ijl} that is well known from plasma physics. Then, instead of Eq. (3.28) we write

$$\delta E^{\alpha,\beta}_{k,i} = G_{k,i,j}\,\delta j^{\alpha,\beta(0)}_{k,i}, \tag{3.49}$$

where $\delta j_{k,i}^{\alpha,\beta(0)}$ is caused by the "zero" fluctuations in the field of the particles,

$$\delta\mathbf{j}^{\alpha,\beta(0)}_{k} = e_{\alpha,\beta}\int \mathbf{v}_{\alpha,\beta}\,\delta f^{\alpha,\beta(0)}_{\mathbf{p},k}\,\frac{d\mathbf{p}}{(2\pi)^3}\,. \tag{3.50}$$

In taking the average $\langle[\mathbf{k}\delta\mathbf{j}]^2\rangle$, we shall use Eq. (3.44). As a result, we immediately obtain Eq. (3.17), which is proportional to $\Phi_{\mathbf{p}\alpha}^{\alpha}$ and $\Phi_{\mathbf{p}\beta}^{\beta}$ [previously we obtained Eq. (3.18) using the currents of the individual particles, i.e., an expression without $\Phi_{p\alpha}{}^{\alpha}$]. The probability obtained in this way is expressed in terms of the nonlinear currents given by Eqs. (3.32) and (3.33).

We have, therefore, proven rigorously that polarization transition bremsstrahlung is a necessary component of the radiation from a plasma and is automatically obtained when a systematic theory of fluctuations is used.

The picture of particles "dressed" in a polarization "cloud" appears in bremsstrahlung, as well as in particle collisions (the Landau–Balescu integral).

Of course, the isolated contribution of δf included in these calculations is not the only one. Nonlinear terms of the type $\delta \mathbf{E}\partial\delta f^{(0)}/\partial \mathbf{p}$, which yield ordinary bremsstrahlung, also appear. When all terms of the required order are included rigorously, this yields the balance equations for $\Phi_{p\alpha}{}^{\alpha}$, $\Phi_{p\beta}{}^{\beta}$, and $N_{\mathbf{k}}{}^{\sigma}$ if $N_{\mathbf{k}} \neq 0$, then there is yet another channel, scattering of photons on particles. In the framework of the balance equations, we must still introduce the scattering probability and include the corresponding terms in these equations [9].

Rigorous fluctuation theory automatically yields bremsstrahlung, as well as scattering, including ordinary Thomson scattering and scattering on polarization clouds, i.e., transition scattering. We have already discussed this in Chapter 1. The results of Rosenbluth and Rostoker [10] on the scattering of electromagnetic waves on electrons with "clouds" and ions with "clouds" in plasmas are fully reproduced in this type of calculation.

It is extremely tedious to include all the terms that yield the Landau–Balescu collision integrals, bremsstrahlung (including polarization bremsstrahlung), and scattering (including transition scattering on polarization clouds). However, the general physical picture obtained from these calculations is perfectly clear. Polarization effects show up in all elementary interactions of particles, both among themselves and with electromagnetic radiation.

Chapter 4

POLARIZATION BREMSSTRAHLUNG OF NONRELATIVISTIC CHARGED PARTICLES ON ATOMS

V. M. Buimistrov

4.1. Elementary Quantum Mechanical Concepts in the Theory of Bremsstrahlung

Ever since Stokes and Sommerfeld [1] explained the "white" x-ray bremsstrahlung spectrum of cathode rays in the anticathode material, bremsstrahlung has been a subject of experimental and theoretical research. The quantum theory of bremsstrahlung was created by Bethe and Heitler [2, 3], Sommerfeld [1], and Sauter [4]. The need for a quantum theory follows from the existence of a maximum frequency for the x rays, ω_{max}, which is determined by the initial (subscript i) kinetic energy T_i of the incident particle and Planck's constant:

$$\omega_{max}=T_i/\hbar. \tag{4.1}$$

Of course, under certain conditions the quantum theory transforms to the classical theory. This happens if the energy of the bremsstrahlung photon is considerably lower than the energy of the incident particle, $\hbar\omega \ll T_{i,f}$ [more precisely, of the final (subscript f) and initial particles]. Then the quantum mechanical transition current can be replaced by the classical particle current and the quantum mechanical formula for the bremsstrahlung intensity becomes the classical formula.

In [1–4], which appeared shortly after the creation of quantum mechanics, and in many subsequent papers, bremsstrahlung is treated as the emission of a photon by a charged incident particle as it is scattered in a specified external field. In the classical theory, an external field is necessary for slowing down the electron;

the intensity of the radiation is proportional to the square of the acceleration. In the quantum theory, a free electron does not radiate because the energy and momentum cannot simultaneously be conserved in an elementary act of photon emission. Because the dispersion laws for electrons and photons are different, the electron cannot transfer to the photon an amount of momentum corresponding to the transferred energy. In the case of a given external field, a suitable third "body" may be an atom, which is assumed to be infinitely heavy and can, therefore, accept an arbitrary amount of excess momentum. In [1–4] the external field is the Coulomb field of a motionless nucleus, which is "bare" or shielded by the electrostatic potential of the atomic electrons (the shielding approximation). This approximation is exact for a "bare" nucleus if we regard it as a point Coulomb charge, i.e., neglect its structure and recoil during the bremsstrahlung process (because of the large mass of the nucleus compared to that of an incident electron). As for an atom or ion, the situation is considerably more complicated.

Unlike the other approximations used in the work of Bethe and Heitler, specifically the Born approximation and perturbation theory in terms of the electromagnetic field, the shielding approximation is not well justified and has a rather intuitive character. The concept of an atomic electron as a "smeared out" electric charge leads to a correct cross section for elastic scattering of an electron on an atom (in the Born approximation and when exchange effects are neglected). Here a rigorous treatment, based on the exact Hamiltonian for the problem, and the shielding approximation yield the same results. The contribution of the atomic electron reduces to introducing a form factor in the scattering cross section which takes the shielding of the charge into account. It probably then seemed natural to apply the shielding approximation to the theory of bremsstrahlung, as well. The approximate nature of this approximation was clear to many authors and was noted, for example, by Ter-Mikaélyan [5]. In fact, in the shielding approximation the incident particle and an atomic electron are not endowed with "equal rights." While the incident particle is treated rigorously, with its kinetic energy and the interaction with the electromagnetic field included in the Hamiltonian for the problem, the corresponding terms for an atomic electron are neglected. Nevertheless, until roughly the mid-1970s, the shielding approximation was regarded as adequate, at least in the x-ray range. This viewpoint was apparently maintained even by those authors who used the exact Hamiltonian for calculating the bremsstrahlung cross section [6–8]. This viewpoint was also reflected in all the textbooks and monographs known to us. Here we should point out that the above remarks apply to processes in which the state of the atom is not changed. If an atom is ionized with simultaneous emission of a photon, then the excess momentum of the incident particle can be transferred to an atomic electron. In this case, of course, the role of the atomic electrons does not merely reduce to shielding. This was noted, for example, by Bethe and Salpeter [2].

In the 1970s several papers [9–13] appeared which showed that in a number of cases, including the x-ray frequency range most characteristic of bremsstrahlung, the shielding approximation is not only quantitatively inaccurate, but also fails to provide a correct physical picture. In this chapter we shall construct a rigorous theory of bremsstrahlung based on the exact Hamiltonian for the problem and demonstrate that some of the results obtained from this theory can be explained through simple physical considerations.

Before proceeding to that discussion, we first recall some elementary quantum mechanical ideas about bremsstrahlung. We are examining a process which involves the scattering of a charged particle on a force center. We shall assume that the interaction of the center with a particle is characterized by a potential energy $V(\mathbf{r})$, where the radius vector of the particle is taken relative to the center of force. As a result of scattering, the momentum and energy of the particle are changed from their initial values p_i, T_i to final values p_f, T_f, and a photon $\hbar\omega$ is emitted simultaneously, that is, in the same single quantum mechanical scattering event. This process is characterized by the differential cross section. If we multiply the cross section by the particle flux, that is, by the number of particles passing through 1 cm^2 per second, then we obtain the number of particles scattered by a single center per second (if this process is bremsstrahlung, then a photon is emitted at the same time). Multiplying this quantity by the density of the centers of force, we obtain the number of scattering events per cm^3 per second. The differential cross section can be represented intuitively in the following way: all particles which pass through an area equal to the differential transverse cross section are scattered in a certain way. When a single atom does the scattering, we might expect that the cross section would be roughly equal to the transverse cross-sectional area of the atom, i.e., be on the order of 10^{-16} cm^2. This is indeed roughly so for elastic scattering. Bremsstrahlung in a static field, however, is a very much less efficient process; for it the transverse cross section is typically smaller by many orders of magnitude. Thus, only a very small fraction of the collisions of an incident particle (electron, proton, etc.) with an atom are accompanied by the emission of a photon.

In the present approximation of a static field, the differential cross section for bremsstrahlung is equal to

$$d\sigma_{st}(\omega) = \frac{e_0^2 p_f 4\pi^2}{\hbar^3 c^3 p_i} |(\mathbf{e}_{\mathbf{k},\sigma}\mathbf{q})|^2 |V(\mathbf{q})|^2 d\Omega_{\mathbf{k}} d\Omega_{\mathbf{p}_f} \frac{d\omega}{\omega}, \tag{4.2}$$

where

$$V(\mathbf{q}) = \int V(\mathbf{r}) \exp(-i\mathbf{q}\mathbf{r})\, d\mathbf{r}/(2\pi)^3, \quad \hbar\mathbf{q} = \mathbf{p}_f - \mathbf{p}_i.$$

The cross section $d\sigma_{st}(\omega)$ describes the emission of a photon with wave vector $\mathbf{k}$ in the frequency interval from ω to $\omega + d\omega$ (the solid angle $d\Omega_{\mathbf{k}}$ is around the direction of $\mathbf{k}$), and the polarization vector $\mathbf{e}_{\mathbf{k},\sigma}$ of the photon corresponds to one of two possible mutually orthogonal polarizations, $\sigma = 1, 2$. The incident particle, with charge e_0 and mass m_0, is scattered into a solid angle $d\Omega_{\mathbf{p}f}$ in momentum space. The cross section $d\sigma_{st}(\omega)$ is proportional to the square of the

Fourier component of the potential energy of the interaction. The argument of the Fourier component is the vector **q** which represents the change in the wave number of the incident particle during the scattering event (or, in a system of units with $\hbar$ = 1, the momentum transfer).

We now discuss briefly how the cross section $d\sigma_{st}(\omega)$ is calculated. The quantum mechanical system of interest to us consists of the incident particle which interacts with the center of force and the quantized electromagnetic field. The Hamiltonian of this system is

$$H = H_p^{(0)} + H_f^{(0)} + H_{p,f} + V(\mathbf{r}). \tag{4.3}$$

Here $H_p^{(0)}$ and $H_f^{(0)}$ are the Hamiltonians of the particle and quantized electromagnetic field, respectively, and $H_{p,f}$ and $V(\mathbf{r})$ describe the interaction of the particle with the electromagnetic field and the force center (specific expressions for the terms in the Hamiltonian (4.3) are given below in Sec. 4.2). The probability (per unit time) of the transition of interest is calculated by standard quantum mechanical methods. Here we are concerned with the term in the transition amplitude that is proportional both to the interaction of the particle with the center of force, $V(\mathbf{r})$, and to the interaction with the quantized field, $H_{p,f}$. Since, as already mentioned, the emission of a bremsstrahlung photon requires that both interactions occur simultaneously, this is the lowest term in the perturbation theory expansion that can describe bremsstrahlung. The cross section is calculated as the transition probability per unit time divided by the flux of incident particles.

The above statement of the problem is fully rigorous for a point Coulomb center, i.e., for a structureless nucleus, provided its recoil is neglected. The bremsstrahlung which occurs during scattering of a charged particle on an atom, however, has been examined in terms of the same formulation. In an atom the nuclear charge is shielded by the atomic electrons. We limit ourselves to the simplest case of a hydrogen atom or a hydrogenlike ion with a nuclear charge of $e_i = Ze$. Then the charge density of the shielding electron cloud is given by $-e|\psi(\mathbf{r})|^2$, where $\psi(\mathbf{r})$ is the wave function of the bound electron in the hydrogen atom. The potential energy of the interaction between the incident particle and the atom is given by

$$V(\mathbf{r}, \mathbf{r}_a) = e_0 Ze/r - e_0 e/|\mathbf{r} - \mathbf{r}_a|. \tag{4.4}$$

Here $\mathbf{r}_a$ is the radius vector of the atomic electrons. The Fourier component of the potential energy is

$$V(\mathbf{q}) = (ee_0/2\pi^2 q^2)\,[Z - F(\mathbf{q})], \tag{4.5}$$

where

$$F(\mathbf{q}) = \int d\mathbf{r} \exp(-i\mathbf{q}\mathbf{r})\,|\psi(\mathbf{r})|^2.$$

The function $F(\mathbf{q})$ is known as the form factor. Since the atomic wave functions $\psi(\mathbf{r})$ have a definite parity, the form factor $F(\mathbf{q})$ is a real quantity. It describes the effect of the atomic electrons on the scattering process. For scattering on a "bare" nucleus, the form factor in the formula for $V(\mathbf{q})$ should be dropped. For scattering on an atom, the form factor $F(\mathbf{q})$ reduces the charge of the nucleus; that is, it describes the shielding of the nucleus by the atomic electrons. Since the form factor depends on the momentum transfer $\mathbf{q}$, the degree of shielding also depends on this quantity.

How can we intuitively conceive shielding in the quantum theory? Strictly speaking, when a wave, rather than a classical particle, is scattered, we cannot introduce the concept of an impact parameter. In the integral (4.5) which defines $F(\mathbf{q})$, the contribution from distances greater than $1/q$ is not important because of the rapid oscillations in the exponent (for any given $\mathbf{q}$). This means that for a given momentum transfer $\mathbf{q}$, the charge which shields the nucleus lies at distances from 0 to $1/q$. Small momentum transfers correspond to large impact parameters. As $q \to 0$, the form factor $F(\mathbf{q}) \to Z$ and the electrons shield the nucleus with their full charge. A large $\mathbf{q}$ corresponds to a small impact parameter $1/q$, for which the form factor $F(\mathbf{q})$ is small and the entire charge of the nucleus acts on the electron as it passes near the nucleus.

Thus, the shielding picture is physically clear and it can be derived rigorously for elastic scattering on an atom in the Born approximation if we proceed from the exact Hamiltonian for this problem.

How do things stand with bremsstrahlung? Unlike elastic scattering, here the shielding picture corresponds to an approximate, rather than the exact, Hamiltonian. A rigorous treatment based on the exact Hamiltonian leads to different results.

4.2. The Total Bremsstrahlung Cross Section Including Polarization Bremsstrahlung

We shall now formulate the bremsstrahlung problem exactly for the collision of a nonrelativistic charged particle with an atom. The Hamiltonian for this problem is

$$H = H_{p,a} + H_f^{(0)} + H_{p,f} + H_{a,f}. \tag{4.6}$$

Here $H_{p,a}$ is the Hamiltonian of the system formed by the atom plus the incident particle that interacts with the atom; $H_f^{(0)}$ is the Hamiltonian of the quantum electromagnetic field; and, the operators $H_{p,f}$ and $H_{a,f}$ describe the interaction of the incident particle and the atomic electrons with the quantized electromagnetic field. The incident particle and the atomic electrons enter the Hamiltonian H on an equal footing. In the following analysis we shall consider only single-photon processes and retain only those terms which are linear in the quantized potential vector **A** of the electromagnetic field in the operators H_{pf} and H_{af}:

$$H_{p,f} = -\frac{e_0}{m_0 c}\hat{\mathbf{A}}(\mathbf{r})\hat{\mathbf{p}}, \qquad H_{a,f} = \frac{e}{mc}\sum_{j=1}^{N}\hat{\mathbf{A}}(\mathbf{r}_j)\hat{\mathbf{p}}_j,$$
$$\hat{\mathbf{A}}(\mathbf{r}) = \sum_{\mathbf{k},\sigma}[\mathbf{A}^{(-)}_{\mathbf{k},\sigma}(\mathbf{r})\hat{a}_{\mathbf{k},\sigma} + \mathbf{A}^{(+)}_{\mathbf{k},\sigma}(\mathbf{r})\hat{a}^{+}_{\mathbf{k},\sigma}],$$
$$\mathbf{k} = \{k_i\}, \quad k_i = (2\pi/L)l_i, \quad l_i = 0, \ \pm 1, \ \pm 2, \quad i = x, \ y, \ z, \tag{4.7}$$
$$\mathbf{A}^{(\mp)}_{\mathbf{k},\sigma} = \delta_{\mathbf{k}}\exp(\pm i\mathbf{kr})\mathbf{e}_{\mathbf{k},\sigma}, \quad \delta_{\mathbf{k}} = (2\pi\hbar c^2/L^3\omega_{\mathbf{k}})^{1/2}, \quad \omega_{\mathbf{k}} = kc.$$

Here $\hat{a}_{\mathbf{k},\sigma}{}^{+}$ and $a_{\mathbf{k},\sigma}$ are the creation and annihilation operators for a photon with wave vector **k** and polarization σ. We assume that the atom remains motionless before and after a collision (i.e., neglect its recoil), and denote the radius vector of each of the N atomic electrons by $\mathbf{r}_j$ and that of the incident particle by **r**. As usual, we introduce an elementary volume in the form of a cube with edges of length L and $V = L^3$, and denote the momentum operator by $\hat{\mathbf{p}}$. Thus, the exact Hamiltonian includes the interaction of both the incident particle (as in the shielding approximation) and the atomic electrons with the quantized electromagnetic field. This ultimately means that their emission during collisions is taken into account.

The probability $dw_{f,i}$ of spontaneous emission of a single photon in a given range of momenta $\hbar\mathbf{k}$ is equal to

$$dw_{f,i} = (2\pi/\hbar)\sum_{\mathbf{k},\varkappa_f}|\langle f|\hat{L}|i\rangle|^2\delta(\varepsilon_f - \varepsilon_i + \hbar\omega), \quad \varkappa_f\hbar = p_f,$$
$$\hat{L} = -\frac{e_0}{m_0 c}\mathbf{A}^{(+)}_{\mathbf{k},\sigma}\hat{\mathbf{p}} + \frac{e}{mc}\sum_{j=1}^{N}\mathbf{A}^{(+)}_{\mathbf{k},\sigma}\hat{\mathbf{p}}_j. \tag{4.8}$$

Here the sum is taken over the final momenta of the photon and incident particle and the δ-function expresses the conservation of energy. The matrix element is calculated from the wave functions Φ_s, the eigenfunctions of the Hamiltonian $H_{a,p}$:

$$(H_a^{(0)} + H_p^{(0)} + V)\Phi_s = \varepsilon_s\Phi_s,$$
$$\langle f|\hat{L}|i\rangle = \int d\mathbf{r}_a d\mathbf{r}\Phi_f^*(\mathbf{r}_a, \mathbf{r})\hat{L}(\mathbf{r}_a, \mathbf{r})\Phi_i(\mathbf{r}_a, \mathbf{r}),$$
$$\mathbf{r}_a = \{\ldots \mathbf{r}_j \ldots\}, \quad d\mathbf{r}_a = d\mathbf{r}_1, \ \ldots, \ d\mathbf{r}_j, \ \ldots, \ d\mathbf{r}_N, \tag{4.9}$$
$$V(\mathbf{r}_a, \mathbf{r}) = \frac{Zee_0}{r} - \sum_{j=1}^{N}\frac{ee_0}{|\mathbf{r} - \mathbf{r}_j|},$$

where $H_a{}^{(0)}$ is the Hamiltonian of the atom; $H_p{}^{(0)}$ is the Hamiltonian of the incident particle; ε_s is the eigenvalue of the Hamiltonian $H_{a,p}$; s is the set of quantum numbers that defines the state of the atom + incident particle system; i and f denote the initial and final states; V is the interaction potential of the atom with the incident particle; and r_a is the set of coordinates of the atomic electrons. We shall solve the problem of the scattering of the incident particle by an atom in the Born approximation; that is, in the formulas for the wave function $\Phi_s^{(1)}$ and matrix elements $\langle f|\hat{L}|i\rangle$ we shall restrict ourselves to first-order terms in V (subscript 1 for $\langle\, |\hat{L}|f|i\rangle$):

$$\langle f|L|i\rangle_1 \equiv \Sigma_i + \Sigma_f = \sum_{s_0 \neq f_0} \frac{\langle f_0|\hat{V}|s_0\rangle\langle s_0|L|i_0\rangle}{\varepsilon_{f_0}^{(0)} - \varepsilon_{s_0}^{(0)}} + $$
$$+ \sum_{s_0 \neq i_0} \frac{\langle f_0|\hat{L}|s_0\rangle\langle s_0|V|i_0\rangle}{\varepsilon_{i_0}^{(0)} - \varepsilon_{s_0}^{(0)}},$$
$$\Phi_s^{(1)} = \Phi_s^{(0)} + \sum_{s_0 \neq s_0'} \frac{\langle s_0|V|s_0'\rangle}{\varepsilon_{s_0}^{(0)} - \varepsilon_{s_0'}^{(0)}} \Phi_{s_0'}^{(0)}, \tag{4.10}$$
$$\Sigma_i = \sum_{s_0 \neq f_0} \frac{\langle f_0|V|s_0\rangle\langle s_0|\hat{L}|i_0\rangle}{\varepsilon_{f_0}^{(0)} - \varepsilon_{s_0}^{(0)}}, \qquad \Sigma_f = \sum_{s_0 \neq i_0} \frac{\langle f_0|\hat{L}|s_0\rangle\langle s_0|V|i_0\rangle}{\varepsilon_{i_0}^{(0)} - \varepsilon_{s_0}^{(0)}}.$$

The matrix elements $\langle s_0|\hat{L}|s_0'\rangle$ and $\langle s_0|V|s_0'\rangle$ are calculated from the eigenfunctions $\Phi_{s_0}^{(0)}$ of the Hamiltonian $H^{(0)} = H_a^{(0)} + H_p^{(0)}$. In Eq. (4.10), we have dropped the matrix element $\langle f_0|\hat{L}|i_0\rangle$, since it describes another process, namely the emission of a photon by the atom. The solution of the unperturbed problem has the following form:

$$H^{(0)}\Phi_{s_0}^{(0)} = \varepsilon_{s_0}^{(0)}\Phi_{s_0}^{(0)}, \quad H^{(0)} = H_a^{(0)} + H_p^{(0)}, \quad s_0 = \{n, \varkappa\} = \{n, \mathbf{p}/\hbar\},$$
$$\mathbf{p} = \hbar\varkappa, \quad T_\mathbf{p} = \mathbf{p}^2/2m_0, \quad \varepsilon_{s_0}^{(0)} = T_\mathbf{p} + E_n, \quad T_\mathbf{p} \ll m_0c^2,$$
$$\varkappa_i = (2\pi/L)\, l_i, \quad l_i = 0, \pm 1, \pm 2, \ldots, \quad \Phi_{s_0}^{(0)} = \varphi_n(\mathbf{r}_a)\,\psi_\mathbf{p}(\mathbf{r}), \tag{4.11}$$
$$\psi_\mathbf{p}(\mathbf{r}) = L^{-3/2}\exp(i\varkappa\mathbf{r}) = \exp(i\mathbf{p}\mathbf{r}/\hbar)\,L^{-3/2}, \quad H_p^{(0)}\psi_\mathbf{p} = T_\mathbf{p}\psi_\mathbf{p},$$
$$H_a^{(0)}\varphi_n(\mathbf{r}_a) = E_n\varphi_n(\mathbf{r}_a).$$

Here $\varphi_n(\mathbf{r}_a)$ and E_n are the wave functions and energies (terms) of the atom; $\psi_\mathbf{p}$ and $T_\mathbf{p}$ are the same for the incident particle; n is the set of quantum numbers of the atom; $\mathbf{p} = \hbar\boldsymbol{\kappa}$ is the momentum of the incident particle; s_0 is the complete set of quantum numbers n and $\mathbf{p} = \hbar\boldsymbol{\kappa}$: $s_0 = \{n, \vec{p}\}$. Using the orthonormal property of the functions $\varphi_n(\mathbf{r}_a)$ and $\psi_\mathbf{p}(\mathbf{r})$, we obtain

$$\langle s|\hat{L}|s'\rangle = (\delta_\mathbf{k}/c)\,[-(e_0/m_0)\,\hbar\,(\varkappa'\mathbf{e}_{\mathbf{k},\sigma})\,\delta_{\varkappa',\varkappa+\mathbf{k}}\delta_{n,n} + (e/m)\langle n|\hat{L}^a|n'\rangle\,\delta_{\varkappa,\varkappa'}], \tag{4.12}$$
$$\hat{L}^a = \sum_j \exp(-i\mathbf{k}\mathbf{r}_j)\,(\mathbf{e}_{\mathbf{k},\sigma}\hat{\mathbf{p}}_j).$$

After Eq. (4.12) is substituted in Eq. (4.10), the sum over the intermediate states becomes much simpler. Now the sum is taken only over the states of the atom:

$$\langle f|\hat{L}|i\rangle_1 = \frac{\delta_\mathbf{k}(2\pi)^3}{L^3c}\left[+\frac{e_0}{m_0}\frac{(\mathbf{q}\mathbf{e}_{\mathbf{k},\sigma})}{\omega}\frac{ee_0}{2\pi^2q^2}(Z\delta_{n_f,n_i} - F_{n_f,n_i}(\mathbf{q})) + \frac{e}{m}\Sigma\right], \tag{4.13}$$

$$F_{n_f,n_i}(\mathbf{q}) = \sum_{j=1}^{N} \langle n_f | \exp(i\mathbf{q}\mathbf{r}_j) | n_i \rangle,$$

$$\dot{\Sigma} = \sum_n \frac{\langle n_f | V(\mathbf{q}, \mathbf{r}_a) | n \rangle \langle n | \hat{L}^a | n_i \rangle}{E_{n_i} - E_n - \hbar\omega} + \sum_n \frac{\langle n_f | \hat{L}^a | n \rangle \langle n | V(\mathbf{q}, \mathbf{r}_a) | n_i \rangle}{E_{n_f} - E_n + \hbar\omega},$$

$$\mathbf{q} = \varkappa_f - \varkappa_i = (\mathbf{p}_f - \mathbf{p}_i)/\hbar.$$

In calculating the first term in Eq. (4.13), we have used the formulas

$$\langle \varkappa_f, n_f | V(\mathbf{r}, \mathbf{r}_a) | \varkappa_i, n_i \rangle = (2\pi)^3 \langle n_f | V(\mathbf{q}, \mathbf{r}_a | n_i \rangle,$$

$$V(\mathbf{q}, \mathbf{r}_a) = \left(\frac{L}{2\pi}\right)^3 \langle \varkappa_f | V(\mathbf{r}, \mathbf{r}_a) | \varkappa_i \rangle = \frac{ee_0}{2\pi^2 q^2} \left(Z - \sum_{j=1}^{N} \exp(i\mathbf{q}\mathbf{r}_j) \right),$$

$$\langle \varkappa_f, n_f | V(\mathbf{r}, \mathbf{r}_a) | n_i, \varkappa_i \rangle = \left(\frac{2\pi}{L}\right)^3 \frac{ee_0}{2\pi^2 q^2} [Z\delta_{n_f,n_i} - F_{n_f,n_i}(\mathbf{q})], \quad (4.14)$$

$$\langle f | \hat{L} | i \rangle_1 = \left(\frac{2\pi}{L}\right)^3 \frac{\delta_{\mathbf{k}}}{c} A_{f\,i}.$$

Equation (4.14) is obtained by substituting the interaction energy $V(\mathbf{r}, \mathbf{r}_a)$ on the left-hand side and carrying out the required integration. In the energy denominators of the sums in Eq. (4.10) the dependence on $\boldsymbol{\kappa}$ in the matrix element vanishes if we note that the first sum is calculated for $\boldsymbol{\kappa} = \boldsymbol{\kappa}_i$ and the second, for $\boldsymbol{\kappa} = \boldsymbol{\kappa}_f$. In the calculations we have used the conservation of energy,

$$E_{n_i} + \frac{p_i^2}{2m_0} = E_{n_f} + \frac{p_f^2}{2m_0} + \hbar\omega, \qquad \mathbf{p}_i = \hbar\varkappa_i, \quad \mathbf{p}_f = \hbar\varkappa_f. \quad (4.15)$$

It has also been assumed that the momentum $\hbar\mathbf{k}$ of the photon can be neglected compared to that of an electron. Nevertheless, this approximation only affects the magnitude of the first term in Eq. (4.13).

We now write down an expression for the differential cross section for bremsstrahlung:

$$d\sigma_{n_f,n_i}(\omega) = 4\pi^2 \frac{p_f}{p_i} \frac{m_0^2}{\hbar^3 c^3} |A_{f\,i}|^2 \omega d\omega \, d\Omega_{\mathbf{k}} \, d\Omega_{\mathbf{p}_f}. \quad (4.16)$$

A photon with frequency ω within the frequency interval $d\omega$ and momentum $\mathbf{k}$ within the solid angle $d\Omega^{\mathbf{k}}$ is emitted as a result of the collision of an incident particle with the atom. Here an incident particle with an initial momentum $\mathbf{p}_i$ is scattered into a solid angle $d\Omega_{\mathbf{p}f}$. The final momentum is determined by the conservation of energy in accordance with Eq. (4.15). In order to obtain Eq. (4.16), we must convert the expression (4.8) for the transition probability in the usual way from a sum to an integral over the range of momenta of the photon and electron of interest to us, $d\mathbf{k}$ and $d\boldsymbol{\kappa}_f$, as well as actually carry out the integration over $\boldsymbol{\kappa}_f$. We then divide the resulting expression by the particle flux p_i/m_0L^3 (where $1/L^3$ is the density of particles in an elementary volume L^3) to obtain the bremsstrahlung cross section.

4.3. Bremsstrahlung of Electrons, Positrons, and Protons

We shall now investigate the expression obtained above for the bremsstrahlung cross section. We shall consider the frequency range characteristic of x rays; that is, we shall assume that the energy of the photon exceeds the ionization energy of any of the electron shells of the atom. In the first approximation we neglect the differences $E_{\eta i,f} - E_n$ in the denominators of the sums in Eq. (4.13) compared to $\hbar\omega$. Then the sums over the intermediate atomic states drop out and we obtain the following expression for the cross section:

$$
\begin{aligned}
d\sigma_{n_f,n_i}(\omega) &= \frac{1}{\pi^2}\frac{p_f}{p_i}\frac{e^2e_0^2m_0^2}{\hbar^3c^3}\frac{(\mathbf{e}_{\mathbf{k},\sigma}\mathbf{q})^2}{q^4}\,|B_{n_f,n_i}|^2\frac{d\omega}{\omega}\,d\Omega_{\mathbf{k}}\,d\Omega_{\mathbf{p}_f},\\
B_{n_f,n_i} &= \frac{e_0}{m_0}\,[Z\delta_{n_f,n_i} - F_{n_f,n_i}(\mathbf{q})] - \frac{e}{m}F_{n_f,n_i}(\mathbf{q}-\mathbf{k}),\\
F_{n_f,n_i} &= \left\langle n_f \left| \sum_j \exp(i\mathbf{q}\mathbf{r}_j) \right| n_i \right\rangle .
\end{aligned}
\tag{4.17}
$$

Taking the sums leading to Eq. (4.17) reduces to calculating the matrix element of the commutator

$$
\begin{aligned}
\Sigma &= -(1/\hbar\omega)\langle n_f|\,[V(\mathbf{q},\mathbf{r}_a),\hat{L}_a]\,|n_i\rangle,\\
[\hat{a},\hat{b}] &= \hat{a}\hat{b} - \hat{b}\hat{a}.
\end{aligned}
\tag{4.18}
$$

Substituting the expressions for $V(\mathbf{q}, \mathbf{r}_a)$ and $\hat{L}^a$ from Eqs. (4.14) and (4.12) in Eq. (4.18), we obtain

$$
\Sigma = -\frac{ee_0}{2\pi^2\omega q^2}(\mathbf{e}_{\mathbf{k},\sigma}\mathbf{q})\,F_{n_f,n_i}(\mathbf{q}-\mathbf{k}). \tag{4.19}
$$

Equation (4.17), in turn, follows from Eqs. (4.16), (4.13), and (4.19).

Equation (4.17) leads to a new physical picture of bremsstrahlung. Let us discuss this question in more detail. At first we shall restrict ourselves to elastic bremsstrahlung ($n_f = n_i$). Then we have

$$
B_{n_i,n_i} = (e_0/m_0)\,[Z - F_{n_i,n_i}(\mathbf{q})] - (e/m)\,F_{n_i,n_i}(\mathbf{q}-\mathbf{k}). \tag{4.20}
$$

The first two terms in Eq. (4.20) describe bremsstrahlung in the shielding approximation: for a given momentum transfer $\mathbf{q}$, the nuclear charge Ze is reduced by an amount $eF_{ni,ni}(\mathbf{q})$ [the function $F_{ni,ni}(\mathbf{q})$ is the form-factor of an atom in the state n_i]. This is precisely the shielding picture: the electron density shields the nucleus as a static charge. The complete theory of bremsstrahlung developed here yields the third term in Eq. (4.20). That term shows up only when an atomic electron is treated as a particle equivalent to the incident particle, i.e., when the kinetic energy of the atom and its interaction with the electromagnetic field are included in the Hamiltonian.

We begin with bremsstrahlung from an electron. Let a bremsstrahlung photon be emitted during scattering of an electron on a hydrogen atom or on a hydrogenlike ion in the ground state ($e_0 = -e$, $m_0 = m$, $n_i = 1s$). For a nonrelativistic electron in the Born approximation, the momentum of the bremsstrahlung photon can be neglected compared to the change in the momentum of the electron ($k \ll q$). Then the second and third terms of Eq. (4.20) compensate one another and the bremsstrahlung takes place as if the incident electron had been scattered on a "bare" nucleus. Therefore, the atomic electron does not shield the nuclear charge, even for large impact parameters where it "should" shield it completely. This new physical picture has a simple explanation. When the conditions

$$p_{f,i}^2/2m_0 \gg I, \quad \hbar\omega \gg I \tag{4.21}$$

(where I is the ionization potential) are satisfied, the atomic electron can be regarded as approximately free. It is known that the amplitude of bremsstrahlung for scattering of particles with the same charge to mass ratio (e_0/m_0) is zero. Thus, the total bremsstrahlung amplitude for scattering of an electron on an atom reduces to the amplitude of bremsstrahlung on the nucleus. Of course, the atomic electron cannot be regarded as entirely free, even when the conditions (4.21) are satisfied, so there is an associated correction. This correction is obtained by noting that there are two terms besides the first in the above expansion in powers of $(E_n - E_{ni,f})/\hbar\omega$. Then for the hydrogen atom we obtain

$$d\sigma_{1s,1s}(\omega) = \frac{1}{\pi^2}\frac{p_f}{p_i}\frac{e^6}{\hbar^3 c^3}\frac{(\mathbf{e}_{\mathbf{k},\sigma}\mathbf{q})^2}{q^4} \times$$
$$\times \left[1 + \frac{4\pi}{3}\left(\frac{\omega_a}{\omega}\right)^2 a_B^3 \,|\,\psi_{1s}(0)\,|^2\right]^2 \frac{d\omega}{\omega}\, d\Omega_{\mathbf{k}}\, d\Omega_{\mathbf{p}_f} \tag{4.22}$$

Here $\psi_{1s}(0)$ is the wave function of the hydrogen atom for $r_a = 0$; $|\psi_{1s}(0)|^2 a_B{}^3 = 1/\pi$; $\omega_a = me^4/\hbar^3$ is the atomic unit of frequency; and, $a_B = \hbar^2/me^2$ is the Bohr radius. This formula is valid for $qa_B \ll 1$, i.e., for sufficiently large impact parameters. In deriving Eq. (4.22), we have used the relation

$$(E_{n'} - E_n)\langle n'\,|\,\hat{B}\,|\,n\rangle = \langle n'\,|\,[H_a^{(0)}\hat{B}]\,|\,n\rangle, \quad \hat{B} = \mathbf{e}_{\mathbf{k}\sigma}\nabla_j. \tag{4.23}$$

Let us compare Eq. (4.22) with the cross section in the shielding approximation:

$$d\sigma_{st}(\omega) = \frac{1}{\pi^2}\frac{p_f}{p_i}\frac{e^6(\mathbf{e}_{\mathbf{k},\sigma}\mathbf{q})^2}{\hbar^3 c^3 q^4}\left[1 - \frac{16}{(4 + q^2 a_B^2)^2}\right]\frac{d\omega}{\omega}\, d\Omega_{\mathbf{k}}\, d\Omega_{\mathbf{p}_f}. \tag{4.24}$$

Scattering on a proton is described by the first term in Eq. (4.24). The influence of the atomic electron reduces to lowering the effective charge, as should happen, given the very meaning of the concept of shielding. For scattering at an angle $\vartheta = 0$, we have $q = 2\omega m/(p_f + p_i)$. Now, when $qa_B \gg 1$, which is compatible with the inequality (4.21), the bremsstrahlung cross section transforms to the cross section for scattering on a nucleus: small values of q correspond to large impact parameters and, therefore, to complete shielding [according to Eq. (4.24), $d\sigma_{st} \to 0$ when $q \to 0$]. The ratio of the cross sections given by Eqs. (4.22) and (4.24) ranges from 500 to 8 when q varies from 0.3 to 1. If $p_i \approx 10$ a.u. and $\omega \approx 3$ a.u., then the angle ϑ varies from 0 to 5°. If we note in addition that under the above conditions (which ensure that the inequality $p_{f,i}^2/2m_0 \gg \hbar\omega$ is satisfied) the intensity of bremsstrahlung during scattering of an electron on a nucleus obeys the classical formula, then we arrive at the conclusion that Eq. (4.24) does not have a correct limiting transition to the classical theory. This shows clearly that the role of the atomic electrons in bremsstrahlung does not reduce to shielding. When $qa_B \gg 1$, Eqs. (4.22) and (4.24) coincide to within $\sim(\omega_a/\omega)^2$ and $1/q^4 a_B^4$, since there is little shielding at sufficiently small impact parameters. In addition, it should be noted that, in this case, the influence of the atomic electron on the bremsstrahlung cross section should be taken into account by using Eq. (4.22) to the extent that it is important at all.

In the above discussion of the limiting transition to the classical theory, we had in mind a simultaneous transition to the case of free particles, when the atomic electron also becomes free. Of course, in the classical theory, as well, it is possible to use the shielding approximation (see Chapters 2 and 3), but a correct classical theory must take polarization bremsstrahlung into account as well. At high frequencies the term describing the shielding and polarization bremsstrahlung cancels out for nonrelativistic incident electrons in the classical theory as it does in the quantum mechanical theory (see Eq. (4.20) with $e_0/m_0 = -e/m$ [2]). In Chapters 2 and 3 the emphasis was on polarization bremsstrahlung as such. Thus, we initially neglected the traditional bremsstrahlung (which corresponds to ignoring the first term in Eq. (4.20); in particular, it becomes small when $m_0 \to \infty$). Then, in Sec. 2.8, ordinary bremsstrahlung was taken into account. For the case in which the incident particle is a nonrelativistic electron, the contributions of the two effects (shielding and polarization bremsstrahlung) to the total bremsstrahlung amplitude are of equal magnitude but opposite in sign [as in Eq. (4.20)]. This happens in the dipole approximation for the photon. The formulas of Sec. 2.8 are more complicated, since they are not restricted to the dipole approximation. In order to establish a closer relationship with the results of Chapters 1 and 2, we shall find the integral cross section $d\sigma_{n_i n_i}(\omega)$ [see Eq. (4.17)], integrated over $d\Omega_k$ and $d\Omega_{pf}$ and summed over the polarizations (in $B_{n_i n_i}$ we restrict ourselves to the first term, $B_{n_i n_i} = e_0 Z/m_0$, and take $e_i = Ze$):

$$\int \sum_\sigma (\mathbf{e}_{\mathbf{k},\sigma} q)^2 \, d\Omega_{\mathbf{k}} = \int \frac{[\mathbf{kq}]}{k^2} d\Omega_{\mathbf{k}} = \frac{8\pi}{3} q^2,$$

$$\int \frac{d\Omega_{\mathbf{p}_f}}{q^2} = \hbar^2 2\pi \int \frac{d\cos(\widehat{\mathbf{p}_i\mathbf{p}_f})}{p_i^2 + p_f^2 - 2p_ip_f\cos(\widehat{\mathbf{p}_i\mathbf{p}_f})} = \frac{2\pi\hbar^2}{p_ip_f}\ln\frac{p_i+p_f}{p_i-p_f} \approx \frac{2\pi\hbar^2}{p_ip_f}\ln\frac{2T_i}{\hbar\omega},$$

$$p_f^2 = p_i^2 - 2m\hbar\omega, \quad p_f \approx p_i - m\hbar\omega/p_i,$$

$$\sum_\sigma \int (d\sigma_{n_in_i}(\omega)/d\Omega_{\mathbf{k}}d\Omega_{\mathbf{p}_f})\, d\Omega_{\mathbf{k}}d\Omega_{\mathbf{p}_f} = \frac{16e_0^4e^2Z^2}{3c^3m_0^2v_0^2\hbar\omega}\ln\frac{2T_i}{\hbar\omega},$$

which corresponds to Eq. (1.4).

We now consider bremsstrahlung from positrons. Obviously the cross section for bremsstrahlung from an electron and a positron is the same in the shielding approximation described by the first term of Eq. (4.20). However, if we include both terms of Eq. (4.20), the situation changes. The second term, naturally, retains its sign when the electron is replaced by a positron, while the first term changes sign. Therefore, the form factors no longer cancel out, but add up. The scattering of a positron on an atomic electron gives a nonzero contribution to the bremsstrahlung (unlike the scattering of an electron on an electron), since the ratio e_0/m_0 differs in sign for an electron and positron. A sharp difference between the previous and the new approaches shows up in the bremsstrahlung of the proton. It is generally accepted that the bremsstrahlung of a proton is less than that of an electron by the reciprocal of the square of their mass ratio. This is indeed true if we limit ourselves to the first ("static") term in Eq. (4.20), which is inversely proportional to the mass m_0 of the incident particle, for both the proton and the electron. Things are considerably different for the second term in the sum: there the factor e_0/m_0 is replaced by e/m. From the standpoint of a physical picture of the radiation, this is quite understandable, since the first term is related to the dipole moment of the incident particle and the second, to the dipole moment of the atomic electrons. The latter ("dynamic") term makes the bremsstrahlung of a proton comparable to the bremsstrahlung of an electron, of course, only at low frequencies.

We now turn to a physical picture of particle scattering which will explain why a proton produces a significant amount of bremsstrahlung. Let an atom be initially at rest in the laboratory coordinate frame and let a proton with velocity v_0 be incident on it. We now transform from the laboratory coordinate system to a system attached to the proton. In that system the atom is approaching a proton that is at rest. The atom, however, consists of a nucleus and electrons. Scattering of a heavy particle (the nucleus) on a proton actually does not produce significant bremsstrahlung, since in the frequency range of interest to us we can neglect the structure of the nucleus. As for the atomic electrons, at sufficiently high frequencies they will radiate as free electrons with the same velocity v_0. For a free electron, the upper limit on the emitted frequency is $mv_0^2/2$. For an electron bound in an atom, this limit is somewhat different, but the emission nonetheless lies in the low-frequency range.

This simple picture (or one like it) has been discussed in various stages from time to time (e.g., by Hayakawa [13] and by Arutyunyan and Tumanyan [14]) and formed the basis of a calculation of one of the types of cosmic γ-radiation by Agaronyan et al. [15]. Nevertheless, the statement that a proton does not radiate is still almost canonical. It appears that the instinctive distrust in such a simple analysis is explained by the fact that it was not based on a calculation that took the binding of the electron to the atom into account. In other words, the analysis was not based, so to say, on first principles. This sort of calculation was done later for bremsstrahlung on atoms [16] and for bremsstrahlung in plasmas [17]. The calculations for atoms show that the simple analysis given above is basically true, but also that the distrust in it is, to some extent, justified.

A rigorous theory yields a result for a multielectron atom that is inconsistent with the idea of independent scattering of a proton on every electron. Under certain conditions the bremsstrahlung cross section is not proportional to the number N of electrons in the atom, but to the square of this number. The emission becomes coherent, similarly to the way that the scattering of light on an atom can become coherent. This analogy is not superficial. In fact, the virtual photons which form the field of the proton are scattered on the atomic electrons to yield real photons. This result is also understandable from a formal point of view. It means that the N amplitudes for scattering on the individual electrons add up, but the bremsstrahlung cross section, which involves the square of the amplitude, is proportional to N^2 (see Chapter 5). This result applies, of course, not only to a proton, but to any charged incident particle.

Thus, the rigorous theory of bremsstrahlung leads to some new results: it shows that in certain cases (see [15], for example), we can limit ourselves to the simple approximation and regard the atomic electrons as free. In other cases (e.g., at photon energies that are sufficiently low compared to the energy of the proton), the complete theory is needed, if only because the atomic electrons can introduce a coherent contribution to the bremsstrahlung.

We now proceed to the question of comparing theory and experiment. At first glance it seems that the experimenters should have "sounded the alarm" on observing emission from protons and other nuclei that was anomalously high compared to the previous theory. In reality the matter was by no means this simple. There was, of course, some interest in continuum x-ray emission from nuclei, but more as a kind of background effect which interfered with the main observations. As an example, we might consider x-ray structural analysis with ionic excitation. The choice of protons and other ions for excitation of the characteristic x-ray spectra of atoms is based on the fact that at high frequencies, all the way up to the initial energy of the proton, the bremsstrahlung is actually weak (here it is correctly described by the shielding theory). Thus, the x-ray continuum background does not interfere with observing the characteristic spectra. The background observed at low frequencies is as expected from the theory discussed above. At the time the background emission was noticed, however, the theory of polarization bremsstrahlung had not yet been created. Then it was assumed, as mentioned above, that the polarization effects in bremsstrahlung are small. Another explanation based on a two-step process was advanced [18]: on colliding with a target (gaseous or solid) atom, the proton knocks out a so-called δ-electron, which in turn emits bremsstrahlung photons as it collides with other atoms. This is at least a two-stage process and may be a multistage process, if, for example, an electron is able to emit several bremsstrahlung photons as it passes through a target. This process is also referred to as a cascade.

Yet another process has been proposed [16, 19] to explain the low-frequency background: bremsstrahlung with simultaneous ionization. Unlike the emission of δ-electrons, this process takes place as a single quantum mechanical event: the electron that is knocked out of an atom is scattered on that same atom and simultaneously emits a photon (in our terminology this is inelastic bremsstrahlung). The analysis has been done for heavy nuclei [19] and for protons and light nuclei [16].

We have, therefore, presented a qualitative explanation of the low-frequency x-ray background. As noted recently by Ishii and Morita [20], however, agreement between theory and experiment has not been obtained at some frequencies. They showed that in this frequency range, which begins roughly at $2mv_0^2$, elastic polarization bremsstrahlung, that is, emission in which the target atom remains in its initial state, predominates. Ishii and Morita derived their general formulas with the same initial assumptions as Buimistrov et al. [16] However, unlike the earlier work [16], they carried out a numerical calculation for an aluminum atom using Slater functions. The theory was compared with an experiment for 1 MeV protons in the frequency range from roughly 2 to 6 keV, in which photons emitted at an angle of 90° to the direction of the proton beam were detected (the maximum emission is observed at this angle). It appeared that in this interval the elastic polarization bremsstrahlung exceeded both radiative ionization and the emission of secondary electrons. At energies above 2 keV, which is roughly equal to $2mv_0^2$ for 1 MeV protons, the emission of secondary electrons falls off quite rapidly. This can be understood on the basis of some fairly simple considerations: the energy of a 1 MeV proton is considerably greater than the binding energy of an electron in the atom, thus the transfer of energy to an electron knocked out of the atom takes place almost as if the electron were free. Because of the conservation of energy and momentum, however, the proton cannot transfer an energy greater than $2mv_0^2$ to a free electron. An energy above $2mv_0^2$ can be transferred to a bound atomic electron, but the number of such electrons decreases rather rapidly with increasing energy and, therefore, the number of photons with energies above 2 keV (i.e., with the maximum energy, almost equal to the initial energy of an electron that has been knocked out and emitted a bremsstrahlung photon) decreases rapidly. Thus, elastic polarization bremsstrahlung appears as a major contributor. It is at least an order of magnitude greater than the other forms of radiation (ionization accompanied by radiation and bremsstrahlung of secondary electrons). Besides radiative ionization, yet another type of inelastic process is evidently possible, bremsstrahlung with simultaneous excitation of an atom to a discrete level. Unfortunately, the contribution of this process has not yet been analyzed in connection with experimental data.

4.4. Bremsstrahlung at Low Frequencies

We now consider frequencies below the ionization energy of an atom ($\hbar\omega < I$). These low frequencies are typical of the optical range if I is the ionization energy of the outer shell of an atom. In the x-ray region these frequencies may be small compared to the ionization energy of an inner shell of the atom or of ions with a high degree of ionization. Low frequencies are of interest in connection with laser breakdown and the amplification of light. Stimulated emission and the inverse bremsstrahlung effect, which is important for laser breakdown, take place in an external electromagnetic field.

Here we shall consider a stimulated process at low frequencies, for which the spontaneous bremsstrahlung cross section can be found easily with the aid of the Einstein relations. In the following we shall use the dipole approximation for the interaction of an incident particle with the electromagnetic field.

The Hamiltonian for the stimulated bremsstrahlung problem now has the form

$$H = H^{(0)} + V + H_{\text{int}}, \tag{4.25}$$

where

$$H^{(0)} = H_a^{(0)} + H_p^{(0)}, \qquad H_{\text{int}} = 2\cos\omega t \hat{L}_E(\mathbf{r}_a, \mathbf{r})$$

and

$$\hat{L}_E = \frac{i\hbar \mathbf{E}_0}{2\omega}\left(\frac{e_0}{m_0}\nabla - \frac{e}{m}\sum_{j=1}^{N}\nabla_j\right).$$

The first term in $\hat{L}_E$ describes the interaction with the electromagnetic field of the incident particle and the second, with the field of the atomic electrons. $\mathbf{E}_0$ is the electric field vector of the wave.

Using the same approximations and repeating the same calculations which led us to Eq. (4.9), we again obtain it, but now the operator $\hat{L}$ is defined by Eq. (4.25). The formula analogous to Eq. (4.13) now has the form

$$\langle f|\hat{L}_E|i\rangle = \left(\frac{2\pi}{L}\right)^3 \frac{i\hbar}{2\omega}\left[\frac{e_0}{m_0}\frac{i(\mathbf{E}_0\mathbf{q})}{\hbar\omega}\langle n_f|V(\mathbf{q}, \mathbf{r}_a)|n_i\rangle - \right.$$

$$-\frac{e}{m}\sum_{n}\frac{\langle n_f|V(\mathbf{q},\mathbf{r}_a)|n\rangle\left\langle n\left|\left(\mathbf{E}_0\sum_{j=1}^{N}\nabla_j\right)\right|n_i\right\rangle}{E_{n_i}-E_n\mp\hbar\omega}-$$

$$-\frac{e}{m}\sum_{n}\frac{\left\langle n_f\left|\left(\mathbf{E}_0\sum_{j=1}^{N}\nabla_j\right)\right|n\right\rangle\langle n|V(\mathbf{q},\mathbf{r}_a)|n_i\rangle}{E_{n_f}-E_n\pm\hbar\omega}\Bigg]. \tag{4.26}$$

Here the sum over the intermediate states of the incident particle has also dropped out (and for the same reason the lower sign denotes absorption of a photon.) Now we can write the differential cross section $d\sigma_{nf,ni}{}^{\text{ind}}(\omega)$ for the stimulated emission of a photon as

$$d\sigma^{\text{ind}}_{n_f,n_i}(\omega)=\frac{m_0^2}{(2\pi)^2\hbar^4}\frac{p_f}{p_i}|L^3\langle f|\hat{L}_E|i\rangle_1|^2\,d\Omega_{p_f}=\frac{E_0^2\pi^2c^3}{\hbar\omega^3}\frac{d\sigma_{n_f,n_i}(\omega)}{d\omega\,d\Omega_{\mathbf{k}}}. \tag{4.27}$$

At the same time the photon is emitted, the incident particle is scattered into the solid angle $d\Omega_{\mathbf{p}f}$ and the state of the atom may change ($n_i\rightarrow n_f$). $d\sigma_{nf,ni}(\omega)/d\omega d\Omega_{\mathbf{k}}$ denotes the cross section for spontaneous emission (4.16).

Again, as at high frequencies, we must calculate the sum (4.26) over the intermediate states of the atom. This is a much more complicated problem for low frequencies (if we exclude resonances) than for high frequencies. Special methods are required to solve it. They have been developed by a number of authors over an extended period [21-24]. A thorough discussion of this problem is given in the book by Rapoport et al. [21], which contains many specific calculations. Here we shall use the following simple approach: turning to the initial sums Σ_i and Σ_f, we shall show that they can be calculated using the formulas [Σ_i and Σ_f are defined by Eq. (4.10) with $\hat{L}=\hat{L}_E$]

$$\Sigma_i=\int d\mathbf{r}\,d\mathbf{r}_a\Phi_f^*V(\mathbf{r},\mathbf{r}_a)\,W_i,\qquad \Sigma_f=\int d\mathbf{r}\,d\mathbf{r}_aW_f^*V(\mathbf{r},\mathbf{r}_a)\,\Phi_i, \tag{4.28}$$

if the auxiliary functions W_i and W_f obey the following differential equations:

$$(H^{(0)}-\varepsilon_f)\,W_i=-\hat{L}_E\Phi_i,\qquad (H^{(0)}-\varepsilon_i)\,W_f=-\hat{L}_E\Phi_f. \tag{4.29}$$

We seek a solution of Eqs. (4.29) in the form of a series in the eigenfunctions of the operator $H^{(0)}$:

$$W_i=\sum_{s_0}a_{s_0}\Phi_{s_0},\qquad W_f=\sum_{s_0}b_{s_0}\Phi_{s_0}. \tag{4.30}$$

Substituting the functions W_i and W_f in Eqs. (4.29), we obtain equations for finding the coefficients a_s and b_s in the usual way. They are soluble only if the eigenfunctions associated with the eigenvalues ε_f and ε_i are orthogonal to the right-hand sides of the corresponding equations, i.e., if $\langle f|L|i\rangle=0$. It has already been pointed out that this condition is satisfied. Evidently, the solution of Eq. (4.29) is determined with the accuracy of the solutions to the corresponding homogeneous equations. We choose the following particular solutions of Eqs. (4.29):

$$W_i = \sum_{s_0 \neq s_{f_0}} \frac{\langle s_0 | \hat{L}_E | i_0 \rangle}{\varepsilon_{f_0} - \varepsilon_{s_0}}, \qquad W_f = \sum_{s_0 \neq s_{i_0}} \frac{\langle s_0 | \hat{L}_E | f_0 \rangle}{\varepsilon_{i_0} - \varepsilon_{s_0}}. \tag{4.31}$$

By direct substitution of Eq. (4.31) in Eq. (4.28) we can confirm that the solutions (4.31) have been chosen correctly. As a result, we obtain the original expression (4.10) for the matrix element $\langle f|\hat{L}_E|i\rangle$.

The problem has, therefore, been reduced to solving Eqs. (4.29) and calculating the integrals (4.28).

The method presented here has a simple physical interpretation. The ordinary theory of quantum transitions involves analyzing transitions, owing to a time-varying perturbation $H_{\rm int}$, among the stationary states of a system described by the Hamiltonian $H^{(0)} + V$. The stationary wave function of such a system can be calculated by first-order perturbation theory in the interaction V:

$$(H^{(0)} - \varepsilon^{(0)}_{i_0,f_0})\, \psi^{(1)}_{i_0,f_0} = -(V(\mathbf{r}, \mathbf{r}_a) - \varepsilon^{(1)}_{i_0,f_0})\, \Phi^{(0)}_{i_0,f_0}. \tag{4.32}$$

Here $\psi^{(1)}_{i_0,f_0}$ and $\varepsilon^{(1)}_{i_0,f_0}$ are the first-order corrections in V to the wave functions $\Phi^{(0)}_{i_0,f_0}$ and energies $\varepsilon^{(0)}_{i_0,f_0}$. If it were possible to solve these equations, then this would be equivalent to taking the sums over the intermediate states of the system. The method proposed above essentially involves replacing Eq. (4.32) by Eq. (4.29). Since the operator $\hat{L}_E$ is much simpler than the operator V, Eqs. (4.29) are simpler than Eqs. (4.32). In Eqs. (4.29) the variables are separable, but not in Eqs. (4.32). Equations (4.29) arise in another physical statement of the problem, which concerns transitions among the nonstationary states of a system described by the Hamiltonian $H^{(0)} + H_{\rm int}$ under the action of a time-independent perturbation V. Of course, both methods lead to the same results. The above "inversion" of the perturbations follows naturally from the variational principle for the nondiagonal matrix elements (see [24]).

Before continuing the calculations, let us make the problem more specific. We are considering elastic bremsstrahlung or the emission of a photon without excitation of the atom. We shall assume that the incident particle is an electron and that it is scattered on a hydrogen atom or a hydrogenlike ion in the ground state ($e_0 = -e$, $m_0 = m$; $\varphi_{n_{i_0}} = \varphi(r_a)$, $n_{i_0} = n_{f_0} = 1s$). We now proceed to solve Eqs. (4.29). The functions $W_{i,f}$ have the following form:

$$W_{i,f} = \left[\tilde{u}^a_{i,f}(\mathbf{r}_a) - \frac{i\alpha\varkappa_{i,f}}{\hbar\omega}\, \varphi(r_a) \right] L^{-3/2} \exp(i\varkappa_{i,f}\mathbf{r}), \tag{4.33}$$

where

$$\alpha = -\frac{ie\hbar}{2m\omega}\, \mathbf{E}_0,$$

and

$$(H_a^{(0)} - E_{n_{i,f}} \pm \hbar\omega)\, \widetilde{u}^a_{i,f}(\mathbf{r}_a) = -\alpha\nabla_a \varphi_{n_i,n_f}(r_a). \tag{4.34}$$

By direct substitution of Eq. (4.33) in Eq. (4.29), we confirm the validity of Eqs. (4.33) and (4.34). A suitable choice of coordinates makes it much easier to solve Eq. (4.34). The z axis of the Cartesian coordinate system is directed along the vector $\mathbf{q}$ and we introduce the spherical coordinates r_a, θ, φ. We now write Eq. (4.34) in spherical coordinates:

$$\begin{aligned}(H_a^{(0)} - E_{n_{i,f}} \pm \hbar\omega)\, \widetilde{u}^a_{i,f}(\mathbf{r}_a) = \\ = -(\partial\varphi(r_a)/\partial r_a)(\alpha_z \cos\theta + \alpha_x \sin\theta\cos\varphi + \alpha_y \sin\theta\sin\varphi)/r_a.\end{aligned} \tag{4.35}$$

We seek a solution of Eqs. (4.35) in the form

$$\widetilde{u}^a_{i,f}(\mathbf{r}_a) = \alpha_z u^a_{i,f}(r_a, \theta) + u'_{i,f}(r_a, \theta, \varphi). \tag{4.36}$$

It can be shown that the functions $u_{i,f}'$, which depend on φ, make no contribution to the integrals $\Sigma_{i,f}$. Thus, it remains only to solve the equations for the functions $u_{i,f}^a$:

$$(H_a^{(0)} - E_{n_{i,f}} \pm \hbar\omega)\, u^a_{i,f} = -(\partial\varphi(r_a)/\partial r_a)\cos\theta. \tag{4.37}$$

We shall perform the following calculations in atomic units. We return to absolute units in the final expressions for the matrix element of the operator $\hat{L}_E$ and the cross section. We seek a solution of Eq. (4.37) in the form

$$u^a_{i,f} = \rho\cos\theta\exp(-\rho/2)\,V(\rho), \quad \rho = 2\beta_{i,f} r_a, \quad \beta_{i,f} = \sqrt{Z^2 \pm 2\omega}. \tag{4.38}$$

Here r_a, ρ, and ω are expressed in atomic units. Substituting $u_{i,f}^a$ in Eq. (4.37), we obtain an equation for $V(\rho)$:

$$\begin{aligned}\rho\frac{d^2}{d\rho^2}V(\rho) + (4-\rho)\frac{d}{d\rho}V(\rho) - \left(2 - \frac{Z}{\beta}\right)V(\rho) = \\ = -\gamma\exp[\pm\omega\rho/\beta(\beta+Z)], \quad \gamma = Z^{5/2}/2\beta^2\sqrt{\pi}.\end{aligned} \tag{4.39}$$

The upper and lower signs refer to the subscripts i and f, respectively, but we shall omit these subscripts in the intermediate calculations. The power series solution of Eq. (4.39) has the form

$$V(\rho) = \gamma\sum_{k=0}^{\infty} a_k\rho^k, \quad a_{k+1} = \frac{(k+2-Z/\beta)\,a_k - b_k}{(k+1)(k+4)}, \quad b_k = \frac{(\pm\omega)^k}{\beta^k(\beta+Z)^k k!}. \tag{4.40}$$

The constant a_0 must be chosen so that the initial functions W_i and W_f can be expanded as a series in the eigenfunctions of the operator $H^{(0)}$. This requirement reduces to certain conditions that the function $V(\rho)$ must satisfy at zero and at infinity.

In determining the constant a_0, it is convenient to transform to a new unknown function $f(\rho)$:

$$f(\rho) = \rho^2 V(\rho) \exp(-\rho/2), \tag{4.41}$$

$$\frac{d^2 f(\rho)}{d\rho^2} + \left(-\frac{1}{4} + \frac{Z}{\beta\rho} + \frac{1/4 - (3/2)^2}{\rho^2}\right) f(\rho) = -\gamma\rho \exp(-Z\rho/2\beta).$$

The general solution of the inhomogeneous equation (4.41) can be expressed in terms of a fundamental system of solutions of the corresponding homogeneous equation:

$$f(\rho) = \frac{\gamma}{\Delta}\left[M(\rho)\int_\rho^\infty W(\rho')\exp\left(-\frac{Z\rho'}{2\beta}\right)\rho' d\rho' + W(\rho)\int_0^\rho M(\rho')\exp\left(-\frac{Z\rho'}{2\beta}\right)\rho' d\rho'\right], \tag{4.42}$$

where

$$\Delta = W\, dM/d\rho - M\, dW/d\rho,\quad W \equiv W_{Z/\beta,3/2}(\rho),\quad M = M_{Z/\beta,3/2}(\rho).$$

The homogeneous equation corresponding to the inhomogeneous equation (4.41) is the well-known Whittaker equation [25]. The functions $M_{Z/\beta,3/2}$ and $W_{Z/\beta,3/2}$ are the Whittaker functions. The limits of integration in Eq. (4.42) are chosen so that the function $f(\rho)$ is finite at zero and falls off at infinity. From Eq. (4.42) we obtain the limit of $f(\rho)$ as $\rho \to 0$ and, therefore, the coefficient a_0, which is given in terms of the hypergeometric function by

$$a_0 = \frac{16}{(2 - Z/\beta)(Z/\beta + 1)^4} F\left(4, 2 - \frac{Z}{\beta}, 3 - \frac{Z}{\beta}; \frac{\mp 2\omega}{(Z+\beta)^2}\right). \tag{4.43}$$

We have, therefore, evaluated the functions W_i and W_f. Substituting these functions in the integrals Σ_i and Σ_f, we obtain

$$\langle f | L_E | i \rangle_1 = \left(\frac{2\pi}{L}\right)^3 \frac{i2\alpha q}{\hbar\omega} \frac{e^2}{2\pi^2 q^2} Z^4 \left[A - \frac{1}{2Z^3} + \frac{8}{(4Z^2 + q^2 a_B^2)^2}\right], \tag{4.44}$$

where

$$A = \frac{\omega}{\omega_a}\left(\sum_k C_k^i B_k^i - \sum_k C_k^f B_k^f\right),$$

and

$$C_k^{i,f} = a_k 2^{k+1} \beta_{i,f}^{k-1} (k+1)!\,(qa_B)^{-3}\left[(Z + \beta_{i,f})^2 + q^2 a_B^2\right]^{-(k+3)/2},$$

$$B_k^{i,f} = (k+3)\, qa_B \cos\left[(k+3)\arctan\frac{qa_B}{Z+\beta}\right] - (Z+\beta)\sin\left[(k+3)\arctan\frac{qa_B}{Z+\beta}\right].$$

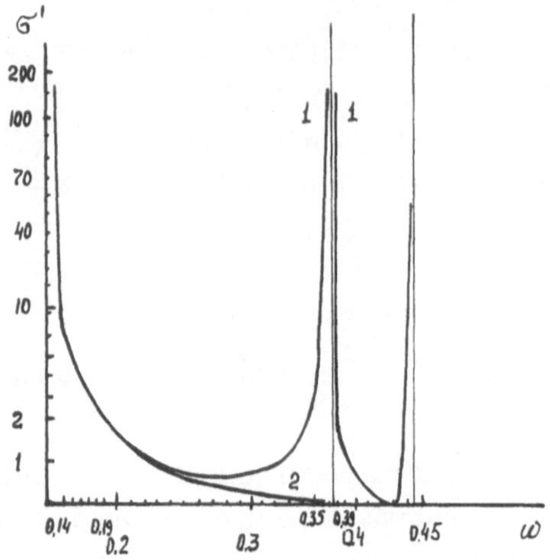

Fig. 4.1. The cross section $\sigma_{1s,1s}(\omega)$ for bremsstrahlung on hydrogen atoms in atomic units (curve 1, $|\mathbf{p}_i| = 3$ and $\mathbf{E}_0 \perp \mathbf{p}_i$) and the cross section for bremsstrahlung calculated in the approximation of a given field (curve 2). (For convenience a nonuniform scale is used; the scale is uniform between the points labelled with numbers on the axes.)

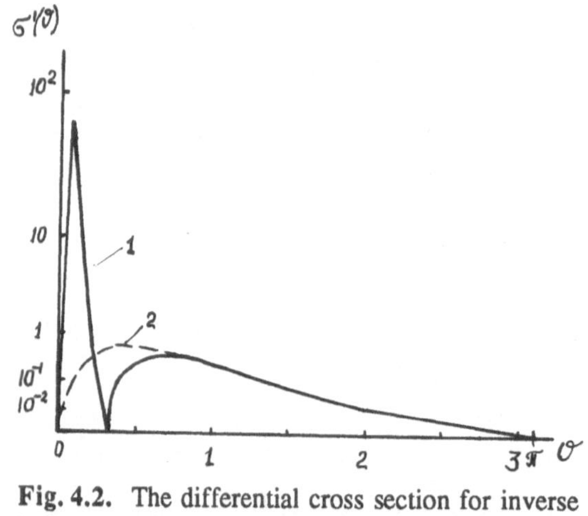

Fig. 4.2. The differential cross section for inverse bremsstrahlung during scattering of an electron on a hydrogen atom [$\sigma' = 10^{-2} \cdot \sigma_{1s,1s}(\omega, \theta)/E_0^2$; $\omega = 0.3$ a.u.; $|\mathbf{p}_i| = 3$ a.u.].

The quantities qa_B and ω/ω_a in the square brackets are given in atomic units. The cross section calculated according to Eq. (4.27) is in absolute units.

The results of some numerical calculations are shown in Figs. 4.1 and 4.2 [10, 26, 27]. Figure 4.1 shows the integral cross section for elastic bremsstrahlung on a ground-state hydrogen atom as a function of the frequency. The ordinate is the quantity

$\sigma' = (10^{-2}/a_B{}^2)\sigma_{1s,1s}(\omega)(e/a_B{}^2E_0)^2$, where $\sigma_{1s,1s}(\omega) = \int(d\sigma_{1s,1s}{}^{ind}/\,d\Omega_{p_f})d\Omega_{p_f}$ is the integral over the entire solid angle calculated according to Eqs. (4.27) and (4.44). The quantities ω/ω_a, $E_0a_B{}^2/e$, and $\sigma_{1s,1s}(\omega)/a_B{}^2$ are given in atomic units. The cross section $\sigma'(\omega)$ (curve 1) evidently differs greatly from the cross section calculated in the shielding approximation (curve 2). The cross section $\sigma'(\omega)$ increases sharply near resonance frequencies. Two resonances are shown in Fig. 4.1 ($\omega/\omega_a \approx \omega_1/\omega_a = 0.37$ and $\omega/\omega_a \approx \omega_2/\omega_a = 0.44$). These correspond to transitions from the ground state of the hydrogen atom into the first two excited states (the singularity at $\omega = 0.4$ corresponds to the "infrared" catastrophe.) Naturally, the curves differ greatly near a resonance, but even outside the resonance region there is a significant difference. Let us consider the behavior at a frequency of $\omega/\omega_a = 0.4$ near the second resonance $\omega/\omega_a = 0.44$ in more detail (the difference $\omega_2/\omega_a - \omega/\omega_a = 0.04$ is considerably greater than the width of the emission line). At $\omega/\omega_a = 0.4$ the cross section σ' is equal to 0.9 in the exact analysis (curve 1) and to 0.1 in the shielding approximation (curve 2). If we try to describe the cross section by a single resonance term in the sum Σ, then we obtain $\sigma' = 1.3$. The need for the exact theory is, therefore, obvious.

We now consider the differential cross section for inverse bremsstrahlung. We shall limit ourselves to the case $\mathbf{p}_i \perp \mathbf{E}_0$. Figure 4.2 (curve 1) shows the cross section as a function of the scattering angle ϑ between $\mathbf{p}_i$ and the final momentum $\mathbf{p}_f$ of the electron (here ϑ and φ are the spherical coordinates of the vector $\mathbf{p}_f$). The ordinate is

$$\sigma'(\vartheta) = 10^{-2}\,\frac{\sigma_{1s,1s}^{ind}(\omega,\vartheta)}{a_B^2}\left(\frac{e}{a_B^2E_0}\right)^2, \tag{4.45}$$

where

$$\sigma_{1s,1s}^{ind}(\omega,\vartheta) = \int_0^{2\pi} d\varphi\,\frac{d\sigma_{1s,1s}^{ind}(\omega)}{d\Omega_{p_f}}.$$

$\sigma_{1s,1s}{}^{ind}(\omega,\vartheta)/a_B{}^2$ is the cross section for scattering of an electron into a solid angle with an aperture of $d\Omega_{p_f}$, integrated over φ (in atomic units). $d\sigma_{1s,1s}{}^{ind}(\omega)/d\Omega_{p_f}$ is given by Eqs. (4.27) and (4.44). It is evident that the angular dependence of $\sigma'(\vartheta)$ (curve 1) is extremely different from that obtained in the shielding approximation (curve 2). Both functions go to zero at the points $\vartheta = 0$ and $\vartheta = \pi$ thanks to their common angular factor $\mathbf{E}_0\mathbf{p}_f$ (for $\vartheta = 0$ and $\vartheta = \pi$ the vector $\mathbf{p}_f$ is parallel or antiparallel to $\mathbf{p}_i$, so that $\mathbf{p}_f \perp \mathbf{E}_0$). The function $\sigma'(\vartheta)$, however, also goes to zero at $\vartheta =$ 0.33. This zero appears for an entirely different reason: the terms corresponding to the shielding approximation in the matrix element are compensated by the sums Σ which appear only in this formulation of the problem. Thus, the two versions of the theory are in distinct disagreement, both qualitatively and quantitatively.

The greatest difference between the traditional and exact formulations of the bremsstrahlung problem shows up in a resonance region. The exact formulation leads to additional terms which represent an infinite sum over the atomic states. When the resonance conditions

$$\hbar\omega = E_n - E_{n_i}, \qquad \hbar\omega = E_{n_f} - E_n, \tag{4.46}$$

are satisfied, the denominator in the corresponding terms of these sums goes to zero.

In the preceding discussion we have limited ourselves to the case in which the frequency of the light is not equal to the resonant frequency, but is close to it, so that there is a detuning Δ off the resonance that is greater than the widths Γ_n of the corresponding lines (Γ_n is the width of the level E_n):

$$\begin{aligned} \Delta &= E_{n_i} + \hbar\omega - E_n, \qquad \hbar\omega \gg |\Delta| \gg \Gamma_n,\ \Gamma_{n_i}, \\ \Delta' &= E_{n_f} - \hbar\omega - E_n, \qquad \hbar\omega \gg |\Delta| \gg \Gamma_n,\ \Gamma_{n_f}. \end{aligned} \tag{4.47}$$

The conditions (4.46) and (4.47) and the subsequent discussion apply to inverse bremsstrahlung (the absorption amplitude is obtained from the emission amplitude by replacing ω by $-\omega$). When conditions (4.47) are satisfied, we can ignore the broadening of the atomic levels and additionally limit ourselves to one term in Eq. (4.26):

$$\langle f | \hat{L}_E | i \rangle_1 \approx -\left(\frac{2\pi}{L}\right)^3 \frac{ie\hbar}{2m\omega\Delta} \langle n_f | V(\mathbf{q}, \mathbf{r}_a) | n \rangle \left\langle n \left| \sum_{j=1}^{N} (\mathbf{E}_0 \nabla_j) \right| n_i \right\rangle \tag{4.48}$$

[for concreteness we assume that the first resonance condition in Eq. (4.46) is satisfied and that the levels E_n and E_{ni} are not degenerate]. Substituting Eq. (4.48) in Eq. (4.27), we obtain the cross section for the resonant bremsstrahlung effect:

$$d\sigma^{\text{ind}}_{n_f, n_i}(\omega) = \frac{(2\pi)^4 p_f e^2}{4\omega^2 p_i \Delta^2 \hbar^2} \left| \langle n_f | V(\mathbf{q}, \mathbf{r}_a) | n \rangle \left\langle n \left| \sum_{j=1}^{N} (\mathbf{E}_0 \nabla_j)\, n_i \right.\right\rangle \right|^2 d\Omega_{p_f}. \tag{4.49}$$

The matrix element $\langle n_f | V(\mathbf{q}, \mathbf{r}_a | n\rangle$ can be expressed in terms of the cross section for inelastic collisions of fast electrons with ions that do not involve photons. In fact, the expression for the cross section of an inelastic electron collision with transitions of the ion from state n to state n_f includes the matrix element

$$\langle n_f | V(\mathbf{p}_i - \mathbf{p}_f', \mathbf{r}_a) | n \rangle,$$

which, as a function of $\mathbf{p}_i - \mathbf{p}_f'$, coincides with the matrix element of interest to us [28]. The only difference is that the final momentum $\mathbf{p}_f'$ is determined from the conservation of energy for inelastic scattering of an electron without the participation of a photon:

$$\frac{p_f'^2}{2m_0}+E_{n_f}=\frac{p_i^2}{2m_0}+E_n. \tag{4.50}$$

At the same time, the final momentum $\mathbf{p}_f$ in Eq. (4.49) is determined from the conservation of energy during scattering with participation of a photon (4.15). Generally speaking, these expressions are different, but if the resonance conditions are satisfied approximately, e.g., if $\hbar\omega \sim E_n - E_{ni}$, then Eq. (4.50) follows from Eq. (4.15). The matrix element $\langle n_f|V|n\rangle$ can then be expressed in terms of the cross section $d\sigma_{nf,n}(\mathbf{p}_f, \mathbf{p}_i)$ for inelastic scattering with transfer of an ion energy $E_n - E_{nf}$, to the electron as follows:

$$|\langle n_f|V|n\rangle|^2=\frac{\hbar^4}{(2\pi)^4 m_0^2}\frac{p_i}{p_f}\frac{d\sigma_{n_f,n}(\mathbf{p}_f,\mathbf{p}_i)}{d\Omega_{\mathbf{p}_f}}. \tag{4.51}$$

Substituting Eq. (4.51) in Eq. (4.49), we obtain ($m_0 = m$)

$$d\sigma^{\text{ind}}_{n_f,n_i}(\omega)=\frac{e^2\hbar^2}{4m^2\omega^2\Delta^2}d\sigma_{n_f,n}(p_f,p_i)\left|\left\langle n\left|\sum_{j=1}^{N}(\mathbf{E}_0\nabla_j)\right|n_i\right\rangle\right|^2. \tag{4.52}$$

If the transition $n_i \to n$ is dipole allowed, then we can express the matrix element $\langle n|\Sigma_{j=1}^N(\mathbf{F}_0\nabla_j)|n_i\rangle^2$ in terms of the oscillator strength $f_{n,ni}{}^Z$ for the transition $n_i \to n$ ($\mathbf{E}_0 \parallel z$):

$$\left|\left\langle n\left|\sum_{j=1}^{N}(\mathbf{E}_0\nabla_j)\right|n_i\right\rangle\right|^2=\frac{m\omega_{n,n_i}}{2\hbar}f^z_{n,n_i}E_0^2. \tag{4.53}$$

In the dipole approximation, therefore, Eq. (4.52) has the form

$$d\sigma^{\text{ind}}_{n_f,n_i}(\omega)=\frac{e^2E_0^2\hbar}{8m\omega\Delta^2}f^z_{n,n_i}\,d\sigma_{n_f,n}. \tag{4.54}$$

It is clear from Eq. (4.54) that the above approximation, in which just the resonant term is retained in the sums (4.26), is valid only for a dipole allowed transition $n_i \to n$. [For a dipole forbidden transition, $f_{n,ni}{}^z = 0$ and in the dipole approximation the resonant term goes to zero. Then it becomes necessary, in general, to take nonresonant terms into account in the sums (4.26).] As for the dipole allowed transitions, a situation in which the resonant term exceeds the nonresonant term is entirely realistic: for comparable values of the numerators, the denominator of the resonant term is much smaller than the denominators of the nonresonant terms. Note that Eq. (4.54) is valid not only when the level E_n is nondegenerate. If the level E_n is degenerate, but all the transitions $n_i \to n$ are dipole forbidden, ex-

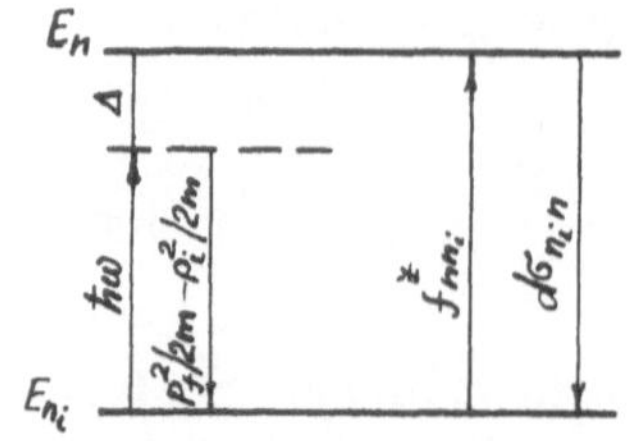

Fig. 4.3. An energy diagram for inverse bremsstrahlung absorption of a photon (the transitions whose probabilities determine the inverse bremsstrahlung cross section are indicated on the right.)

cept one, then the sums (4.26) can also be approximated by the resonant term alone. For example, let the frequency ω be close to the frequency of the transition of a hydrogenlike ion from its ground state to the first excited state. Then a transition from the state n_i is dipole allowed only into one of the four states belonging to the term E_n (into the state with $l = 1$ and $m = 0$). Equation (4.54) is valid in this case if f_{n,n_i}^z is the oscillator strength of the allowed transition.

During a collision of an atom (or ion), electron, and photon, the states of the electron and ion may change simultaneously (and usually do). First let us consider the case when only the state of the electron changes. As in inverse bremsstrahlung for an electron in a given external field, in this case the energy of the photon is transferred entirely to the electron. The cross section for the process is given by Eq. (4.54) with $n_i = n_f$:

$$d\sigma^{\text{ind}}_{n_i,n_i}(\omega) = \frac{e^2E_0^2\hbar}{8m\omega\Delta^2}\, d\sigma_{n_i,n} f^z_{n,n_i}. \tag{4.55}$$

Therefore, the cross section for (inverse) bremsstrahlung absorption of a photon is given in terms of the oscillator strength of the transition $n_i \to n$ (Fig. 4.3). This should not be interpreted as a real excitation of the atom followed by a transition into an intermediate state n (such excitation is generally impossible in our approximation, where the widths of the atomic levels are equal to zero and the detuning off the resonance is $\Delta \neq 0$).

In deriving Eq. (4.55) we have neglected all the nonresonant terms, including the first term of Eq. (4.26). It turns out that this term has a simple physical meaning: it describes conventional inverse bremsstrahlung for an electron in a given external field of an atom or ion. Let us consider just the case of an ion. The matrix element $\langle n_i|V|n_i\rangle$ is the potential energy of the interaction between an incident electron and the ion, averaged over the wave function of the ion. We can take the field of the bound electrons in the ion into rough account by treating the ion as a point charge $(Z - N)e$ (which is correct for $a_B q \ll 1$ [2, 3]). Then the contribution of the first term of Eq. (4.26), alone, to the cross section is given by

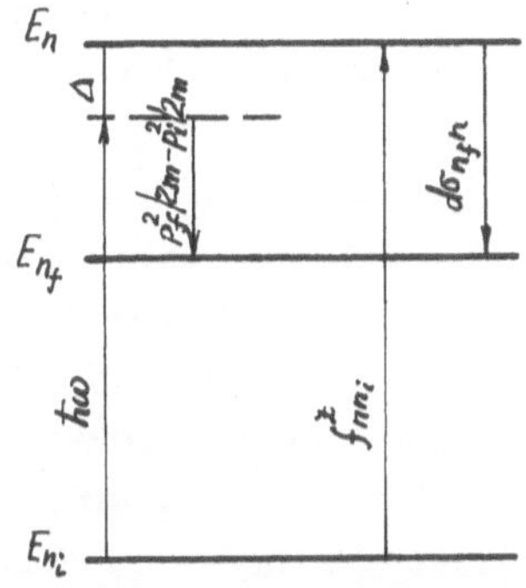

Fig. 4.4. An energy diagram for a transition with simultaneous excitation of an atom and an incident electron (the transitions whose probabilities determine the inverse bremsstrahlung cross section are indicated on the right.)

$$d\sigma_{st}^{ind}(\omega) = \frac{(Z-N)^2 e^6}{(\hbar\omega)^4 q^4} \frac{p_f}{p_i} (\mathbf{E}_0\mathbf{q})^2 \, d\Omega_{\mathbf{p}_f}. \tag{4.56}$$

This is the standard formula for the cross section for absorption of a photon during scattering of an electron in the Coulomb field of a point charge $(Z-N)e$.

It remains to compare the resonance term, which approximates the sum in Eq. (4.26), with the last term in this formula [i.e., Eq. (4.53) or (4.56)]. We can simplify Eq. (4.55) if we use an approximate formula for the inelastic scattering cross section $d\sigma_{n_i,n}$. Since the main contribution to the total cross section is from collisions with $qa \ll 1$, we restrict ourselves to comparing the differential cross sections for small q. When $qa \ll 1$, Eq. (4.55) takes the form

$$d\sigma_{n_i,n_i}^{ind}(\omega) = \frac{e^6 p_f (\mathbf{E}_0\mathbf{q})^2}{4(\hbar\omega)^2 \Delta^2 q^4 p_i} (f_{n,n_i}^z)^2 \, d\Omega_{\mathbf{p}_f}. \tag{4.57}$$

Here, as above, we have assumed that the sums (4.26) contain only a single resonance term. In order to simplify the calculation, we consider the case in which the quantum numbers n_i correspond to an s-state of the valence electron and the quantum numbers n correspond to a p-state. Then only the term $q_z\langle n|z|n_i\rangle$ remains in the matrix element $\mathbf{q}\langle n|\mathbf{r}|n_i\rangle$ which arises in the course of expanding $\exp(i\mathbf{qr})$ in the cross section $d\sigma_{n_i,n}$. We now consider the ratio of the cross sections (4.57) and (4.56):

$$\frac{d\sigma_{n_i,n_i}^{ind}(\omega)}{d\sigma_{st}^{ind}(\omega)} = \frac{(f_{n,n_i}^z)^2}{4(Z-N)^2} \left(\frac{\hbar\omega}{\Delta}\right)^2. \tag{4.58}$$

For $Z - N = 1, f_{n,n_i}Z \sim 10^{-1}$ and $\Delta/\hbar\omega \sim 10^{-2}$, the cross-section ratio is 25. Consequently, we can limit ourselves to examining just the resonance term in the matrix element (4.26).

Bremsstrahlung and inverse bremsstrahlung must, therefore, be treated rigorously as a many-particle problem that also takes the effect of the electromagnetic field on the bound electrons in the atom into account. The shielding approximation yields a cross section (4.56) which is considerably smaller than the cross section (4.57) obtained by solving the many-particle problem.

In the framework of the many-particle problem we can also examine collisions in which the states of an electron and an ion change simultaneously. As a result of such a collision, the electron acquires an energy that is smaller or greater than the energy of the photon. In the first case, the ion is excited at the same time the incident electron gains energy owing to the energy of a single photon (Fig. 4.4). The cross section for such processes in the dipole approximation is given by Eq. (4.54) for $n_i \neq n_f$. On comparing the bremsstrahlung cross sections with and without excitation, we obtain

$$\frac{d\sigma^{\text{ind}}_{n_f,n_i}(\omega)}{d\sigma^{\text{ind}}_{n_i,n_i}(\omega)} = \frac{d\sigma_{n_f,n}}{d\sigma_{n_i,n}} . \tag{4.59}$$

It is clear from Eq. (4.59) that if the cross section $d\sigma_{nf,n}$ is greater than the cross section $d\sigma_{ni,n}$, then a collision with excitation of the ion may be more probable that one without excitation.

The low-frequency condition, therefore, encompasses the resonance situation when the frequency of the emitted (or absorbed) photon is close to the frequency of an atomic transition. The bremsstrahlung cross section increases sharply near resonances, so that resonance frequencies are of special interest. In turn, the resonant case breaks up naturally into two cases: the first occurs when the detuning from the resonance is much greater than the atomic line width and the second, when the detuning is comparable to the line width. We have only examined the first case. It is simpler because it does not require an investigation of the specific mechanisms for broadening of the resonance line. Nevertheless, even here we can ask the following question. We know that there is a process in which an atom can be excited by an electron beam (inelastic collision of an electron with an atom) and "afterwards" emit a resonance photon. At the same time, during bremsstrahlung a photon is emitted "simultaneously" with the scattering of an incident particle (e.g., an electron). Is there a real physical difference between these ("cascade" and "direct") processes? Evidently, there is, at least when collisions with other atoms, electrons, etc. take place between the times the atom is excited and deexcited. Then the concepts of "simultaneously" and "afterwards" have a real physical content. Is is possible to imagine an experiment, even an imaginary one, which would allow us to distinguish these processes? Yes.

Let us consider one such experiment. It concerns the inverse bremsstrahlung effect. An electron in a given external field with frequency ω can acquire an energy $\hbar\omega$ through inverse bremsstrahlung and its energy becomes equal to $T_i + \hbar\omega$. If an electron beam is sufficiently monochromatic, then for a fixed frequency ω there will be a peak in the electron energy distribution at $T_i + \hbar\omega$ with a spread that corresponds to the spread in energy of the initial electron beam. As the frequency ω approaches a resonance, the magnitude of this peak will increase. An additional peak will develop in the electron energy distribution owing to those electrons which lose energy through inelastic collisions with atoms and excitation of atoms to the level corresponding to this resonance. This peak will be displaced relative to the inverse bremsstrahlung peak by an amount equal to the detuning Δ from the resonance.

It is important to note that this sort of experiment has actually been carried out, but, unfortunately, only in the nonresonance region [29]. These experiments were aimed at studying multiphoton processes during bremsstrahlung, which had previously been examined theoretically [30]. It would be very interesting to carry out this type of experiment in the resonance region as well.

A rigorous examination of the bremsstrahlung problem, therefore, leads to a new physical picture. Electron and positron bremsstrahlung in long-range collisions, proton bremsstrahlung at low frequencies, and resonance bremsstrahlung show up only in an exact formulation of the problem where the atomic electrons and the incident particle enter on an equal basis in the Hamiltonian for the system. Like the shielding picture, the rigorous analysis is physically intuitive, but it involves scattering on particles (the nucleus and atomic electrons), rather than on a static potential.

The concept of an incident particle which carries a field of virtual photons with it is very useful in understanding the nature of resonant bremsstrahlung [5]. Then polarization bremsstrahlung can be regarded as the scattering of these photons by atomic electrons. This analogy works: it facilitates an understanding of the physics of the phenomenon, especially in a resonance region. From this point of view, elastic bremsstrahlung is resonant scattering of virtual photons on an atom when the initial and final states of the atom coincide. Inelastic bremsstrahlung is analogous to resonant Raman (combination) scattering where the initial and final states of the atom differ. In both cases, the radiation is not associated with the actual filling of an intermediate resonance level. In the theory of resonant Raman scattering, the situation which we are examining, where the detuning from the resonance satisfies the inequality (4.47), is referred to as preresonant. A clear distinction has been drawn [31] between resonant Raman scattering in this region and the luminescence associated with actual filling of an intermediate level. The case of an exact resonance, when the detuning from the resonance is on the order of or smaller than the level width, is more complicated. It requires further study.

Calculating the sums over intermediate states for many-electronic atoms for arbitrary $\mathbf{q}$, when the sums do not reduce to the polarizabilities, and evaluating the accuracy of these calculations are also of great current interest, especially in connection with the experiments of Ishii and Morita [20]. The accuracy with which transition probabilities are calculated is well known to be substantially less than the accuracy to which the energies of atoms can be calculated [24].

We have already mentioned the close relationship between polarization dynamic bremsstrahlung and the polarization transition bremsstrahlung introduced and studied by Akopyan and Tsytovich [17, 32]. It is interesting that both effects can be regarded as transition radiation [33, 34] where the inhomogeneity is a cloud of atomic electrons in the first case and the Debye cloud surrounding an ion in a plasma in the second.

Polarization bremsstrahlung arises in a three-body collision between an atom, electron, and photon. As a whole, collisions of this type form a more general class of processes than bremsstrahlung. For example, when an atom collides simultaneously with an electron and a photon, the atom may be excited even if the energy of the photon is less than the excitation energy of the atom. The electron supplies the missing energy [35]. The theory of this type of process [35] is essentially the same as the theory of polarization bremsstrahlung. They obey the same formulas as inelastic bremsstrahlung. The earlier results [35] have been greatly extended [36]. In particular, it has been suggested [36] that the scattered electrons should be tracked in order to identify the process by which an atom is excited in three-body collisions.

Chapter 5

POLARIZATION BREMSSTRAHLUNG OF RELATIVISTIC CHARGED PARTICLES ON ATOMS

V. A. Astapenko, V. M. Buimistrov, and Yu. A. Krotov

5.1. Introductory Comments

As noted previously, the theory of bremsstrahlung of relativistic charged particles on atoms was developed in the classical papers of Bethe and Heitler [1, 2]. Even in that early work, as well as in the work of Landau and Rumer [3] and Wheeler and Lamb [4] it was noted that the role of the atomic electrons in bremsstrahlung of relativistic particles is not merely to shield the nucleus. The essence of the problem was clearly formulated in 1969 by Ter-Mikaélyan [5], who wrote, "besides the emission on the nucleus, we must take the emission on the atomic electrons into account. Up to now, the bonding of the electrons has not been included in any analysis of this problem by the exact methods of quantum electrodynamics because of its complexity."

The contribution of atomic electrons to the bremsstrahlung of relativistic particles has usually been taken into account in two ways. On one hand, the atom is replaced by a fixed central potential created by the nucleus and the cloud of atomic electrons that shields it, just as in the nonrelativistic bremsstrahlung problem. For the differential (with respect to the scattering angles of the incident particle) cross section, this reduces to an effective reduction in the nuclear charge Ze by the magnitude of the form factor, which is usually calculated in the Hartree–Fock model or, for atoms with high Z, in the statistical Fermi–Thomas model. In the integral cross section this leads to the concept of the shielding radius (in the Fermi–Thomas model $R_a \sim Z^{-1/3}a_B$), which "cuts off" the lower limit of integration with respect to the momentum transfer and yields the well-known logarithm in the bremsstrahlung cross section.

On the other hand, besides the emission on the shielded nucleus at relativistic incident particle energies, the emission on the atomic electrons was also taken into account. The latter usually reduced to replacing the factor Z^2 in the integral cross section by $Z(Z + 1)$ and was explained as a supplement to the emission on the nucleus that is proportional to Z^2 by emission on "fastened" unit charges (Z of them) which replace the atomic electrons.

In fact, when $\omega \gg m$ ($\hbar = 1$) both the method of equivalent quanta [6, 7] and the Born approximation [8] yield an expression for the cross section for bremsstrahlung of a fast electron on a fixed nucleus that coincides with the bremsstrahlung cross section for an electron on a "fastened" unit charge.

The emission on the atomic electrons was first taken into account in this fashion by Landau and Rumer [3], who noted the importance of a correction that was proportional to Z in the theory of cosmic ray showers.

For large momentum transfers and in the soft ($\omega \ll m$) region of the bremsstrahlung spectrum, however, one must take the emission from the atomic recoil electrons into account [8]. In addition, it has been noted [5, 9] that including the bonding of the atomic electrons for low momentum transfer is not equivalent to the shielding effect, since the atomic electrons can acquire energy, be excited into the discrete spectrum, and, therefore, lead to scattering of the incident particle and to bremsstrahlung at impact parameters which exceed those for the atom.

Later on, Wheeler and Lamb [4] took the sum of the Bethe–Heitler cross sections over the final states of the atomic electrons and obtained a term proportional to Z, but with a different logarithmic term for the shielding. This was actually a refinement of Landau and Rumer's result [3].

The results of [3, 4] were then supplemented [10] by including exchange and large momentum transfers, which, incidentally, did not lead to significant corrections.

All the above remarks referred to traditional bremsstrahlung, in which the bremsstrahlung photon is emitted by the incident particle and a scatterer is necessary only to accept the excess momentum.

After the nonrelativistic theory of bremsstrahlung on atoms was supplemented in the 1970s by including a polarization term, the contribution of the dynamic polarization of the atom to the bremsstrahlung from a relativistic particle was studied in detail [11–13]. This showed that the polarization term in the bremsstrahlung of a relativistic particle on an atom has some additional features not seen in the nonrelativistic case.

In this chapter we shall use the methods of quantum electrodynamics to calculate the polarization bremsstrahlung of relativistic particles (see Chapter 9 as well). In the rest of this chapter, we take $\hbar = c = 1$.

5.2. Bremsstrahlung Amplitude of a Relativistic Particle on an Atom

We now consider the collision of a relativistic particle [charge e_0, mass m_0, and energy $\varepsilon_i = (\mathbf{p}_i^2 + m_0^2)^{1/2}$] in the state $|p_i\rangle$ with an atom in the state $|n_i\rangle$ with energy $E_{n_i} \equiv E_i$. As a result of the collision, the incident particle enters the state $|p_f\rangle$ with energy $\varepsilon_f = (\mathbf{p}_f^2 + m_0^2)^{1/2}$ and emits a bremsstrahlung photon with frequency ω and wave vector $\mathbf{k}$, while the atom goes into the state $|n_f\rangle$ with energy $E_{n_f} \equiv E_f$. The conservation of energy has the form

$$\varepsilon_i + E_{n_i} = \varepsilon_f + E_{n_f} + \omega. \tag{5.1}$$

Here we shall assume for simplicity that the atom has one electron. We also assume that the nucleus of charge Ze is motionless. The incident particle obeys the Dirac equation. The generalization to the case of a nonrelativistic many-electron atom is given in Sec. 5.3.

Let us also assume that the velocities of the incident particle before and after the collision obey the Born condition: $Z|ee_0| \ll v_{f,i}$. Then the incident particle is described by a plane wave, unlike the exact solution of the Dirac equation in the external field of the nucleus (the Furry representation), which is needed to describe the atomic electron. When the incident particle is an electron, this makes it possible to neglect the exchange terms in the amplitude, as well.

We shall use standard perturbation theory for the scattering operator. In the lowest order with respect to the interaction of the incident particle with the atom and with respect to the interaction of the incident particle and the atomic electron with the electromagnetic field, we obtain a diagram expression for the amplitude of this process (Fig. 5.1) [11, 12].

Here the single lines correspond to the wave functions and propagator of the incident particle, and the double lines, to the atomic electron in the field of the nucleus. A wavy line denotes an electromagnetic field: the photon propagator and the wave function of a photon, $A_{\mathbf{k},\sigma}{}^*$. The analytic expression for the amplitude of this process has the form

$$\begin{aligned} M_{fi} = M_{fi}^{st} + M_{fi}^{p} = &-\frac{4\pi e_0^2 e}{q^2} A^{\mu *}_{\mathbf{k},\sigma} [Z g^{0\nu} \langle f|i\rangle - j^{\nu}_{fi}(\mathbf{q})] \times \\ &\times \frac{\bar{u}_f}{\sqrt{2\varepsilon_f}} \left[\gamma_\nu \frac{\gamma p_2 + m_0}{p_2^2 - m_0^2} \gamma_\mu + \gamma_\mu \frac{\gamma p_1 + m_0}{p_1^2 - m_0^2} \gamma_\nu\right] \times \\ &\times \frac{u_i}{\sqrt{2\varepsilon_i}} + \frac{4\pi}{q_1^2} A^*_{\mathbf{k},\sigma,\nu} e^2 e_0 \sum_n \left[\frac{j^{\nu}_{nf}(\mathbf{k})\, j^{\mu}_{ni}(\mathbf{q}_1)}{-E_n + E_f + \omega \pm i0} + \right. \\ &\left. + \frac{j^{\mu}_{nf}(\mathbf{q}_1)\, j^{\nu}_{ni}(\mathbf{k})}{E_i - E_n - \omega \pm i0}\right] \frac{\bar{u}_f \gamma_\mu u_i}{2\sqrt{\varepsilon_f \varepsilon_i}}, \end{aligned}$$

$$q_1 = p_f - p_i,\quad q = q_1 + k,\quad p_2 = p_f + k,\quad p_1 = p_i - k. \tag{5.2}$$

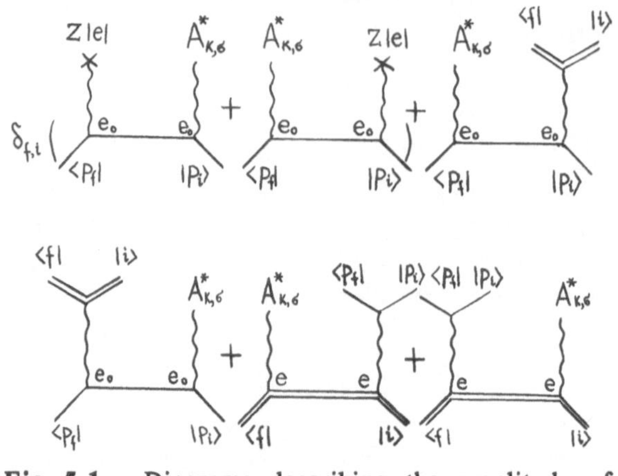

Fig. 5.1. Diagrams describing the amplitude of bremsstrahlung on an atom in the third order of perturbation theory.

Here we have used the following notation:

$$A_{\mathbf{k},\sigma} = \sqrt{\frac{2\pi}{\omega}}\, e_{\mathbf{k},\sigma}, \qquad |p\rangle = \psi_p(x) = \frac{u(p,s)}{\sqrt{2\varepsilon}} \exp(-ipx),$$

$$j^{\mu}_{n_1 n}(\varkappa) = \langle n_1 | \gamma^{\mu} \exp(-i\varkappa\mathbf{r}) | n \rangle, \qquad a = a^{\mu} = \{a^0, \mathbf{a}\},$$

$$ab = a^{\mu} b_{\mu} = a^0 b_0 - \mathbf{ab}, \qquad \mu, \nu = 0, 1, 2, 3.$$

The metric, normalization, and notation in Eq. (5.2) are similar to those in [8]: $g_{\mu\nu}$ is the metric tensor whose nondiagonal elements are zero and whose diagonal elements are given by $g_{00} = -g_{11} = -g_{22} = -g_{33} = 1$; γ^{μ} are the Dirac matrices; the normalization of the bispinors, $\bar{u}u = u^{+}\gamma^{0}u = 2m_0$, corresponds to normalization of the wave function $\psi_p(x)$ of the incident particle to one particle in an elementary region of unit volume; the wave function $A_{\mathbf{k},\sigma}$ of the photon is also normalized to one photon in an elementary volume; $e_{\mathbf{k},\sigma}$ is the polarization 4-vector which obeys a three-dimensional transverse gauge in the laboratory coordinate system; $e_{\mathbf{k},\sigma} = \{0, \mathbf{e}_{\mathbf{k},\sigma}\}$; $\mathbf{k}\mathbf{e}_{\mathbf{k},\sigma} = 0$; $n, f,$ and i are the set of quantum numbers which determine the stationary state of the atom; and the sum over the intermediate states extends over both the positive ($i0$) and negative ($-i0$) energy spectra of the atomic electron.

Let us analyze the basic formulas corresponding to Fig. 5.1 and Eq. (5.2). Note that, in the case where the atom does not change its state ($|i\rangle = |f\rangle$), the first four terms of Fig. 5.1 and the corresponding first term of Eq.(5.2) give the classical result of Bethe and Heitler [1, 2]: bremsstrahlung of a relativistic electron in the static field of the nucleus and the field of the atomic electron that shields it. If we include all the possible final states of the atomic electron in these amplitude terms, then we obtain the result of Wheeler and Lamb [4]. The Fourier transform of the time component of the transition current 4-vector when $f = i$, i.e., $j_{ii}{}^{0}(\mathbf{q})$, gives the usual form factor for the charge shielding. The spatial component $\mathbf{j}_{fi}(\mathbf{q})$ is the current (magnetic) shielding term and can be important during bremsstrahlung involving the excitation of deep-lying atomic shells.

It should be noted that a rigorous electrodynamic approach to relativistic bremsstrahlung on atoms leads to new results, even in the traditional static part of the amplitude for this process. These include the possibility of a change in the state of the atom during bremsstrahlung and the appearance of a current correction in the form factor caused by the spatial components of the transition 4-current for the atomic electrons.

The last two terms in Fig. 5.1 and the corresponding second term in Eq. (5.2) describe the emission of a bremsstrahlung photon by an atomic electron during a collision of an incident particle with the atom. These terms show up if the atomic electron is treated as an equivalent dynamic particle interacting with the electromagnetic field. We refer to their contribution to the total bremsstrahlung as polarization bremsstrahlung, since it is determined by the dynamic polarization of the atom in the fields of the incident particle and the bremsstrahlung photon.

One characteristic feature of the polarization term in the amplitude of bremsstrahlung on an atom is the presence of sums over the virtual states of an atomic electron with resonant energy denominators. Thus, besides the resonance in the electron spectrum of the atomic state when $\omega = \varepsilon_n^{(+)} - \varepsilon_f$, the relativistic problem has a resonant denominator in the positron part of the sum when $\omega = \varepsilon_i - \varepsilon_n^{(-)} \approx 2m$. Here we are not considering such high frequencies, although, in principle, it is necessary to include the contribution of the dynamic polarization of the atom to pair production from γ-ray quanta in the field of the atom, which process is cross-invariant to bremsstrahlung.

The total bremsstrahlung cross section also includes an interference contribution from static and polarization bremsstrahlung. As will be apparent from the following, however, that contribution is small for a relativistic incident particle.

It is interesting to trace the two limiting cases contained in Eq. (5.2). Let us first suppose that there is no nucleus ($Z = 0$). In this case the first two diagrams of Fig. 5.1 vanish, while in the remaining four the double lines describing an atomic electron in the field of the nucleus must be replaced by ordinary lines (free electron). Then the diagrams of Fig. 5.1 become the well-known graphical representation of quantum electrodynamics for the bremsstrahlung of an incident particle on a free electron. Here the first pair of diagrams describes the contribution of the incident particle to the bremsstrahlung as it is scattered on the electron and the second, the contribution to the bremsstrahlung from the recoil electron. When the incident particle is an electron and we consider frequencies $\omega \gg m$, we obtain the standard result of quantum electrodynamics to the effect that the emission from the recoil electron can be neglected, while the fast electron radiates on the slow free electron as on a stationary unit charge. This, in particular, explains the result of Landau and Rumer [3]. Note that in order to obtain this limiting transition, we must include the possibility of exciting the atom into discrete as well as continuum states in the diagrams of Fig. 5.1 and Eq. (5.2).

In the other limiting case there is no atomic electron. The last four diagrams of Fig. 5.1 disappear and the amplitude refers to the trivial case of bremsstrahlung on a "bare" nucleus.

The 4-transfer momentums q_1 and q enter the static and polarization terms of the bremsstrahlung amplitude in different ways and, in general, have both space and time components. **q** coincides with the momentum transfer in Chapter 3:

$$q = \{E_f - E_i;\ \mathbf{p}_f - \mathbf{p}_i + \mathbf{k}\},$$

$$|\mathbf{q}|_{\min} \approx \omega/\gamma^2 \qquad \varepsilon_{f,i} \gg \omega, \quad \gamma = \frac{\varepsilon_i}{m_0} \gg 1;$$

$$q_1 = \{\varepsilon_f - \varepsilon_i;\ \mathbf{p}_f - \mathbf{p}_i\}, \qquad |\mathbf{q}_1|_{\min} \approx \omega\,(1 + 1/\gamma^2)^{1/2}.$$

Here $|\mathbf{q}|$ is independent of the scattering angle in the range $\omega/\gamma^2\varepsilon \gg \vartheta \geq 0$, where $\cos\vartheta = (\mathbf{p}_i\mathbf{p}_f)/p_ip_f$, and remains of order $|\mathbf{q}|_{\min}$ within that range. At the same time this range is significantly wider for $|\mathbf{q}_1|$: $\omega/\gamma\varepsilon \gg \vartheta \geq 0$. The range of scattering angles $\vartheta \gg \omega/\gamma^2\varepsilon$ is characterized by the fact that the dependence of **q** on **k** can be neglected and we can take $\mathbf{q} \sim \mathbf{q}_1 = \mathbf{p}_f - \mathbf{p}_i$ with good accuracy.

We shall show later that the amount of momentum transferred to the atom during the bremsstrahlung process determines the probability of the atomic electron remaining in its initial state or being excited. Here we note that in a number of papers on polarization bremsstrahlung, the channels involving excitation of the bound electron were not examined, although excitation of the atom may be of decisive importance in some cases.

It is clear from the diagrams of Fig. 5.1 and Eq. (5.2) that the traditional static (Bethe–Heitler) and polarization terms in the amplitude depend in different ways on the mass and charge of the incident particle (this was mentioned in Chapter 1). Indeed, $M_{f,i}^{st} \sim e_0^2 e/m_0$, while $M_{f,i}^{P} \sim e_0 e^2/m$, and the static bremsstrahlung falls off as the mass of the incident particle approaches infinity, while the polarization term remains finite. This means, for example, as noted above, that the bremsstrahlung of a proton on an atom, which is practically equal to zero in the shielding approximation, is comparable to the bremsstrahlung from an electron within a certain frequency range when the polarization of the atom is included. Changing the sign of the incident particle does not change the static amplitude (the shielding approximation for bremsstrahlung is sensitive to the sign of the charge of the incident particle only in higher orders of perturbation theory) and changes the sign of the polarization amplitude. This leads to a change in the sign of the interference term in the total cross section for bremsstrahlung on an atom.

Let us consider the case of a nonrelativistic atomic electron ($Z \ll 137$, $|E_{f,i} - m| \ll m$). If, in addition, $\omega \ll m$, then Eq. (5.2) can be transformed to a form which contains only nonrelativistic characteristics of the atomic electron.

In fact, for $Z \ll 137$, we have

$$j^{\mu}_{f,n}(\mathbf{q}_1) = \{j^0_{f,n}(\mathbf{q}_1), \mathbf{j}_{f,n}(\mathbf{q}_1)\} = \int d\mathbf{r}\varphi_f^* \{1, \alpha\} \varphi_n \exp(-i\mathbf{q}_1\mathbf{r}) \approx$$

$$\approx \left\{ \int d\mathbf{r}\varphi_f^* \exp(-i\mathbf{q}_1\mathbf{r}) \varphi_n; \quad \int d\mathbf{r}\varphi_f^* \left[\exp(-i\mathbf{q}_1\mathbf{r}) \frac{1}{2m}(-i\nabla) \right.\right.$$

$$\left.\left. + \frac{1}{2m}(-i\nabla) \times \exp(-i\mathbf{q}_1\mathbf{r}) \right] \varphi_n \right\}.$$

This corresponds to a formal decomposition of the atomic bispinors into large (~1) and small ($\sim v_a$) spinors and subsequent neglect of the spin corrections.

In the polarization term of the amplitude (5.2), therefore, the transition to the nonrelativistic description in the sum over the intermediate states with a positive energy reduces to replacing the relativistic expressions for the transition currents by the nonrelativistic ones. The sum over the intermediate states can be transformed if we assume that the primary contribution to it is from states whose energies obey the inequality $||E_n^{(-)}| - m| \ll m$. Given that $|E_{f,i} - m| \ll m$ and $\omega \ll m$, the energy denominators in the terms with a negative energy can be replaced by $2m$. Furthermore, by applying the projection operator $(m - \hat{H}_a)/2m$ (where $\hat{H}_a$ is the atomic Hamiltonian) to the space of wave functions with negative energy, we can extend the sum over the entire energy spectrum of the atomic electron. To do this, we assume that

$$(m - \hat{H}_a)/2m = (1 - \gamma^0)/2, \quad \gamma^0 | n^{\pm} \rangle = \pm | n^{\pm} \rangle.$$

Then we have

$$\sum_{E_n<0} \approx \frac{1}{2m} \left\langle f \left| \exp(-i\mathbf{q}\mathbf{r}) \left(\gamma^{\mu} \frac{(1-\gamma^0)}{2} \gamma^{\nu} + \gamma^{\nu} \frac{(1-\gamma^0)}{2} \gamma^{\mu} \right) \right| i \right\rangle.$$

Using the commutation relation $\gamma^{\mu}\gamma^{\nu} + \gamma^{\nu}\gamma^{\mu} = 2\delta^{\mu\nu}$, we obtain

$$\sum_{E_n<0} \approx \frac{\delta^{\mu\nu}}{m} \langle f | \exp(-i\mathbf{q}\mathbf{r}) | i \rangle = \frac{\delta^{\mu\nu}}{m} j^{(0)}_{fi}(\mathbf{q}).$$

The polarization term in the amplitude for a nonrelativistic atomic electron, therefore, has the form

$$M^p_{fi} = \frac{4\pi e_0 e^2}{q_1^2} A^{\nu *}_{\mathbf{k},\sigma} \sum_{E_n>0} \left[\frac{j^{\nu}_{fn}(\mathbf{k})\, j^{\mu}_{ni}(\mathbf{q}_1)}{\omega_{fn} + \omega + i0} + \frac{j^{\mu}_{fn}(\mathbf{q}_1)\, j^{\nu}_{ni}(\mathbf{k})}{\omega_{in} - \omega + i0} + \frac{\delta^{\mu\nu}}{m} j^0_{fi}(\mathbf{q}) \right] \frac{\bar{u}_f \gamma_{\mu} u_i}{2\sqrt{\varepsilon_f \varepsilon_i}} \cdot \quad (5.3)$$

In the case of a nonrelativistic incident particle, Eq. (5.3) yields the well-known result from the nonrelativistic theory of polarization bremsstrahlung (see Chapter 4):

$$M^p_{fi} = \frac{4\pi e_0 e^2}{q_1^2} \sqrt{\frac{2\pi}{\omega}} \sum_n \left\{ \frac{\mathbf{e}^*_{\mathbf{k},\sigma} \mathbf{j}_{fn}(\mathbf{k})\, j^0_{ni}(\mathbf{q}_1)}{\omega_{fn} + \omega + i0} + \frac{j^0_{fn}(\mathbf{q}_1)\, \mathbf{e}_{\mathbf{k},\sigma} \mathbf{j}_{ni}(\mathbf{k})}{\omega_{in} - \omega + i0} \right\}. \quad (5.4)$$

In order to obtain Eq. (5.4) from Eq. (5.3), one must set (neglecting spin effects)

$$\bar{u}_f\gamma^\mu u_i/2\sqrt{\varepsilon_i\varepsilon_f} \approx \{1, \mathbf{v}_0\}, \ |\mathbf{q}_1| \ll |\mathbf{p}_{i,f}|,$$

i.e., replace the transition current of the incident particle with the classical current and retain terms of lower order in the ratio of the velocity of the incident particle to the speed of light.

5.3. Bremsstrahlung Amplitude of Relativistic Incident Particles on Nonrelativistic Many-Electron Atoms

A rigorous quantum electrodynamic treatment of the polarization bremsstrahlung of a relativistic incident particle on a many-electron atom is made more complicated by the need to include the interaction among the atomic electrons in the relativistic formalism, as well as by the problem of summing over the negative-energy states in a many-electron system. In the meantime, the problem can be greatly simplified for nonrelativistic atomic electrons if we use the nonrelativistic Hamiltonian from the very beginning and the incident particle is replaced by the electromagnetic field which it produces.

We now justify this type of substitution. Let the field operator of the free incident particle, $\hat{\varphi}(x)$ ($x = \{t, \mathbf{r}\}$), satisfy the Dirac equation

$$(\gamma p - m_0)\,\hat{\varphi}(x) = 0. \tag{5.5}$$

We shall assume that the following Dirac equation is valid for the operator $\hat{\psi}(x)$ of the electron–positron field of the atomic electrons:

$$[\gamma(p + eA^{\text{ext}}(x) + e\hat{A}^{\text{a.e}}(x)) - m]\,\hat{\psi}(x) = 0. \tag{5.6}$$

Here $A^{\text{ext}}(x)$ is the potential of the external field of the nucleus and $\hat{A}^{\text{a.e}}(x)$ is the operator for the electromagnetic field created by the atomic electrons, which satisfies the Maxwell equation

$$\partial^\nu\partial_\mu\hat{A}^{\text{a.e}\,\mu}(x) - \partial^\mu\partial_\mu\hat{A}^{\text{a.e}\,\nu}(x) = 4\pi e\hat{j}^\nu(x), \tag{5.7}$$

where $\hat{j}^\nu(x) = \hat{\bar{\psi}}(x)\gamma^\nu\hat{\psi}(x)$ is the current operator for the atomic electrons.

It is thus assumed that the interaction between the atomic electrons is included in $\hat{\psi}(x)$. We write the state vector of the system of fields (of the atomic electrons, incident particle, electromagnetic field) in the form of the product $|\Phi_j\rangle = |j\rangle|\varphi_j\rangle\cdot|n_{\mathbf{k},\sigma}\rangle$, where $|j\rangle$ is the state vector of the atomic electrons interacting with one another, $|\varphi_j\rangle$ is the state vector of the free incident particle, and $|n_{\mathbf{k},\sigma}\rangle$ is the state vector of the electromagnetic field (where $\mathbf{k}$ and σ denote the wave vector and polarization index of the photon). The equation for $|\Phi\rangle$ in the interaction representation has the form

$$i\partial|\Phi\rangle/\partial t = \int d\mathbf{r}\,[e_0\hat{J}^{\nu}(x) - e\hat{j}^{\nu}(x)]\,\hat{A}_{\nu}(x)\,|\Phi\rangle,$$

where

$$\hat{J}^{\nu}(x) = \hat{\bar{\varphi}}(x)\,\gamma^{\nu}\hat{\varphi}(x).$$

From this we obtain the standard equation

$$\hat{S} = T\exp\left\{-i\int dx\hat{A}_{\nu}(x)\,[e_0\hat{J}^{\nu}(x) - e\hat{j}^{\nu}(x)]\right\} \tag{5.8}$$

for the scattering operator, where T is the chronological ordering symbol.

The amplitude of the polarization bremsstrahlung in the lowest order of perturbation theory is given by the third term in the expansion of $\hat{S}$ ($x_i \equiv i$):

$$\hat{S}_3 = (-i)^3 e^2 e_0 \int d1\,d2\,d3T\,\{\hat{A}_{\nu}(1)\,\hat{j}^{\nu}(1)\,\hat{A}_{\mu}(2)\,\hat{j}^{\mu}(2)\hat{A}_{\lambda}(3)\,\hat{J}^{\lambda}(3)\}. \tag{5.9}$$

Here we have brought in similar terms which originate in a permutation of the integration variables. Subsequently we shall assume that there is no exchange between the incident particle and the atomic electrons. Using the commutativity of the corresponding operators, we can rewrite Eq. (5.9) in the form

$$\hat{S}_3 = (-i)^2 \int d1\,d2\hat{A}_{\nu}(1)\,T\,\{e^2\hat{j}^{\nu}(1)\,\hat{j}^{\mu}(2)\}\int d3e_0 D_{\mu\lambda}(2,3)\,\hat{J}^{\lambda}(3), \tag{5.10}$$

where $D_{\mu\lambda}(2, 3) = iT\langle 0|\hat{A}_{\mu}(2)\hat{A}_{\lambda}(3)|0\rangle$ is the photon propagator.

One uncoupled A-operator remains in Eq. (5.10). This corresponds to a change in the electromagnetic field for a single photon.

Taking the matrix elements of $\hat{S}_3$ with respect to the initial $|\Phi_i\rangle = |i\rangle\,|\varphi_i\rangle|0\rangle$ and final $|\Phi_f\rangle = |f\rangle\,|\varphi_f\rangle|1_{\mathbf{k},\sigma}\rangle$ states of the system, we obtain

$$S^{p}_{3,fi} = (-i)^2 \int d1\,d2A^{*}_{\mathbf{k},\sigma,\nu}L^{\nu\mu}_{fi}(1,2)\,A^{(0)}_{\mu,fi}(2). \tag{5.11}$$

Here

$$L^{\nu\mu}_{fi}(1,2) = e^2\,\langle f|\,T\,\{\hat{j}^{\nu}(1)\,\hat{j}^{\mu}(2)\}\,|i\rangle \tag{5.12}$$

is the relativistic analog of the scattering tensor for the electromagnetic field of the atom [7], and

$$A^{(0)}_{\mu,fi}(2) = -e_0\int d3D_{\mu\nu}(2,3)\,\langle\varphi_f|\,\hat{J}^{\nu}(3)\,|\varphi_i\rangle \tag{5.13}$$

is the 4-potential of the virtual photon created by the incident particle during the scattering process $|\varphi_i\rangle \rightarrow |\varphi_f\rangle$. Note that the potential of the virtual photon, $A_f^{(0)}$, could be found from Maxwell equations (5.7) if we inserted the matrix element $\langle\varphi_f|\hat{J}^\mu(3)|\varphi_i\rangle$ for the transition current of the incident particle on the right-hand side.

Equation (5.11) for the amplitude of the polarization bremsstrahlung makes it possible to interpret it as the scattering (conversion) of a virtual photon $A_{fi}^{(0)}$ on the atomic electrons into a real bremsstrahlung photon.

It is easy to show that the same expression for the polarization bremsstrahlung amplitude can be obtained (for fixed $|\varphi_i\rangle$ and $|\varphi_f\rangle$) from another form of the interaction Hamiltonian,

$$V' = -e \int d\mathbf{r} \, \{\hat{A}_\nu(x) + \hat{A}^{(0)}_{fi,\nu}(x)\} \, \hat{j}^\nu(x). \tag{5.14}$$

Here the incident particle is replaced by the electromagnetic field $A_{fi}^{(0)}$ which it produces and is, therefore, excluded from the discussion as a dynamic degree of freedom. The field $A_{fi}^{(0)}$ can be regarded as a specified external field defined by Eq. (5.13). Then the amplitude of the polarization bremsstrahlung is obtained in the standard way in second-order perturbation theory. After calculating the matrix element between $|\Phi_i\rangle$ and $|\Phi_f\rangle$, we find the amplitude to be

$$S'^{p}_{2,fi} = (-i)^2 \int d1 \, d2 A^{*}_{\mathbf{k}\sigma,\nu}(1) \, \langle f | T\{e^2 \hat{j}^\nu(1) \, \hat{j}^\mu(2)\} | i \rangle \, A^{(0)}_{\mu,fi}(2). \tag{5.15}$$

A comparison of Eqs. (5.11) and (5.15) shows that $S_{3,fi}{}^{p} = S_{2,fi}{}'^{p}$.

The polarization bremsstrahlung amplitude can, therefore, be calculated (for fixed initial and final states of the incident particle) by replacing the incident particle by the field which it creates in accordance with Eq. (5.13). Then in the case of nonrelativistic atomic electrons being considered here, the only relativistic degree of freedom (the incident particle) will be eliminated and we can use the nonrelativistic formalism for calculating the amplitude of the process.

Note that replacing a particle by its field is widely used, even for calculating Bethe–Heitler bremsstrahlung by the method of equivalent photons, where in the rest system of the incident particle the field of the atom (shielded nucleus) is replaced by equivalent photons which are Compton scattered on the incident particle to form the bremsstrahlung photons.

We now calculate the polarization bremsstrahlung amplitude for a nonrelativistic multi-electron atom ($Z \ll 137$) by replacing the incident particle with its field while neglecting "incident particle ↔ atomic electron" exchange. We shall use an axial gauge for the electromagnetic potential ($A_0 = 0$). The nonrelativistic Hamiltonian for the perturbation of the atomic electrons by the electromagnetic field has the form

$$V = \frac{e}{2m} \sum_{j=1}^{N} \{\hat{\mathbf{p}}_j \cdot \hat{\mathbf{A}}(\mathbf{r}_j, t) + \hat{\mathbf{A}}(\mathbf{r}_j, t)\,\hat{\mathbf{p}}_j + e\hat{\mathbf{A}}^2(\mathbf{r}_j, t)\}, \qquad \mathbf{p}_j = -i\nabla_j, \tag{5.16}$$

where $\hat{\mathbf{A}} = \hat{\mathbf{A}}^{ph} + \mathbf{A}_{fi}^{(0)}$ is the total vector potential and $\hat{\mathbf{A}}^{ph}$ describes the photon field ($kx = \omega t - \mathbf{kr}$):

$$\hat{\mathbf{A}}^{ph}(x) = \sum_{\mathbf{k},\sigma} \sqrt{\frac{2\pi}{\omega}} \{\mathbf{e}_{\mathbf{k},\sigma} \hat{c}_{\mathbf{k},\sigma} \exp(-ikx) + \\ + \mathbf{e}^*_{\mathbf{k},\sigma} \hat{c}^+_{\mathbf{k},\sigma} \exp(ikx)\}, \quad \omega = |\mathbf{k}|. \tag{5.17}$$

Here $\mathbf{e}_{\mathbf{k},\sigma}$ is the unit polarization vector of the photon; $c_{\mathbf{k},\sigma}^+$ and $c_{\mathbf{k},\sigma}$ are the creation and annihilation operators; and, $A_{fi}^{(0)}$ is given by Eq. (5.13) (this is the external field created by the incident particle).

Passing to the interaction representation $\hat{V}_{\text{int}} = \exp(iH_a t) \cdot V \exp(-iH_a t)$ (the photon field is already written in the interaction representation), we have

$$\hat{S} = T \exp\left\{-i \int_{-\infty}^{\infty} dt \hat{V}_{\text{int}}(t)\right\} \tag{5.18}$$

for the scattering operator. The first and second terms in the expansion of S contribute to the polarization bremsstrahlung amplitude in the lowest order of perturbation theory (second order in the charge e). The zeroth order term in this expansion (unity) corresponds to an unchanged state of the system. In the first-order term ($\sim V_{\text{int}}$) the component containing $\hat{\mathbf{A}}^2$ contributes to the process and in the second-order term ($\sim \hat{V}_{\text{int}}^2$) the component containing $\hat{\mathbf{p}}\hat{\mathbf{A}} + \hat{\mathbf{A}}\hat{\mathbf{p}}$ contributes. According to the physical picture of polarization bremsstrahlung, we must include the components which contain the mixed product of $\hat{\mathbf{A}}^{ph}$ and $\mathbf{A}_{fi}^{(0)}$. Thus, the matrix element for the process can be written in the form

$$S_{fi}^p = S_{fi}^{(1)} + S_{fi}^{(2)}.$$

Here

$$S_{fi}^{(1)} = -i \left\langle \Phi_f \left| \int_{-\infty}^{\infty} dt \exp(iH_a t) \frac{e^2}{2m} \sum_{j=1}^{N} 2\hat{\mathbf{A}}^{ph}(\mathbf{r}_j, t) \mathbf{A}_{fi}^{(0)}(\mathbf{r}_j, t) \exp(-iH_a t) \right| \Phi_i \right\rangle, \tag{5.19}$$

with $|\Phi_j\rangle = |j\rangle|n_{\mathbf{k},\sigma}\rangle$, since the incident particle has already been taken into account in $\mathbf{A}_{fi}^{(0)}$. From Eq. (5.19) we find

$$S_{fi}^{(1)} = -2\pi i \delta(\varepsilon_f + E_f + \omega - \varepsilon_i - E_i) \sqrt{\frac{2\pi}{\omega}} \mathbf{e}^*_{\mathbf{k},\sigma} \mathbf{A}_{fi}^{(0)}(q_1) \times \\ \times \left\langle f \left| \sum_{j=1}^{N} \exp(-i\mathbf{q}\mathbf{r}_j) \right| i \right\rangle \frac{e^2}{m}, \tag{5.19a}$$

where $\mathbf{A}_{fi}^{(0)}(q_1)$ is the spatial-temporal transform of the field of the incident particle calculated for the 4-vector $q_1 = \{\varepsilon_f - \varepsilon_i;\ \mathbf{p}_f - \mathbf{p}_i\}$. We neglect spin effects. By analogy, for $S_{fi}^{(2)}$ we have the expression

$$S_{fi}^{(2)} = -\frac{1}{2}\left\langle \Phi_f \left| T \iint\limits_{-\infty}^{\infty} dt\, dt' \hat{V}_{\text{int}}(t)\, \hat{V}_{\text{int}}(t') \right| \Phi_i \right\rangle. \tag{5.20}$$

After some simple transformations, $S_{fi}^{(2)}$ reduces to the form

$$S_{fi}^{(2)} = -e^2 \sqrt{\frac{2\pi}{\omega}}\, 2\pi\delta(\varepsilon_f + E_f + \omega - \varepsilon_i - E_i)\, e^{*}_{\mathbf{k},\sigma,l} A^{(0)}_{fi,s}(q_1) \times$$
$$\times \left\langle f \left| \int\limits_{-\infty}^{\infty} d\tau \exp(i\omega\tau)\, T \hat{j}^l(\mathbf{k}, \tau)\, \hat{j}^s(\mathbf{q}_1) \right| i \right\rangle. \tag{5.20a}$$

Here

$$\mathbf{j}^l(\varkappa, \tau) = \exp(iH_a\tau) \frac{1}{2m} \sum_{j=1}^{N} \{p_j^l \exp(-i\varkappa\mathbf{r}_j) + \exp(-i\varkappa\mathbf{r}_j)\, p_j^l\} \exp(-iH_a\tau)$$

is the spatial Fourier transform of the current operator in the interaction representation.

Adding $S_{fi}^{(1)}$ and $S_{fi}^{(2)}$, we obtain the polarization bremsstrahlung amplitude in the form

$$S_{fi}^{p} = 2\pi i \sqrt{\frac{2\pi}{\omega}}\, (q_1^0)^2\, \delta(\varepsilon_f + E_f + \omega - \varepsilon_i - E_i)\, e^{*}_{\mathbf{k},\sigma,l} A^{(0)}_{fi,s}(q_1) \times$$
$$\times \langle f | \hat{c}^{ls}(k, \mathbf{q}_1) | i \rangle; \qquad q_1^0 = \varepsilon_f - \varepsilon_i. \tag{5.21}$$

In Eq. (5.21) $\hat{c}^{ls}(\mathbf{k}, \mathbf{q}_1)$ is the operator for scattering of the electromagnetic field on the atom in the nonrelativistic (for the atomic electrons) approximation, which can be written in the following form:

$$\hat{c}^{ls}(\mathbf{k}, \mathbf{q}_1) = \frac{e^2}{m(q_1^0)^2}\left[im \int\limits_{-\infty}^{\infty} d\tau \exp(i\omega\tau) \times \right.$$
$$\left. \times T\{\hat{j}^l(\mathbf{k}, \tau)\, \hat{j}^s(\mathbf{q}_1, 0)\} - \delta^{ls}\hat{n}(\mathbf{q}) \right], \quad \hat{n}(\mathbf{q}) = \sum_{j=1}^{N} \exp(-i\mathbf{q}\mathbf{r}_j). \tag{5.22}$$

If we normalize the initial relativistic expression (5.12), which yields Eq. (5.22), then we can say that the first term in the square brackets of Eq. (5.22) originates in the sum over the positive part of the spectrum of the atomic electrons and describes the scattering of the electromagnetic field on the current of the atomic electrons. The second term in Eq. (5.22) shows up after taking the sum over negative energy states and describes the scattering of the field on the charge of the atomic electrons [cf. Eq. (5.3)].

We now write the matrix element $c_{fi}^{ls}(\mathbf{k}, \mathbf{q}_1)$ as a sum over the intermediate states of the atomic electrons:

$$c_{fi}^{ls}(\mathbf{k}, \mathbf{q}_1) = \frac{e^2}{m(q_1^0)^2}\Big\{-\delta^{ls} n_{fi}(\mathbf{q}) + + m\sum_n\Big[\frac{j_{fn}^l(\mathbf{k})\, j_{ni}^s(\mathbf{q}_1)}{\omega_{fn}+\omega+i0} + \frac{j_{fn}^s(\mathbf{q}_1)\, j_{ni}^l(\mathbf{k})}{\omega_{in}-\omega+i0}\Big]\Big\}. \tag{5.23}$$

In particular, in the case of a spherically symmetric state $|i\rangle$ for $f = i$ and $\mathbf{k} = \mathbf{q}_1 = 0$, Eq. (5.23) yields the standard expression for the dipole polarizability of an atom,

$$c_{ii}^{ls}(\mathbf{q}, \mathbf{k}\to 0) \to \alpha(\omega)\,\delta^{ls} = \delta^{ls}(e^2/m)\sum_n \frac{f_{in}}{\omega_{in}^2-\omega^2}, \tag{5.24}$$

where $f_{in} = (2/3)m\omega_{ni}|\langle i|\mathbf{r}|n\rangle|^2$ is the oscillator strength of the transition $i \to n$. In Eqs. (5.22)-(5.24) it is understood that $\Delta = |\omega - \omega_{f(i)n}| \gg \Gamma_{f(i)n}$, where $\Gamma_{f(i)n}$ is the linewidth of the $n \to f, i$ transition. In the opposite case, we must include the linewidths of the corresponding transitions in these formulas. It is quite evident that Eq. (5.21) for the polarization bremsstrahlung amplitude corresponds to an interpretation of this process as the scattering of the field of the incident particle on the atomic electrons to produce a bremsstrahlung photon.

We now calculate the amplitude of the "traditional" bremsstrahlung (owing to the emission of a photon by the incident particle), including the possible excitation of atomic electrons. We again use the interpretation of bremsstrahlung as a process in which a virtual photon is scattered into a real photon. Now the virtual photons are created by the atom (nucleus and bound electrons). The virtual photons created by a stationary atom and nonrelativistic atomic electrons are mostly longitudinal. In this case it is convenient to use a Coulomb gauge for the electromagnetic potential (div $\mathbf{A} = 0$), since then we need only consider its time component. In a Coulomb gauge the spatial components describe the transverse part of the field and are small in this case. According to Eq. (5.13), the time component of the potential of a virtual photon created by the atom is given by

$$A_{0fi}(1) = -\int d1' D_{00}(1, 1')\langle f|\hat{J}^0(1')|i\rangle. \tag{5.25}$$

Here $\hat{J}^0(1) = ZE\delta(\mathbf{r}_1 - \mathbf{r}_0) - e\sum_{j=1}^{N}\delta(\mathbf{r}_1 - \mathbf{r}_j)$ is the operator for the charge density of the atom in the coordinate representation ($\mathbf{r}_0$ is the radius vector of the atom). According to the standard rules it is easy to obtain the amplitude for the bremsstrahlung, which, for brevity, can be referred to as static:

$$S_{fi}^{\mathrm{st}} = -i2\pi\sqrt{(2\pi/\omega)}e_0^2 e_{\mathbf{k}\sigma,\nu}^{*} T^{\nu}(p_{f,i};k)\,A_{fi}^0(q)\times \\ \times\delta(\varepsilon_f + E_f + \omega - \varepsilon_i - E_i). \tag{5.26}$$

Here we introduce the following notation:

$$T^{\nu} = \frac{\bar{u}_f}{\sqrt{2\varepsilon_f}}\left\{\gamma^{\nu}\frac{p_f\gamma + \gamma k + m_0}{(p_f + k)^2 - m_0^2}\gamma^0 + \gamma^0\frac{\gamma p_i - \gamma k + m_0}{(p_i - k)^2 - m_0^2}\gamma^{\nu}\right\}\frac{u_i}{\sqrt{2\varepsilon_i}} \tag{5.27}$$

and

$$A_{fi}^0(\mathbf{q}) = (4\pi/\mathbf{q}^2)\{\delta_{fi}Ze - en_{fi}(\mathbf{q})\}. \tag{5.28}$$

Physically, Eq. (5.28) describes the shielded potential of the nucleus and Eq. (5.27) describes the scattering of the electromagnetic field on the incident particle.

The total bremsstrahlung amplitude for a relativistic incident particle on a nonrelativistic atom ($Z \ll 137$) when the polarization mechanism and the possibility of exciting the atomic electrons are taken into account, therefore, has the form (neglecting exchange and spin effects):

$$S_{fi}^{\mathrm{Br}} = S_{fi}^{\mathrm{st}} + S_{fi}^{p}, \tag{5.29}$$

where S_{fi}^{p} and S_{fi}^{st} are given by Eqs. (5.21) and (5.26), respectively.

5.4. Cross Section for Total Bremsstrahlung on a Nonrelativistic Atom

Proceeding from the above expression for the amplitude, we now write down an expression for the total spectral cross section for bremsstrahlung:

$$\frac{d\sigma^{\mathrm{Br}}(\omega)}{d\omega} = \frac{\varepsilon_i}{|\mathbf{p}_i|}\sum_{f,\sigma}\omega^2\int\frac{d\Omega_{\mathbf{k}}}{(2\pi)^3}\frac{d\mathbf{q}}{(2\pi)^3}\lim_{T\to\infty}\frac{|S_{fi}^{\mathrm{Br}}(\sigma;\mathbf{p}_{f,i};\mathbf{k})|^2}{T}. \tag{5.30}$$

Here $d\Omega_{\mathbf{k}}$ is the solid angle around the direction $\mathbf{k}$ and T is a parameter with the significance of time. The sum is taken over the polarizations of the emitted photon (σ) and the final states of the atom (f). We also assume that the incident particle satisfies the Born approximation and the initial state of the atom is nondegenerate.

Given the explicit form of S_{fi}^{Br}, Eq. (5.30) can be rewritten as

$$\frac{d\sigma^{\mathrm{Br}}}{d\omega} = \frac{\varepsilon_i}{|\mathbf{p}_i|}\sum_{f,\sigma}\omega^2\int\frac{d\Omega_{\mathbf{k}}}{(2\pi)^3}\frac{d\mathbf{q}}{(2\pi)^3}2\pi\delta(q_1^0 + \omega + \omega_{fi})\times$$

$$\times \frac{2\pi}{\omega} \left| e^{*}_{\mathbf{k}\sigma,l} \left\{ e_0^2 T^l (p_{f,i}; k) \frac{4\pi}{q^2} (Ze\delta_{fi} - en_{fi}(\mathbf{q})) + \right.\right.$$

$$\left.\left. + (q_1^0)^2 c_{fi}^{ls}(\mathbf{k}, \mathbf{q}_1) A_{fi,s}^{(0)}(q_1) \right\} \right|^2 = d\sigma^{st}/d\omega + d\sigma^{p}/d\omega + d\sigma^{int}/d\omega. \quad (5.31)$$

The last term in Eq. (5.31), $d\sigma^{int}/d\omega$, describes the interference of the static and polarization terms of the bremsstrahlung. $T^l(p_{j,i}; k)$ and $c_{fi}^{ls}(\mathbf{k}, \mathbf{q}_1)$ are given by Eqs. (5.27) and (5.23).

In the following we shall assume that $|\mathbf{q}_1| \ll |\mathbf{p}_{f,i}|$, i.e., that the motion of the incident particle is only weakly perturbed during bremsstrahlung. Then, for $A_{fi}^{(0)}$ [see Eq. (5.13)] we have the expression

$$\mathbf{A}^{(0)}(\mathbf{q}) = \frac{4\pi e_0}{q^0} \frac{\mathbf{v}q^0 - \mathbf{q}}{(q^0)^2 - \mathbf{q}^2}, \quad q^0 = \mathbf{q}\mathbf{v}_0. \quad (5.32)$$

Here $\mathbf{v}_0$ is the velocity of the incident particle. In this approximation

$$\mathbf{T} = \mathbf{q}_i / m_0 \gamma (\omega - \mathbf{k}\mathbf{v}_0), \quad \gamma = \varepsilon_i / m_0. \quad (5.33)$$

The result (5.31) for the cross section for bremsstrahlung on an atom is the most general expression. When the internal degrees of freedom of the incident particle and atomic nucleus are neglected, it provides a rigorous description of the contribution of the atomic electrons to the bremsstrahlung process. After several simple transformations we obtain

$$\frac{d\sigma^{st}}{d\omega} = \frac{\varepsilon_i}{|\mathbf{p}_i|} \omega^2 \int \frac{d\Omega_{\mathbf{k}}}{(2\pi)^3} \frac{d\mathbf{q}}{(2\pi)^3} \frac{2\pi}{\omega} \int dt \exp[it(\omega + q_1^0)] \times$$
$$\times \sum_\sigma |\mathbf{e}^{*}_{\mathbf{k},\sigma}\mathbf{T}|^2 \frac{2\pi}{\mathbf{q}^2} e_0^4 e^2 \langle i | (Z - \hat{n}(-\mathbf{q}))(Z - \hat{n}(\mathbf{q}, t)) | i \rangle. \quad (5.34)$$

for the static bremsstrahlung cross section from Eq. (5.31). If it is possible to neglect the excitation energy of the atomic electrons compared to the frequency ω, then we can set $\hat{n}(\mathbf{q}, t) \sim \hat{n}(\mathbf{q}, 0)$ in Eq. (5.34) and get

$$\frac{d\sigma^{st}}{d\omega} = \frac{\varepsilon_i}{|\mathbf{p}_i|} \omega^2 \int \frac{d\Omega_{\mathbf{k}}}{(2\pi)^3} \frac{d\mathbf{q}}{(2\pi)^3} 2\pi\delta(q_1^0 + \omega) [\mathbf{n}\mathbf{T}]^2 \frac{2\pi}{\omega} \frac{2\pi}{\mathbf{q}^2} e_0^4 e^2 \langle i || Z - \hat{n}(\mathbf{q})|^2 | i \rangle, \quad \mathbf{n} = \mathbf{k}/k. \quad (5.35)$$

In deriving Eq. (5.35), we have used the equality $\Sigma_\sigma e_{\mathbf{k}\sigma,l}^{*} e_{\mathbf{k}\sigma,s} = \delta_{ls} - n_l n_s$. Equation (5.35) coincides with the result of Wheeler and Lamb [4], who first took rigorous account of the contribution from excitation of the atomic electrons to static bremsstrahlung.

In the case of a heavy incident particle ($m_0 \gg m$), the first term under the absolute value sign in Eq. (5.31) can be neglected compared to the second, since $|T| \propto 1/m_0$, while $\mathbf{A}^{(0)}(\mathbf{q})$ and $\hat{c}^{ls}(\mathbf{k}, \mathbf{q}_i)$ are independent of the incident particle. Then the total cross section for bremsstrahlung on an atom reduces to the cross section for polarization bremsstrahlung, for which Eq. (5.31) gives

$$\frac{d\sigma^p}{d\omega} = \frac{\varepsilon_i}{|\mathbf{p}_i|}\,\omega^2 \int \frac{d\Omega_{\mathbf{k}}}{(2\pi)^3}\,\frac{d\mathbf{q}}{(2\pi)^3}\,\frac{2\pi}{\omega}\,(\delta_{ls} - n_l n_s)\,(q_1^0)^4 A^{(0)}_{fi,s'}(q_1) \times$$
$$\times A^{(0)}_{fi,l'}(q_1) \int dt \exp(iq^0 t)\, \langle i |\, \hat{c}^{sl'*}(\mathbf{k}, \mathbf{q}_1)\, \hat{c}^{ls'}(\mathbf{k}, \mathbf{q}_1, t)\, | i \rangle. \qquad (5.36)$$

Here we have set

$$\mathbf{A}^{(0)} = \mathbf{A}^{(0)*}, \qquad \hat{c}^{ls}(\mathbf{k}, \mathbf{q}_1, t) = \exp(iH_a t)\,\hat{c}^{ls}(\mathbf{k}, \mathbf{q}_1)\exp(-iH_a t). \qquad (5.37)$$

Therefore, the polarization bremsstrahlung cross section summed over all the final states of the atomic electrons is expressed in terms of the correlation function of the operator for scattering of the electromagnetic field on the atom.

We now consider polarization bremsstrahlung without atomic excitation (elastic polarization bremsstrahlung). The cross section is given by the component with $f = i$ in the second term under the absolute value sign in Eq. (5.31):

$$\frac{d\sigma^p_{ii}}{d\omega} = \frac{\omega^2}{v_0} \int \frac{d\Omega_{\mathbf{k}}}{(2\pi)^3}\,\frac{d\mathbf{q}}{(2\pi)^3}\,\frac{2\pi}{\omega}\,(\delta_{ls} - n_l n_s)\,(q_1^0)^4 A^{(0)}_h(q_1)\, A^{(0)}_r(q_1) \times$$
$$\times 2\pi\delta(q_1^0 + \omega)\, \langle i | \hat{c}^{lh}(\mathbf{k}, \mathbf{q}_1) | i \rangle \langle i | \hat{c}^{sr*}(\mathbf{k}, \mathbf{q}_1) | i \rangle, \quad v_0 = |\mathbf{p}_i|/\varepsilon_i. \qquad (5.38)$$

Initially let $\omega \ll p_a v_0$ ($p_a \sim Z^{1/3} m e^2$ is the characteristic atomic momentum.) Then the main contribution to this process will be from the absolute values of $|\mathbf{q}_1| \ll p_a$ allowed by the conservation of energy. Otherwise ($|\mathbf{q}_1| \gg p_a$), polarization bremsstrahlung with excitation and ionization of the atom must predominate. Thus, in this case we can use the dipole approximation in $\mathbf{q}_1$ and $\mathbf{k}$ for the scattering tensor:

$$c^{lh}_{ii}(\mathbf{k}, \mathbf{q}_1) \rightarrow \delta^{lh}\alpha_i(\omega)\,\theta(p_a - |\mathbf{q}_1|),$$

and, instead of Eq. (5.38), we obtain

$$\frac{d\sigma^p_{ii}}{d\omega} \approx \frac{1}{v_0}\,\omega^2 \int \frac{d\Omega_{\mathbf{k}}}{(2\pi)^3}\,\frac{d\mathbf{q}}{(2\pi)^3}\,\frac{2\pi}{\omega}\,[\mathbf{n}\mathbf{A}^{(0)}(q_1)]^2 \times$$
$$\times \delta(q^0)\,\theta(p_a - |\mathbf{q}_1|)\,|\omega^2\alpha_i(\omega)|^2, \qquad p_a v_0 > \omega. \qquad (5.39)$$

For the frequency and angular distribution of elastic polarization bremsstrahlung in the spectral interval considered here, this yields [cf. Eqs. (1.3) and (2.127)]

$$\frac{d\sigma^p_{ii}(\omega, \theta)}{d\omega} = \frac{2e_0^2}{v_0^2}\,\frac{d\omega}{\omega}\left|\omega^2\alpha_i(\omega)\right|^2 (1 + \cos^2\theta)\sin\theta d\theta \ln\frac{\gamma p_a v}{\omega}. \qquad (5.40)$$

Here we have neglected terms of order unity compared to the large logarithm.

Equation (5.40) has two important consequences (see Chapters 1 and 2 and [11–13]):

a) as opposed to the static case, the polarization bremsstrahlung of a ultrarelativistic particle ($\gamma \gg 1$) at frequencies $\omega < p_a v$ is not narrowly directed, but has a dipole character; and

b) the cross section for polarization bremsstrahlung increases logarithmically with the energy of the incident particle in the ultrarelativistic limit when $\omega < p_a v$.

These characteristics of the polarization bremsstrahlung of a relativistic incident particle have an intuitive physical interpretation. The logarithmic growth in the cross section for polarization bremsstrahlung with the energy of the incident particle is related to the spatial structure of the electromagnetic field of a relativistic incident particle. The spatial distribution of the potential of this field at a frequency of ω is given by the formula

$$A_\omega^{(0)} \sim \exp\left[i\frac{\omega}{v_0}(z - v_0 t) - i\frac{\omega\rho}{\gamma v_0}\right], \tag{5.41}$$

where z and ρ are the coordinates of the field.

We thus obtain a transverse dimension of $\rho_{max} \sim \gamma v_0/\omega$ for the field, so that $|\mathbf{q}_\perp|_{min} \sim \omega/\gamma v_0$. Then the second property of the polarization bremsstrahlung follows from the formula for the spectral cross section for polarization bremsstrahlung, $d\sigma^p(\omega) \propto \ln[|\mathbf{q}_\perp|_{max}/|\mathbf{q}_\perp|_{min}]$. Note that in the case of static bremsstrahlung on a neutral atom the maximum dimension of the field that is scattered on the incident particle into a bremsstrahlung photon is determined by the dimensions of the atom.

We now consider elastic polarization bremsstrahlung at frequencies $m \gg \omega \gg I$ (where I is the ionization potential of atom). In this case we can use the asymptotic expression for the scattering operator,

$$\hat{c}^{ls}(\mathbf{k}, \mathbf{q}_1) \approx -\frac{e^2}{m(q_1^0)^2}\hat{n}(\mathbf{q})\left\{\delta^{ls} + \frac{q_1^l q_1^s}{2m\omega}\right\}, \quad m \gg \omega \gg I. \tag{5.42}$$

Equation (5.42) is obtained by expanding the matrix element c_{fi}^{ls} (5.23) as a power series in $|\omega_{jn}|/\omega$ ($j = f, i$), since the terms in this sum over intermediate states with $|\omega_{jn}| > \omega$ make a small contribution to c_{fi}^{ls} for $\omega \gg I$. Substituting Eq. (5.42) in Eq. (5.38), we find

$$\frac{d\sigma_{ii}^p}{d\omega} = \frac{\omega^2}{v_0}\int \frac{d\Omega_\mathbf{k}}{(2\pi)^3}\frac{d\mathbf{q}}{(2\pi)^3}\frac{2\pi}{\omega}\delta(q^0)\left(\frac{e^2}{m}\right)^2 \times$$
$$\times |n_{ii}(\mathbf{q})|^2\left[\mathbf{n}, \left(\mathbf{A}^{(0)}(\mathbf{q}_1) + \frac{\mathbf{q}_1(\mathbf{q}_1 A^{(0)}(\mathbf{q}_1))}{2m\omega}\right)\right]^2, \quad m \gg \omega \gg I. \tag{5.43}$$

In order to simplify the calculations we assume that $\gamma \gg 1$. Then the field of the incident particle is essentially transverse and $\mathbf{q}_1\mathbf{A}^{(0)}(q_1) = 0$.. Let us use the exponential shielding approximation for calculating the spectral cross section. Then

$$n_{ii}(\mathbf{q}) = N/(1 + \mathbf{q}^2/p_a^2). \tag{5.44}$$

Here N is the number of atomic electrons (for a neutral atom we have $N = Z$).

Using Eq. (5.44) for the spectral cross section for polarization bremsstrahlung and taking $\alpha_i \to \alpha_a = -Ne^2/m\omega^2$, after integrating over $d\Omega_\mathbf{k}$ and $d\mathbf{q}$ [also see Eq. (9.22); compare with Eqs. (2.111) and (2.112), given Eqs. (1.3) and (2.132)], for three frequency ranges we find that

$$\frac{d\sigma_{ii}^p}{d\omega} = \frac{16}{3} N^2 \frac{e^4 e_0^2}{m^2\omega} \ln \frac{\gamma p_a}{\omega}, \qquad p_a \gg \omega \gg I; \tag{5.45}$$

$$\frac{d\sigma_{ii}^p}{d\omega} = 2N^2 \frac{e^4 e_0^2}{m^2\omega} \left(\frac{p_a}{\omega}\right)^2 \ln \gamma, \qquad \gamma^2 p_a \gg \omega \gg p_a; \tag{5.46}$$

and

$$\frac{d\sigma_{ii}^p}{d\omega} = 4N^2 \frac{e^4 e_0^2}{m^2\omega} \left(\frac{p_a}{\omega}\right)^2 \left(\frac{\gamma^2 p_a}{\omega}\right)^2, \qquad m \gg \omega \gg \gamma^2 p_a. \tag{5.47}$$

Equations (5.45) and (5.46) are insensitive to the specific form of shielding. The spectral cross section (5.45) can be derived from Eq. (5.40), since the dipole approximation is still valid in this range and we can write the polarizability in the form $\alpha(\omega) = -Ne^2/m\omega^2$, which is valid for $\omega \gg I$. The frequency range for Eq. (5.46) is characteristic of ultrarelativistic incident particles ($\gamma \gg 1$). In that range, $\mathbf{q}_1$ can be compensated as a result of $\mathbf{k}$ in the expression for $n_{ii}(\mathbf{q})$ with $|\mathbf{q}_1 + \mathbf{k}| \ll p_a$. This, however, occurs only for small photon emission angles:

$$\theta \lessdot p_a/\omega \approx \lambda/R_a, \qquad R_a \approx 1/p_a.$$

In fact, in order for $\mathbf{q}_1$ to be compensated, it is necessary that $|\mathbf{q}_1| \sim |\mathbf{q}_{1,\parallel}| \sim |\mathbf{k}|$; then $|\mathbf{q}| \sim 2\omega \sin\theta/2 < p_a$ ($\omega \gg p_a$), which also implies a condition on the maximum emission angle. In the frequency range for Eq. (5.46), therefore, polarization bremsstrahlung acquires a directionality ($\theta < \lambda/R_a$) and the momentum of the photon must be taken into account in calculating the cross section. In the range for Eq. (5.47) (if it exists), it is always true that $|\mathbf{q}| \gg p_a$ and polarization bremsstrahlung is strongly suppressed. This already follows from the form of $n_{ii}(\mathbf{q}) \sim p_a^2/\mathbf{q}^2$. Physically, this means that for the large momentum transfers to the atom ($|\mathbf{q}| \gg p_a$) characteristic of this frequency range, the inelastic channels for polarization bremsstrahlung, which are accompanied by excitation and ionization of the atomic electrons, will predominate.

We note that this "elastic" polarization bremsstrahlung is coherent over the atomic electrons and that its cross section is proportional to the square of the number of electrons (N^2 or Z^2). The last point can be explained by the fact that, in this case, where the state of the atomic electrons does not change, the electronic charge remains localized in the atom and manifests itself as a charge Ne on a single particle (for $\lambda > R_a$). The amplitude of its interaction with the electromagnetic field, therefore, is proportional to Ne, while the cross section is proportional to N^2e^2.

We now return to the total (including excitation of the atomic electrons) cross section for polarization bremsstrahlung given by Eq. (5.36). It is not possible to obtain an explicit general expression for the spectral cross section for polarization bremsstrahlung. We shall examine some common cases which are of practical importance.

Let the frequency ω be such that the main contribution to the differential cross section relative to $\mathbf{q}_1$ is from $|\mathbf{q}_1| \ll p_a$. Then the dipole approximation is valid and we can take the integral over $d\mathbf{q}$ using the explicit form of $\mathbf{A}^{(0)}(\mathbf{q})$ (5.32). Given the spherical symmetry of the $|i\rangle$ state, after making a number of transformations we obtain the following expression for the spectral cross section for polarization bremsstrahlung when $\omega_{si} < p_a v_0 - \omega$:

$$\frac{d\sigma_{fi}^p}{d\omega} = \frac{16e_0^2}{9v_0^2} \sum_{m,l,f} \omega^3 \langle f | \hat{c}_{ml}(\omega) | i \rangle \langle f | \hat{c}_{ml}(\omega) | i \rangle^* \ln \frac{\gamma p_a v_0}{\omega + \omega_{fi}} . \qquad (5.48)$$

Here a sum over l and m is understood and $\langle i|\hat{c}_{ml}(\omega)|i\rangle = \alpha_i(\omega)\delta_{lm}$ is the dynamic polarizability of the atom in the state $|i\rangle$. Note that the term with $f = i$ in Eq. (5.48) gives the spectral cross section for elastic polarization bremsstrahlung, which also follows from Eq. (5.40) (after integration over $d\theta$). For states $|f\rangle$ such that $\omega_{fi} > p_a v_0 - \omega$, this dipole approximation in q is not valid.

We now consider the case of a resonance. Let the frequency ω be such that two energy levels of the discrete spectrum with $E_n > E_f$ obey the inequality $\omega \gg |\omega - \omega_{nf}| \gg \Gamma_{nf}$. Then, in the expression for the matrix element $\langle f|\hat{c}_{ml}(\omega)|i\rangle$ we can isolate a single resonance term which makes the major contribution and neglect the imaginary part of the scattering tensor compared to the real part. A single resonance term also remains in the sum over f in Eq. (5.48).

After taking the sum over the projections of the total angular momentum of the resonant states ($|f\rangle$, $|n\rangle$) and noting that $J_i = 0$ (where J is the total angular momentum quantum number), for the resonant bremsstrahlung cross section we find

$$\frac{d\sigma_{fi}^p}{d\omega} = \frac{4e_0^2e^4}{3v_0^2m^2} \left(\frac{\omega}{\Delta}\right)^2 \frac{f_{in}}{\omega_{ni}} (2J_f + 1) f_{fn} \ln \frac{\gamma p_a v_0}{\omega + \omega_{fi}} , \qquad (5.49)$$

where $\Delta = \omega - \omega_{nf}$ and $\omega \gg |\Delta| \gg \Gamma_{fn}$.

For $f = i$, Eq. (5.49) yields an expression for "elastic" resonant bremsstrahlung. We now calculate the cross section for polarization bremsstrahlung with excitation (including ionization) of the atom for $m \gg \omega \gg I$.

Substituting the expression (5.42) for $\hat{c}^{lh}$ in this range in Eq. (5.36), we find

$$\frac{d\sigma^p}{d\omega} = \frac{2\pi\omega^2}{v_0} \int \frac{d\Omega_{\mathbf{k}}}{(2\pi)^3} \frac{d\mathbf{q}}{(2\pi)^3} \frac{2\pi e^4}{\omega m^2} \left[\mathbf{n}, \left(\mathbf{A}^{(0)}(q_1) + \frac{\mathbf{q}_1 (\mathbf{q}_1 \mathbf{A}^{(0)}(q_1))}{2m\omega}\right)\right]^2 S_{ii}(\mathbf{q}). \quad (5.50)$$

Here we have introduced the quantity

$$S_{ii}(q) = \frac{1}{2\pi} \int_{-\infty}^{\infty} dt \exp(iq^0 t) \langle i | \hat{n}(-\mathbf{q}) \hat{n}(\mathbf{q}, t) | i \rangle, \tag{5.51}$$

which we shall refer to as the dynamic form-factor of the atom in accordance with the terminology used for describing these effects in a medium. For simplicity we shall set $\mathbf{q}_1 \sim \mathbf{q}$. in the following and thereby neglect the terms of order $(p_a/\omega)^2$ compared to unity when $\omega \gg p_a$ [Eq. (5.46)]. When $\omega \ll p_a$, the dipole approximation is valid, so that the absolute values $|\mathbf{q}|$ and $|\mathbf{k}|$ can be neglected compared to p_a. As can be seen from Eq. (5.50), in order to calculate the spectral cross section for polarization bremsstrahlung we must know the explicit form of the functional dependence of S_{ii} on $\mathbf{q}$. As we are only concerned with the fundamental points in the following discussion, we shall use the simplest approximation for $S_{ii}(\mathbf{q})$ in the calculations:

$$S_{ii}(q) \approx \{\theta(|\mathbf{q}| - p_a)\,\delta(q^0 + \mathbf{q}_1^2/2m)\,N + \theta(p_a - |\mathbf{q}|)\,\delta(q^0)\,N^2\}. \tag{5.52}$$

This formula can be obtained by noting the explicit forms of the operators $\hat{n}(\mathbf{q})$ and $\hat{n}(\mathbf{q}, t)$; i.e.,

$$\hat{n}(\mathbf{q}) = \sum_{j=1}^{N} \exp(-i\mathbf{q}\mathbf{r}_j)$$

and

$$\hat{n}(\mathbf{q}, t) = \exp(i\hat{H}_{at} t)\,\hat{n}(\mathbf{q}) \exp(-i\hat{H}_{at} t),$$

and using the fact that $|\mathbf{r}_j| \leq 1/p_a$. Then, for $|\mathbf{q}| < p_a$, the exponents in the definition of $\hat{n}(\mathbf{q})$ can be replaced approximately by unity (noting that $\mathbf{q}_1 \sim \mathbf{q}$). This yields the second term in Eq. (5.52). In order to find the first term, we use the formula

$$\exp(i\hat{H}_{at} t) \sum_j \exp(-i\mathbf{q}\mathbf{r}_j) = \sum_j \exp(-i\mathbf{q}\mathbf{r}_j) \exp\left\{it\left[U(\mathbf{r}_j) + \sum_{j'} \frac{1}{2m} (\hat{\mathbf{p}}_{j'} - \delta_{jj'}\mathbf{q})^2\right]\right\}. \tag{5.53}$$

Here $U(\mathbf{r}_j)$ is the potential energy of the interaction of the atomic electrons with the nucleus and with one another. This last equality is a consequence of the fact that $\exp(-i\mathbf{q}\mathbf{r}_j)$ is a displacement operator in momentum space. Since $|\mathbf{q}| > p_a$, we can neglect p_j compared to $|\mathbf{q}|$ in the exponent on the right-hand side of this equation. We then obtain the approximation

$$\langle i | \hat{n}(-\mathbf{q}) \hat{n}(\mathbf{q}, t) | i \rangle \approx \left\langle i \left| \sum_{jj'} \exp[i\mathbf{q}(\mathbf{r}_j - \mathbf{r}_{j'})] \right| i \right\rangle \times$$
$$\times \exp(i\mathbf{q}^2 t/2m), \qquad |\mathbf{q}| > p_a. \tag{5.54}$$

Since $|\mathbf{q}| > p_a$, after averaging over $|\,i\,\rangle$, the contribution of the terms with $j \neq j'$ is small in view of the oscillations in the exponent. Thus, the terms with $j = j'$ remain, and after integrating over dt we arrive at the first term in the curly brackets of Eq. (5.52). Here the first term in the curly brackets corresponds to large momentum transfers, when the binding energy of an atomic electron with the nucleus is not enough to hold the electron in orbit. As a result, ionization takes place with an energy expenditure of order $\mathbf{q}^2/2m$. Here the contribution of the various atomic electrons to the process is incoherent and is proportional to their number ($\sim N$). The second term in Eq. (5.52) refers to small $|\mathbf{q}|$, when the atom remains unexcited, with an overwhelming probability, during the bremsstrahlung process. In this case the emission of a bremsstrahlung photon by all the atomic electrons takes place coherently, which corresponds to the factor N^2 in the formula for S_{ii}. The emission is coherent because, when $|\mathbf{q}| < p_a$, the phase of the electromagnetic interaction leading to polarization bremsstrahlung varies weakly over a distance on the order of the radius of the atom ($\sim 1/p_a$), so that all the atomic electrons make contributions with the same sign to the process.

It is easy to find the spectral cross section for polarization bremsstrahlung in this approximation from Eqs. (5.52) and (5.50) [see Eq. (9.22), as well]:

$$\frac{d\sigma^p}{d\omega} = \frac{16e_0^2e^4}{3m^2v_0^2}\Big\{\theta(p_a v_0 - \omega)\left[N^2 \ln\frac{\gamma p_a v_0}{\omega} + N \ln\frac{m_0 v_0}{p_a}\right] + \\ + \theta(\omega - p_a v_0) N \ln\frac{\gamma m_0 v_0^2}{\omega}\Big\}, \quad m \gg \omega \gg I, \quad \varepsilon_i \gg \omega. \tag{5.55}$$

This formula is also valid for a nonrelativistic incident particle (but one that satisfies the Born approximation); however, to examine it separately without taking the static bremsstrahlung into account would be meaningful only in those cases where interference between the polarization and conventional bremsstrahlung can be neglected (for example, with heavy or relativistic incident particles). In the case of nonrelativistic incident electrons or positrons, it is important to include this interference. Thus, for an incident electron with small momentum transfer ($|\mathbf{q}| < p_a$) and $\omega \gg I$, the atom is "stripped"; i.e., the emission from the incident and atomic electrons compensate one another [the first term in the square brackets of Eq. (5.55) vanishes]. An analogous effect should also occur with large momentum transfer ($|\mathbf{q}| > p_a$), when atomic electrons are knocked out of the atom during a collision with an incident electron. In the latter case, the emission from the two types of electrons will be compensated for all $\omega \ll m$, and not only for $p_a v_0 \gg \omega \gg I$, as when $|\mathbf{q}| < p_a$, since for large $|\mathbf{q}|$ ($|\mathbf{q}| > p_a$) the atomic electrons can be regarded as free in the course of the bremsstrahlung emission, regardless of frequency, provided the nonrelativistic description of the process is still valid ($\omega \ll m$, $v \ll 1$), while free nonrelativistic electrons cannot emit a dipole photon. As a result of this compensation, the second or third term vanishes (depending on the frequency). For a nonrelativistic incident positron, the total bremsstrahlung is enhanced because of the interference of the amplitudes of the polarization and static bremsstrahlung [11].

Equation (5.55) has an intuitive interpretation. When $\omega < p_a v_0$ (the term in square brackets), polarization bremsstrahlung occurs both without excitation of the atom (for $|\mathbf{q}| < p_a$) and with ionization of the atom (if $|\mathbf{q}| > p_a$). In the first case, the polarization bremsstrahlung is coherent (over the atomic electrons the cross section for the process is proportional to N^2), while in the second it is incoherent (the cross section is proportional to N). If $\omega > p_a v_0$, then only the case with $|\mathbf{q}| > p_a$ is allowed by the conservation of energy and momentum. Then the polarization bremsstrahlung is mostly accompanied by ionization of the atom and the cross section is proportional to N.

The first and third terms of Eq. (5.55) were first obtained by Krotov [11], who used a Fermi–Thomas model for the atom.

It is important to emphasize that the total cross section for polarization bremsstrahlung, including excitation of the atom, given by Eq. (5.55) allows a limiting transition to the case $Z = 0$. This limit in Eq. (5.55) corresponds to $p_a = 0$. Then the term in the square brackets vanishes, since frequencies $\omega < p_a v_0$ drop out. The last term of Eq. (5.55) remains and the result is the same as the equation [8, 11] for the emission from a slow free recoil electron during a collision with a relativistic charged particle, which should happen according to the physics of the process. Note that this limit does not occur in the cross section for elastic polarization bremsstrahlung (5.40), since this process becomes fundamentally inelastic in the absence of a nucleus (the atomic electron gains the excess momentum and changes its state).

We now compare the integral (over the scattering and emission angles) cross sections for polarization and static bremsstrahlung. These cross sections have their simplest form in the quasiclassical ($E_{f,i} \gg \omega$) and ultrarelativistic ($\gamma \gg 1$) limits at frequencies above the ionization potential of the atom.

Thus, for an incident electron at frequencies $p_a v_0 > \omega \gg I$ the main contributions to polarization and static bremsstrahlung are from the elastic terms in the cross sections (we assume that $Z, N \gg 1$):

$$\frac{d\sigma^{p}_{it}}{d\omega} = \frac{16N^2e^6}{3m^2\omega}\ln\frac{\gamma p_a}{\omega} \tag{5.56a}$$

and

$$\frac{d\sigma^{st}_{ii}}{d\omega} = \frac{16Z^2e^6}{3m^2\omega}\ln\frac{m}{p_a}\,, \tag{5.56b}$$

which differ only in the logarithmic factors, although the two emission processes have substantially different directional diagrams.

Let us now write down the cross sections for inelastic polarization and inelastic static bremsstrahlung in the interval where the main contribution to the polarization part of the bremsstrahlung is from processes involving ionization of the atom [see Eq. (9.22), as well]:

$$\frac{d\sigma^{p}_{\text{inel}}}{d\omega} = \frac{16Ne^6}{3m^2\omega}\ln\frac{\varepsilon}{\omega} \tag{5.57a}$$

and

$$\frac{d\sigma_{\text{inel}}^{\text{st}}}{d\omega} = \frac{16Ne^6}{3m^2\omega} \ln \frac{m}{p_a}, \quad \gamma \gg \sqrt{\frac{\omega}{p_a}}. \tag{5.57b}$$

Therefore, when $p_a v_0 \ll \omega \ll m$ the cross section for elastic static and elastic polarization bremsstrahlung differ only in the factor under the logarithm, while the corresponding inelastic terms in the cross sections are close in magnitude all the way up to $\omega \sim m$.

When $\omega \gg m$, the spatial part of the 4-momentum transferred to the atom is large ($|\mathbf{q}_1|_{\min} \gg p_a$) and the atomic electrons can be regarded, with good accuracy, as free. This leads to the well-known result of quantum electrodynamics according to which the emission from the recoil electron is a factor of ω/m smaller than from the fast electron [8, 14]. Thus, the contribution of the polarization term to the total bremsstrahlung cross section for an electron on an atom at high ($\omega \gg m$) frequencies is negligible compared to the static term.

The above remarks are all true for the case of bremsstrahlung by an ultrarelativistic positron on an atom, when the sign of the polarization term in the amplitude changes. As for an electron, however, because of the different dependences on the emission angle for the static and polarization terms, their interference can be neglected, so that the total cross section for bremsstrahlung by an ultrarelativistic particle can be written as the sum of two terms, polarization and static.

If the relativistic particle is heavy, for example, a proton, then the elastic and inelastic static terms in the total cross sections (5.56b) and (5.57b) are inversely proportional to the square of the mass of the proton, while the polarization terms (5.56a) and (5.57a) do not change. This leads, as in the nonrelativistic problem, to anomalously large bremsstrahlung by a proton on an atom in the low-frequency part of the spectrum. It is interesting to note that the importance of including the bremsstrahlung of fast protons on electrons in a medium has been pointed out previously in [15–18], where the atomic electrons were regarded as free because of the high energy of the bremsstrahlung photons and the correspondingly large momentum transfer. Including the emission from protons and nuclei on the free electrons in a medium is important for interpreting observations of cosmic radiation in the x-ray and γ-ray regions [18].

We now examine the term which describes the interference between the polarization and static bremsstrahlung in the bremsstrahlung cross section. It follows from an analysis of the angular dependences that this interference is small for an ultrarelativistic incident particle. Thus, here we shall assume that the incident particle is nonrelativistic ($v \ll 1$), but, as before, describable in the Born approximation. We shall neglect excitation of the atom in the course of bremsstrahlung. Then, given Eqs. (5.32) and (5.33), from Eq. (5.31) we can obtain

$$\frac{d\sigma_{ii}^{\text{int}}}{d\omega} \approx \frac{32e_0^2\omega^3}{3v_0^2} \int\limits_{\omega/v_0}^{|\mathbf{q}_1|_{\max}} \frac{|e_0 e|}{m_0\omega} \operatorname{Re}\{c_{ii}(\omega, |\mathbf{q}_1|)\}(Z - n_{ii}(\mathbf{q}_1)) \frac{d|\mathbf{q}_1|}{|\mathbf{q}_1|}. \tag{5.58}$$

In deriving Eq. (5.58) we used the fact that that when $v \ll 1$, $|\mathbf{q}_1| \geq \omega/v_0 \gg |\mathbf{k}|$. We note that only the real part of the diagonal matrix element of the scattering operator $c_{ii}(\omega, |\mathbf{q}_1|)$ contributes to the interference. Let us take the simplest approximation for c_{ii} as a function of $|\mathbf{q}_1|$:

$$c_{ii}(\omega, \ |\mathbf{q}_1|) \approx \theta(p_1 - |\mathbf{q}_1|)\alpha_i(\omega). \tag{5.59}$$

Here p_1 is the characteristic momentum of the atomic electrons which provide the major contribution to the polarizability of the atom at the given frequency ω. We also assume that the field of the atom is negligibly small for $|\mathbf{q}_1| < p_{\min}$, where $p_{\min}$ is the characteristic momentum of the electrons in the outer shell. Then, we have the approximation

$$\frac{d\sigma_{ii}^{\text{int}}(\omega)}{d\omega} \approx \frac{32\,|e_0^3 e\,|}{3m_0 v_0^2}\,\omega^2\,\mathrm{Re}\,\alpha_i(\omega) \int_{p_*}^{p_1} \frac{(Z - n_{ii}(\mathbf{q}_1))}{|\mathbf{q}_1|}\, d\,|\mathbf{q}_1|, \tag{5.60}$$

where $p_* = \max\{p_{\min}, \omega/v_0\}$. It follows from Eq. (5.60) that the interference term can be significant if the largest contribution to the polarizability is from an inner shell of the atom. Then the cross sections for polarization and static bremsstrahlung are comparable. This happens, for example, with the bremsstrahlung of electrons in xenon at frequencies near the ionization potential of the 4*f*-subshell. The emission spectrum of xenon with an electron beam passing through it was recorded by Verkhovtseva et al. [19] who observed a 20-eV shift of the peak from the value calculated without taking interference into account. They explained [19] the discrepancy in terms of the possibility that the velocity of the beam electrons might not have been high enough for the Born approximation to work "well." On the other hand, the shift may be caused by the neglected interference term in the total bremsstrahlung cross section. If the incident particle is heavy or ultrarelativistic, then the expected magnitude of the shift must be small because the interference term in the total bremsstrahlung cross sections is small for these cases. For an ultrarelativistic incident particle, the theory leads to an additional interesting possibility: the shift in the bremsstrahlung frequency peak relative to the ionization potential of the corresponding subshell of the atom depends strongly on the emission angle because of the considerable difference in the directional diagrams for polarization and static bremsstrahlung in the ultrarelativistic limit.

In conclusion we note that the dynamic approach to the theory of bremsstrahlung on many-electron atoms can be used to obtain a more general expression for the cross section for this process. It takes into account the possibility of exciting bound electrons, the nondipole nature of the electromagnetic interaction for a relativistic incident particle, and the interference between static and polarization bremsstrahlung [20].

Interpreting bremsstrahlung as a process in which virtual photons are scattered into real photons on charged particles offers a simple way to obtain an expression for the total cross section (summed over the final states of the atom), which in special cases is expressed in terms of the dynamic and static atomic form factors.

In the directional diagram for bremsstrahlung of relativistic electrons and positrons on an atom, including the polarization contribution leads to dipole emission which, in the interval $p_a v > \omega > I$, is comparable (in an integral sense) with the narrowly directed static bremsstrahlung, as "elastic" bremsstrahlung predominates at these frequencies. In the interval $p_a v < \omega$, the main contribution to polarization bremsstrahlung is from processes involving atomic excitation and ionization. The substantial difference between the directional diagrams for static and polarization bremsstrahlung in the ultrarelativistic case makes it possible to neglect their interference.

The inelastic terms in the total cross section for bremsstrahlung on a many-electron atom are proportional to the number of atomic electrons for both polarization and static bremsstrahlung, so that these terms are important in bremsstrahlung processes on light atoms.

In the case of a heavy incident particle, the contribution of polarization bremsstrahlung is much greater than that of static bremsstrahlung in the frequency range $\omega < m_0^2/m$ because the former is independent of the mass of the incident particle.

Polarization bremsstrahlung at frequencies $\omega > m$ is small compared to static bremsstrahlung. Their contributions to the total bremsstrahlung cross section are comparable at frequencies $\omega < m$. Then the polarization bremsstrahlung depends on the energy of the incident particle much more strongly than does the static bremsstrahlung, so that it determines the energy dependence of the total bremsstrahlung.

Further investigation of this problem is of particular interest for the cases when the incident particle velocities do not satisfy the Born approximation, when the atomic electrons are relativistic, and when spin and exchange corrections must be included in the cross section. A detailed experimental study of the polarization bremsstrahlung of relativistic particles is also necessary and should be compared with the available theoretical results.

Chapter 6

POLARIZATION BREMSSTRAHLUNG ON IONS AND ATOMS IN PLASMAS

V. A. Astapenko, V. M. Buimistrov, Yu. A. Krotov, and V. N. Tsytovich

6.1. Introductory Comments

We now examine the emission from incident particles that obey the Born approximation in a plasma [1, 2]. We shall assume that the plasma ions are partially ionized and have an electronic structure. The Debye electron "cloud" of fully ionized ions in the plasma effectively replaces the bound electrons of the atom. When a complicated many-electron atom is partially ionized in a plasma, some of its electrons form a "skeleton" of bound electrons, while the remainder are the passing (unbound) electrons in its Debye "cloud." As a rule, the characteristic length R_a within which the bound electrons are concentrated is much shorter than the Debye length r_d (i.e., $R_a \ll r_d$). Thus, we must distinguish the radiation for different impact parameters: $\rho < R_a$, $R_a < \rho < r_d$, and $\rho > r_d$. As we showed in Chapter 1, for incident particles with given impact parameter ρ, the emitted frequencies are less than v_0/ρ. Thus, collisions with impact parameters $\rho < R_a$ will contribute to frequencies greater than v_0/R_a, collisions with impact parameters $\rho < r_d$ will contribute to frequencies greater than v_0/R_a, and, finally, collisions with impact parameters less than v_0/ω will contribute to frequencies less than v_0/r_d . In the last case, interference must occur between ordinary bremsstrahlung and polarization bremsstrahlung on the bound and free electrons. The difference between the bound and free electrons is not so large (in a certain sense and under certain conditions). For the bound electrons, the Coulomb field of the nucleus of the ion must, of course, be taken into exact account, rather than through perturbation theory [3]. On the other hand, the free electrons in the Debye cloud have high energies (temperatures) by definition, so that for them the Coulomb field of the ion is a perturbation. In the meantime, for the bound electrons, as well, the field of the nucleus of the ion will be a perturbation at the high bremsstrahlung frequencies corresponding to energies greater than the ionization potential, when the polarizabilities of the free and bound electrons are practically the same. Here we shall assume that the electrons in the Debye cloud are weakly perturbed by the ion. The qualitative picture of polarization bremsstrahlung for an ion with an electronic structure located in a plasma was described in Chapters 1 and 2. It was shown that one must add the polarizabilities of the plasma and bound electrons in the amplitude for polarization bremsstrahlung. In the following we shall present a rigorous quantitative quantum mechanical theory of polarization bremsstrahlung which simultaneously includes the plasma and bound electrons by using the methods of quantum electrodynamics.

6.2. Cross Section for Polarization Bremsstrahlung on Ions in Plasmas

For calculating the bremsstrahlung of ions in plasmas including the contribution of the bound electrons, we shall formulate the problem in a way that is characteristic for polarization bremsstrahlung on an atom. We shall consider the emission of a transverse photon by the electrons (delocalized and bound) of the medium during inelastic scattering of an incident particle that obeys the Born approximation. We make one simplifying approximation that is valid for a nonrelativistic plasma with $\omega \gg \omega_{pe}$ and $v_0 \gg v_{Te}$: we neglect scattering of virtual photons into real photons on the polarization charge around the incident particle. Then the latter only serves as a "supplier" of virtual photons and, as demonstrated in Chapter 5, it can be replaced by the appropriate electromagnetic field (assuming that the initial and final states of the incident particle are specified) in accordance with the formula

$$A^{(0)}_{fi,\mu}(1) = -e_0 \int d1' D^{(0)}_{\mu\nu}(1, 1') J^{\nu}_{fi}(1'). \tag{6.1}$$

Here e_0 is the charge of the incident particle and $J_{fi}^{\nu}(1')$ is the transition current density of the incident particle in the coordinate representation. We assume further that the motion of the incident particle is only weakly perturbed in the bremsstrahlung process, i.e., $|\mathbf{q}_1| = |\mathbf{p}_f - \mathbf{p}_i| \ll |\mathbf{p}_{i,f}|$ (where $\mathbf{p}_{i,j}$ are the initial and final momenta of the incident particle), so that we have

$$J^{\nu}_{f,i} = v^{\nu}_0 \tag{6.2}$$

for the transition current density, where v_0^{ν} is the 4-velocity of the incident particle. In Eq. (6.2) the normalization volume is set equal to unity.

It must be noted that the electromagnetic field in the medium is "dressed," i.e., the effect of the polarization of the medium on this field must be taken into account. This can be done formally by replacing the vacuum photon propagator $D_{\mu\nu}^{(0)}(1, 1')$ in Eq. (6.1) by the photon propagator in the medium, whose Fourier transform in an isotropic medium with an axial gauge ($A_0 = 0$) is given by [4]

$$D_{mn}(q) = \frac{4\pi}{(q^0)^2}\left\{\frac{q_m q_n}{\mathbf{q}^2 \varepsilon_q^l} + \frac{(q^0)^2}{(q^0)^2 \varepsilon_q^t - \mathbf{q}^2}\left(\delta_{mn} - \frac{q_m q_n}{\mathbf{q}^2}\right)\right\}, \tag{6.3}$$

where $n, m = 1, 2, 3$, while ε_q^l and ε_q^t are the longitudinal and transverse components of the dielectric permittivity tensor taken on the 4-vector $q = \{q^0, \mathbf{q}\}$. Note that the propagator (6.3) coincides with the linear Green function for Maxwell's equations to within a constant factor. Equation (6.3) for the photon propagator in a medium can be obtained rigorously using standard techniques [5]:

$$D_{mn}^{-1}(q) = D_{mn}^{(0)-1}(q) - P_{mn}(q)/4\pi. \tag{6.4}$$

Introducing the definitions [6]

$$\varepsilon_q^l = 1 - q_n q_m P_{nm}(q)/\mathbf{q}^2 (q^0)^2, \tag{6.5a}$$

and

$$\varepsilon_q^t = 1 - P_{nm}(q)(\delta_{nm} - q_n q_m/\mathbf{q}^2)/2(q^0)^2, \tag{6.5b}$$

for $P_{nm}(q)$ we then have

$$P_{nm}(q) = (q^0)^2\left[\delta_{nm} - \varepsilon_q^l \frac{q_n q_m}{\mathbf{q}^2} - \varepsilon_q^t\left(\delta_{nm} - \frac{q_n q_m}{\mathbf{q}^2}\right)\right]. \tag{6.6}$$

Using Eq. (6.4), we now come to a definition of $D_{nm}(q)$ with $\varepsilon_q^{l,t}$ given by Eqs. (6.5). The polarization operator P_{nm} for nonrelativistic particles with $|\mathbf{q}| \ll m$ can be written in the form

$$P_{nm}(q) = 4\pi\left\{\delta_{nm}\sum_\alpha \frac{e_\alpha^2 n_\alpha}{m_\alpha} + \sum_s \frac{\left\langle 0\left|\sum_\alpha e_\alpha \hat{j}_\alpha^n(\mathbf{q})\right|s\right\rangle}{\omega_{0s} + q^0 + i0} \times \right.$$

$$\left. \times \left\langle s\left|\sum_{\alpha'} e_{\alpha'} \hat{j}_{\alpha'}^m(\mathbf{q})\right|0\right\rangle + \frac{\left\langle 0\left|\sum_{\alpha'} e_{\alpha'} \hat{j}_{\alpha'}^m(\mathbf{q})\right|s\right\rangle}{\omega_{0s} - q^0 + i0}\left\langle s\left|\sum_\alpha e_\alpha \hat{j}_\alpha^n(q)\right|0\right\rangle, \right. \tag{6.7}$$

where α is a subscript representing the type of particle; n_α is the density of these particles; $\hat{j}_\alpha^l(\mathbf{q}) = (1/2m_\alpha)\sum_i (\hat{p}_{i\alpha}^l e^{-i\mathbf{q}\mathbf{r}_{i\alpha}} + e^{-i\mathbf{q}\mathbf{r}_{i\alpha}}\hat{p}_{i\alpha}^l)$ is the Fourier transform of the current density of particles of type α; and $|0, s\rangle$ are the many-particle wave functions of the ground and excited states of the system. Equation (6.7) is valid when the temperature $T = 0$. When $T > 0$, Eq. (6.7) must be averaged over the initial state. If the interaction of the different types of particles can be neglected in the zeroth approximation, then $P_{lm}(q) = \Sigma_\alpha P_{lm}{}^\alpha(q)$ and we obtain the following expression from Eqs. (6.5) and (6.7) for the components of the dielectric permittivity of particles of type α, $\varepsilon_q{}^{l,t(\alpha)}$:

$$\varepsilon_q^{l(\alpha)} = 1 - \frac{4\pi e_\alpha^2}{m_\alpha (q^0)^2 \mathbf{q}^2} \left\{ m_\alpha \sum_s \frac{2\omega_{0s}^3 \,|\, n_{0s}^{(\alpha)}(\mathbf{q})\,|^2}{\omega_{0s}^2 - (q^0)^2} + \mathbf{q}^2 n_\alpha \right\}, \tag{6.8a}$$

and

$$\varepsilon_q^{t(\alpha)} = 1 - \frac{2\pi e_\alpha^2}{m_\alpha (q^0)^2 \mathbf{q}^2} \left\{ m_\alpha \sum_s \frac{|\,[q j_{0s}^{(\alpha)}(\mathbf{q})]\,|^2}{\omega_{0s}^2 - (q^0)^2} + 2\mathbf{q}^2 n_\alpha \right\}. \tag{6.8b}$$

Here ω_{0s} are the excitation energies of the system of particles and $\hat{n} = \Sigma_i e^{-i\mathbf{q}\mathbf{r}_i}$ are the Fourier components of the density operator for the particles.

Since quantum mechanical averaging is understood in Eqs. (6.7) and (6.8) (rather than averaging over a physically infinitely small volume), the condition $|\mathbf{q}| \ll n^{1/3}$ drops out and the $\varepsilon_q{}^{l,t}$ are defined for all values of q for which the nonrelativistic treatment of the medium is still valid.

Since the medium is always microscopically inhomogeneous, we note that the "dressed" photon propagator must also include contributions from photon scattering processes that involve a change in the 4-vector q that were not included in Eq. (6.3) and will not be considered here because of their smallness.

The potential of a virtual photon created by the incident particle during scattering ($p_i \to p_f$) is, therefore, equal to

$$\mathbf{A}_{fi}^{(0)}(x) = \mathbf{A}^{(0)}(q) \exp[i(q^0 t - \mathbf{q}\mathbf{r})], \qquad A_m^{(0)}(q) = -e_0 D_{mn}(q) v_{0n}. \tag{6.9}$$

The total electromagnetic field that excites the plasma electrons is made up of $\mathbf{A}^{(0)}$ and the field of the free photons, for which the quantum mechanical vector potential is given by

$$\hat{\mathbf{A}}^{\mathrm{ph}}(x) = \sum_{k,\sigma} \sqrt{4\pi \Big/ \frac{\partial}{\partial\omega} \omega^2 \varepsilon_k^t \Big|_{\omega=\omega_k^t}} \{\mathbf{e}_{\mathbf{k}\sigma} \hat{a}_{\mathbf{k}\sigma} \times$$
$$\times \exp[i(\mathbf{k}\mathbf{r} - \omega_k^t t)] + \mathbf{e}_{\mathbf{k}\sigma}^* \hat{a}_{\mathbf{k}\sigma}^t \exp[-i(\mathbf{k}\mathbf{r} - \omega_k^t t)]\}. \tag{6.10}$$

Here σ and $\mathbf{k}$ are the subscripts representing the polarization and wave vector of a photon; $\mathbf{e}$ is the polarization unit vector; $a_{\mathbf{k}\sigma}{}^+$ and $a_{\mathbf{k}\sigma}$ are the photon creation and annihilation operators; and, $\omega_k{}^t$ is the frequency of a transverse photon with wave vector $\mathbf{k}$ which is given by the dispersion law for an isotropic medium as

$$\omega_k^t = |\mathbf{k}| / \sqrt{\varepsilon_k^t}. \tag{6.11}$$

We now write down an explicit expression for ε_k^t including the bound electrons in the ions. Neglecting the interaction among the bound and plasma electrons, we obtain

$$\varepsilon_k^t = 1 - \omega_{pe}^2/\omega^2 + 4\pi n_i \alpha_i(\omega). \tag{6.12}$$

Here $\alpha_i(\omega)$ is the dynamic polarizability of the ions (which are assumed to be identical) at the frequency ω and n_i is the ion density. In deriving Eq. (6.12) from Eq. (6.8b) we have used the condition $\omega \gg |\mathbf{k}|v_{Te}$ (where v_{Te} is the thermal speed of the plasma electrons, which are assumed to be in thermal equilibrium). We neglect the Doppler effect in the contribution of the bound electrons to ε_k^t.

In the following we shall be interested in the emission of transverse photons of frequency $\omega \gg \omega_{pe}$, so that we set $\varepsilon_k^t \sim 1$ in the nonresonant configurations. In addition, we shall omit the factor $(2)^{1/2}(\partial\omega^2\varepsilon_{\mathbf{k}}^t/\partial\omega)^{-1/2}$ in Eq. (6.10) in the following calculations, given that after taking the square of the modulus of the amplitude for this process it cancels out a similar factor in the expression for the density of photon states:

$$\frac{d\mathbf{k}}{(2\pi)^3} = \frac{\partial\omega^2\varepsilon_{\mathbf{k}}^t}{2\partial\omega}\,\frac{\omega d\omega\, d\Omega_{\mathbf{k}}}{(2\pi)^3}, \tag{6.13}$$

where $d\Omega_{\mathbf{k}}$ is the element of solid angle around the direction of $\mathbf{k}$ or $\mathbf{n} = \mathbf{k}/|\mathbf{k}|$. The Hamiltonian of the perturbation of the nonrelativistic plasma electrons by the electromagnetic field (using an axial gauge) has the form

$$\begin{aligned} \hat{V} &= \frac{e}{2m}\sum_j \{\hat{\mathbf{p}}_j\hat{\mathbf{A}}(\mathbf{r}_j, t) + \hat{\mathbf{A}}(\mathbf{r}_j, t)\hat{\mathbf{p}}_j + e\hat{A}^2(\mathbf{r}_j t)\}, \\ \mathbf{p}_j &= -i\nabla_j, \qquad \hat{\mathbf{A}} = \hat{\mathbf{A}}^{(0)} + \hat{\mathbf{A}}^{\mathrm{ph}}. \end{aligned} \tag{6.14}$$

Here the sum is taken over both the bound and plasma electrons.

In the second order of perturbation theory with respect to the interaction $\hat{\mathbf{p}}\hat{\mathbf{A}} + \hat{\mathbf{A}}\hat{\mathbf{p}}$ and in the first order with respect to the interaction $\hat{\mathbf{A}}^2$, the amplitude of the process of interest to us which contains the "crossover" terms $\mathbf{A}^{(0)}A^{ph}$ has the form

$$\begin{aligned} M_{fi}(\mathbf{k}, \sigma, q) = \sqrt{2\pi}\,\frac{e^2}{m}\Bigg\{ m\sum_s \Bigg[&\frac{\langle f|\,\mathbf{e}_{\mathbf{k}\sigma}^*(\hat{\mathbf{j}}_{\mathbf{k}}^p + \hat{\mathbf{j}}_{\mathbf{k}}^b)\,|s\rangle\langle s|\,(\hat{\mathbf{j}}_{\mathbf{q}_1}^p + \hat{\mathbf{j}}_{\mathbf{q}_1}^b)\,\mathbf{A}_{\mathbf{q}_1}^{(0)}\,|i\rangle}{\Omega_{fs} + \omega + i\Gamma_{fs}/2} + \\ + &\frac{\langle f|\,\mathbf{A}_{\mathbf{q}_1}^{(0)}(\hat{\mathbf{j}}_{\mathbf{q}_1}^p + \hat{\mathbf{j}}_{\mathbf{q}_1}^b)\,|s\rangle\langle s|\,(\hat{\mathbf{j}}_{\mathbf{k}}^p + \hat{\mathbf{j}}_{\mathbf{k}}^b)\,\mathbf{e}_{\mathbf{k}\sigma}^*\,|i\rangle}{\Omega_{is} - \omega + i\Gamma_{is}/2}\Bigg] + \\ &+ \mathbf{e}_{\mathbf{k}\sigma}^*\mathbf{A}_{\mathbf{q}}^{(0)}\langle f|\,\hat{n}^p(\mathbf{q}) + \hat{n}^b(\mathbf{q})\,|i\rangle\Bigg\}, \end{aligned} \tag{6.15}$$

where $\hat{\mathbf{j}}_{\mathbf{k}}^{p,b}$ and $\hat{n}_{\mathbf{k}}^{p,b}$ are the Fourier components of the current density and charge density operators defined above for the plasma (superscript p) and bound (superscript b) electrons; $|f, s, i\rangle$ are the many-particle wave functions; Ω_{ij} are the perturbation energies, including the Doppler effect; and Γ_{ij} are the linewidths of the $i \to j$ transitions. We assume that the subsystems formed by the bound electrons of each ion interact weakly with the plasma electrons, as well as with the electrons in neighboring ions, so that the wave functions of the bound electrons of an isolated ion are determined only by the parameters of this ion. Let us suppose that the plasma electrons interact with the ions as with point charges ($\mathbf{q}_1 = \mathbf{q} - \mathbf{k}$).

Given the above remarks, we write the wave function of the plasma in the form

$$\psi_s(\mathbf{r}_l, \mathbf{R}_j, \rho_\alpha^j) = \Phi_s(\mathbf{r}_l, \mathbf{R}_j) \prod_j \psi_s^{(j)}(\rho_\alpha^j). \tag{6.16}$$

Here $\mathbf{r}_l$ is the radius vector of the lth plasma electron; $\mathbf{R}_j$ is the radius vector of the center of mass of the jth ion; $\rho_\alpha{}^j$ is the radius vector of the αth electron belonging to the jth ion relative to the center of mass of the ion; Φ_s is the wave function of the interacting plasma electrons and ions; and $\psi_s{}^{(j)}$ are the wave functions of the electron subsystem of the jth ion. We shall assume that the systems of functions Φ_s and $\psi_s{}^{(j)}$ (for each j) are orthonormal and form a complete set. For simplicity, in the following we shall assume that the subsystem of bound electrons does not change its state in the course of bremsstrahlung; i.e., $\psi_{si}{}^{(j)} = \psi_{sf}^{(j)} = \psi_1 e^{i\varphi_{1j}}$, where φ_{1j} is the phase of the electron wave function of the jth ion in the ground state ψ_1.

Substituting the approximation (6.16) in Eq. (6.15), we find

$$\begin{aligned} M_{fi}^{(1,1)}(\mathbf{k}, \sigma, \mathbf{q}) = \sqrt{2\pi} \Big\{ \frac{e^2}{m} \Big\langle \Phi_f \Big| \sum_l e^{-i\mathbf{q}\mathbf{r}_l} \Big| \Phi_i \Big\rangle \times \\ \times \left[\mathbf{e}^{\bullet}_{\mathbf{k}\sigma} \mathbf{A}^{(0)}_{\mathbf{q}_1} + \frac{(\mathbf{e}^{\bullet}_{\mathbf{k}\sigma}\mathbf{q}_1)(\mathbf{q}_1\mathbf{A}^{(0)}_{\mathbf{q}_1})}{2m\omega} \right] - \mathbf{e}^{\bullet}_{\mathbf{k}\sigma,h} \mathbf{A}^{(0)}_{\mathbf{q}_1 m} (q^0)^2 \times \\ \times \Big\langle \Phi_f \Big| \sum_j e^{-i\mathbf{q}_1\mathbf{R}_j} c_{hm}^{(1,1)j}(\mathbf{k}, \mathbf{q}_1, \omega - \mathbf{k}\mathbf{v}^j) \Big| \Phi_i \Big\rangle \Big\}. \end{aligned} \tag{6.17}$$

Here $c_{hm}{}^{(1,1)}(\mathbf{k}, \mathbf{q}_1)$ is the diagonal matrix element of the scattering operator for the electromagnetic field on the electrons of the jth ion, which can be written in the following form that includes the width of the line corresponding to the transition (see Chapter 5):

$$\begin{aligned} c_{hm}^{(1,1)}(\mathbf{k}, \mathbf{q}_1, \omega - \mathbf{k}\mathbf{v}^j) = -\frac{e^2}{m}(q_1^0)^{-2} \Big\{ \delta_{hm} \langle 1 \,|\, \hat{n}^{(j)}(\mathbf{q}) \,|\, 1 \rangle + \\ + m \sum_s \left[\frac{j_{1s}^h(\mathbf{k})\, j_{s1}^m(\mathbf{q}_1)}{\omega - \mathbf{k}\mathbf{v}_f^j + \omega_{1s} + i\Gamma_{1s}/2} + \frac{j_{1s}^m(\mathbf{q}_1)\, j_{s1}^h(\mathbf{k})}{\omega_{1s} - \omega + \mathbf{k}\mathbf{v}_i^j + i\Gamma_{1s}/2} \right] \Big\}. \end{aligned} \tag{6.18}$$

In deriving Eq. (6.17) from Eq. (6.15), we have used the orthonormality and completeness of the system of functions Φ_s, $\psi_s{}^{(j)}$. In obtaining the second term of Eq. (6.15) for the energy denominators we have taken

$$\Omega_{fs} = \omega_{1s} - \mathbf{k}\mathbf{v}_f^j, \qquad \Omega_{is} = \omega_{1s} + \mathbf{k}\mathbf{v}_i^j,$$

where ω_{1s} is the frequency of the electronic excitation of the ion and $\mathbf{kv}_f{}^j$ and $\mathbf{kv}_i{}^j$ are the excitation energies of the ion associated with the motion of its center of mass. In this approximation, we can take the sum over the entire set of functions Φ_s, which leads to the appearance of the factors $e^{-i\mathbf{q}\mathbf{R}_j}$ in Eq. (6.17). Including $\mathbf{kv}_f{}^j$ and $\mathbf{kv}_i{}^j$ in the energy denominators corresponds to the appearance of the Doppler effect. For nonrelativistic plasma electrons, this is important in the resonant frequency regions for the bremsstrahlung photons, or those ω which satisfy $|\Delta| = |\omega - \omega_{1s}| \leq kv_{Ti}$ (where v_{Ti} is the ion thermal speed). For $T = 300$ K and $Z \sim 10$, where Ze is the nuclear charge of an ion, the inequality $\Delta \leq 10^{-5}\omega$ must be satisfied. Otherwise (large detunings $|\Delta|$ from the resonance), we can neglect the Doppler effect and simplify Eq. (6.17). We note that in deriving the first term of Eq. (6.17), we have used the approach described in Chapter 5 for obtaining an asymptotic form for the scattering tensor when $\omega \gg I$ (where I is the ionization potential of an atom). In the present case, this corresponds to neglecting the excitation energy of the plasma electrons when a momentum $\mathbf{k}$ is absorbed compared to the photon frequency ω, i.e., $\omega \gg |\mathbf{k}|v_{Te}$ (where $v_{Te} \ll 1$). Given these remarks, we rewrite Eq. (6.17) in the form

$$M_{fi}^{(1,1)}(\mathbf{k}, \sigma, \mathbf{q}) = \sqrt{2\pi}\left\{\frac{e^2}{m} n_{fi}^{(e)}(\mathbf{q})\, \mathbf{e}^*_{\mathbf{k}\sigma}\mathbf{A}^{(0)}(\mathbf{q}_1) - \right.$$
$$\left. - \mathbf{e}^*_{\mathbf{k}\sigma,h}\mathbf{A}_m^{(0)}(\mathbf{q}_1)\, c_{hm}^{(1,1)}(\omega, \mathbf{k}, \mathbf{q}_1)\, n_{fi}^{(i)}(\mathbf{q})\right\}, \tag{6.19}$$

where

$$n_{fi}^{(e)}(\mathbf{q}) = \left\langle \Phi_f \left| \sum_l e^{-i\mathbf{q}\mathbf{r}_l} \right| \Phi_i \right\rangle; \quad n_{fi}^{(i)}(\mathbf{q}) = \left\langle \Phi_f \left| \sum_j e^{-i\mathbf{q}\mathbf{R}_j} \right| \Phi_i \right\rangle$$

are the matrix elements of the Fourier transforms of the density operators for the plasma electrons and ions. In Eq. (6.19) we have neglected the second term in the square brackets of Eq. (6.17) compared to the first. That is possible because, as will be shown below, the main contribution to this term of the bremsstrahlung amplitude is from $|\mathbf{q}| < (2m\omega)^{1/2}$ $(\omega \gg \omega_{pe})$.

We now calculate the differential cross section for bremsstrahlung in a plasma (owing to the interaction of the plasma electrons with photons) summed over all the final states of the delocalized particles (the plasma electrons and ions) and averaged over their initial states (we assume that the plasma is in thermodynamic equilibrium):

$$d\sigma_{11}(\mathbf{k}, \mathbf{q}) = \frac{2\pi}{v_0} \sum_{f,i,\sigma} \delta(\varepsilon_f - \varepsilon_i + \omega + \mathbf{q}_1\mathbf{v})\, w(i)\, |M_{fi}^{(1,1)}(\mathbf{k}, \sigma, \mathbf{q})|^2 \frac{\omega\, d\omega\, d\Omega_{\mathbf{k}}\, d\mathbf{q}}{(2\pi)^6}, \tag{6.20}$$

where $w(i) = \exp(-\varepsilon_i/T)/\Sigma_s \exp(-\varepsilon_s/T)$; and ε_f, ε_s, and ε_i are the energies of the delocalized plasma particles. Equation (6.20) has been obtained in accordance with the standard rules [7] with $v \gg v_{Te}$. Substituting Eq. (6.19) in Eq. (6.20) and summing over f and i, we obtain

$$d\sigma_{1,1}(\mathbf{k}, \mathbf{q}) = \frac{(2\pi)^2}{v_0}(\delta_{hl} - n_h n_l) A_m^{(0)}(\mathbf{q}_1) A_n^{(0)}(\mathbf{q}_1) \times$$
$$\times \frac{\omega d\omega d\Omega_k d\mathbf{q}}{(2\pi)^6}\left\{\delta_{hm}\frac{e^4}{m^2} S^{(e)}(q) + (q^0)^4 S^{(i)}(q) \times\right.$$
$$\times c_{hm}^{(1,1)}(\omega, k, q)\, c_{ln}^{(1,1)*}(\omega, k, q) - 2(q^0)^2 (e^2/m_e)\,\delta_{ln} \times$$
$$\left.\times \operatorname{Re}(c_{hm}^{(1,1)})\, S^{(e,i)}(q)\right\}; \quad \mathbf{q}_1 = \mathbf{q} - \mathbf{k},\ q^0 = \mathbf{q}\mathbf{v}_0 + \omega - \mathbf{k}\mathbf{v}_0. \tag{6.21}$$

Here $S^{(e)}(q)$, $S^{(i)}(q)$, and $S^{(ei)}(q)$ are the dynamic form factors of the plasma electrons and ions and the mixed electron–ion dynamic form factor, respectively. They are defined by the following formulas:

$$S^{(e)}(q) = \frac{1}{2\pi}\int_{-\infty}^{\infty} dt e^{iq^0 t}\langle \hat{n}^{(e)}(\mathbf{q}, t)\, \hat{n}^{(e)}(-\mathbf{q}, 0)\rangle, \tag{6.22a}$$

$$S^{(i)}(q) = \frac{1}{2\pi}\int_{-\infty}^{\infty} dt e^{iq^0 t}\langle \hat{n}^{(i)}(\mathbf{q}, t)\, \hat{n}^{(i)}(-\mathbf{q}, 0)\rangle, \tag{6.22b}$$

and

$$S^{(ei)}(q) = \frac{1}{2\pi}\int_{-\infty}^{\infty} dt e^{iq^0 t}\langle \hat{n}^{(e)}(\mathbf{q}, t)\, \hat{n}^{(i)}(-\mathbf{q}, 0)\rangle, \tag{6.22c}$$

where

$$\hat{n}^{(e)}(\mathbf{q}) = \sum_l e^{-i\mathbf{q}\mathbf{r}_l};\quad \hat{n}^{(i)} = \sum_j e^{-i\mathbf{q}\mathbf{R}_j};\quad \hat{n}(\mathbf{q}, t) = e^{i\hat{H}t}\, \hat{n}(\mathbf{q})\, e^{-i\hat{H}t};$$

and $\hat{H}$ is the Hamiltonian of the delocalized particles. The angle brackets denote both quantum mechanical and statistical averaging in accordance with the equation

$$\langle \ldots \rangle = \mathrm{Sp}\,(e^{-\hat{H}/T} \ldots)/\mathrm{Sp}\, e^{-\hat{H}/T},$$

where the trace (Sp) is taken over the entire set of functions Φ_s. Note that the thermodynamic average is important here, since the delocalized particles have a continuous energy spectrum. In deriving Eq. (6.21), we have taken the sum over the photon polarizations ($\Sigma_\sigma e_{\mathbf{k}\sigma,m}{}^* e_{\mathbf{k}\sigma,n} = \delta_{mn} - n_m n_n$ for unpolarized radiation), used the integral representation of the δ-function

$$2\pi\delta(\omega) = \int_{-\infty}^{\infty} d\tau e^{i\omega\tau},$$

and made use of the completeness of the system Φ_s and of the fact that the mixed dynamic form factor is real, i.e.,

$$S^{(ei)}(q) = S^{(ei)*}(q) = S^{(ie)}(q).$$

We now comment on Eq. (6.21). Three terms, proportional to $S^{(e)}(q)$, $S^{(i)}(q)$, and $S^{(ei)}(q)$, respectively, can be distinguished in it. The first term describes the scattering of the electromagnetic field of the incident particle $[\mathbf{A}^{(0)}(\mathbf{q})]$ into a bremsstrahlung photon on the plasma electrons; the second describes the same process on the bound electrons of the ions; and the third is an interference term. The dynamic form factors introduced here are the Fourier transforms of the spatial-temporal density-density correlation functions of the different plasma components [8]. They characterize the effectiveness with which the plasma absorbs energy-momentum through its different components. It is important to emphasize that when they are defined in this way, the dynamic form factors include the interactions among the plasma electrons and ions, such as their mutual shielding. This follows from the assumption that the wave function $\Phi_s(\mathbf{r}^l, \mathbf{R}_j)$ in the definition of the dynamic form factors (6.22) takes this interaction into account. We note that the dynamic form factor arises in a number of other problems (energy loss by fast particles in a medium, scattering of light in matter, x-ray diffraction). This quantity has been well studied, so it is extremely convenient to introduce the dynamic form factor for calculating the bremsstrahlung cross section.

The cross section (6.21) describes the emission of photons during scattering of an incident particle on all the charges in the plasma within a unit volume, so that it corresponds to Eq. (3.18). In this regard it also differs from the bremsstrahlung cross section discussed above which assumed a collision between two isolated particles and from which it is also easy to find the emission from a unit volume of plasma (see Chapter 3). In the formulation of the problem used here, those bremsstrahlung processes involving the transfer of excess energy-momentum to the collective excitation of the medium, such as to plasmons, are taken into account in the framework of a unified formalism. In order to avoid misunderstandings, we point out that the dimensions of the quantity (6.21) determined in this way are inverse length, so that here the cross section for the process on a single particle has been multiplied by the density. We shall, however, retain the name "cross section" for the quantity (6.21), with the understanding that it is to be multiplied by the volume $L^3 = 1$ (to maintain the required dimensions).

6.3. Dynamic Form Factors of the Plasma Components

For specific calculations, therefore, one requires knowledge of the explicit dependence of the dynamic form factors (6.22) on the 4-vector q. For this purpose we express them in terms of some characteristics of the noninteracting particles, namely the components of the dielectric permittivity of the plasma described in [9]. The basis of this method is the fact that the interaction of the plasma electrons and ions is weak and can be taken into account through perturbation theory. Technically, the dynamic form factor will be calculated using the fluctuation dissipation theorem, which relates the dynamic form factor to another quantity, the linear response function of the plasma components to an external field. Thus, for a plasma electron we have

$$S^{(e)}(q) = [\pi e^2(\exp(-q^0/T))-1]^{-1}\mathrm{Im}\, F_{ee}(q), \tag{6.23}$$

where $F_{ee}(q)$ is the linear response function of the electron component to a fictitious external potential that acts only on the plasma electrons. In accordance with [9], we introduce a second linear response function to an external potential, F_{ei}, which describes the response of the electron component of the plasma to a fictitious external potential that acts only on the ions. Note that in defining the fictitious potential, the term "interaction" was meant to refer to a direct interaction of the external potential with one or another component, which can then be transferred to the other component through Coulomb collisions. Here for convenience we shall use the Coulomb gauge for the electromagnetic potential. Let an external potential $\varphi^{\text{ext}}(q)$ act on the plasma. Then the electron charge density induced in the plasma [we are only considering delocalized particles for which the dynamic form factors (6.22) are determined] is given in terms of the above response functions by

$$\rho_\alpha(q) \equiv \langle\hat{\rho}_\alpha(q)\rangle = e_\alpha\langle\hat{n}_\alpha(q)\rangle, \qquad \rho_l(q) = [F_{ee}(q) + F_{ei}(q)]\,\varphi^{\text{ext}}(q). \tag{6.24}$$

The functions $F_{ee}(q)$ and $F_{ei}(q)$, therefore, account for the interaction among the different plasma components. For a weak interaction they can be expressed in terms of the characteristics of noninteracting particles. For this purpose we introduce another response function $\alpha_a(q)$ – the response function of particles of type a to the total potential in the plasma. It includes the effect on the plasma particles of the induced potential $\varphi_a^{\text{ind}}(q)$ owing to the redistribution of the charged particles under the action of the external potential. It is important that $\alpha_a(q)$ can be determined approximately in terms of the characteristics of noninteracting particles, since the interaction is taken into account by introducing an induced potential. With the aid of $\alpha_a(q)$ the induced charge of the ath component can be expressed in terms of the total potential as

$$\rho_\alpha(q) = \alpha_\alpha(q)\,\varphi^{\text{tot}}(q). \tag{6.25}$$

By using the momentum distribution function of the particles, $\Phi_p{}^{\alpha}$, we can obtain the following expression for $\alpha_{\alpha}(q)$ [9] (here $L = 1$) according to perturbation theory:

$$\alpha_{\alpha}(q) = e_{\alpha}^{2} Q_{\alpha}(q), \qquad Q_{\alpha}(q) = \sum_{\mathbf{p}} \frac{\Phi_{\mathbf{p}}^{\alpha} - \Phi_{\mathbf{p}+\mathbf{q}}^{\alpha}}{q^0 + \mathbf{p}^2/2m - (\mathbf{p}+\mathbf{q})^2/2m + i0}. \tag{6.26}$$

In the following we shall need to know the imaginary part of the function $Q_{\alpha}(q)$. In the case of a Maxwellian distribution, Eq. (6.26) yields

$$\operatorname{Im} Q_{\alpha}(q) = \pi (e^{-q^0/T} - 1) n_{\alpha} \frac{\exp[-(q^0)^2/2\,^2 v_{T\alpha}^2]}{\sqrt{2\pi}\,|\mathbf{q}|\, v_T}. \tag{6.27}$$

From the definition of the longitudinal part of the dielectric permittivity,

$$\tilde{\varepsilon}_q^{l(\alpha)} = 1 - \frac{\varphi^{\text{ind}(\alpha)}(q)}{\varphi^{\text{tot}(\alpha)}(q)}, \qquad \varphi^{\text{tot}} = \varphi^{\text{ext}} + \varphi^{\text{ind}}, \quad \varphi^{\text{ind}(\alpha)}(q) = \frac{4\pi}{\mathbf{q}^2} \rho_{\alpha}(q)$$

and the expression (6.25) for $\rho_{\alpha}(q)$, we find

$$\tilde{\varepsilon}_q^{l(\alpha)} = 1 - \frac{4\pi}{\mathbf{q}^2} \alpha_{\alpha}(q), \tag{6.28}$$

where the tilde above the symbol for the dielectric permittivity means that the contribution of the bound electrons has been neglected in determining this quantity. We express $F_{\alpha\beta}$ in terms of α_{α} (and, thus, in terms of $\tilde{\varepsilon}^{l(\alpha)}$). To do this, we introduce a fictitious external potential $\varphi^{\text{ext}*}$ which acts only on the electrons. Then, according to the definition, we have

$$\rho_e^{*}(q) = F_{ee}(q)\, \varphi^{\text{ext}*}(q). \tag{6.29a}$$

On the other hand, $\rho_e{}^*$ can be expressed in terms of α_e as

$$\rho_e^{*}(q) = \alpha_e(q)\, [\varphi^{\text{ext}*} + \varphi^{\text{ind}*}]. \tag{6.29b}$$

The induced potential $\varphi^{\text{ind}*}$ is determined from the Poisson equation:

$$\varphi^{\text{ind}*}(q) = (4\pi/\mathbf{q}^2)\, [\rho_e^{*}(q) + \rho_i^{*}(q)]. \tag{6.30}$$

We still have to find $\rho_i{}^*$:

$$\rho_i^{*} = \alpha_i \varphi^{\text{ind}*}, \tag{6.31}$$

which is determined by noting that, according to its definition, the fictitious external potential $\varphi^{\text{ext}*}$ acts directly only on the electrons. Here the asterisk above a quantity means that it is defined for the fictitious external potential $\varphi^{\text{ext}*}$. Solving the system of Eqs. (6.29)-(6.31), we find

$$F_{ee}(q)=\alpha_e(q)\,[1-(4\pi/\mathbf{q}^2)\,\alpha_i(q)\,]\,[1-(4\pi/q^2)\,(\alpha_e(q)+\alpha_i(q))\,]^{-1}. \tag{6.32}$$

Substituting Eq. (6.32) into the right-hand side of the fluctuation dissipation theorem (6.23) and using the definition (6.28) of the dielectric permittivity, we find a formula for the desired dynamic form factor for the electrons:

$$S^{(e)}(q)=\left[\pi e^2\left(\exp\left(-\frac{q^0}{T}\right)-1\right)\right]^{-1}\left\{\left|\frac{1-\dfrac{4\pi}{q^2}\alpha_i(q)}{\tilde{\varepsilon}^l_q}\right|^2\times\right.$$
$$\left.\times\operatorname{Im}\alpha_e(q)+\left(\frac{4\pi}{\mathbf{q}^2}\right)^2\left|\frac{\alpha_e(q)}{\tilde{\varepsilon}^l_q}\right|^2\operatorname{Im}\alpha_i(q)\right\} \tag{6.33}$$

Using Eqs. (6.26) and (6.27), we can rewrite Eq. (6.33) as

$$S^{(e)}(q)=\frac{n_e}{\sqrt{2\pi}\,|\mathbf{q}|}\left\{\frac{1}{v_{Te}}\left|\frac{\tilde{\varepsilon}^{l(i)}_q}{\tilde{\varepsilon}^l_q}\right|^2\exp\left[-\frac{(q^0)^2}{2\mathbf{q}^2v^2_{Te}}\right]+\right.$$
$$\left.+\frac{Z_i}{v_{Ti}}|(1-\tilde{\varepsilon}^{l(e)}_q)/\tilde{\varepsilon}^l_q|^2\exp[-(q^0)^2/2\mathbf{q}^2v^2_{Ti}]\right\}. \tag{6.34a}$$

In a completely analogous fashion, for $S^{(i)}(q)$ and $S^{(ei)}(q)$ we find

$$S^{(i)}(q)=\frac{n_i}{\sqrt{2\pi}\,|\mathbf{q}|}\left\{\frac{1}{v_{Ti}}|\tilde{\varepsilon}^{l(e)}_q/\tilde{\varepsilon}^l_q|^2\exp[-(q^0)^2/2\mathbf{q}^2v^2_{Ti}]+\right.$$
$$\left.+\frac{1}{Z_iv_{Te}}|(1-\tilde{\varepsilon}^{l(i)}_q)/\tilde{\varepsilon}^l_q|^2\exp[-(q^0)^2/2\mathbf{q}^2v^2_{Te}]\right\}; \tag{6.34b}$$

and

$$S^{(ei)}(q)=\frac{n_e}{\sqrt{2\pi}\,|\mathbf{q}|}\left\{|(1-\tilde{\varepsilon}^{l(i)}_q)/\tilde{\varepsilon}^l_q|^2\frac{1}{Z_iv_{Te}}\exp[-(q^0)^2/2\mathbf{q}^2v^2_{Te}]+\right.$$
$$\left.+\frac{1}{v_{Te}}|(1-\tilde{\varepsilon}^{l(e)}_q)/\tilde{\varepsilon}^l_q|^2\exp[-(q^0)^2/2\mathbf{q}^2v^2_{Te}]\right\}. \tag{6.34c}$$

We now give a physical interpretation of these expressions for the dynamic form factors. As an example we consider the dynamic form factor for the electrons, $S^{(e)}(q)$. It is conveniently represented in the form

$$\begin{aligned}&S^{(e)}(q)=|\delta n_e|^2_q\,|1^{\mathrm{ef}}_{e,q}|^2+|\delta n_i|^2_q\,|Z^{\mathrm{ef}}_{i,q}|^2,\\&|\delta n_{e,i}|^2_q=(n_{e,i}/\sqrt{2\pi}\,|\mathbf{q}|\,v_{Te,i})\exp[-(q^0)^2/2\mathbf{q}^2v^2_{Te,i}],\\&1^{\mathrm{ef}}_{e,q}=\tilde{\varepsilon}^{l(i)}_q/\tilde{\varepsilon}^l_q,\quad Z^{\mathrm{ef}}_{i,q}=Z(1-\tilde{\varepsilon}^{l(e)}_q)/\tilde{\varepsilon}^l_q.\end{aligned} \tag{6.35}$$

Here $|\delta n_{ei}|_{\mathbf{q}}^2$ denote the squares of the thermal fluctuations in the electron and ion components at a frequency of q^0 and a wave vector of $\mathbf{q}$; $eZ_{i,q}{}^{\text{ef}}$ is the effective electronic charge that shields a plasma ion and is defined along the 4-vector q; and $e1_{e,q}{}^{\text{ef}}$ is the effective charge of a plasma electron, including the shielding effect of the other electrons. Using this representation for the dynamic form factor of the electrons, we can interpret it as the sum of the squares of the thermal fluctuations in two types of electron density. The first term in Eq. (6.35) describes the fluctuations associated with isolated electrons shielded by other electrons. The second term applies to the electronic charge that shields the fluctuations in the ion component. An analogous interpretation of the dynamic form factor for the ions is also possible. The mixed electron–ion dynamic form factor $S^{(ei)}(q)$ is the sum of fluctuations in the charges of both plasma components which shield the charges of opposite sign. The zeroes in the longitudinal part of the dielectric permittivity in Eqs. (6.34) correspond to charge fluctuations associated with the appearance of collective excitations in the plasma. In this case, the excess energy-momentum that develops during bremsstrahlung is transferred to collective excitations of the medium.

6.4. Polarization Bremsstrahlung on Free Electrons and Ions

We first consider the bremsstrahlung caused by the conversion of a virtual photon of the field of the incident particle on plasma electrons. The cross section is given by the term proportional to $S^{(e)}(q)$ in Eq. (6.21). We write it including the explicit expression for $S^{(e)}(q)$ (6.34a) as

$$d\sigma^p(\mathbf{k},\mathbf{q}) = \frac{(2\pi)^2 e^4}{v_0 m^2}\left[\frac{\mathbf{k}}{|\mathbf{k}|}\mathbf{A}^{(0)}(q)\right]^2 \Big\{ n_e |\tilde{\varepsilon}_q^{l(i)}/\tilde{\varepsilon}_q^l|^2 \times$$
$$\times \frac{1}{\sqrt{2\pi} q v_{Te}} \exp[-(q^0)^2/2q^2 v_{Te}^2] + n_i Z_i^2 |(1-\tilde{\varepsilon}_q^{l(e)})/\tilde{\varepsilon}_q^l|^2 \frac{1}{\sqrt{2\pi} q v_{Ti}} \times$$
$$\times \exp[-(q^0)^2/2q^2 v_{Ti}^2]\Big\} \frac{\omega\, d\omega\, d\Omega_k\, d\mathbf{q}}{(2\pi)^6}. \tag{6.36}$$

The terms in curly brackets in Eq. (6.36) describe the scattering of a virtual photon into a real photon on the plasma electrons. The first term is related to the transfer of energy-momentum q to the subsystem of the plasma electrons and the second describes the transfer of q to the ions through the interaction of the field of the incident particle with the plasma electrons, which is a consequence of the Coulomb interaction of the electrons and ions of the plasma. This, in particular, is indicated by the exponents, one of which is related to the thermal speed of the electrons, while the other is related to that of the ions.

We now examine the part of the cross section (6.36) that is determined by the second term in the curly brackets. It coincides with the cross section for polarization (transition) bremsstrahlung during a collision of a fast incident particle with a point plasma ion at frequencies $\omega \gg \omega_{pe}$. This process has been studied in [2]. In order to confirm this, we note that if we neglect the recoil, an ion at rest obeys the equation

$$\lim_{v_{Ti}\to 0} \frac{1}{\sqrt{2\pi}\,|\mathbf{q}|\,v_{Ti}} \exp\left[-(q^0)^2/2\mathbf{q}^2 v_{Ti}^2\right] = \delta(q^0), \tag{6.37}$$

where $q^0 = \mathbf{q}\mathbf{v}_0 + \omega - \mathbf{k}\mathbf{v}_0$. In this case, $q^0 \gg |\mathbf{q}|v_{Te}$ and $\omega \gg \omega_{pe}$, so that scattering of virtual photons on the electron polarization charge around the incident particle can be neglected. In fact, the magnitude of this charge is determined by the value of $|(1 - \tilde{\varepsilon}_q^{l(e)})/\tilde{\varepsilon}_q^{l}|$ which is small in this case; i.e., $|(1 - \tilde{\varepsilon}_q^{l(e)})/\tilde{\varepsilon}_q^{l}| \sim \omega_{pe}^2/\omega^2 \ll 1$, where $\tilde{\varepsilon}_q^{l(e)}) \sim \tilde{\varepsilon}_q^{l} \sim 1 - \omega_{pe}^2/\omega^2$. As a result, the transition bremsstrahlung is caused only by scattering of the field of the incident particle on the electron polarization charge around the ion, which has been included in Eq. (6.36). Taking these remarks into account and using the explicit form of $\mathbf{A}^{(0)}(q)$ from Eqs. (6.9) and (6.3) (noting that $\varepsilon_q^t = \varepsilon_q^l = \varepsilon_q$ in this case), we find that the cross section

$$d\sigma_i^p(\mathbf{k}, \mathbf{q}) = \frac{(2\pi)^2 e^4 [\mathbf{k}, (\mathbf{q}_1 - q_1^0\mathbf{v})]^2}{v_0 m^4 \mathbf{k}^2 (q^2 - (q^0)^2 \varepsilon_q)^2} \left(\frac{4\pi e_0}{q_1^0}\right)^2 \delta(q^0) Z_i^2 \,|(1 - \tilde{\varepsilon}_q^{l(e)})/\tilde{\varepsilon}_q^{l}|^2 \frac{\omega d\omega d\Omega_{\mathbf{k}} d\mathbf{q}}{(2\pi)^6}, \tag{6.38}$$

where

$$q^0 = \mathbf{q}\mathbf{v}_0 + \omega - \mathbf{k}\mathbf{v}_0, \quad q_1^0 = q^0 - \omega, \quad \mathbf{q}_1 = \mathbf{q} - \mathbf{k},$$

coincides with the cross section for polarization (transition) bremsstrahlung of a fast particle on an ion at rest in a plasma for $\omega \gg \omega_{pe}$ given by the matrix element (21) of [2]. In deriving Eq. (6.38) from Eq. (6.36), we have used Eq. (6.37) and the result has been divided by the ion density [compare the results of Chapter 2 with Eq. (3.18)]. The agreement of Eq. (6.38) with the results of [2] is not surprising, since Eq. (6.38) is a purely classical result and should correspond to a special case of the results of Chapters 2 and 3.

The cross section $d\sigma^p$ obtained here for $\omega \gg \omega_{pe}$ is, therefore, the same as the cross section for polarization (transition) bremsstrahlung on a plasma ion at rest. We note that without the limiting transition (6.37) and dividing by $n_i = \int \Phi_{\mathbf{p}}^{(i)} d\mathbf{p}/(2\pi)^3$, it would describe transition bremsstrahlung on an ensemble of plasma ions in thermodynamic equilibrium.

We now examine the cross section corresponding to the first term in the curly brackets of Eq. (6.36) while including the term in square brackets in Eq. (6.17) that was dropped earlier:

$$d\sigma_e^p(\mathbf{k}, \mathbf{q}) = \frac{(2\pi)^2}{v_0 k^2} \left[\mathbf{k}, \left(\mathbf{A}^{(0)}(\mathbf{q}_1) + \frac{\mathbf{q}_1 (\mathbf{q}_1 \mathbf{A}^{(0)}(\mathbf{q}_1))}{2m\omega}\right)\right] \times$$

$$\times \frac{e}{m} n_e \frac{1}{\sqrt{2\pi}\,|\mathbf{q}|\,v_{Te}} \exp\left[-(q^0)^2/2q^2 v_{Te}^2\right] \times$$

$$\times \frac{\omega d\omega d\Omega_{\mathbf{k}} d\mathbf{q}}{(2\pi)^6} |\tilde{\varepsilon}_q^{l(i)}/\tilde{\varepsilon}_q^{l}|^2, \quad \mathbf{q}_1 = \mathbf{q} - \mathbf{k},\ q_1^0 = q^0 - \omega,\ q^0 = \mathbf{q}\mathbf{v}_0 + \omega - \mathbf{k}\mathbf{v}_0. \tag{6.39}$$

Recall that Eq. (6.39) describes bremsstrahlung owing to the conversion of virtual photons of the field of an incident particle on isolated shielded plasma electrons. The adjective "isolated" in this case means that the excess energy (momentum q) is transferred to a single plasma electron.

From the explicit forms of $\tilde{\varepsilon}_q{}^l$ for $|q^0| \ll |\mathbf{q}|v_{Te}$ and $|q^0| \gg |\mathbf{q}|v_{Te}$, i.e.,

$$\tilde{\varepsilon}_q^l \simeq 1 + \frac{1}{q^2 r_d^2} - \frac{\omega_{pi}^2}{(q^0)^2}, \quad r_d = \frac{v_{Te}}{\omega_{pe}}, \quad \omega_{pi} = \sqrt{\frac{4\pi n_i e^2}{m_i}}, \tag{6.40a}$$

and

$$\tilde{\varepsilon}^{l(e)} \approx 1 + 1/q^2 r_d^2 \tag{6.40b}$$

it follows that the cross section (6.39) is small when $|\mathbf{q}| < r_d^{-1}$. This corresponds to the well-known fact of plasma physics that long-wavelength Coulomb perturbations are shielded. If the factor $\tilde{\varepsilon}_q{}^{l(i)}/\tilde{\varepsilon}_q{}^l$ is written in the form

$$\tilde{\varepsilon}_q^{l(i)}/\tilde{\varepsilon}_q^l = 1 - (\tilde{\varepsilon}_q^{l(e)} - 1)/\tilde{\varepsilon}_q^l, \tag{6.41}$$

then the first term in Eq. (6.41) describes the scattering of the field of an incident particle on a "bare" (unshielded) ion and the second, on the polarization electron charge surrounding the ion. In the terminology of [1, 2], the first type of emission is traditional bremsstrahlung on the plasma electrons, while the second is polarization (transition) bremsstrahlung on them. The total cross section is the superposition of the two. Note that for $|q^0| \ll |\mathbf{q}|v_{Te}$, the interference between the two leads to a sharp reduction in the cross section. This is explained by the fact that the polarization electronic charge around an isolated electron has the opposite sign, so that the scattering amplitudes for the field of the incident particle on an electron and on the electron polarization charge associated with it will be close in absolute value and have opposite signs if $|\mathbf{q}| \to 0$.

The situation is different for scattering of a virtual photon on a plasma ion: scattering on the charge of the ion itself can be neglected in view of its large mass, so that the scattering on the electron polarization cloud surrounding the ion is uncompensated. These remarks make it clear why a plasma ion (ion plus its electron "cloud") radiates when $|\mathbf{q}| < 1/r_d$, although for these $|\mathbf{q}|$ its effective charge is negligibly small: only charges with small masses "show up" in the radiation.

The term proportional to $S^{(e)}$ in Eq. (6.21), therefore, describes the superposition of polarization (transition) bremsstrahlung on the plasma electrons and ions and traditional bremsstrahlung on the plasma electrons. These mechanisms for bremsstrahlung have been calculated [1, 2] under more general assumptions in the approximation of structureless ions (with no bound electrons) on the basis of a classical analysis (see Chapters 2 and 3).

We note that the contribution of the bound electrons to the dielectric permittivity (6.12) is unimportant for transition bremsstrahlung in a rarefied plasma (the criterion for "rarefied" will be explained below). In fact, the cross section for polarization (transition) bremsstrahlung contains the factor $|(\varepsilon_q^{l(e)} - 1)/\varepsilon_q^l|^2 \sim (1 + |\mathbf{q}|^2 r_d^2)^{-2}$, so it is small when $|\mathbf{q}| \gg 1/r_d$. Given that $|\mathbf{k}| \ll |\mathbf{q}|$ for polarization (transition) bremsstrahlung, the last inequality can be rewritten in the form $|\mathbf{q}_1| \geq 1/r_d$. Furthermore, since the conservation of energy implies that $|\mathbf{q}| \geq \omega/v_0$, polarization (transition) bremsstrahlung is small for $\omega > v_0/r_d$. Let us consider a specific, simple example. For a rarefied plasma with $r_d \sim 10^{-5}$ cm and $v_0 \sim 1$, the region in which polarization bremsstrahlung on the Debye "cloud" is significant will be determined by the inequality $\omega < 10^{15}$ Hz. For these frequencies, the polarizability of an ion can be set equal to its static limit $\alpha_i(\omega) \underset{\omega\to 0}{\to} R_i^3$, where $r_i \sim 1$ Å is the average radius of an ion. Consequently, the contribution of the bound electrons to the dielectric permittivity (6.12) can be neglected, since the inequality $\omega_{pe}^2/\omega^2 \gg n_i R_i^3$ is usually satisfied for frequencies $\omega < v_0/r_d$. Thus, this approach for calculating the bremsstrahlung owing to the plasma electrons yields the same result as the theory of [1, 2] for frequencies $\omega \gg \omega_{pe}$ ($v_0 \gg v_{Te}$).

6.5. Polarization Bremsstrahlung on Bound Electrons

We now consider bremsstrahlung in plasmas caused by the conversion of the field of an incident particle into a bremsstrahlung photon on the bound electrons near the corresponding ions. We did not include this emission above. For rarefied plasmas at frequencies $v/r_d > \omega$, where bremsstrahlung on the Debye shielding clouds is important, the ions can, as a rule, be treated as points. Over a wider frequency range ($m > \omega$), however, the contribution of the bound electrons near the ions must be taken into account. If we extend the discussion still further and generalize the analysis to frequencies $\omega > m$, then we must take the structure of the nucleus into account in the bremsstrahlung [10]. In this last case, however, bremsstrahlung in a plasma can be regarded as taking place on isolated ions.

The cross section for bremsstrahlung owing to the interaction of bound electrons with photons is given by the term proportional to the ion dynamic form factor $S^{(i)}(q)$ in Eq. (6.21):

$$d\sigma^p_{i,\text{bound}}(\mathbf{k},\sigma) = \frac{2\pi\omega d\omega}{v_0\mathbf{k}^2}[\mathbf{k}\mathbf{A}^{(0)}(\mathbf{q}_1)]^2 (q_1^0)^4 \quad |c(\omega,\mathbf{k},\mathbf{q}_1)|^2 \frac{n_i d\Omega_{\mathbf{k}} d\mathbf{q}}{\sqrt{2\pi}\,|\mathbf{q}|\,(2\pi)^6}\left\{|\tilde{\varepsilon}_q^{l(e)}/\tilde{\varepsilon}_q^l|^2 \frac{1}{v_{Ti}}\times\right.$$
$$\left.\times\exp[-(q^0)^2/2q^2v_{Ti}^2] + |(1-\tilde{\varepsilon}_q^{l(i)})/\tilde{\varepsilon}_q^l|^2\frac{1}{v_{Te}Z_i}\exp[-(q^0)^2/2q^2v_{Te}^2]\right\}. \tag{6.42}$$

In writing down this equation, we have used the approximation $c_{hm}^{(1,1)} = c(\omega, \mathbf{k}, \mathbf{q}_1)\delta_{h,m}$, which is valid for the centrally symmetric state $|1\rangle$ when $|\mathbf{q}| < (2m\omega)^{1/2}$. It will be clear from the following that it is precisely this range of momentum transfers from the incident particle which is of interest. The cross section (6.42) for bremsstrahlung on bound electrons does not include the possible excitation of the electronic subsystem of the ion during the bremsstrahlung.

A plasma has a twofold influence on this process compared to the case of bremsstrahlung on an isolated ion in a vacuum. The ions in a plasma shield one another. This effect is described by the factor in the first term in the curly brackets of Eq. (6.42). The interaction between the ions and plasma electrons can lead to the transfer of excess energy-momentum from the bound to the plasma electrons, which is described by the second term in the curly brackets of Eq. (6.42). In addition, the interaction of the electromagnetic field of the incident particle with the plasma causes "dressing" of the corresponding virtual photon. This is reflected in the form of the photon propagator (6.3) and leads to a change in the dispersion formula (6.11) for a real photon in a medium. This influence of the medium has been taken into account in Eq. (6.42). In a rarefied plasma, shielding of the ions and the transfer of excess energy-momentum from the bound to the plasma electrons can be neglected. In fact, it is easy to show that these effects are important for $|\mathbf{q}_1| \sim |\mathbf{q}| \sim 1/r_d$, as this follows from the form of the ion dynamic form factor (6.34b) and from the expression for $\tilde{\varepsilon}_q{}^l$ in the low-frequency limit (6.40a) and (6.40b). It follows from the conservation of energy-momentum, however, that momentum transfers in these amounts are possible for frequencies $v_0/r_d > \omega$, when the polarizability of the bound electrons is small. Thus, when $1/r_d > |\mathbf{q}|$, the corresponding level of bremsstrahlung is low. The polarizability of the bound electrons becomes significant when $\omega > \omega_{\min}{}^{(i)}$ (where $\omega_{\min}{}^{(i)}$ is the minimum frequency for excitation of the electrons in the ions), but then $|\mathbf{q}_1| \sim |\mathbf{q}| > 1/r_d$, since $\omega_{\min}{}^{(i)} \gg v/r_d$ for a rarefied plasma, and in Eq. (6.42) we can set $|\tilde{\varepsilon}_q{}^{l(e)}/\tilde{\varepsilon}_q{}^l|^2 = 1$ and $|(1 - \tilde{\varepsilon}_q{}^{l(i)})/\tilde{\varepsilon}_q{}^l|^2 = 0$. Given these remarks and the explicit form of $\mathbf{A}^{(0)}(q_1)$ for $n_i = 1$, we obtain

$$d\sigma^P_{i,\text{bound}}(\mathbf{k}, \mathbf{q}) = \frac{2\pi}{v_0}(4\pi e_0)^2 \frac{d\omega d\Omega_{\mathbf{k}} d\mathbf{q}}{\omega k^2 (2\pi)^6} \delta(\omega + \mathbf{q}\mathbf{v}_0 - \mathbf{k}\mathbf{v}_0) \times$$
$$\times |\omega^2 c(\omega, \mathbf{k}, \mathbf{q}_1)|^2 [\mathbf{k}, (\omega\varepsilon(\omega)\mathbf{v}_0 - \mathbf{q})]^2 [\omega^2\varepsilon(\omega) - \mathbf{q}_1^2]^{-2}, \quad \mathbf{q}_1 = \mathbf{q} - \mathbf{k}. \qquad (6.43)$$

This yields the following formula for the spectral cross section at frequencies $\omega < v_0/R_i$ [cf. Eq. (2.129)]:

$$d\sigma^P_{i,\text{bound}}(\omega) = \frac{16e_0^2}{3v_0^2}\frac{d\omega}{\omega} |\omega^2\alpha_i(\omega)|^2 \ln\frac{1}{\omega R_i (1/v_0^2 - \varepsilon(\omega))^{1/2}}. \qquad (6.44)$$

In deriving this equation we have used the approximate equality $c(\omega, \mathbf{k}, \mathbf{q}_1) \approx \theta(1/R_i - |\mathbf{q}_1|)\alpha_i(\omega)$, which follows from the fact that when $|\mathbf{q}_1| > R_i^{-1}$, the contribution of the bound electrons to the scattering tensor becomes incoherent (see Chapter 5) while its coherent part, $c(\omega, \mathbf{k}, \mathbf{q}_1)$, is small.

Let us compare $d\sigma_{i,\text{bound}}{}^P$ with the cross section for transition bremsstrahlung on ions at rest in the frequency range $\omega < \{v_0/r_d, \omega_{\min}{}^{(i)}\}$. The formula for the latter cross section follows from Eq. (27) of [2], as well as from Eq. (2.106) of this book with $e_i = Z_i e$ and Eq. (6.38):

$$d\sigma^P_i(\omega) = \frac{16e_0^2 e^4 Z_i^2}{3v_0^2 m^2}\frac{d\omega}{\omega} \ln\frac{\gamma v_0}{r_d(\omega^2 + \gamma^2\omega_{pe}^2)^{1/2}}. \qquad (6.45)$$

We approximate (at these frequencies) the polarizability of the ion by its low-frequency limit $\alpha_i(\omega) \sim R_i^3 \sim (me^2)^{-3}Z^{-1}$. Then, neglecting the difference in the logarithmic factors in the cross sections, we find the ratio η_ω of the cross sections to be

$$\eta_\omega = \frac{d\sigma^p_{i,\text{bound}}}{d\sigma^p_i} \approx \frac{\omega^4}{Z_i^2 Z^2 m^4 e^{16}}, \quad \omega \ll v_0/r_d,\ \omega^{(i)}_{\min}. \tag{6.46a}$$

It follows from Eq. (6.46a) that polarization bremsstrahlung on bound electrons is small compared to the emission on free electrons ($\eta_\omega \ll 1$) for frequencies

$$\omega \ll \sqrt{ZZ_i}\, me^4 \approx \omega^{(i)}_{\min}. \tag{6.46b}$$

If $v/r_d \ll \omega_{\min}{}^{(i)}$, then the emissions on plasma and bound electrons are important in different frequency intervals and interference between the two can be neglected. We now calculate the plasma densities (n_*) for which this interference is important. For this purpose, we note that polarization bremsstrahlung on bound electrons plays a role when $\omega \geq \omega_{\min}{}^{(i)} \sim me^4(ZZ_i)^{1/2}$. The emission on plasma electrons has an upper bound frequency of order $v_0(4\pi ne^2/T_e)^{1/2}$. Thus, for $v \sim 1$ we find $n_* \sim m^2e^6T_eZZ_i/4\pi$, which for $T_e \sim me^4Z_i^2$, and given that $a_B = (me^2)^{-1}$ and $e^2 = 1/137$, gives

$$n_* \approx ZZ_i^3/(137)^2 a_B^3 \approx 10^{19} ZZ_i^3,$$

where n_* is in cm^{-3}. Finally, the above-mentioned criterion for a "rarefied" plasma, in which the interference between the emissions on bound and plasma electrons can be neglected, is $n \ll n_*$.

We now write down an expression for $d\sigma_{i,\text{bound}}{}^p(\omega)$ at frequencies $v_0/R_i \gg \omega \gg I$ (where I is the ionization potential of an ion). For the dynamic polarizability of an ion we then have $\alpha(\omega) = -(e^2/m\omega^2)(Z - Z_i)$, where $Z - Z_i = N$ is the number of bound electrons in the ion (and Ze is the nuclear charge of the ion). For a relativistic incident particle we find the following expression for the cross section:

$$d\sigma^p_{i,\text{bound}} = \frac{16 e_0^2 e^4 (Z - Z_i)^2}{3m^2} \frac{d\omega}{\omega} \ln \frac{\gamma}{R_i(\omega^2 + \gamma^2\omega_p^2)^{1/2}}, \tag{6.47}$$

where

$$Z - Z_i = N; \quad v_0/r_i > \omega > I; \quad \omega_p = \sqrt{4\pi n_i Z e^2/m}.$$

At frequencies $\omega > v_0/R_i$, the cross section $d\sigma_{i,\text{bound}}{}^p(\omega)$ falls off sharply and the main contribution to this form of bremsstrahlung will be from processes involving the ionization of the bound electrons (see Chapter 5).

It is interesting to compare Eq. (6.47) and the formula (6.45) for the cross section for bremsstrahlung on the "cloud" of a plasma ion. These expressions are the same if we make the substitutions $Z - Z_i \to Z_i$, $R_i \to r_d$, and $\omega_{pe} \to \omega_p$. From this it is clear that polarization bremsstrahlung on bound electrons at $\omega \gg I$ is completely analogous to polarization bremsstrahlung on the "cloud" of a plasma ion, except that in this case, the role of the polarization "cloud" is played by an electron "cloud" consisting of the bound electrons of the ion. This correspondence is explained by the fact that when $\omega \gg I$, the bound electrons interact with the electromagnetic field as free electrons, since their eigenfrequencies are very much lower than the frequency of the field. The coupling with the nucleus shows up only in the spatial localization of the electrons in the ion. An analogous situation also exists for the plasma electrons in the Debye sphere surrounding an ion. If, however, $\omega < I$, then the behavior of the bound electrons is quantum mechanical and polarization bremsstrahlung on them has features which differ from polarization (transition) bremsstrahlung on polarization "clouds." In particular, if the frequency of a bremsstrahlung photon is close to one of the eigenfrequencies of the bound electrons, then the cross section for polarization bremsstrahlung on them increases sharply and resonant bremsstrahlung takes place. The analogy between these mechanisms for bremsstrahlung can also be seen in the fact that when $\omega > v_0/r_d$ ($\omega > v_0/R_i$), where large momentum transfers $|\mathbf{q}| > 1/r_d$ ($|\mathbf{q}| > 1/R_i$) are important, the polarization (transition) bremsstrahlung on a plasma ion is small and processes in which the excess energy-momentum is transferred to individual electrons (Compton bremsstrahlung on plasma electrons) predominate. This corresponds to ionization of bound electrons during polarization bremsstrahlung on an ion (atom), when the excess energy-momentum is transferred to the electrons rather than to the nucleus (see Chapter 5).

Polarization (transition) bremsstrahlung on a plasma ion is "elastic" polarization bremsstrahlung on the plasma electrons, while the corresponding Compton bremsstrahlung represents "inelastic" polarization bremsstrahlung on the plasma electrons.

Equation (6.44) differs from the corresponding expression for polarization bremsstrahlung on an ion (atom) in a vacuum only in the part under the logarithm. It follows from this, in particular, that polarization bremsstrahlung of ions in a plasma on bound electrons is not suppressed by the density effect, as in the case of polarization bremsstrahlung on plasma electrons [2]. Both these facts have the same explanation: the photon is emitted by nonrelativistic particles – plasma electrons and bound electrons. Thus, the increased phase velocity of the electromagnetic wave in a plasma is not reflected in the probability of the radiative process. The change in the logarithm in Eq. (6.44) compared to bremsstrahlung in a vacuum is related to the "dressing" of the field of the incident particle in the medium. Here it should be noted that for structured plasma ions, in the frequency range $I \gg \omega \gg \omega_{min}^{(i)}$, the dielectric permittivity $\varepsilon(\omega)$ (6.12) may be even greater than unity. Then the phase velocity of a photon decreases and we should see an effect opposite to the density effect for Bethe–Heitler bremsstrahlung and a factor greater than unity appears in front of the logarithm in the cross section. For polarization bremsstrahlung on bound electrons, this means that as a result of "dressing," the field of the incident particle increases and is not attenuated. Formally, this condition shows up as an increase in the logarithm in Eq. (6.44), which goes to infinity when $v^{-2} = \varepsilon(\omega)$. This means that the scheme we have used for calculating the bremsstrahlung fails when $v^{-2} = \varepsilon(\omega)$, and the process under consideration is the scattering of Cerenkov radiation on bound electrons. For a nonrelativistic incident particle, the medium does not affect polarization bremsstrahlung in the case of a rarefied plasma, and this follows from Eq. (6.44). In particular, interference between the polarization bremsstrahlung on plasma electrons and that on bound electrons does not occur. In fact, the term proportional to the mixed dynamic form factor $S^{(ei)}(q)$ in the total cross section for polarization bremsstrahlung (6.21), which describes this interference, is small. This is explained by the fact that for the momentum transfers $|\mathbf{q}| > \omega_{\min}^{(i)}/v_0 > 1/r_d$, characteristic of polarization bremsstrahlung on bound electrons, the interaction between the plasma electrons and ions is weak, so that energy-momentum exchange does not occur between them ($|\tilde{\varepsilon}_q^{l(e)} - 1| \ll \tilde{\varepsilon}_q^{l}$). Thus, conversion of the field of an incident particle into a real photon on the plasma electrons leads to excitation of the latter. The same process on bound electrons (for $v/R_i > \omega$) is accompanied by the transfer of excess energy-momentum to the ions. Therefore, these processes are uncorrelated and there is no interference between them. We note also that the emission determined by Eqs.(6.45) and (6.47) occurs in different frequency intervals.

Polarization bremsstrahlung on bound electrons in a rarefied plasma is practically identical to the same process on isolated ions (atoms) in a vacuum, and its interference with polarization bremsstrahlung on the plasma electrons can be neglected.

6.6. Polarization Bremsstrahlung in Degenerate Plasmas

In any discussion of polarization bremsstrahlung in real degenerate plasmas, two circumstances must be kept in mind. First, in this case the approximation (6.16) for the wave function is not correct for the excited states of bound electrons, since overlapping of the wave functions of the electronic subsystems of different ions becomes important. Second, as a rule the ions in a degenerate electron plasma cannot be regarded as free and the ion and mixed dynamic form factors generally have a different form. Here we calculate the polarization bremsstrahlung in a plasma with degenerate electrons using a simple model.

We shall assume that the bremsstrahlung frequency of interest to us is such that the contribution of the intermediate states of the bound electrons to this process can be neglected. Then the bremsstrahlung cross section is determined by the wave function of the ground state of the subsystem of bound electrons, when the approximation (6.16) for the wave function can be assumed to be correct in the zeroth approximation. We shall consider the case in which the polarizability of the ion component associated with the motion of the center of mass of the ions is negligible, i.e., we set $|1 - \tilde{\varepsilon}_q^{l(i)}| \ll |\tilde{\varepsilon}_q^{l}|$. Here, however, we assume that $\operatorname{Im} \tilde{\varepsilon}^{l(i)} \neq 0$. We shall use an expression for the static dielectric permittivity of the plasma electrons which coincides with Eq. (6.40b) if the electron Debye radius r_d is replaced by the Fermi–Thomas radius r_{FT}, which describes the shielding effect of the degenerate electrons. Then, neglecting the ion recoil, in place of Eqs. (6.34) for the dynamic form factors we obtain

$$S^{(e)}(q) = (\delta n_e)_q^2 |\tilde{\varepsilon}_q^l|^{-2} + Z_i^2 n_i \delta(q^0)(1 + r_{TF}^2 \mathbf{q}^2)^{-2}, \tag{6.48a}$$

$$S^{(i)}(q) = n_i \delta(q^0), \tag{6.48b}$$

and

$$S^{ei}(q) = Z_i n_i \delta(q^0)(1 + r_{TF}^2 \mathbf{q}^2)^{-2}. \tag{6.48c}$$

The first term in Eq. (6.48a) refers to fluctuations in the electron density which are unrelated to the presence of the ions. We shall neglect it in the following discussion. We are considering the frequency interval $v_0/R_i > \omega \gg I$. The following approximation for the scattering tensor is valid in that interval:

$$c^{lh}(\omega, \mathbf{k}, \mathbf{q}_1) = -\frac{(Z - Z_i)\, e^2 \delta^{lh}}{m\omega^2 (1 + \mathbf{q}^2 R_i^2)}. \tag{6.49}$$

We shall use the fact that for an electron density $n_e \sim 10^{23}\ \text{cm}^{-3}$, the quantities R_i and r_{FT} are of similar orders of magnitude, so that we can set $r_{FT} \sim R_i \sim r$ (where r is some average radius such that $r_{FT} > r > R_i$). Given these remarks, we find the cross section per ion for polarization bremsstrahlung on both types of electrons (delocalized and bound) to be ($v_0/R_i \gg \omega \gg I$)

$$d\sigma_\Sigma^p(\mathbf{k}, \mathbf{q}) = \delta(q^0) \frac{\omega d\omega}{v_0 k^2} [\mathbf{k}\mathbf{A}^{(0)}(\mathbf{q}_1)]^2 \frac{e^4}{m^2 (1 + \mathbf{q}^2 R_i^2)^2} \{Z_i^2 + (Z - Z_i)^2 + $$
$$+ 2Z_i (Z - Z_i)(1 + \mathbf{q}^2 r_d^2)^{-1}\}\, d\Omega_{\mathbf{k}} d\mathbf{q}\, (2\pi)^{-6},\ q^0 = \omega + \mathbf{q}\mathbf{v}_0 - \mathbf{k}\mathbf{v}_0. \tag{6.50}$$

For the spectral polarization bremsstrahlung cross section in this frequency interval we have

$$d\sigma_\Sigma^p(\omega) = \frac{16}{3} Z^2 \frac{e^4 e_0^2}{m^2} \frac{d\omega}{\omega} \ln \frac{\gamma}{r(\omega^2 + \gamma^2 \omega_p^2)^{1/2}}, \tag{6.51}$$

where

$$\omega_p = (4\pi e^2 n_i Z/m)^{1/2}.$$

This expression has an intuitive physical meaning. At frequencies $I \ll \omega \ll v/R_i$, the bound electrons reradiate virtual photons into real photons as free charges localized around a radius of R_i. Therefore, their dynamic properties in radiation are the same as the properties of plasma electrons and the two types of electrons become indistinguishable. The main contribution to polarization bremsstrahlung is from small momentum transfers $|\mathbf{q}| < 1/r_d \leq 1/R_i$ (permitted by the conservation of energy). Thus, all the electronic charge within a Debye sphere, including both plasma and bound electrons, can be regarded as a point charge. As a result, the cross section $d\sigma_\Sigma{}^p$ is proportional to Z^2e^2, the square of the total electronic charge that shields the nucleus of the ion in the plasma. In this case, therefore, the interference between polarization bremsstrahlung on bound and plasma electrons is very important and, since the amplitudes of these processes add coherently, the total cross section increases: $d\sigma_\Sigma{}^p \propto \{Z_i + (Z - Z_i)]^2$.

Evaluating polarization bremsstrahlung for the degenerate electron plasma in a metal is much more complicated because of the presence of a periodic structure which affects both the electronic spectrum and the form of the ion dynamic form factor. It is clear, however, that in this case the polarization bremsstrahlung on bound and delocalized electrons must be taken into account simultaneously.

6.7. Comparison with the Classical Theory of Transition Bremsstrahlung

We now return to the correspondence between the theory of polarization (transition) bremsstrahlung [2, 4] for an arbitrary medium and the results for polarization bremsstrahlung which follow from the above analysis. For two colliding particles α and (0), the classical theory of polarization bremsstrahlung yields the following matrix element for the process [see Eq. (3.33)]:

$$M_i^{\alpha(0)} = 2e_\alpha e_0 S_{ijl}(k,q)\, G_{ls}(q)\, G_{jn}(-q_1)\, v_{0,s} v_{\alpha,n} =$$
$$= \frac{2}{(4\pi)^2} S_{ijl}(k,q)\, E_j^{(0)}(-q_1)\, E_l^{(\alpha)}(q) \propto \varepsilon_{ij}^{nl}(k,q)\, E_j^{(0)}(-q_1), \qquad (6.52)$$

where S_{ijl} is a nonlinear current coefficient (nonlinear susceptibility) which can be calculated for an arbitrary medium. Thus, although the matrix element (6.52) for this process was derived classically in Chapter 3, this result is easily generalized to the case in which the polarization current that emits the photon is created by bound electrons. In fact, the quantum mechanical features of their motion are taken into account in the tensor S_{ijl}, which can be known independently. For bremsstrahlung owing to bound electrons, however, Eq. (6.52) describes another type of radiation than that which was calculated here, when the particles α entered into the composition of the medium. Then the radiating polarization current is determined by the linear susceptibility of the medium. Thus, the amplitude (6.19) can be written in the form

$$M_{fi} = e_h^* M_h^{(p)}, \quad M_h^{(p)} = \frac{1}{4\pi}\left(\delta_{hm} - \varepsilon_{hm}^{(fi)}(\omega, \mathbf{k}, \mathbf{q})\right) A_m^{(0)}(\mathbf{q}), \qquad (6.53a)$$

with

$$\varepsilon_{hm}^{(fi)}(\omega, \mathbf{k}, \mathbf{q}) = \delta_{hm} - 4\pi \frac{e^2}{m\omega^2} n_{fi}^{(e)} \delta_{hm} + 4\pi c_{hm}^{(1,1)}(\omega, \mathbf{k}, \mathbf{q}) n_{fi}^{(i)}(\mathbf{q}_1), \qquad (6.53b)$$

where $\varepsilon_{hm}^{(fi)}$ is the analog of the nonlinear dielectric susceptibility tensor of the medium and takes the microinhomogeneity and the possible excitation of the material into account.

Thus, in this case the polarization current which produces the bremsstrahlung is the linear response of the medium to the field of the incident particle (0); the second particle α, which takes up the excess energy-momentum, enters $\varepsilon_{hm}^{(fi)}$ implicitly.

The correspondence between Eqs. (6.52) and (6.53a) occurs when the influence of the particles α (which are part of the medium) on the radiating electrons can be regarded as a perturbation. This is just what happens for bremsstrahlung owing to plasma electrons.

We note that the expression for $\varepsilon_{hm}^{(fi)}$ in a plasma can be generalized to the case of an arbitrary medium, so that Eq. (6.53a) for the amplitude of the polarization bremsstrahlung remains in force. The quantity $(1/4\pi)(\delta_{hm} - \varepsilon_{hm}^{fi}(k, q))$ describes the conversion of a photon with 4-momentum $q = \{q^0, \mathbf{q}\}$ into a photon with $k = \{\omega, \mathbf{k}\}$ on the electrons of the medium, analogously to the scattering tensor for an atom. Here $\mathbf{q}$ and $\mathbf{k}$ are the wave vectors of the virtual and real photons. Thus, the cross section for polarization bremsstrahlung in a medium cannot be written in the form

$$d\sigma^p(\omega) = A\,|1 - \varepsilon(\omega)|^2 d\omega/\omega \sim n^2,$$

where A is a required factor, n is the particle density in the medium, and $\varepsilon(\omega)$ is the dielectric permittivity at a frequency of ω. In fact, parts of the medium with linear dimensions $l \leq |\mathbf{q}|^{-1}$ make a coherent contribution to the process. Thus, after averaging over a volume $L^3 \gg l^3$, the interference terms disappear and the intensity of the total polarization bremsstrahlung is equal to the sum of the intensities of the radiation from each elementary emitter. As a result, $d\sigma^{(p)}(\omega) \sim n d\omega/\omega$ (we are not considering periodic media). This situation is analogous to the scattering of light in matter: it can be illustrated by considering the cross section (6.21). Let the frequency ω be such that only the emission from bound electrons need be included and let $c_{hm}(\omega, \mathbf{k}, \mathbf{q}_1) = \delta_{km}\theta(1/R_i - }\mathbf{q}_1|)\alpha(\omega)$; then, we have

$$d\sigma^p(\mathbf{k}, \mathbf{q}) = \frac{2\pi}{v_0 k^2}[\mathbf{k}\mathbf{A}^{(0)}(\mathbf{q}_1)]^2 (q_1^0)^4 |\alpha(\omega)|^2 S^{(i)}(q) \frac{\omega d\omega d\Omega_k d\mathbf{q}}{(2\pi)^6}. \qquad (6.54)$$

Here the quantity $S^{(i)}(q) = (1/2\pi)\int_{\infty}^{\infty} dt e^{iq^0 t} \langle \hat{n}^{(i)}(\mathbf{q}, t)\hat{n}^{(i)}(-\mathbf{q})\rangle$ describes the degree of coherence of the contribution of the different ions to the process. When $|\mathbf{q}| \neq 0$, it

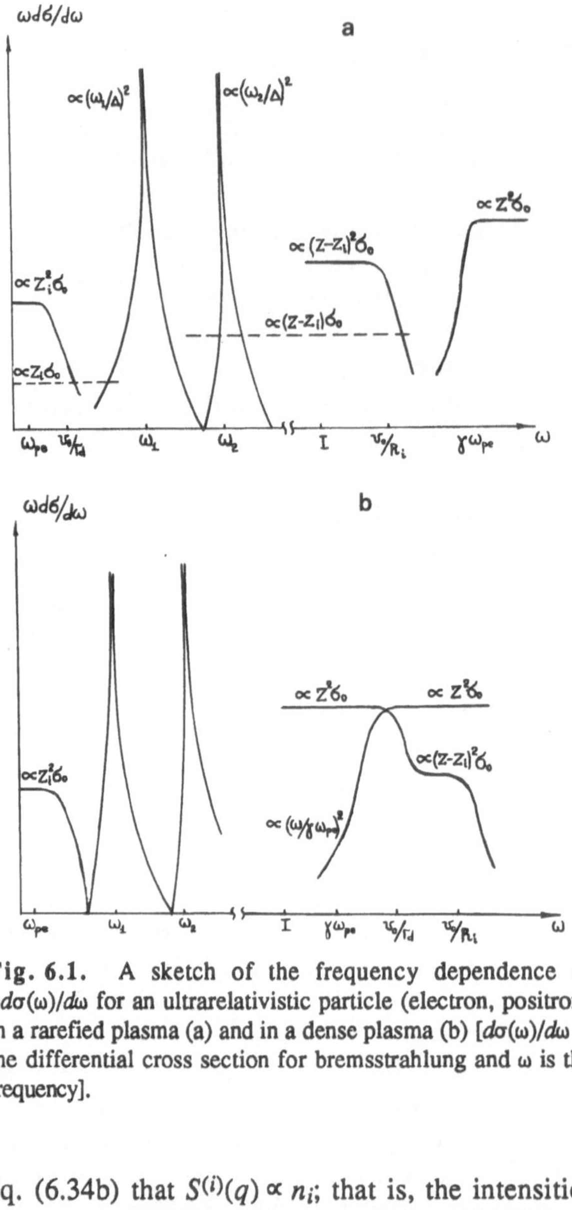

Fig. 6.1. A sketch of the frequency dependence of $\omega d\sigma(\omega)/d\omega$ for an ultrarelativistic particle (electron, positron) in a rarefied plasma (a) and in a dense plasma (b) [$d\sigma(\omega)/d\omega$ is the differential cross section for bremsstrahlung and ω is the frequency].

follows from Eq. (6.34b) that $S^{(i)}(q) \propto n_i$; that is, the intensities of the bremsstrahlung on each ion add up. When $|\mathbf{q}| = 0$, on the other hand, we have $S^{(i)}(q) = \delta(q^0)n_i^2$ and the contribution of the different ions to polarization bremsstrahlung is completely coherent. Note that Eq. (6.34b) is not applicable in this case. We must proceed from the definition of $S^{(i)}(q)$.

In sum, when the nonlinear response function $\varepsilon_{hm}{}^{fi}(\omega, \mathbf{k}, \mathbf{q})$ is used for polarization bremsstrahlung, it is important to include the correct dependence on $\mathbf{k}$ and $\mathbf{q}$.

Polarization bremsstrahlung on the bound electrons of ions is important for frequencies $\omega_{\min}{}^{i} < \omega < v_0/R_i$ and is not suppressed by the density effect for relativistic particles.

6.8. General Picture of Bremsstrahlung

We now offer a characterization of the total bremsstrahlung in plasmas including the traditional (Bethe–Heitler) and polarization mechanisms. We shall assume that the incident particle is an electron.

For a rarefied plasma with $\omega_{\min}{}^{(i)} \gg v_0/r_d$, the picture is especially simple in the ultrarelativistic limit, when $\gamma \gg v_0/R_i\omega_{pe}$. In this, the simplest case, the different mechanisms for bremsstrahlung predominate in different frequency intervals, so that they do not interfere with one another. The corresponding frequency dependence of the quantity $\omega d\sigma/d\omega$ (where σ is the total bremsstrahlung cross section per ion) is shown in Fig. 6.1a. For concreteness, we have assumed that $Z/2 > Z_i > 1$ and $\sigma_0 = e^6/m^2$. The smooth curves represent transitional and elastic bremsstrahlung, while the dashed curves represent inelastic polarization bremsstrahlung (inelastic polarization radiation on bound electrons is not shown for $\omega < I$). The magnitudes of the corresponding cross sections have been written down neglecting the logarithmic factor. We can distinguish three frequency intervals in Fig. 6.1: (a) $v_0/r_d \gg \omega \gg \omega_{pe}$, (b) $v_0/R_i > \omega > \omega_{\min}{}^{(i)}$, and (c) $\omega \gg \gamma\omega_{pe}$, in which polarization bremsstrahlung on plasma electrons, polarization bremsstrahlung on bound electrons, and traditional (Bethe–Heitler) bremsstrahlung, respectively, predominate. As noted above, the reason for the reduction in the polarization radiation on plasma electrons with increasing frequency for $\omega > v_0/r_d$ is that at these frequencies the Debye cloud emits a bremsstrahlung photon incoherently. The increase in the cross section for polarization bremsstrahlung on bound electrons when $\omega > \omega_{\min}{}^{(i)}$ is related to an increase in their polarizability in this frequency range compared to the static limit. The decrease in the cross section for polarization bremsstrahlung on bound electrons ("elastic") when $\omega > v_0/R_i$ is caused by the incoherence with which the short-wavelength photons are emitted by the cloud of bound electrons, analogously to case (a). Maxima are observed in the cross section for polarization radiation on bound electrons at frequencies close to the excitation frequencies of the atom. Traditional bremsstrahlung predominates in interval (c) and is suppressed by the density effect for $\omega < \gamma\omega_{pe}$.

We now consider the case of a dense plasma with $\omega_{\min}{}^{(i)} < v_0/r_d$, when interference should occur between polarization bremsstrahlung on the plasma and that on the bound electrons. This case is illustrated in Fig. 6.1b, for which it is also assumed that $\gamma \gg v_0/R_i\omega_{pe}$. Interference between the two mechanisms for polarization bremsstrahlung can lead to an increase or a decrease (even to a zero level) in the total polarization bremsstrahlung, depending on the sign of the amplitude of the emission from the plasma and bound electrons. If $v_0/r_d > \omega > I$, then "constructive" interference takes place and the polarization bremsstrahlung cross section is proportional to the total electronic charge around the nucleus of an ion in the plasma, which is made up of both plasma and bound electrons, i.e., to the square of the nuclear charge. Note that, although both polarization and traditional bremsstrahlung occur at frequencies $\omega < \gamma\omega_{pe}$, they do not interfere because these mechanisms are important in different ranges of the scattering angle of the incident particles and, in the relativistic case, the angular distribution of the bremsstrahlung photons is also different.

For a heavy incident particle, traditional bremsstrahlung is negligibly small and only polarization bremsstrahlung remains.

In conclusion, we note that polarization bremsstrahlung will be richer and more complicated in a condensed medium, especially one that is periodic [11].

Chapter 7

POLARIZATION BREMSSTRAHLUNG ON MANY-ELECTRON ATOMS AND IN ATOMIC COLLISIONS

M. Ya. Amus'ya

7.1. Introductory Comments

We have already presented the physics of polarization bremsstrahlung in detail in Chapter 1. Here, before proceeding to a concrete discussion of the results concerning many-electron atoms, it is worthwhile to emphasize once again a number of important properties of polarization bremsstrahlung. The internal structure of the particles is important for polarization bremsstrahlung. This structure makes it possible for both the particles of which the medium is made and the particles which are created because of the variability induced by the polarization to become polarized and radiate. In such a generalized formulation, any structured particles can, in principle, emit any other particles (provided this is allowed by the conservation laws). In an interaction involving a structured particle it is natural to distinguish elastic polarization bremsstrahlung, when the colliding particles do not change their internal state, from bremsstrahlung with excitation or inelastic bremsstrahlung. Polarization of structured particles is also possible during a collision process if the particles are electrically neutral on the whole, but have different charges and masses, or interact in different ways with an incident particle. In all these cases, a variable dipole moment is induced in a structured particle during the collision process.

Since the ability of a heavy particle to deform or polarize a partner is generally no less than and, perhaps, even greater than that of a light particle, we should expect that polarization bremsstrahlung caused by a heavy particle would be no less than that caused by a light particle (having the same velocity) but at least of the same magnitude.

Polarization bremsstrahlung is extremely important and shows up in collisions between various kinds of partners: electrons on atoms and nuclei, atoms on atoms, electrons, mesons, and nucleons on atoms and ions, and nuclei on one another.

Just 10 years ago it was generally accepted [1] that bremsstrahlung is inherent only in charged particles. Now, however, it is certain that in all cases where particle collisions will be accompanied by real or virtual excitation, polarization bremsstrahlung will make a significant contribution to the total bremsstrahlung spectrum. Obviously, the contribution from the discrete excitation of a target atom to the narrow lines in the bremsstrahlung spectrum produced by collisions of electrons with atoms was discussed some time ago [2]. For resonant particles whose energies coincide with those of the levels of an atom, however, bremsstrahlung actually proceeds through a two-step mechanism: initially an atomic level is actually excited by electron impact and then, after a significant time on an atomic scale, the atom goes into the ground state, emitting a photon.

An incident electron, however, can shift an atom into a short-lived (virtual or real) state that does not correspond to discrete energy levels of the atom, but to its continuum. This state will decay shortly after (in fact, simultaneously with) its formation, resulting in a single polarization bremsstrahlung process. Polarization bremsstrahlung as a single process in which the entire spectrum of excitations, including virtual excitation, rather than individual excitations of the target (and only real excitations) is important was first examined for the case of fast electrons on hydrogen [3] and for slow electrons on the noble gas atoms [4, 5]. Subsequently, an expression was obtained for the differential (with respect to the scattering angle for a fast electron) bremsstrahlung cross section including the contribution of polarization radiation for arbitrary atoms [6]. It has been shown that, because of the contribution from the "atomic" component, the bremsstrahlung spectrum as a function of the frequency ω of the emitted photon should have peaks whose locations and intensities are directly related to the intensities of the maxima in the dipole dynamic polarizability of a target atom. For small scattering angles of fast incident electrons, the contribution of polarization bremsstrahlung in the neighborhood of the peaks is considerably greater than that of ordinary bremsstrahlung.

For incident electron energies of tens and hundreds of electron volts, the contribution of polarization bremsstrahlung is immense when the emitted frequency is close to the ionization potential of one of the atomic shells. Polarization bremsstrahlung becomes especially important if the electron in the final state is slow and is scattered resonantly in the field of a virtually excited target atom. This mechanism has been used to explain the sharp increase in the photon yield at the edge of the spectrum of solid barium, lanthanum, and cerium irradiated by electrons with energies close to the threshold for the $3d^{10}$-shells of these elements [7].

The formula for the total bremsstrahlung cross section for fast electrons obtained by Zon [8] and Amus'ya et al. [9] has been used to interpret [9] experimental data on photon emission during the interaction of electrons with metallic lanthanum at frequencies close to the ionization potential of its $4d^{10}$-subshell [10]. The bremsstrahlung cross section (including polarization bremsstrahlung) has been obtained for Ar, Xe, and La atoms [11, 12] by treating the incident electron in the Born approximation.

The experimentally observed emission peak at energies close to the ionization potential of the $4d^{10}$-subshell of xenon for incident electrons with energies of about 1 keV [13] has been explained with the aid of these calculations.

Besides predicting the maxima associated with the dependence of the dipole dynamic polarizability of target atoms on the photon frequency, these studies of the bremsstrahlung spectrum of fast electrons have predicted the "stripping" of electronic subshells (see [11, 14] and Chapter 4).

The contribution of polarization bremsstrahlung at moderate frequencies may be greater for relativistic electrons [15–17]. This feature is explained by the fact that the interaction of the incident electron with the atoms is stronger: the virtual photon with which they undergo exchange becomes almost real and the process approaches a resonant process in which the photon emitted by the incident electron is absorbed by the atoms, thereby causing polarization of the atomic electron shells with subsequent emission of a bremsstrahlung photon by the latter. The probability of resonance processes is especially large.

As noted in Chapter 4, the total bremsstrahlung spectrum is different for electrons and positrons moving at high velocities [18, 19], unlike the scattering cross section, which is independent of the sign of the charge of fast particles (the same effect appears in plasmas during interactions of particles "dressed" in polarization clouds). The difference between the bremsstrahlung spectra at high frequencies disappears for relativistic electrons and positrons [15].

Polarization bremsstrahlung, as already noted, is important for ions (both negatively and positively charged) scattered on atoms, as well as for electrons.

A heavy charged particle polarizes a target atom at least as efficiently as an electron. Thus, for example, at certain frequencies the polarization bremsstrahlung of a proton is of the same order of magnitude as the polarization bremsstrahlung of an electron with the same velocity. Since even a neutral structured particle (an atom) polarizes a target, its bremsstrahlung is of the same order as the bremsstrahlung of an electron [20]. When two atoms moving at velocities large compared to the intraatomic velocities collide, radiation is produced as a result of the polarization of one atom by the static field of the other. At distances on the order of the radius of an atom or less, the field created by an atom is far from zero, so that the polarization it induces in the other atom is significant and fully comparable to the field created by an incident charged particle. This means that the polarization bremsstrahlung and, therefore, the total bremsstrahlung spectrum from collisions of an electron, heavy charged particle, or neutral atom with a target atom is of the same order for equal velocities of the colliding particles [21].

On the whole the mechanism for production of radiation during collisions of neutral atoms is close to that which leads to intense emission from heavy particles in plasmas, which, shielded by the free electrons, essentially form a unique kind of neutral atoms [22].

Bremsstrahlung produced in collisions involving structured particles can be accompanied by the excitation or ionization of the particles, i.e., it may be inelastic. The secondary electrons produced during ionization of colliding particles are a source of radiation, which may be even more intense than that of the heavy collision partners. However, it appears [23] that in some frequency ranges (and these are extremely important) the contribution from elastic bremsstrahlung will predominate and the effect of the inelastic processes will be small (see Chapters 6 and 7). We note that for high frequencies it is possible to obtain the bremsstrahlung cross section including the contributions from elastic and inelastic processes. In this chapter we shall discuss the polarization bremsstrahlung of many-electron atoms in detail, describing the atoms in terms of its polarization under the influence of an incident particle and the radiation in terms of the self-consistent Hartree–Fock approximation for the field. For this purpose we shall use the appropriate Feynman diagram techniques and employ the atomic system of units.

7.2. Many-Body Diagram Techniques for Bremsstrahlung

We shall use the apparatus of many-body quantum theory and Feynman diagrams [24]. For concreteness we discuss the bremsstrahlung of electrons on atoms. One can examine the bremsstrahlung of nucleons on nuclei in an analogous fashion (see Chapter 8). It is simplest just to consider atoms with closed shells as a target, since the simplest excitations for them involve the transition of an electron from an initially occupied level to a free one (including continuum states in this case), i.e., the creation of an electron–hole pair.

The graphical symbols in conventional use include the following: a continuous line with an arrow to the right denotes both an incident electron and an atomic electron in a free level of the target atom; a line with an arrow pointing to the left denotes a vacancy. The wave functions of the electron and hole are assumed to be calculated in the Hartree–Fock self-consistent field approximation, which is taken as an initial approximation for describing the bremsstrahlung process to be discussed in this chapter. A vertical wavy line denotes the Coulomb interaction between the incident and atomic particles (i.e., an electron and a nucleus). A dashed line represents an emitted or absorbed photon.

At low electron energies it is more convenient to consider the inverse bremsstrahlung absorption of photons, rather than the direct bremsstrahlung process. In the language of diagram techniques, the amplitude of electron bremsstrahlung is rep-

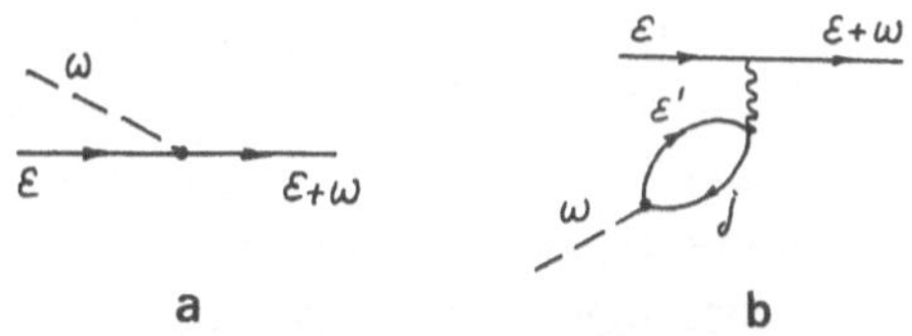

Fig. 7.1. Simplest diagrams describing the amplitude of ordinary bremsstrahlung absorption (a) and polarization bremsstrahlung absorption (b).

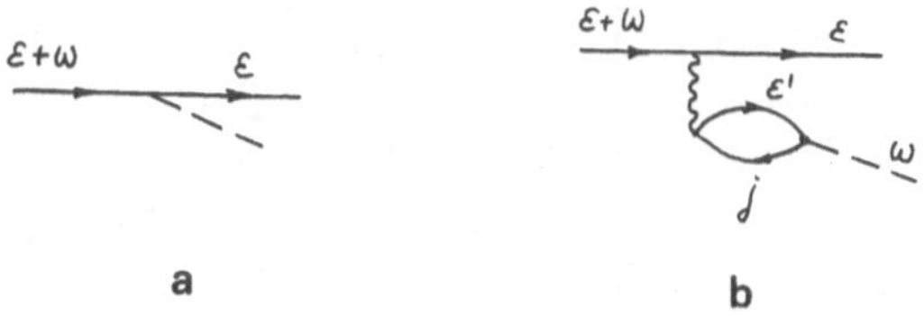

Fig. 7.2. Simplest diagram describing the amplitude of bremsstrahlung.

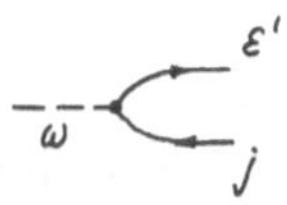

Fig. 7.3. Simplest diagram describing the absorption of a photon by an atom.

resented by the diagram of Fig. 7.1a. Here the point represents the interaction operator of the electron with the photon.

The simplest example of a diagram representing the amplitude of polarization bremsstrahlung absorption is shown in Fig. 7.1b. This diagram represents the absorption of a photon when an atomic electron undergoes a transition into an excited state, i.e., an electron–hole pair (real or virtual) is created, and then, as it returns to its initial state, the electron transfers energy to the accelerated electron.

The diagrams similar to these for bremsstrahlung (emission) are shown in Fig. 7.2. The amplitude of atomic bremsstrahlung (of which an example is shown in the diagram of Fig. 7.2b) represents a process in which the slowed-down electron is scattered by a target atom with excitation (real or virtual) of an electron in the atom, while the latter emits a photon after returning to its initial state.

As can be seen from the following discussion, polarization bremsstrahlung is important if $\omega \geq I$, where I is the ionization potential of the atom. We shall be interested in precisely these values of ω. In this frequency range it is important to include the interaction of the electron and the vacancy, which create a photon as they undergo annihilation. This part of the diagram, which is illustrated in Fig. 3, describes the photoionization amplitude. It is, therefore, natural to include the interaction of the electron ε' and hole j in the random phase approximation with exchange [25], which yields good agreement with experiment, just as in the case of the photoionization of the outer and intermediate shells. In the language of diagrams, the random phase approximation with exchange includes an infinite sequence (Fig. 7.4) that intuitively describes the photoionization process as evolving in time: first a photon is absorbed and then one, two, etc., excitation events occur in which an electron–hole pair undergoes a transition from one state to another. In our figures, the time is regarded as increasing from left to right.

A technique has been developed for summing an infinite series of these diagrams, which, besides the terms shown in Fig. 7.4, includes all the exchange terms obtained by replacing one or several of the elements which describe the annihilation of a pair followed by the creation of a pair (Fig. 7.5a) by an element that corresponds to scattering of an electron on a hole (Fig. 7.5b).

This technique is also used to account for the so-called time-reversed diagrams, in which the order of the interaction between a photon and (or) electron is switched in time. For example, to the first order with respect to the electron interaction, the diagram of Fig. 7.6b is included along with the diagram shown in Fig. 7.6a.

The corrections to the random phase approximation with exchange do yield significant changes in the dipole dynamic polarizability of a target atom, as well as in the photoabsorption (and photoemission) amplitude [25]. They usually reduce the latter at the ionization threshold relative to its value for purely Coulomb wave functions of the atomic electrons. We shall take these corrections into account in our calculations of polarization bremsstrahlung.

For intermediate shells, especially those near their thresholds for photoionization, it is important that other corrections related to the readjustment of the outer electron shells during ionization (real or virtual processes) be taken into account, as well as the corrections to the random phase approximation with exchange.

This readjustment means that the photoelectron ε'' moves in the self-consistent Hartree–Fock field of an ion with a vacancy j', and not simply in the field of the vacancy j' (see the diagrams of Fig. 7.4). Figure 7.7a shows diagrams which take this effect into account (here j'' refers to the vacancies in the outer electronic subshells of the atom).

This readjustment usually reduces the photoionization amplitude in the near-threshold region and yields satisfactory agreement with experimental data [26] for intermediate shells. Near the thresholds for the inner shells, besides the readjustment one must include the possible decay of the vacancy, either by an Auger mechanism, i.e., with emission of an electron, or by radiative decay, with the emission

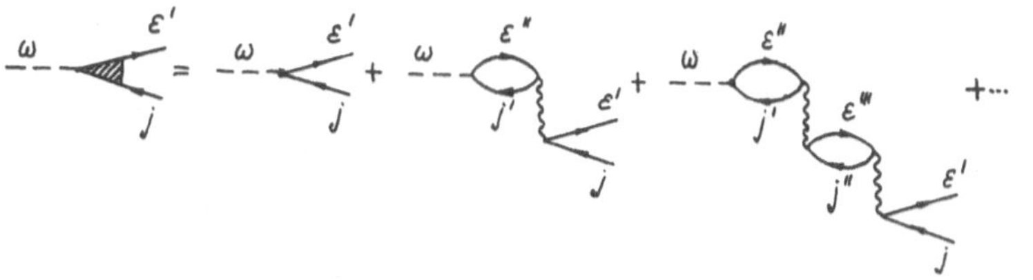

Fig. 7.4. Sequence of diagrams describing the amplitude of photoabsorption in the random phase approximation.

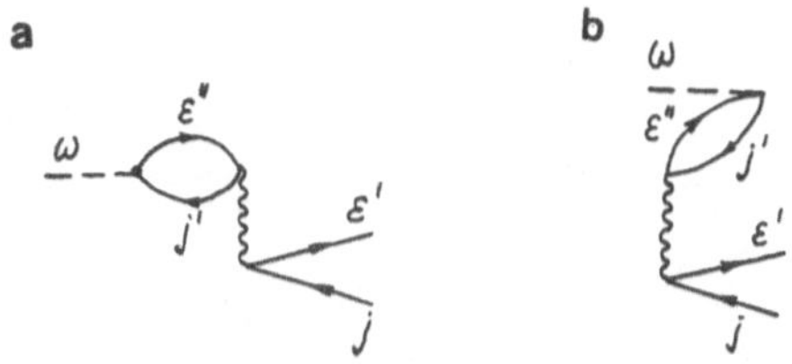

Fig. 7.5. Simplest diagrams for the annihilation and creation of an electron–hole pair (a) and for scattering of an electron on a hole (b).

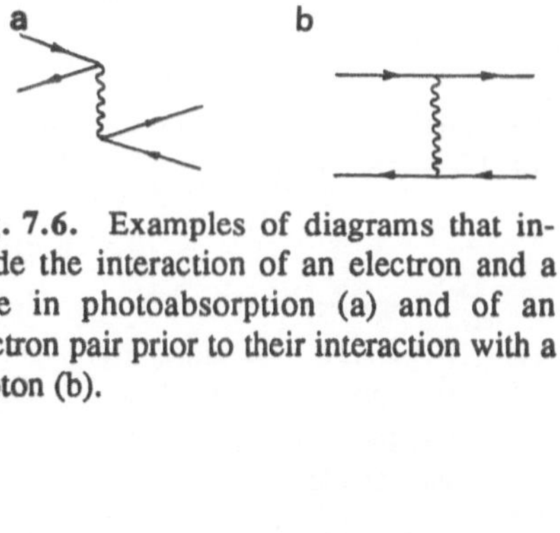

Fig. 7.6. Examples of diagrams that include the interaction of an electron and a hole in photoabsorption (a) and of an electron pair prior to their interaction with a photon (b).

of a photon. Including decay, as opposed to the readjustment, leads to an increase in the photoionization amplitude near the threshold [27]. Satisfactory agreement with experiment is obtained for the inner shells only when both effects are taken into account.

If the energy of an incident electron in its final or initial state is low, then it may be necessary to include radiationless polarization of the target atom by the incident electron. In the language of diagrams, the simplest example of a suitable cor-

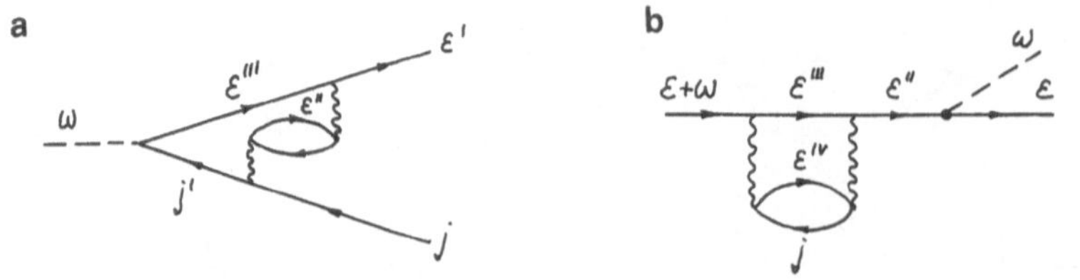

Fig. 7.7. Examples of diagrams that include the interaction of an electron and a hole in photoionization through excitation of the other electrons (a) and through the radiationless polarization of a target atom during bremsstrahlung (b).

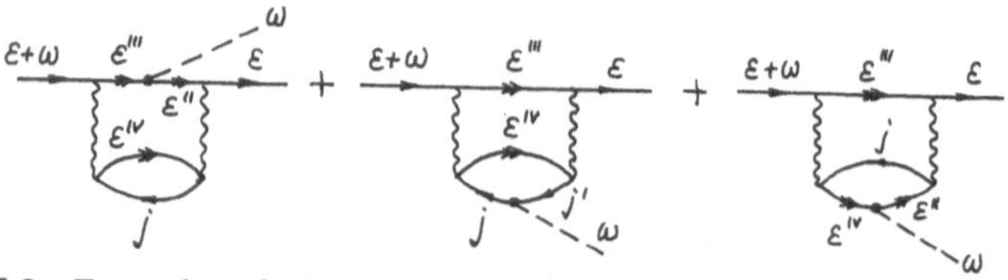

Fig. 7.8. Examples of diagrams that include the emission of a photon by a compound "electron + atom" system.

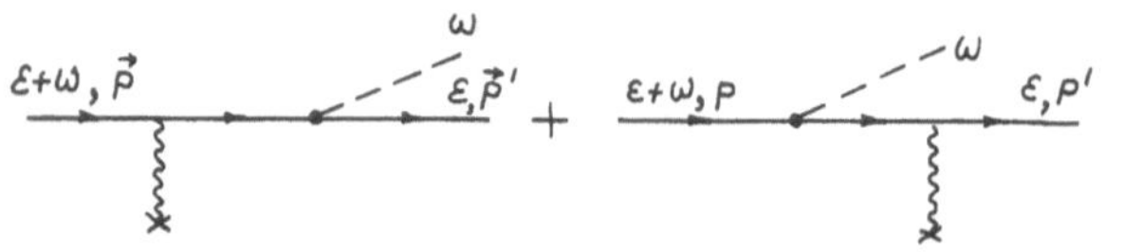

Fig. 7.9. Diagrams describing the bremsstrahlung of fast electrons in the field of an atom.

Fig. 7.10. Notation for the propagation of any structureless particle other than an electron.

rection is illustrated in Fig. 7.7b. It is understood that exchange may also occur between the ε^{III} and ε^{IV} electrons in the intermediate state of the diagram.

In order to describe radiationless polarization of an atom by a slow electron it is sufficient to calculate the corresponding potential by second-order perturbation theory in terms of the interelectronic interaction, as is confirmed by the agreement between computational and experimental data on elastic scattering [25].

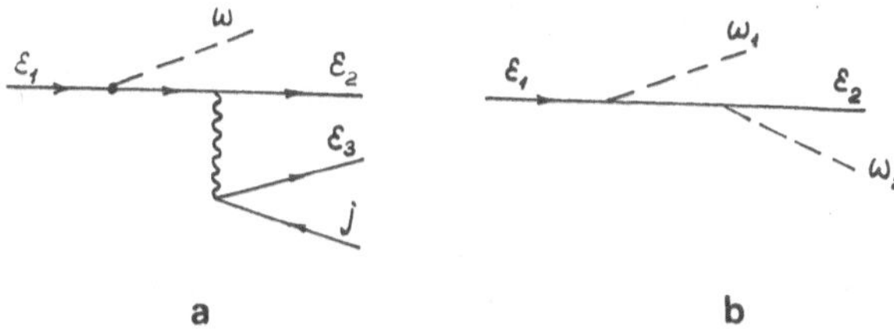

Fig. 7.11. Examples of diagrams that describe inelastic bremsstrahlung.

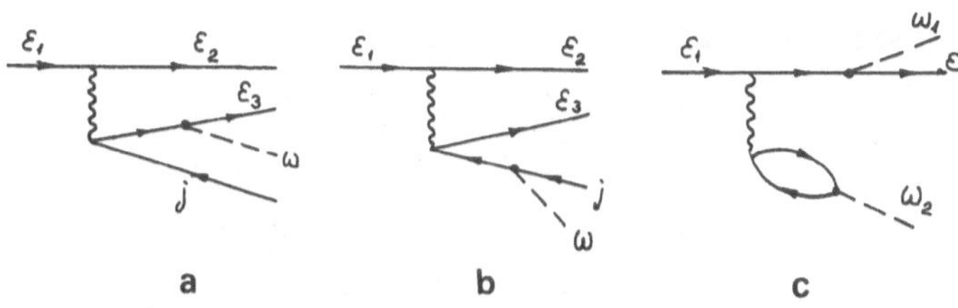

Fig. 7.12. Examples of diagrams that describe inelastic processes accompanying polarization bremsstrahlung.

The intermediate state in Fig. 7.7b ($\varepsilon^{\mathrm{III}}$, $\varepsilon^{\mathrm{IV}}$, and j denote two excited electrons and a hole) may be long-lived and quasistable. In this case, not only is the radiation from the electron or atom important, but so is the radiation from this state itself (the electrons or vacancies that form it), a compound "electron + atom" system. The simplest examples of diagrams for these processes are shown in Fig. 7.8. Quasistationarity (or discreteness of the level to which the electron is excited) of an intermediate state is denoted by a double arrow.

We note that, in general, a sum must be taken over all the intermediate states (of electrons and holes) in each diagram.

On the whole, radiationless polarization of a target atom by an incident electron means that, besides being subject to the short-range Hartree–Fock potential, the electron will also be acted on by the long-range, so-called polarization potential, which at large distances r from an atom has the form $\alpha_d/2r^4$, where α_d is the dipole polarizability of the atom.

If the energy of the electron in the final state is small and ω is close to the ionization potential of one of the atomic shells, then one must include the interaction of the incident electron with the excited atom in diagrams for polarization bremsstrahlung similar to those shown in Fig. 7.2b.

Fig. 7.13. Notation for the propagation of a structured incident particle.

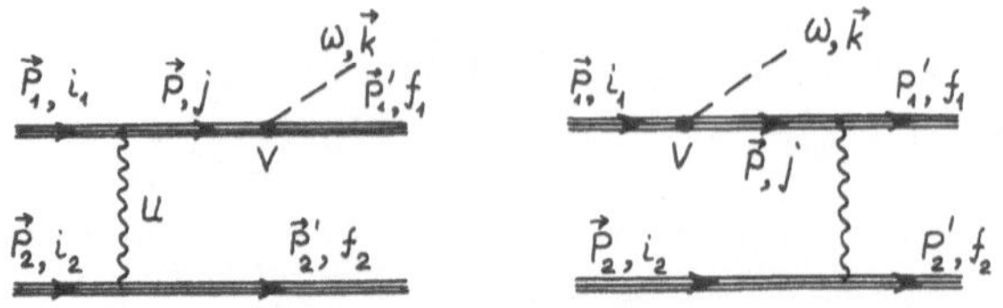

Fig. 7.14. Diagrams describing bremsstrahlung of fast structured particles.

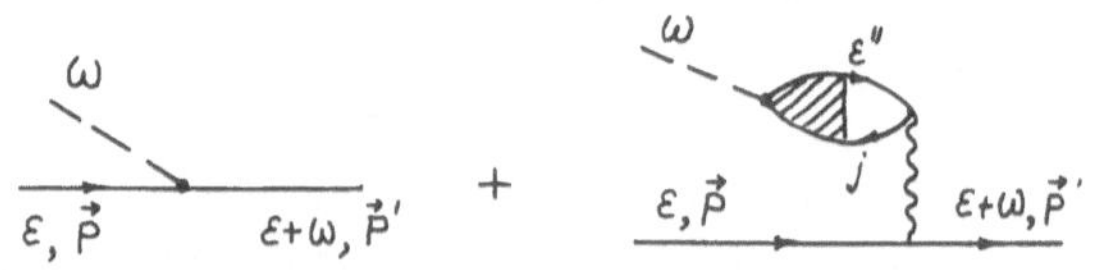

Fig. 7.15. Diagrams depicting the bremsstrahlung amplitude of slow electrons.

It is sufficient to treat the interaction of a fast electron with an atom by perturbation theory in the lowest order with respect to the potential of the atom, so that the amplitude of ordinary bremsstrahlung is described by two diagrams (Fig. 7.9), where **p** and **p**' are the momenta of the electron in its initial and final states. In this figure the wavy line with a cross at its end represents the interaction of the incident electron with the total static field of the atom. At high energies, the wave function of the incident electron must be considered to be a plane wave. In the absence of an external field, the creation or emission of a photon by a free electron is forbidden by the conservation of energy and momentum.

Diagram techniques can also be used to describe bremsstrahlung from any structureless particle, and not just electrons. We shall represent these "other" particles by a double line (Fig. 7.10).

Until now we have only considered diagrams for elastic bremsstrahlung. Inelastic bremsstrahlung differs in that either the incident particle or an atomic electron knocks out an electron from the target atom or emits a second bremsstrahlung photon.

The diagram of Fig. 7.11 serves as an example of inelastic electron bremsstrahlung. Besides the emission of a bremsstrahlung photon, the diagram of

Fig. 7.11a describes the knocking of an electron with energy ε_3 out of an atomic subshell in the lowest approximation. The diagram of Fig. 7.11b describes the creation of two photons by a single electron. For a fast incident electron, it is sufficient to include its interaction with the target atom leading to ionization in the lowest order of perturbation theory with respect to the interelectronic interaction, as shown in the diagram of Fig. 7.11a.

Inelastic processes can also accompany "atomic" bremsstrahlung, and the diagrams of Fig. 7.12 include some examples.

The diagrams of Fig. 7.12a, b describe the creation of an electron–hole pair by an incident electron followed by the emission of a bremsstrahlung photon by one of the particles of the pair.

It seems reasonable that in any diagram describing a real physical process (elastic or inelastic bremsstrahlung), energy should be conserved. As an example, for the diagrams of Fig. 7.12a, b this means that

$$\varepsilon_1=\varepsilon_2+\varepsilon_3+\omega+\varepsilon_j.$$

In the diagrams shown above, the indices ε (ε') and j (j') denote all the quantum numbers of the states of the electrons and holes, respectively: not only the energy or principal quantum number, but also the angular or total momentum, spin, and their projections.

We shall represent a structured incident particle by a triple line (Fig. 7.13) and an interaction between it and another structured particle (including actually pairwise interactions of the elements of both structured particles), by a vertical wavy line. The bremsstrahlung amplitude of fast colliding structured particles is shown in the diagrams of Fig. 7.14, to which we should add two others in which a photon ω is emitted by the second particle. In the diagram, $\mathbf{p}_1$, $\mathbf{p}_2$, $\mathbf{p}_1'$, and $\mathbf{p}_2'$ are the momenta of the colliding particles at the beginning and end of the process, respectively, while i_1 (i_2) and f_1 (f_2) denote the internal states of the structured particles (atoms or ions).

Between the emission of a photon and the interaction, the momentum $\mathbf{p}$ of a particle and its internal state j may differ from the initial values.

The diagrams shown in this section are used to compose analytic expressions in accordance with certain rules known as the correspondence rules [24, 25] and we shall be making use of this relationship.

7.3. Bremsstrahlung of Slow Electrons on Atoms

In accordance with Figs. 7.1 and 7.4, bremsstrahlung absorption by slow electrons is described by a set of diagrams [4]. The corresponding analytic expression for the amplitude $F_{\varepsilon,\varepsilon+\omega}$ is

$$F_{\varepsilon,\varepsilon+\omega} = \langle \varepsilon | (\mathbf{ed}) | \varepsilon + \omega \rangle + 2 \sum_{\substack{j \leq F \\ \varepsilon'' > F}} \langle \varepsilon j | u | \varepsilon + \omega, \varepsilon'' \rangle \times$$

$$\times \frac{(\varepsilon'' + I_j) \langle \varepsilon'' | (\mathbf{e}D(\omega)) | j \rangle}{[\omega^2 - (\varepsilon'' + I_j)^2]}. \tag{7.1}$$

Here $\mathbf{e}$ is the polarization vector of the photon; $\mathbf{d} = \mathbf{r}$ (or $i\nabla/\omega$) is the dipole moment operator of the electron; and I_j is the ionization potential of the jth subshell. The sum is taken over the states that are occupied in the target atom ($j \leq F$) and the free states, including an integral over the continuum. The amplitude $D(\omega)$, which summarizes the sequence of diagrams in Fig. 7.4, is found by solving the integral equation in the random phase approximation with exchange. In Eq. (7.1), $\langle \varepsilon i|$ and $|\varepsilon + \omega, \varepsilon'')$ is the set of direct and exchange Coulomb matrix elements for the interaction between an atomic electron and a free electron. In Eq. (7.1), $\langle \varepsilon|$, $\langle i|$, etc. denote the wave functions of the electrons and vacancies.

The differential cross section for bremsstrahlung absorption, $d\sigma_{l,l'}^{abs}(\omega, \Omega_{\mathbf{k}})$, is related to the amplitude $F_{\varepsilon,\varepsilon+\omega}$ by the formula [4] (in atomic units)

$$d\sigma_{ll'}^{abs}(\omega, \Omega_{\mathbf{k}}) = \frac{\omega}{3cp^3p'} | F_{\varepsilon,l;\varepsilon+\omega,l'} |^2 \, d\omega d\Omega_{\mathbf{k}}. \tag{7.2}$$

Here $\mathbf{p}$, $\mathbf{p}'$, l, and l' are the momentum and angular momentum of the accelerated (because it has absorbed a photon) electron in the initial and final states, respectively, and $d\Omega_{\mathbf{k}}$ is the solid angle within which the photon leaves.

Since the photon corresponds to a dipole transition, l and l' differ by unity. In Eq. (7.2) the integral is taken over all directions of the momentum $\mathbf{p}'$.

For very small ω, the first term predominates in Eqs. (7.1) and (7.2). For low electron energies, distances that are large compared to the size of the atom make an important contribution to the matrix element $\langle \varepsilon|(\mathbf{ed})|\varepsilon + \omega\rangle$, so that $\varphi_\varepsilon(\mathbf{r})$ and $\varphi_{\varepsilon+\omega}(\mathbf{r})$ may be replaced by their asymptotic values. If ε and $\varepsilon + \omega$ are small, then we can limit ourselves to the term for the s-phase $\delta_0(l = 0)$, alone, in the field of the neutral atom, so that the asymptotic form of the radial part of the wave function $\varphi_\varepsilon(r)$ is $\sqrt{2/\pi}(1/r)\sin(pr + \delta_0)$ for $l = 0$ and $\sqrt{(p/r)}J_{l+1/2}(pr)$ for $l \neq 0$, where $J_{l+1/2}(pr)$ is the Bessel function of half-integer order corresponding to free motion.

Integrating over $\mathbf{r}$, we obtain

$$\langle \varepsilon | (\mathbf{ed}) | \varepsilon + \omega \rangle = \frac{4\pi pp'}{\omega^2} \left[\frac{(\mathbf{ep}')}{p} \sin \delta_0 - \frac{(\mathbf{ep})}{p'} \sin \delta_0' \right] \tag{7.3}$$

for the bremsstrahlung absorption amplitude of an electron in the field of an atom, where $p = \sqrt{2\varepsilon}$; $p' = \sqrt{2(\varepsilon + \omega)}$; and δ_0 and δ_0' are the phases of the free electron in its initial and final states, respectively.

As $\omega \to 0$, the second term in Eq. (7.1) ceases to depend on ω at all, which is also confirmed by the dominance of ordinary bremsstrahlung at low ω. In evaluating the second term of Eq. (7.1), we shall use the fact that the wave function of the free electron is "smeared out" over a volume much greater than that of the atom. If scattering does not distort the wave function of the free electron too much at these distances, then the wave function can be replaced by a plane wave. As a result, the second polarization term in Eq. (7.1) is equal to $-8\pi pp'[(\mathbf{eq})/q^2]\,\alpha_d(\omega)$ ($\mathbf{q} = \mathbf{p}' - \mathbf{p} + \mathbf{k} \approx \mathbf{p}' - \mathbf{p}$), so that the total amplitude for bremsstrahlung is (a similar expression has been obtained by Zon [28])

$$F_{\varepsilon,\varepsilon+\omega} = \frac{4\pi pp'}{\omega}\left[\frac{1}{\omega}\left(\frac{(\mathbf{ep}')}{p}\sin\delta_0 - \frac{(\mathbf{ep})}{p'}\sin\delta_0'\right) + 2\omega\frac{(\mathbf{eq})}{q^2}\alpha_d(\omega)\right], \tag{7.4}$$

where $-\mathbf{q} = \mathbf{p} - \mathbf{p}'$ and $\alpha_d(\omega)$ is the dipole polarization of the atom, which is given by

$$\alpha_d(\omega) = \sum_{\substack{\varepsilon' > F \\ j \leqslant F}} \frac{2(\varepsilon' + I_j)\langle j|(\mathbf{e}r)|\varepsilon'\rangle\langle\varepsilon'|(\mathbf{e}D(\omega))|j\rangle}{(\varepsilon' + I_j)^2 - \omega^2}. \tag{7.5}$$

It follows from Eq. (7.5) that for low ω, $\alpha_d > 0$ and depends only weakly on the frequency. If ω is considerably higher than $p^2/2$ and p is small, then $\mathbf{ep}' \approx -\mathbf{eq}$ and $\sin\delta_0/p \sim a$, where a is the scattering length for the electron on an atom [29], so that Eq. (74) becomes much simpler and has the form

$$F_{\varepsilon,\varepsilon+\omega} = [4\pi pp'(\mathbf{ep}')/\omega][a/\omega + \alpha_d(\omega)]. \tag{7.6}$$

For negative scattering lengths, which correspond to an effective attraction of the accelerated electron to the atom, $F_{\varepsilon,\varepsilon+\omega}$ goes to zero for $\omega < I$, where $\alpha_d(\omega) > 0$ at the point $\omega_{min} \approx |a|/\alpha_d(0)$, which, for example, is 0.26 Ry for argon [$a = -1.4$, $\alpha_d(0) = 10.7$] and 0.4 Ry for xenon [$a = -5.7$, $\alpha_d(0) = 28$].

For positive scattering lengths corresponding to effective repulsion of the accelerated electron from the atom, $F_{\varepsilon,\varepsilon+\omega}$ does not go to zero when $\omega < I$. When $\omega > I$ the polarizability of the atom becomes a complex quantity, whose real part changes sign with increasing ω and only $\operatorname{Re} F_{\varepsilon,\varepsilon+\omega}$ can go to zero. In this case, however, according to Eq. (7.5) the nonzero imaginary part $\operatorname{Im} F_{\varepsilon,\varepsilon+\omega}$ is proportional to $\sigma_\gamma(\omega)$:

$$\operatorname{Im}\alpha_d(\omega) = (c/4\pi\omega)\sigma_\gamma(\omega), \tag{7.7}$$

where c is the speed of light and $\sigma_\gamma(\omega)$ is the photoionization cross section of the atom.

With the aid of Eqs. (7.2) and (7.6), the cross section for bremsstrahlung absorption can be written in the form

$$\sigma^{abs}(\omega) = \sigma^{abs(0)}(\omega)\,[1 + (\omega/a)\alpha_d(\omega)]^2, \tag{7.8}$$

where $\sigma^{abs(0)}$ corresponds only to the first term in the amplitude (7.1). It is clear that the contribution of polarization (atomic) bremsstrahlung absorption, which is given by the term $\eta \equiv (\omega/a)\alpha_d(\omega)$ and is small for $\omega\alpha_d(\omega)/a \ll 1$, increases rapidly with increasing ω. If $|(\omega/a)\alpha_d(\omega)| \geq 1$, then polarization (atomic) bremsstrahlung predominates.

The quantity $\alpha_d(0)$ can be estimated fairly well by using the formula $\alpha_d(0) \sim N_{ext}/I^2$, where N_{ext} is the number of outer electrons and I is the ionization potential of the atom. The scattering length $|a| \sim \langle r \rangle$, while $I \propto \langle r \rangle^{-1}$, so that $|\eta| = |(\omega/a)\alpha_d(\omega)| \sim N_{ext}\omega/I$, and when $\omega \sim I$ the contribution of polarization bremsstrahlung is already large, with $\eta \sim N_{ext} > 1$.

Note that when $a < 0$, the cross section $\sigma^{abs}(\omega)$ goes to zero at ω_{min}, while when $a > 0$ it is equal to

$$\sigma^{abs}(\omega_0) = \sigma^{abs(0)}(\omega_0)\, c^2\sigma_\gamma^2(\omega_0)/(4\pi a)^2. \tag{7.9}$$

at the point $\omega_0 = a/\mathrm{Re}\,\alpha_d(\omega_0)$.

A negative scattering length is obtained for complicated atoms only if the polarization of the atom by an incident electron is taken into account, that is, outside the framework of the Hartree–Fock approximation.

Amus'ya et al. [4] have made an accurate numerical calculation of the amplitude (7.1) and cross section (7.2), integrated over the directions of the momentum p', for electrons with energies of $\varepsilon = 0.01$ and 0.09 Ry that have been accelerated as a result of the absorption of a photon in the field of argon and xenon atoms [4]. According to an estimate based on Eq. (7.6), the amplitudes of ordinary (electronic) and polarization (atomic) bremsstrahlung have opposite signs when $\omega \sim I$. The results of the calculations for Ar are shown in Fig. 7.16. The increase in σ^{abs} as $\omega \to I_{3p\to3d}$ (where $I_{3p\to3d}$ is the excitation energy of the resonance level) is caused by the fact that the denominator of Eq. (7.1) goes to zero.

7.4. Bremsstrahlung of Fast (but Nonrelativistic) Electrons

The bremsstrahlung amplitude of fast electrons can be calculated in the Born approximation with the aid of the diagrams in Figs. 7.1b, 7.2, and 7.4 [6], where the wave function of the incident electron is treated as a plane wave. An analytic expression for the bremsstrahlung amplitude can be obtained either from the diagrams of Fig. 7.17 by using the correspondence rules or directly from the analytic expression (7.1) if the wave function of the free electron in the second term is replaced by plane waves and the first-order correction in the interaction of the fast electron with the static field of the atom,

$$W(\mathbf{r}) = -Z/r + \int (1/|\mathbf{r}-\mathbf{r}'|)\rho(\mathbf{r}')\,d\mathbf{r}' \tag{7.10}$$

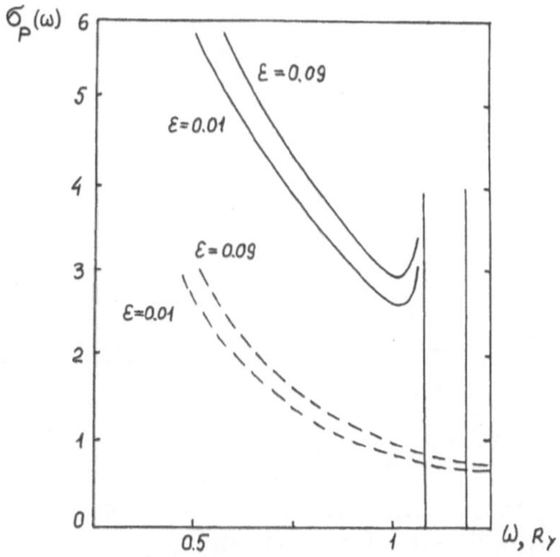

Fig. 7.16. Cross section for absorption of a photon by a slow electron on Ar [4] including polarization bremsstrahlung (1) and ordinary bremsstrahlung (2).

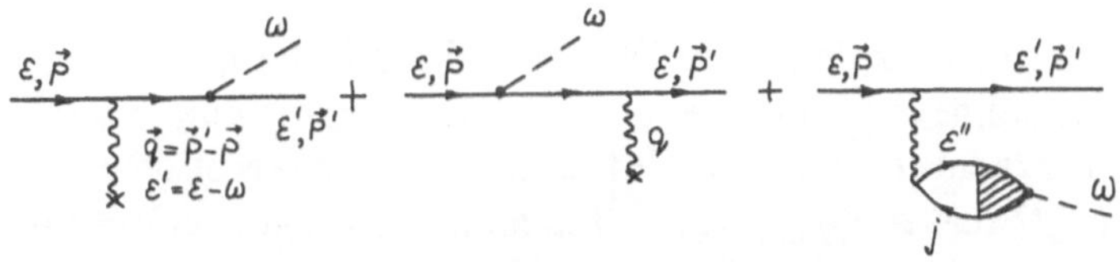

Fig. 7.17. Diagrams depicting the bremsstrahlung amplitude of fast electrons.

is included in the first term. Here Z is the nuclear charge and $\rho(\mathbf{r})$ is the electron density.

As a result, we obtain the following expression for the bremsstrahlung amplitude:

$$F_{fq}(\omega) = -\frac{W(q)\,(\mathbf{eq})}{\omega} + i\,\frac{4\pi\omega}{q^2} \times$$
$$\times \sum_{\substack{\varepsilon'' > F \\ j \leqslant F}} \frac{2\,\langle j \,|\, \exp(+i\mathbf{qr}) \,|\, \varepsilon'' \rangle \langle \varepsilon'' \,|\, (\mathbf{eD}(\omega)) \,|\, j \rangle\, (\varepsilon'' + I_j)}{(\varepsilon'' + I_j)^2 - \omega^2}. \tag{7.11}$$

Here

$$W(q) = \int W(r) \exp[i\mathbf{qr}]\, d\mathbf{r} = \left[-Z + \int \rho(\mathbf{r}) \exp[i\mathbf{qr}]\, d\mathbf{r}\right]/q^2;$$
$$\mathbf{q} = \mathbf{p}' - \mathbf{p};\ \omega = (p^2 - p'^2)/2;\ k \ll p;\ \mathbf{q}_1 = \mathbf{q} - \mathbf{k} \approx \mathbf{q};$$

and $J_1(qr)$ is the spherical Bessel function. As in Eq. (7.1), the summation in Eq. (7.11) includes an integral over the continuum states ε".

We now introduce the generalized polarizability of the atom,

$$\alpha_d(\omega, q) = \frac{-i}{(\mathbf{eq})} \sum_{\substack{\varepsilon'' > F \\ j \leq F}} \frac{2 \langle j \,|\, \exp(i\mathbf{qr}) \,|\, \varepsilon'' \rangle \langle \varepsilon'' \,|\, (\mathbf{eD}(\omega)) \,|\, j \rangle (\varepsilon'' + I_j)}{(\varepsilon'' + I_j)^2 - \omega^2}, \tag{7.12}$$

which transforms to $\alpha_d(\omega)$ when $\mathbf{q} \to 0$ ($\mathbf{q}_1 = \mathbf{q} - \mathbf{k} \sim \mathbf{q}$).

Using Eq. (7.12), we can write the amplitude (7.11) as

$$F_{pq}(\omega) = -(\mathbf{eq}) \left[W(q)/\omega + (4\pi\omega/q^2)\alpha_d(\omega, q) \right]. \tag{7.13}$$

The differential cross section for emission of a bremsstrahlung photon with energy ω and momentum transfer $|\mathbf{q}|$ within the interval $d\omega dq$, summed over the polarizations of the photon and integrated over all the angles at which it can emerge, is related to the amplitude $F_{pq}(\omega)$ by the formula (for converting from atomic to Gaussian units, see the list of notation)

$$d\sigma(\omega, q) = \frac{d\omega dq}{16\pi^4 c^3 p^2} \int |F_{pq}(\omega)|^2 \omega q d\Omega_{\mathbf{k}} d\Omega_{\mathbf{q}}, \tag{7.14}$$

where $d\Omega_{\mathbf{q}}$ is the solid angle for the directions of the vector $\mathbf{q}$.

Let us examine $F_{pq}(\omega)$ in the region of small momentum transfers, $q < R_a^{-1}$, where R_a is the radius of the atom. It follows from conservation of the energy $\omega = [p^2 - (\mathbf{p} - \mathbf{q})^2]/2$ that $q > q_{\min} = \omega/p$. Expanding the exponent in Eq. (7.11) in a series in $\mathbf{qr}$ and retaining the first nonvanishing term, we obtain

$$F_{pq}(\omega) \approx -(\mathbf{eq}) \left[\frac{W(q)}{\omega} + \frac{4\pi\omega}{q^2} \alpha_d(\omega) \right] = -\frac{4\pi(\mathbf{eq})}{q^2\omega} [-Z + Q(\mathbf{q}) + \omega^2 \alpha_d(\omega)], \tag{7.15}$$

where $Q(\mathbf{q}) = \int \rho(r) \exp(-i\mathbf{qr}) d\mathbf{r}$ is the form factor of the atom, which equals the number N of electrons in the atom for small q.

For small q and δ_0, the expressions in the square brackets of Eqs. (7.4) and (7.15) differ only by factors which have no influence on the relative contributions of polarization (atomic) and ordinary (electronic) bremsstrahlung.

On approximating the potential of the atom $W(r)$ by $W(r) = -[Z \exp(-\mu r)]/r$ (where $\mu \sim (I)^{1/2}$, and Z is the nuclear charge) and noting that $\alpha_d(\omega)$ is greater than $\alpha_d(0) \sim N_{\text{ext}}/I^2$ up to the ionization threshold, we find the ratio η of the contributions to the bremsstrahlung amplitude from polarization and ordinary emission to be

$$\eta(q) = \frac{4\pi\omega^2 \alpha_d(\omega)}{q^2 W(q)} \approx \frac{\omega^2}{Iq^2} \frac{N_{\text{ext}}}{Z}. \tag{7.16}$$

For $q \sim q_{\min}$, this yields $\eta = p^2 N_{\text{ext}}/IZ$ for fast electrons with $p^2 \gg I$.

At very high ω in excess of the ionization potentials of all the atomic shells, the second term in Eq. (7.11) becomes much simpler, since ($\varepsilon'' + I_j$) can be neglected compared to ω in the denominator.

Using the formula $(\varepsilon'' + I_j)\langle\varepsilon''|(\mathbf{eD}(\omega))|j\rangle = -\langle\varepsilon''|(\mathbf{e}\nabla)|j\rangle$ [25] and the completeness of the functions $\varphi_{\varepsilon''}(r)$, we now take the sum over ε" in Eq. (7.11). Integrating $\langle j|[\exp(i\mathbf{qr})](\mathbf{e}\nabla)|j\rangle$ by parts, we obtain the equality

$$\langle j|e^{i\mathbf{qr}}(\mathbf{e}\nabla)|j\rangle = +\frac{i}{2}(\mathbf{eq})\langle j|e^{i\mathbf{qr}}|j\rangle,$$

Since

$$\sum_{j\leqslant F}\langle j|e^{i\mathbf{qr}}|j\rangle = Q(q),$$

for high ω, we arrive at the following extremely simple expression for $F_{pq}(\omega)$:

$$F_{pq}(\omega) = -\frac{\mathbf{eq}}{\omega}\left[W(q) - \frac{4\pi Q(q)}{q^2}\right] = +\frac{4\pi(\mathbf{eq})Z}{\omega q^2}. \tag{7.17}$$

According to Eq. (7.17) polarization bremsstrahlung completely unshields the nucleus, and the bremsstrahlung of the electron at high ω reduces to its bremsstrahlung on the nucleus [14]. This happens only when $p^2 \gg \omega \gg I_{1s}$. Of course, unshielding owing to the contribution from atomic bremsstrahlung is important if $q < R_a^{-1}$; otherwise, for $q \gg R_a^{-1}$ the form factor $Q(q)$ is small.

The electron shells of an atom are well separated from one another in terms of energy. Thus, the above calculations provide good accuracy only when applied to the outer electrons (i.e., to the electrons with an ionization potential much lower than ω) alone. As a result, instead of Eq. (7.17) we obtain

$$F_{pq}(\omega) = -\frac{4\pi(\mathbf{eq})}{\omega q^2}\left[-Z + Q^{(\mathrm{ext})}(q) + \omega^2\alpha_d^{(\mathrm{ext})}(\omega, q)\right], \tag{7.18}$$

where $Q^{\mathrm{ext}}(q)$ and $\alpha_d^{\mathrm{ext}}(\omega, q)$ are the form factor and generalized polarizability, respectively, of the shells whose ionization potentials are greater than or on the order of ω. It is clear that as ω increases, there is a gradual, shell-by-shell deshielding of the nucleus, which is referred to as stripping the electronic shells [11, 12]. Electrons with ionization potentials less than ω do not contribute physically to the bremsstrahlung: at these ω, the electrons can be regarded as free and a system of free electrons does not have a dipole moment, so it cannot emit a dipole photon.

On the whole, the contribution of polarization bremsstrahlung for fast electrons is large, especially when q is small. Since $W(q)$ for the atoms is negative, while $\alpha_d(\omega) > 0$ when $\omega < I$, a very broad maximum that goes to zero appears in the bremsstrahlung cross section given by Eq. (7.14). When $\omega > I$ the polarizability can be either negative or positive and, as noted above, has an imaginary part that is proportional to the photoionization cross section. Figure 7.18 shows the calculated cross section (7.14) for an electron on Ar [6, 19]. A broad minimum can be seen at $\omega = 0.55$ Ry, as well as a strong peak at $\omega = 2$ Ry owing to the presence of polarization bremsstrahlung. The dependence of the cross section on ω beyond the ionization threshold of the $3p$-subshell is fairly complicated.

The bremsstrahlung of fast electrons on ions is also determined by Eqs. (7.11) and (7.14). The contribution of ordinary (electronic) bremsstrahlung is greater in this case, while, since the polarizability of ions is smaller than that of atoms, the contribution of the polarization term to the bremsstrahlung amplitude is smaller. For an ion with $N < Z$, $W(q)$ is negative, but it is impossible to say in advance whether the equality $F_{pq}(\omega) = 0$ might hold for some value of ω up to the ionization threshold. This will depend on the sign of $\xi = [-z + \omega^2 \alpha_d(\omega)]$, where $z = Z - Q(0)$ is the degree of ionization. If $\xi > 0$ is possible when $\omega \leq I$, which clearly depends on the polarizability, then the cross section for bremsstrahlung on an ion will have a deep minimum, as in the case of bremsstrahlung on an atom.

Of course, $\xi > 0$ near the first discrete excited level of an ion, where $\alpha_d(\omega)$ is very large. Here, however, we are not concerned with bremsstrahlung at frequencies close to the energies of the level of an ion (or of a target atom), which was discussed in [2].

For a negative ion the situation is somewhat different [30, 19]. Here we write the bremsstrahlung amplitude $F_{pq}{}^{(-)}(\omega)$ as the sum of $F_{pq}(\omega)$ and $\Delta_{pq}{}^{(-)}(\omega)$, which, according to Eq. (7.11), is given by

$$\Delta_{pq}^{(-)}(\omega) = -\frac{4\pi\,(\mathbf{eq})}{\omega^2 q^2}\left[Q^{(-)}(q) + \omega^2 \alpha_d^{(-)}(\omega)\right], \tag{7.19}$$

where $Q^{(-)}(q)$ and $\alpha_q{}^{(-)}(\omega)$ are an additional form factor and the additional polarizability of the negative ion $A^{(-)}$ compared to that of the atom, with $Q^{(-)}(0) = 1$. When $\omega < \varepsilon_c$ (where ε_c is the electron affinity) and $q < (r^{(-)})^{-1} \sim (\varepsilon_c)^{1/2}$ (where $r^{(-)}$ is the radius of the negative ion), the correction $\Delta_{pq}{}^{(-)}(\omega)$ will predominate in the amplitude, since in this region we have $W(0) = 0$ and $\alpha_d{}^{(-)} \gg \alpha_d$.

With increasing q, the quantity $Q^{(-)}(q)$ decreases, while remaining positive. Since $W(q) < 0$, at some q determined by the equality $Q^{(-)}(q) = (4\pi/3)q^2 \int \rho(r) r^4 dr$, which is obtained by expanding $W(q)$ in powers of q, the contribution of ordinary (electronic) bremsstrahlung to the amplitude $F_{pq}{}^{(-)}(\omega)$ goes to zero and the emission corresponds entirely to polarization bremsstrahlung.

When $\omega > \varepsilon_c$, the polarizability $\alpha_d{}^{(-)}$ changes sign and compensates the contribution of $Q^{(-)}(q)$, so that bremsstrahlung on a negative ion $A^{(-)}$ becomes equal to the bremsstrahlung on an atom A. When $q \gg 1/r^{(-)}$, the contribution of the correction $A^{(-)}$ is negligible, as can be seen by comparing Eqs. (7.19) and (7.18). This happens because $q \gg 1/r^{(-)}$ corresponds to distances at which the additional electron is actually absent.

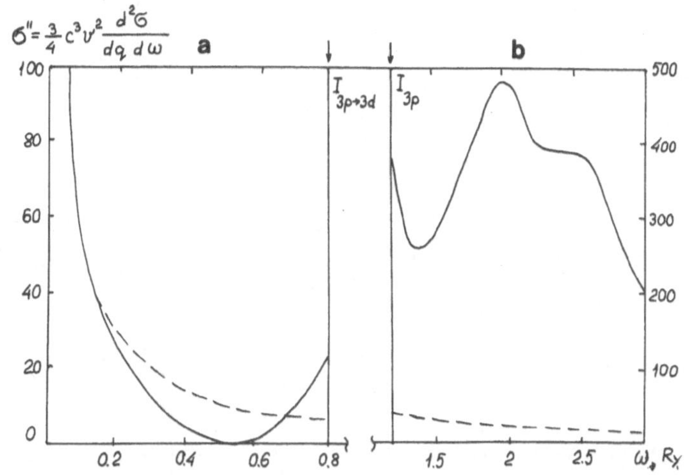

Fig. 7.18. The differential cross section for bremsstrahlung σ" of fast electrons on Ar [6] including polarization bremsstrahlung (1) and ordinary bremsstrahlung (2) (q = 0.5 a.u., $\sigma'' = (3/4)c^3v^2 d\sigma(\omega, q)/d\omega dq$).

7.5. Bremsstrahlung Spectrum of Fast Electrons

As shown in Sec. 7.3, the effect of polarization bremsstrahlung is quite noticeable for small q. In order to study bremsstrahlung experimentally in this case, however, one must not only detect the photon or the energy lost by the fast electron, but also the small momentum transfer, which is rather difficult. Thus, we now clarify the role of polarization bremsstrahlung in the formation of the total emission spectrum $d\sigma(\omega)$ obtained from Eq. (7.14) by integrating over q. It is especially easy to obtain an expression for $d\sigma(\omega)$ in the so-called logarithmic approximation [8, 9].

To do this we use Eq. (7.14). The momentum transfer q is limited to the range $\omega/p < q < 2p$. The first term in Eq. (7.13) is large, on the order of Z for $q > 1/R_Z$, where R_Z is the region of the atom outside which roughly Z electrons can be found. According to the Fermi–Thomas model, $R_Z \sim Z^{-1/3}$. The second term is large when the momenta are small. For a given frequency ω, this means $q < q_\omega \sim R_\omega^{-1}$, where R_ω is the radius of the shell that makes the largest contribution to the polarizability $\alpha_d(\omega)$ for this ω. Taking the square of Eq. (7.13) and integrating the square of the first and second terms with respect to q in those regions of q where either of them is large, we obtain

$$d\sigma(\omega) \approx \frac{16Z^2 d\omega}{3c^3p^2\omega} \ln 2pR_Z + \frac{16}{3}\frac{\omega^3 d\omega}{c^3p^2} |\alpha_d(\omega)|^2 \ln \frac{p}{\omega R_\omega} \tag{7.20}$$

The logarithms in Eq. (7.20) are assumed to be much larger than unity. Thus, in deriving Eq. (7.20) we have dropped the terms that do not contain a large logarithm, including the interference term formed by the product of the first and second terms in $F_{pq}(\omega)$. This is known as the logarithmic approximation. We note that $\ln p$ in the first term of Eq. (7.20) comes from the integral over large $q \sim p$, while in the second it comes from the integral over low $q \sim \omega/p$. For just this reason, the total nuclear charge [when $q \sim p$ the form factor $Q(q)$ is negligibly small] and the simple (rather than the generalized) polarizability $\alpha_d(\omega)$ enters in Eq. (7.20).

At very high frequencies the formula for the polarizability is simpler. Using the sum rule for the dipole strengths of the atomic oscillators, we find from Eq. (7.5) that $\alpha_d(\omega) = -N/\omega^2$. Then, to the same logarithmic accuracy, Eq. (7.20) transforms to

$$d\sigma = \frac{16}{3}\frac{Z^2 d\omega}{c^3 p^2 \omega}\ln 2p^2/\omega \tag{7.21}$$

the bremsstrahlung spectrum for an electron on a bare nucleus.

We now evaluate the comparative roles of polarization (atomic) and traditional (electronic) bremsstrahlung at different frequencies. Near the strong peaks in the photoionization cross section $\sigma_\gamma(\omega)$, it can be seen from Eq. (7.5) that $\operatorname{Re}\alpha_d(\omega)$ is much smaller than $\operatorname{Im}\alpha_d = c\sigma_\gamma(\omega)/4\pi\omega$. We can, therefore, reach an important and extremely general conclusion from Eq. (7.20): a strong peak in $\sigma_\gamma(\omega)$ corresponds to an analogous maximum in the bremsstrahlung spectrum.

We now obtain an estimate for the relative contribution $\eta(\omega)$ of polarization bremsstrahlung,

$$\eta(\omega) = \frac{d\sigma^p}{d\sigma^t} \approx \left(\frac{c\omega_{\max}\sigma_\gamma^{\max}}{4\pi Z}\right)^2 = \eta \quad \text{with } \omega \approx \omega_{\max}, \tag{7.22}$$

where $\omega_{\max}$ is the photon energy at the peak in the cross section $\sigma_\gamma(\omega)$.

As an example, in Mn we have $\omega_{\max}$ = 1.9 a.u. for the $3p$-shell, while η = 1.5. For the $4d^{10}$-shell of Xe and La, we have $\omega_{\max}$ = 3.6 and 4 a.u., while η = 0.4 and 3.2, respectively. The value of η shows that the contribution of polarization bremsstrahlung may be considerably greater than that of conventional (electronic) bremsstrahlung.

The contribution of polarization bremsstrahlung remains large for high ω. Thus, polarization and ordinary (electronic) bremsstrahlung make contributions of the same order to Eq. (7.21), so that $\eta(\omega) \to 1$ with increasing ω.

We note that $d\sigma^p(\omega)$ is more sensitive to the interaction between an excited atomic electron and a vacancy described by the sequence of diagrams in Fig. 7.4 than is the photoionization cross section $\sigma_\gamma(\omega)$, since $d\sigma^p(\omega) \sim |\sigma_\gamma(\omega)|^2 d\omega$.

Even for large p, it is desirable to improve the logarithmic approximation since, for example, $\ln pR_Z = 5$ even for $p = c$ (i.e., it is not very large). For this purpose, the bremsstrahlung spectrum $d\sigma(\omega)$ has been calculated [11, 12] using Eqs. (7.13) and (7.14). This yielded the expression

$$d\sigma(\omega) = \frac{16}{3}\frac{d\omega}{\omega}\frac{1}{c^3 p^2}\int_{\omega/p}^{2p} |C(\omega, q)|^2 \frac{dq}{q}, \tag{7.23}$$

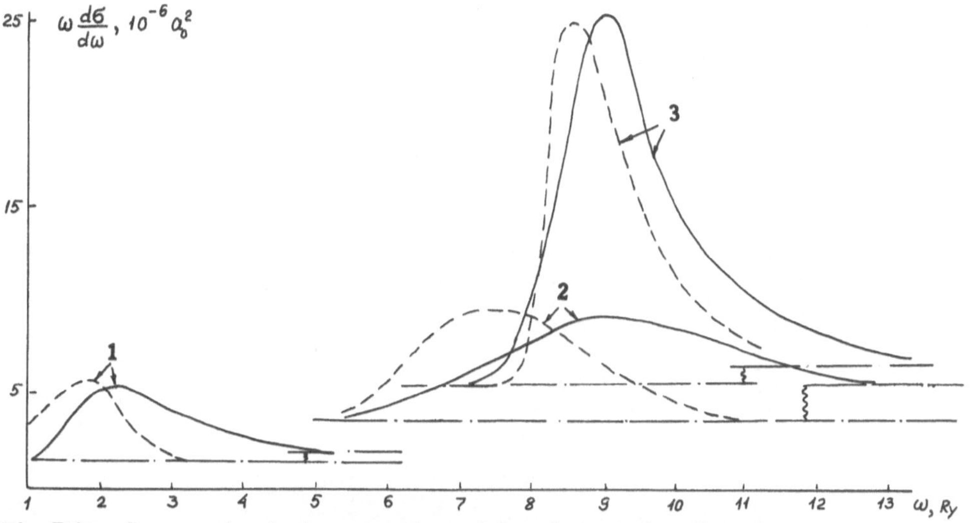

Fig. 7.19. Cross section for bremsstrahlung of fast electrons (ε = 50 keV) on Ar (1), Xe (2), and La (3) [11, 12]. Smooth curves $\omega d\sigma(\omega)/d\omega$ (in units of $10^{-6}a_B{}^2$); dashed curves $\omega\sigma_\gamma(\omega)$ (normalized to $\omega d\sigma(\omega)/d\omega$ at the peak).

where

$$C(\omega, q) = -Z + Q(q) + \omega^2\alpha_d(\omega, q). \tag{7.24}$$

Equation (7.12) was used in this calculation [11, 12] of $\alpha_d(\omega, q)$, which also included the above-mentioned readjustment (Fig. 7.7) of the outer electron shells during virtual creation of vacancies j in an intermediate shell. Specific numerical calculations have been carried out using Hartree–Fock wave functions for the electrons in the ground and excited states of argon, xenon, and lanthanum at frequencies ω close to the ionization potential of the many-electron and comparatively easily ionized polarizing $3p$-shell of Ar and $4d$-shell of Xe and La [12]. The resulting bremsstrahlung cross sections $[\omega d\sigma(\omega)/d\omega]$ for 10-keV electrons are shown in Fig. 7.19. It is clear that the contribution of polarization bremsstrahlung is large, even greater than indicated by the estimates of Eq. (7.22). It is important to note that, although $\alpha_d(\omega)$ is determined by $\sigma_\gamma(\omega)$, because of the dispersion relation between $\mathrm{Re}\,\alpha_d(\omega)$ and $\mathrm{Im}\,\alpha_d(\omega)$, $\alpha_d(\omega)$ is not proportional to $\sigma_\gamma(\omega)$ as a whole. This may lead to a difference in the details of the shape, as well as in the location of the peak, of the photoabsorption and photoemission cross sections $\sigma_\gamma(\omega)$ and $\omega d\sigma(\omega)/d\omega$. This is apparent from a comparison of $\omega d\sigma(\omega)/d\omega$ and $\sigma_\gamma(\omega)$, which matches $\omega d\sigma(\omega)/d\omega$ at its peak in Fig. 7.19.

A calculated $d\sigma(\omega)/d\omega$ profile is compared with the emission spectrum of La bombarded by 0.5-keV electrons in Fig. 7.20. This electron energy, of course, can-

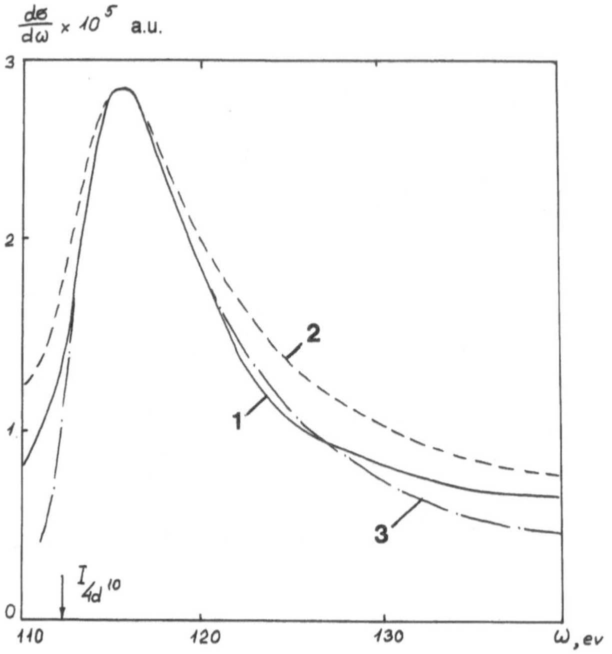

Fig. 7.20. A comparison of measured and calculated emission spectra (in units of $10^{-6}a_B{}^2$) of electrons on lanthanum [9, 10] (the peaks of the curves are matched in height): (1) calculation; (2) experiment, and (3) photoionization cross section.

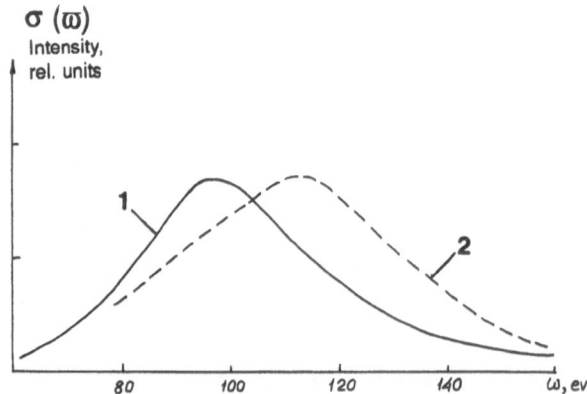

Fig. 7.21. Comparison of the calculated (1) and measured (2) emission spectra of electrons on xenon [13]. (The peaks of the curves are matched in height.)

not be regarded as sufficiently high. The experimental curve (in relative units) matches the calculated curve at the peak. The clear agreement between the shapes of the curves suggests that the peak in the emission spectrum of La [10] is explained by polarization bremsstrahlung [9]. Analogous data for Xe [13] are compared in Fig. 7.21. Unfortunately, the measured emission cross section was not calibrated absolutely, and the electron energy is not (as in [10]) sufficiently high to justify the Born approximation, i.e., the use of Eq. (7.23). As can be seen from Fig. 7.21, however, the experiment does demonstrate the existence of a peak in the emission spectrum that clearly corresponds to the peak given by Eq. (7.23), which is relatively even stronger. Here we note that, as can be seen from Figs. 7.19 and 7.20, $\omega d\sigma(\omega)/d\omega$ is much larger above the ionization threshold than below it. This is a manifestation of the shell-by-shell deshielding or "stripping" of the electronic shells [11, 12] in the total bremsstrahlung cross section mentioned above [see Eq. (7.18)].

The "stripping" effect can be taken roughly into account in a fairly simple way. To do this let us consider the contribution to Eq. (7.13) of a certain shell i (with an ionization potential I_i) and study how $F_{pq}(\omega)$ changes with increasing ω in the interval $\Delta\omega_i$ from $\omega < I_i$ to $\omega > I_i$. Here it is assumed that $\omega \gg I_{i+1}$ and $\omega \ll I_{i-1}$, where I_{i+1} and I_{i-1} are the ionization potentials of the outer and inner shells, respectively, relative to the ith shell. As an estimate we can assume that the contribution of $\alpha_d(\omega, q)$ of the ith shell to $C(\omega, q)$ given by Eq. (7.24) is $\omega^2 Q_i(q)/I_i^2 \ll Q_i$ (if $\omega \ll I_i$) up to its ionization threshold, while substantially beyond the threshold this contribution is much greater and equals $-Q_i(q)$, where $Q_i(q)$ is the form factor for the ith shell. Thus, as ω passes through the region of the ionization potential of the ith shell ($\Delta\omega \leq I_i$), $C(\omega, q)$ changes by a large amount $-Q_i(q)$, which equals N_i for small q. Neglecting the contribution of the shells inside the ith shell to $\alpha_q(\omega, q)$, we find the "jump" in $\omega d\sigma(\omega)/d\omega$ when ω changes by $\Delta\omega_i$ to be

$$\Delta\left(\omega\frac{d\sigma(\omega)}{d\omega}\right) \approx \frac{32}{3}\frac{1}{c^3p^2}\int_{I_i/p}^{2p} Q_i(p)\Big[Z-\sum_{l<i} Q_l(q)\Big]\frac{dq}{q}. \tag{7.25}$$

The main contribution to Eq. (7.25) is from small q [$Q_i(q) \sim N_i$], so that $\Delta[\omega d\sigma(\omega)/d\omega] \sim N_i$. The shell-by-shell "stripping" of the atom in bremsstrahlung is illustrated schematically in Fig. 7.22. Each step $\Delta[\omega d\sigma(\omega)/d\omega]$ in the curve corresponds to the ionization threshold of the next shell. The polarization (atomic) bremsstrahlung for all the subshells is taken into approximate account by including the form factors of only the deepest shells in the amplitude, i.e., by using the formula

$$\left(\omega\frac{d\sigma(\omega)}{d\omega}\right)_i = \frac{16}{3}\frac{1}{c^3p^2}\int_{I_i/p}^{2p}\Big[Z-\sum_{l<i} Q_l(q)\Big]^2\frac{dq}{q}. \tag{7.26}$$

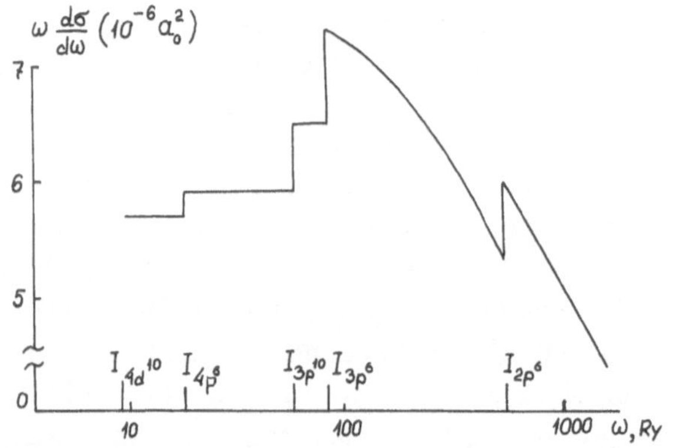

Fig. 7.22. A schematic illustration of the cross section for bremsstrahlung of electrons on lanthanum atoms [11] (in units of $10^{-6}\,a_B{}^2$).

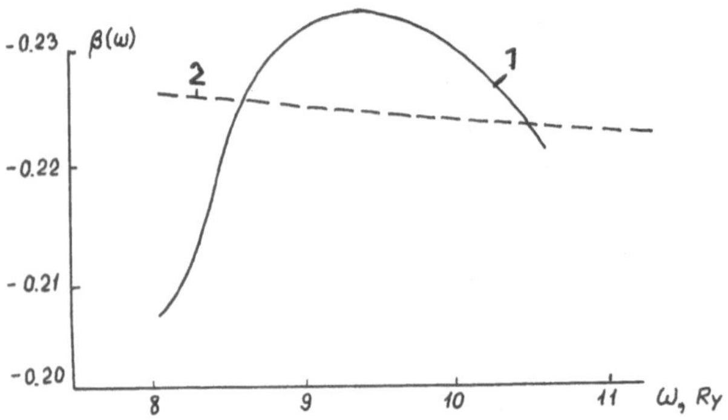

Fig. 7.23. The angular anisotropy parameter for bremsstrahlung on lanthanum atoms including (1) polarization bremsstrahlung and (2) ordinary bremsstrahlung.

The lower limit of integration, ω/p, is replaced by I_i/p, which does not cause a significant difference if I_i/p and ω/p are much less than r_i^{-1}, where r_i is the radius of the ith shell. As a result, $\omega d\sigma(\omega)/d\omega$ is constant for each subshell. With increasing ω, the lower limit ω/p becomes greater than r_i^{-1}, so that $Q_i(q)$ in Eq. (7.26) should be neglected and $[\omega d\sigma(\omega)/d\omega]_i$ begins to fall with increasing ω because of the increase in the lower limit ω/p . This happens for the $3p$-subshell in Fig. 7.22.

Actually, of course, the bremsstrahlung cross section does not change by jumps, but increases monotonically (or nonmonotonically) according to Eq. (7.23) on going from $\omega < I_i$ to $\omega > I_i$. In this frequency interval there is also a region where $\omega d\sigma(\omega)/d\omega$ undergoes extremely rapid changes associated with the poles in $\alpha_d(\omega, q)$ at the discrete excitation energies of the atom.

One important characteristic of the bremsstrahlung spectrum is the angular distribution of the photons with respect to the direction of the incident electron flux.

The anisotropy in the emission of bremsstrahlung photons is determined by the factor (**eq**) in Eq. (7.13). On switching from the directions of **q** (over which the integral must be taken) and **e** to the directions of **p** and **k** (photon momentum), after some trigonometric transformations [11] we obtain

$$d\sigma(\omega, \Omega_k) = \frac{d\Omega_k}{4\pi} d\sigma(\omega)\left[1 - \beta(\omega) P_2(\cos(\widehat{\mathbf{kp}}))\right], \tag{7.27}$$

where

$$\beta(\omega) = \left(\frac{d\sigma(\omega)}{d\omega}\right)^{-1} \frac{8}{3\omega^3 p^2 c^3} \int_{\omega/p}^{2p} |C(\omega, q)|^2 \left[\frac{3}{4p^2}\left(\frac{2\omega}{q} + q\right)^2 - 1\right] \frac{dq}{2q}. \tag{7.28}$$

Equation (7.27) is valid for $\omega \ll p^2/2$. If $C(\omega, q)$ is independent of (as for a purely Coulomb field) or depends only weakly on q, then we can set $\beta(\omega) \approx -1/2$ to logarithmic accuracy. Calculations show, however, that $\beta(\omega)$ differs substantially from this value and depends significantly on the frequency, as is illustrated by Fig. 7.23, which shows the calculated $\beta(\omega)$ [11] for La near the threshold for ionization of the $4d^{10}$-subshell. It can be seen that polarization bremsstrahlung shows up extremely weakly in $\beta(\omega)$.

The bremsstrahlung is linearly polarized and the degree of polarization, like the angular distribution, can be expressed in terms of the $\beta(\omega)$ defined by Eq. (7.28). It has been shown [31] that the angular distribution of polarization bremsstrahlung is given by the following formulas:

$$d\sigma^{\parallel}(\omega, \Omega_k) = \frac{d\Omega_k}{8\pi} d\sigma(\omega)\left[1 + \beta(\omega)\left(2 - 3\cos^2(\widehat{\mathbf{kp}})\right)\right], \tag{7.29a}$$

and

$$d\sigma^{\perp}(\omega, \Omega_k) = \frac{d\Omega_k}{8\pi} d\sigma(\omega)\left[1 - \beta(\omega)\right]. \tag{7.29b}$$

Here $d\sigma^{\parallel}(\omega, \Omega_k)$ and $d\sigma^{\perp}(\omega, \Omega_k)$ are the differential cross sections for emission with **e** directed perpendicular and parallel, respectively, to the plane formed by **p** and **k**.

7.6. Bremsstrahlung of Relativistic Electrons: Spectrum and Angular Distribution of the Emitted Photons

The bremsstrahlung of relativistic electrons including the polarization of a target atom has been examined by Amus'ya et al. [15] and by Astapenko et al. [16]. The amplitude of the bremsstrahlung is determined from diagrams (see Fig. 7.17) in which Dirac plane waves can be used for the wave function of the fast electron.

The atomic electrons, of course, should be described in the nonrelativistic approximation of Hartree–Fock functions. The interaction operator of the fast electron with the electromagnetic field should be the relativistic $\mathbf{e}\boldsymbol{\gamma} = \hat{\mathbf{e}}$ (where $\boldsymbol{\gamma} = \{\gamma^1, \gamma^2, \gamma^3\}$ are the Dirac matrices) instead of (**er**) [or (**ep**)]. The interaction of the fast electron with the atomic electrons also changes significantly. We shall use a Coulomb gauge for the electromagnetic field. Then the interaction of the incident electron with the static field of a target atom (or ion) will obey the formula $[-Z + Q(p)]/q^2$, as in the nonrelativistic case. Two terms, however, should be included in the interaction of the incident electron with an atomic electron: the first is a purely Coulomb term proportional to $1/q^2$ and the second is caused by exchange of a "transverse" photon by the electrons and is described by the expression

$$\left(\delta_{ij} - \frac{q_{1,i}q_{1,j}}{q_1^2}\right)\frac{1}{\omega^2 - c^2 q_1^2}, \quad \mathbf{q}_1 = \mathbf{q} - \mathbf{k} = \mathbf{p}' - \mathbf{p},$$

where q_i and q_j denote the projections of the momentum transfer $\mathbf{q} = \{q_x, q_y, q_z\}$.

The amplitude of the ordinary (electronic) bremsstrahlung is well known [1] and is given by [cf. Eq. (5.2)]

$$F_{pq}^{(t)}(\omega) = \frac{4\pi}{q^2}(Z - Q(\mathbf{q}))\,\bar{u}(\mathbf{p}', s')\left\{\hat{e}\,\frac{[(\hat{p}' + \hat{k}) + c]\,\gamma^0}{[(\mathbf{p}' + \mathbf{k})^2 - (\varepsilon' + \omega)^2/c^2 - c^2]} + \right.$$

$$\left. + \gamma^0\,\frac{(\hat{p} - \hat{k}) + c}{[(\mathbf{p} - \mathbf{k})^2 - (\varepsilon - \omega)^2/c^2 - c^2]}\,\hat{e}\right\} u(\mathbf{p}, s). \tag{7.30}$$

Here $u(\mathbf{p}, s)$ is the bispinor amplitude; in the nonrelativistic limit the upper components of this amplitude approach unity; s, s' and $\varepsilon, \varepsilon'$ are the projection of the spin and the energy of the fast electron in the initial (and final) state; **k** is the momentum of the photon; $\hat{a} = a_\mu\gamma^\mu$ (where $\mu = 0, 1, 2, 3$); γ^μ are the Dirac matrices; $\mathbf{q} = \mathbf{p}' - \mathbf{p} + \mathbf{k}$, and $\mathbf{q}_1 = \mathbf{p}' - \mathbf{p} = \mathbf{q} - \mathbf{k}$.

The formula for $F_{pq}^{(p)}$ is extremely complicated [15], so it is not given in full here. It can, however, be greatly simplified and, as in the nonrelativistic case, expressed in terms of the dynamic polarizability of the atom for small q_1. As can be seen from Eq. (7.12), the momentum q_1 shows up in the polarization (atomic) bremsstrahlung amplitude through the matrix elements $\langle j|e^{i\mathbf{q}_1\mathbf{r}}|e\rangle$ [in Eq. (7.12), $q_1 \sim q$], which differ greatly from zero only for $q_1 \leq R_j^{-1}$. The smallest momentum transfer $q_{1,\min}$ is given by

$$q_{1,\min} = \sqrt{(\varepsilon/c)^2 - c^2} - \sqrt{[(\varepsilon - \omega)/c]^2 - c^2} \approx \omega/v_0 \tag{7.31}$$

for $\omega/\varepsilon \ll 1$, where v_0 is the velocity of the incident electron.

The inequality $q_{1,\min} \leq R_j^{-1}$ is satisfied up to extremely high (on an atomic scale) emission frequencies ω ($\omega < cR_j^{-1}$), which for an atom Z are of order cZ. Replacing the wave functions of the incident electron by plane waves is legitimate if $Z/c \ll 1$. Thus, the inequality $q_{1,\min} < R^{-1}$ means that the contribution of polarization (atomic) bremsstrahlung is important only for $\omega < cZ < c^2$. In this range of q

and ω, the polarization bremsstrahlung amplitude becomes much simpler and takes the form

$$F_{pq}^{(p)}(\omega) = + \frac{4\pi}{\mathbf{q}_1^2} \left[+ b^0 (\mathbf{e}\mathbf{q}_1) + \mathbf{q}_1^2 \left((\mathbf{e}\mathbf{b}) - \frac{(\mathbf{b}\mathbf{q}_1)(\mathbf{e}\mathbf{q}_1)}{q_1^2} \right) \omega \quad \frac{c}{\omega^2 - c^2 q_1^2} \right] \omega \alpha_d(\omega), \quad (7.32)$$

where $\mathbf{q}_1 = \mathrm{q} - \mathbf{k} = \mathbf{p}' - \mathbf{p}$ and $b^\mu = \overline{u}(\mathbf{p}', s')\gamma^\mu u(\mathbf{p}, s)$. The amplitude (7.32) was obtained in the dipole approximation, which is valid if $k = \omega/c \ll R_a^{-1}$.

We now consider the bremsstrahlung spectrum. It is made up of the contributions from the the polarization and electronic amplitudes, together with their interference. The contribution of the latter to the bremsstrahlung spectrum is small, because, as noted in the preceding section, the main contributions from polarization and electronic bremsstrahlung correspond to different ranges of the momentum q_1 [see the derivation and discussion of Eq. (7.20)]. Electronic bremsstrahlung on a neutral atom occurs in the range $R_a^{-1} < q_1 < c$, while polarization bremsstrahlung occurs in $q_{\min} < q_1 < R_a^{-1}$.

The spectrum, angular distribution, and polarization characteristics of ordinary (electronic) bremsstrahlung have been discussed in detail by Berestetskii et al. [1] It is known, for example, that in the ultrarelativistic case

$$d\sigma^{(t)}(\omega) = (16/3)(Z^2 d\omega / c^5 \omega)(\ln cR_a - 1/2) \quad (7.33)$$

for a neutral atom.

For comparing Eq. (7.33) with the spectrum (7.20), which contains an increasing logarithmic term, we note that the upper limit $q_{1,\max} = 2p$ in Eq. (7.33) is inaccessible because of the increasingly virtual character of the relativistic electron in the intermediate state of the diagrams (see Fig. 7.21), which "cut" q_1 off at $q_{1,\max} = c$.

The contribution of polarization bremsstrahlung is given by Eq. (7.32) with the aid of Eq. (7.14), where p^2 must be replaced by v^2. Then, after integrating over dq, to logarithmic accuracy we obtain

$$d\sigma^{(p)}(\omega) = \frac{16}{3} \frac{d\omega}{\omega c^3 v^2} |\omega^2 \alpha_d(\omega)|^2 \ln \frac{v_0 \varepsilon}{\omega R_a c^2}. \quad (7.34)$$

This implies that the polarization bremsstrahlung predominates in the relativistic case with increasing ε. For large ω, the polarizability is given by $\alpha_d(\omega) \sim -Z/\omega^2$, so that

$$\eta = \frac{d\sigma^{(p)}(\omega)}{d\sigma^{(t)}(\omega)} \approx \ln \varepsilon / \omega R_a c \gg 1.$$

The Coulomb gauge used in deriving Eq. (7.32) makes it possible to separate the contributions of the scalar (purely Coulomb) and vector (from the exchange of "transverse" photons) interactions, i.e., the first and last terms in Eq. (7.32). This corresponds to writing the logarithm in Eq. (7.34) in the form of the sum

$$\ln \frac{\varepsilon}{\omega R_a c} = \ln \frac{\varepsilon}{c^2} + \ln \frac{v_0}{\omega R_a}. \tag{7.35}$$

Here we point out that the second term in Eq. (7.35) coincides with the second logarithm in Eq. (7.20), where it follows from dimensional considerations that p must be replaced by v in an arbitrary system of units. The first term (the contribution of "transverse" photons to polarization bremsstrahlung) clearly becomes absolutely dominant for $\varepsilon \gg c^2$.

The angular distribution of bremsstrahlung photons is different in the polarization and traditional parts of the total bremsstrahlung spectrum (this was first noted by Hubbard and Rose [17] for bremsstrahlung of electrons on nuclei). When $v_0 \to c$, all the traditional bremsstrahlung is concentrated in a narrow range of angles $\theta < c^2/\varepsilon$. At the same time, with the aid of Eqs. (7.14) and (7.32) we obtain

$$d\sigma^{(p)}(\omega, \Omega_k) = \frac{d\omega d\Omega_k}{\omega \pi c^3 v^2} |\omega^2 \alpha_d(\omega)|^2 (1 + \cos^2\theta) \ln \frac{v_0 \varepsilon}{\omega R_a c^2}. \tag{7.36}$$

The θ dependence in Eq. (7.36) is the angular distribution of the emission from a rotating atomic dipole polarized by an incident electron. The overall angular distribution of the bremsstrahlung is, therefore, characterized by a smooth dependence (7.36) on the angle for $\theta > c^2/\varepsilon$ and a sharp anisotropy in the region $\theta < c^2/\varepsilon$. For such small angles $d\sigma(\omega, \Omega_k)$ is very large, on the order of $(32Z^2\varepsilon^2/3c^9\omega) \times \ln(c/Z^{1/3}) d\omega d\Omega_k$ according to Eq. (7.33).

The presence of the difference $(\omega^2 - c^2 q_1^2)$ in the denominator of Eq. (7.32) causes a singularity in the angular distribution of the fast electrons. With the aid of Eqs. (7.14) and (7.32) we obtain

$$d\sigma^p(\omega, q_1) = \frac{16}{3} \frac{d\omega dq_1}{\omega c^3 v^2} |\omega^2 \alpha_d(\omega)|^2 \frac{q_1 c^2}{c^2 q_1^2 - \omega^2}. \tag{7.37}$$

Since $q_{1,\min} = \omega/v$ and v are close to c, for small q_1 (i.e., very small scattering angles for the incident electron) the polarization bremsstrahlung cross section is especially large. The rise in the cross section (7.37) in this region is limited by the fact that $v < c$ always; however, the cross section $d\sigma^{(p)}(\omega, q_1)$ itself is very large for $q_1 = q_{1,\min}$ and equals $(16/3)\,(d\omega dq_1/\omega^2 c^3 v)(\varepsilon^2/c^4)|\omega^2 \alpha_d(\omega)|^2$.

The bremsstrahlung spectra for nonrelativistic and relativistic electrons differ greatly. For $v \to c$ and high ω, the deshielding effect mentioned above is replaced by the opposite effect. In fact, on replacing $\alpha_d(\omega)$ by $(-Z/\omega^2)$, in the ultrarelativistic case $(v \to c)$ we obtain

$$d\sigma(\omega) = (16/3)\,(Z^2 d\omega/c^5\omega) \ln \varepsilon/\omega. \tag{7.38}$$

for the total bremsstrahlung spectrum from Eqs. (7.33) and (7.34). This expression differs by a factor of Z^2 from the bremsstrahlung cross section for an electron on an ultrarelativistic electron (see [1]). When $\omega \gg I$, the atomic electrons can be regarded as free. The total bremsstrahlung amplitude in this case is made up of three parts:

$$F_{pq}^{(\omega)} = ZF_{pq}^{(1)}(\omega) + ZF_{pq}^{(2)}(\omega) + F_{pq}^{(3)}(\omega), \tag{7.39}$$

where $F_{pq}^{(1)}(\omega)$, $F_{pq}^{(2)}(\omega)$, and $F_{pq}^{(3)}(\omega)$ are the amplitudes for emission of a photon by an incident electron on an atomic electron, by an atomic electron as it interacts with the incident electron, and by the incident electron on the nucleus. Since the bremsstrahlung amplitude of a fast electron on a free particle at rest is independent of its mass in the ultrarelativistic case, but is determined solely by its charge, we have $ZF_{pq}^{(1)} + F_{pq}^{(3)}(\omega) = 0$ and $F_{pq}(\omega) = ZF_{pq}^{(2)}(\omega)$. It turns out, therefore, that the total bremsstrahlung spectrum in the ultrarelativistic case is determined solely by the emission from free atomic electrons, as opposed to the nonrelativistic case, where the total bremsstrahlung spectrum at high ω is the same as the bremsstrahlung on a bare nucleus and the contribution of the atomic electrons is insignificant.

7.7. Bremsstrahlung of Heavy Charged Particles

Polarization bremsstrahlung plays an especially important role if the incident particle is heavy with a mass $m_0 \gg 1$. The emission from such particles is m_0^2 times smaller than that from an electron [1] (here we are using atomic units, with $e = \hbar = m = 1$). At the same time, the polarization of a target atom and the associated polarization bremsstrahlung produced by a heavy particle are certainly not less than those induced by an electron with the same velocity [18, 19]. Thus, the intensity of the total bremsstrahlung spectrum from an electron and a heavy particle should be of the same order for $\omega \geq I$.

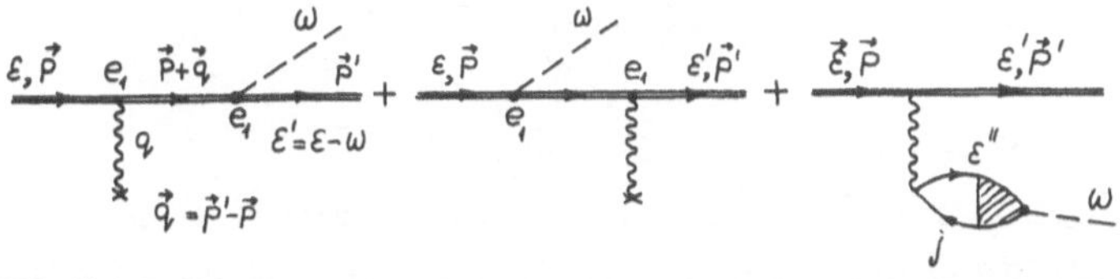

Fig. 7.24. Diagrams describing the bremsstrahlung amplitude of fast heavy particles on atoms.

For simplicity we shall assume that the incident particle of charge e_0 and mass m_0 is fast, i.e., that its velocity is large compared to that of the atomic electrons. In this case we can limit ourselves to the lowest-order Born approximation and treat the wave function of the incident particle as a plane wave. Using the notation of the diagrams of Fig. 7.10, the bremsstrahlung amplitude for a fast heavy particle can be represented (by analogy with Fig. 7.17b) by the set of diagrams in Fig. 7.24.

As above, we restrict ourselves to dipole photons and, therefore, neglect their momenta ($\mathbf{q}_1 = \mathbf{p}' - \mathbf{p} = \mathbf{q} - \mathbf{k} \sim \mathbf{q}$).

The amplitude represented by the diagrams of Fig. 7.24 can be written analytically in the form [cf. Eq. (7.13)]

$$F_{pq}(\omega) = \frac{4\pi e_0^2}{m_0 q^2}[Z - Q(q)]\frac{(\mathbf{eq})}{\omega} - \frac{4\pi e_0}{q^2}\omega\alpha_d(\omega, q)(\mathbf{eq}). \qquad (7.40)$$

In the diagrams of Fig. 7.24 we have noted where the interaction is determined by the charge e_0, so that it is understandable that the first term contains $e_0{}^2$ and the second, $-e_0$. Here e_0 is the charge on the incident particle expressed in units of the electronic charge.

The differential cross section for bremsstrahlung emission of a photon with energy ω and momentum transfer $|\mathbf{q}|$ in the interval $d\omega dq$, summed over the polarizations of the photon and integrated over all directions of emission, is related to the amplitude $F_{pq}(\omega)$ by a formula that differs from Eq. (7.14) in having p^2 replaced by v^2 in the denominator. Because $m_0 \gg 1$, polarization (electronic) bremsstrahlung predominates, even for small ω up to large momenta. According to Eq. (7.40), the ratio η of the polarization and ordinary bremsstrahlung amplitudes is given by

$$\eta(\omega) = \frac{4\pi\omega^2\alpha_d(\omega)\, m_0}{e_0^2 q^2 W(q)} \approx \frac{\omega^2}{Iq^2}\frac{N_{\max} m_e}{Ze_0} \sim \frac{\omega^2 m_0}{q^2 I}. \qquad (7.41)$$

The estimate (7.41) was made (as in Sec. 7.4) for an incident electron and differs from the earlier estimate only by a factor of m_0. It follows from this that $\eta > 1$ up to $q < \omega(m_0/I)^{1/2}$, i.e., within the range $\omega/v_0 < q < \omega(m_0)^{1/2}/v_a$, where v_a is the characteristic velocity of the atomic electrons. This upper bound on q is meaningful if $\omega(m_0)^{1/2}/v_a < R_a^{-1}$, i.e., for low frequencies on the order of $\omega < I(m_0)^{1/2}$. For large q the parameter η is given by

$$\eta = \omega^2\alpha_d(\omega, q)\, m_0/e_0 Z, \qquad (7.42)$$

which falls off rapidly with increasing q for $q > R_a^{-1}$.

In a way similar to that used to obtain an expression for the bremsstrahlung spectrum (7.23) from Eqs. (7.13) and (7.14), for heavy particles we find

$$d\sigma(\omega) = \frac{16}{3}\frac{d\omega}{c^3 v_0^2 \omega}\int\limits_{\omega/v_0}^{2p}\left|\left[\frac{Z - Q(q)}{m_0}e_0^2 - e_0\omega^2\alpha_d(\omega, q)\right]\right|^2\frac{dq}{q}. \qquad (7.43)$$

A closed expression for the bremsstrahlung spectrum is easy to obtain from Eq. (7.43) in the logarithmic approximation, where it is similar to Eq. (7.20):

$$d\sigma(\omega) = \frac{16Z^2e_0^4 d\omega}{3c^3v_0^2m_0^2\omega}\ln 2v_0m_0R_Z + \frac{16e_0^2 d\omega}{3c^3v_0^2\omega}|\omega^2\alpha_d(\omega)|^2 \ln\frac{v_0}{\omega R_\omega}, \tag{7.44}$$

which differs from Eq. (7.20) in having p replaced by v_0 in the second term. Previously, in Eq. (7.20) no distinction was made between p and v_0, since $m_e = 1$. It makes sense to retain both terms in Eq. (7.44) only for low frequencies, where these terms are of the same order, i.e., for

$$\omega \sim \omega_R = (Ze_0^2/m_0\alpha_d(0))^{1/2}. \tag{7.45}$$

When $\omega \ll \omega_R$, ordinary bremsstrahlung, i.e., emission by the incident particle itself, dominates. It is interesting that the interference term in Eq. (7.43) makes a contribution to the total bremsstrahlung spectrum which is given roughly by

$$\Delta d\sigma^{\text{int}}(\omega) \approx -\frac{32d\omega\,(\omega^2\alpha_d(\omega))}{3\omega c^3v_0^2m_0}e_0^3, \tag{7.46}$$

which differs from the geometric mean of the first and second terms in Eq. (7.44) only by a factor of –2 and by the absence of the large logarithm. Thus, the interference term can be neglected (at least for purposes of estimates) for all ω. For $\omega \gg 1$, we have

$$d\sigma(\omega) \approx \frac{16Z^2e_0^2 d\omega}{3c^3v_0^2\omega}\ln\frac{v_0}{\omega R_a}, \tag{7.47}$$

which is m_0^2 times greater than the contribution of ordinary bremsstrahlung. In Eq. (7.47) it is assumed that, although ω is much greater than the ionization potential of most atomic shells, the relation $q_{\min} \sim \omega/v < R_a^{-1}$ also is true [the latter equation is possible if $v_0^2 > (\omega R_a)^2$]. If this inequality is true only for some of the atomic shells (the inner ones) and is violated for the outer shells, then, neglecting terms of order $1/m_0$ in the amplitude $F_{pq}(\omega)$, we can write $d\sigma(\omega)$ in the form

$$d\sigma(\omega) = \frac{16}{3}\frac{d\omega}{\omega}\frac{e_0^2}{c^3v_0^2}\int_{\omega/v_0}^{1/R_b}\left(\frac{N_{\text{ext}}I^2}{q^4} + N_{\text{in}}\right)^2\frac{dq}{q} \approx \frac{16}{3}e_0^2\frac{N_{\text{in}}^2 d\omega}{c^3v_0^2\omega}\ln\frac{v_0}{\omega R_b}. \tag{7.48}$$

Here R_b is the radius of the shells containing the N_{in} inner electrons for which $\omega/v_0 < 1/R_b$. The terms containing N_{ext} have been dropped, because $q_{\min} = \omega/v_0 > 1/R_{\text{ext}} \gg 1/R_{\text{in}}$ and, therefore, $I/q^2 \ll 1$.

Effects similar to the "stripping" of electron shells in bremsstrahlung of electrons on atoms show up in the cross section for bremsstrahlung of heavy particles, which, according to Eq. (7.43), actually has an entirely polarization character. In fact, as ω increases from values lying considerably below the threshold I_{med} for ionization of an intermediate shell to those considerably above I_{med}, the cross section $\omega d\sigma(\omega)/d\omega$ increases discontinuously and $\Delta(\omega d\sigma(\omega)/d\omega)$ is proportional to the number N_{med} of electrons in the intermediate subshell. To convince ourselves of this, we write the term $\omega^2\alpha_d(\omega, q)$ in the bremsstrahlung amplitude (in the range $q_{min} < q < R_a^{-1}$) in the form

$$\omega^2\alpha_d(\omega) \approx (-N_{ext} + \omega^2\alpha_d^{(med)}(\omega)), \tag{7.49}$$

where we have legitimately left out the contribution of the inner shells to $\alpha_d(\omega)$ and included the contribution of the outer shells containing a total of N_{ext} electrons. The contribution of $\omega^2 \operatorname{Im} \alpha_d^{(med)}(\omega) \sim (c/4\pi)\omega\sigma_\gamma(\omega)$ for $\omega \sim I_{med}$ is important only beyond the threshold, i.e., for $\omega \geq I_{med}$, and falls rapidly with increasing ω. Thus, if ω is much higher than I_{med}, then we can neglect the contribution of $\omega^2 \operatorname{Im} \alpha_d^{(med)}(\omega)$ to the amplitude. Up to the threshold, $I_{med}\, \alpha_d^{(med)}(\omega)$ is of the same order as N_{med}/I_{med}^2, while significantly beyond the threshold, it is of order $(-N_{med}/\omega^2)$. Thus, the expression for $\omega^2\alpha_d(\omega)$ is enhanced by additional term $2N_{med}/I_{med}^2$ and the increase in the cross section, $\Delta(\omega d\sigma(\omega)/d\omega)$, is given roughly by

$$\Delta\left(\omega\frac{d\sigma(\omega)}{d\omega}\right) \approx 4\frac{16e_0^2}{3c^3v_0^2} N_{ext}N_{med} \ln\frac{2pv_0}{I_{med}}. \tag{7.50}$$

In deriving this formula we have set $\omega = I_{med}$.

Since the polarizability $\alpha_{med}(\omega)$ increases while remaining positive as ω approaches I_{med}, it is possible in principle that $\omega d\sigma(\omega)/d\omega$ may go to zero near the thresholds for several intermediate many-electron shells. Beyond the threshold I_{med}, as a rule we must also include the contribution of $\operatorname{Im} \alpha_d(\omega)$ where the photoionization cross section is large, since the contribution of $\operatorname{Im} \alpha_d(\omega)$ is most important in the cross section $\omega d\sigma(\omega)/d\omega$. The increase in $\alpha_d(\omega)$ up to the threshold and the large cross section $\sigma_\gamma(\omega)$ beyond the threshold I_{med} means that as ω varies from values lying significantly below I_{med} to values considerably above I_{med}, $\omega d\sigma(\omega)/d\omega$ may pass through a maximum.

The above discussion is correct if $q_{min} = \omega/v_0 < R_a^{-1}$. Otherwise, when $q_{min} \gg R_{ext}^{-1}$, but $q_{min} < R_{med}^{-1}$, where R_{ext} and R_{med} are the radii of the outer and intermediate subshells, respectively, the situation changes. Instead of Eq. (7.49), for $\omega^2\alpha_d(\omega)$ we have

$$\omega^2\alpha_d(\omega, q) \approx \left(-\frac{N_{ext}I^2}{q^4} + \omega^2\alpha_d^{(med)}(\omega)\right). \tag{7.51}$$

Since $q_{min} \gg R_{ext}^{-1}$, we have $q^2 \gg I$ and the first term may be considerably smaller than the second, so that it can be neglected. In this case, $\Delta(\omega d\sigma(\omega)/d\omega)$ becomes small and even negative, since as ω increases from below the threshold I_{med}

to considerably above this threshold, the absolute value of $\omega^2\alpha_d(\omega, q)$ varies little, while the lower limit of integration in Eq. (7.43) increases with rising ω.

For frequencies ω considerably greater than ωR [see Eq. (7.45)], we can also make a fairly simple estimate of the bremsstrahlung spectrum when the heavy particle is slow. It has been shown [32] that for incident particle velocities in the range $1/m_0 \ll v_0^3 \ll 1$, the ionization of a target atom can be described by replacing the wave function of the incident particle by a plane wave. Neglecting ordinary bremsstrahlung for $\omega \gg \omega_R$, we now estimate the contribution of polarization bremsstrahlung while again assuming that the wave function of the heavy particle is a plane wave. For frequencies ω such that $\omega/v_0 < 1/R_a$, Eq. (7.47) provides a valid estimate. The range of frequencies $\omega_R \ll \omega \ll v_0/R_a$ may be compatible with the condition $m_0^{-1/3} \ll v_0 \ll 1$, which can be seen most easily for the case of bremsstrahlung of a proton on a hydrogen atom, where $\omega_R \sim 1/2m_0^{-1/2} \sim 0.01$, $R_a = 1$, and $m_0^{-1/3} \sim 0.1$. Thus, the conditions $0.01 \ll \omega \ll v_0$ and $0.1 \ll v_0 \ll 1$ are compatible. For $\omega > v_0 R_a^{-1}$ the bremsstrahlung intensity is low. Equation (7.47) is also valid for purposes of estimates over the entire range of incident particle velocities v_0, from low to high.

On the whole, as can be seen here, the bremsstrahlung spectra of heavy charged particles and electrons are similar at equal velocities. For heavy particles, however, the emission from so-called secondary electrons that are removed from the atom as it interacts with the heavy particle is extremely important. We shall discuss this question in more detail in Sec. 7.10.

7.8. Emission during Collisions of Two Atomic Particles

Polarization bremsstrahlung makes a significant contribution to the total emission spectrum resulting from collisions of two particles with internal structures. We shall refer to these particles as atomic particles below. They can be neutral (atoms) or charged (ions). If both collision partners have masses $m_1, m_2 \gg 1$, then the ordinary bremsstrahlung of both can be neglected. In this case, by ordinary bremsstrahlung we mean radiation by the static field of one of the colliding particles in the static field of the other, without real or virtual excitation of their internal degrees of freedom. Polarization bremsstrahlung in this case refers to bremsstrahlung resulting from the polarization of at least one of the particles during the collision process. In principle, it is not necessary that the mass of the atomic particle be large; it could be on the order of the electron mass, as in positronium. This somewhat special case will be discussed in Chapter 8.

We now consider collisions of neutral particles. They have a static field $W(r)$ [see Eq. (7.10)]. For neutral particles $W(r)$ falls off with increasing r at large distances at least as fast as $1/r^3$. This interaction causes dipole polarization of the collision partner (by displacing its electron cloud and nucleus in opposite directions) and, therefore, bremsstrahlung [20, 21]. We shall refer to the displacement of the nuclei themselves as "recoil."

It is natural that the bremsstrahlung intensity should rise with increasing polarizability and higher static potential of the collision partner.

The elastic bremsstrahlung amplitude for fast particles is described by a set of four diagrams, two of which are shown in Fig. 7.14, while the other two differ from Fig. 7.14 in that the second particle radiates. In this section we shall consider particle collisions in which the particles are in their ground states both before and after the collision ($i_1 = f_1 = 0$ and $i_2 = f_2 = 0$), while they can be in excited states ($j = n$) in between (after the interaction or emission of a photon). We shall denote the total interaction of the two particles, including the interactions of the two nuclei, of the two groups of electrons, and of the electrons of each of the collision partners with the nucleus of the other, by $\hat{u}$.

The operator describing the interaction with the electromagnetic field can be written most simply in the form of a "velocity" [21]

$$\hat{V}^{(g)} = -\sum_{j=1}^{N_g} \frac{1}{c} e^{-i\mathbf{kr}_j} (\mathbf{ep}_j) + \frac{Z_g}{m_g c} e^{-i\mathbf{kr}_n} (\mathbf{ep}_n), \quad g = 1, 2, \tag{7.52}$$

where the sum is taken over the coordinates of the electrons in the atom, $\mathbf{r}_n$ and $\mathbf{p}_n$ denote the coordinates and momentum of the nucleus, and the indices 1 and 2 refer to the first and second atomic particles. In order to simplify this extremely cumbersome and complicated problem involving radiation during a collision of two structured particles, we shall restrict ourselves to considering only long-wavelength (dipole) photons, which allows us to neglect their momentum $\mathbf{k}$. Essentially, this is not a very severe limitation, since the dipole approximation is valid even for frequencies above the ionization potentials of the inner shells. In the problem which we shall discuss, it is natural to separate the coordinates of the center of mass and the total momenta $\mathbf{P}$ of the colliding particles. Then we have

$$V^{(g)} \approx (Z_g - N_g) \frac{(\mathbf{eP}_g)}{m_g c} - \sum_{j=1}^{N_g} \frac{1}{c} (\mathbf{ep}_{jg}), \quad g = 1, 2, \tag{7.53}$$

instead of Eq. (7.52), where $\mathbf{p}_{jg}$ denotes the momentum of an electron in the center of mass system of the atomic particle. Since most of the mass of the particle is concentrated in the nucleus, however, the coordinates (and momentum) of an electron in the center of mass and nuclear systems are close to one another.

The matrix element for the interaction of two colliding particles, $\hat{u}$ (see Fig. 7.14), can be written in the form [21]

$$\langle \mathbf{P}_1, s_1; \mathbf{P}_2, s_2 | \hat{u} | \mathbf{P}_1', s_1'; \mathbf{P}_2' s_2' \rangle = (2\pi)^3 \delta(\mathbf{P}_1 + \mathbf{P}_2 - \mathbf{P}_1' - \mathbf{P}_2') \frac{4\pi}{q^2} S^{(1)}_{s_1, s_1'}(\mathbf{q}) S^{(2)}_{s_2, s_2'}(-\mathbf{q}), \tag{7.54}$$

where s_1, s_2 and s_1', s_2' denote the internal states of the first (1) and second (2) particles, respectively, before and after the collision and $\mathbf{P}_1$, $\mathbf{P}_2$ and $\mathbf{P}_1'$, $\mathbf{P}_2'$ denote their total momenta. In Eq. (7.54) we have $\mathbf{q} = \mathbf{P}_1' - \mathbf{P}_1 = -(\mathbf{P}_2' - \mathbf{P}_2)$ and

$$S^{(g)}_{s_g, s_g'}(\mathbf{q}) = \left\langle s_g \right| - Z_g e^{-i\mathbf{q}\mathbf{r}_n} + \sum_{j=1}^{N_g} e^{-i\mathbf{q}\mathbf{r}_j} \left| s_g' \right\rangle . \tag{7.55}$$

The main contribution to the total bremsstrahlung spectrum is from small q, at least from those not much greater than $1/R_a$ [21]. The relatively large mass of the nuclei ensures that their recoil makes a comparatively small contribution to the total bremsstrahlung spectrum, so that it can be neglected. As a result, in place of Eq. (7.55) we obtain

$$S^{(g)}_{s_g, s_g'}(\mathbf{q}) \approx - Z_g \delta_{s_g, s_g'} + \left\langle s_g \left| \sum_{j=1}^{N_g} e^{-i\mathbf{q}\mathbf{r}_j} \right| s_g' \right\rangle, \tag{7.56}$$

where the second term is a generalized expression for the form factor $Q(\mathbf{q})$ and coincides with it when $\langle s_g| = \langle s_g'| = \langle 0|$ [see Eq. (7.15)].

An analytic expression for the bremsstrahlung amplitude F can be written down in the form

$$F = F_1 + F_2,$$

where the first and second terms describe the emission of a photon by the first and second particles. Using Eqs. (7.53) and (7.56), taking the integral over the momenta $\mathbf{P}$ of the intermediate state, and summing over the excited intermediate states, we find the amplitude of bremsstrahlung during collisions of atomic particles to be

$$F^{(1)}_{pq}(\omega) = \frac{4\pi\,(\mathbf{e}\mathbf{q})}{q^2\omega}\left[+\frac{(Z_1 - N_1)}{m_1}(Z_1 - Q^{(1)}(q)) - \omega^2 \alpha_d^{(1)}(\omega, q)\right](Z_2 - Q^{(2)}(q)). \tag{7.57}$$

An expression for $F^{(2)}$ can be found from Eq. (7.57) by exchanging the indices 1 and 2. It is very important that the momentum $-q$ transferred from the first particle to the second be included in $F_{pq}^{(2)}(\omega)$ with the opposite sign; for this amplitude $\mathbf{q}$ is the acquired momentum and the momentum transfer is $(-\mathbf{q})$. As a result, the total bremsstrahlung amplitude is equal to

$$\begin{aligned} F_{pq}(\omega) = &- \frac{4\pi\,(\mathbf{e}\mathbf{q})}{q^2\omega}\Big\{[Z_1 - Q^{(1)}(q)]\,[Z_2 - Q^{(2)}(q)] \times \\ &\times \left(\frac{Z_2 - N_2}{m_2} - \frac{Z_1 - N_1}{m_1}\right) + \omega^2\,[\alpha_d^{(1)}(\omega, q)\,(Z_2 - Q^{(2)}(q)) - \\ &- \alpha_d^{(2)}(\omega, q)\,(Z_1 - Q^{(1)}(q))]\Big\}. \end{aligned} \tag{7.58}$$

For neutral particles $Z_1 = N_1$ and $Z_2 = N_2$, so that only the polarization radiation of one particle that has been polarized by the second particle makes a contribution to the bremsstrahlung and the role of the distributed charge is played by the $Z_g - Q^{(g)}(q)$. The terms with $\alpha_d^{(2)}(\omega, q)$ and $\alpha_d^{(1)}(\omega, q)$ appear in Eq. (7.58) with different signs, so that the induced dipole moments of the collision partners are oppositely oriented. As a function of ω the bremsstrahlung amplitude reproduces the structure of the dipole polarizabilities $\alpha_d^{(g)}(\omega, q)$, and with maxima where they occur in the $\alpha_d^{(g)}(\omega, q)$.

If we restrict ourselves for neutral particles to small (compared to $1/R_a^{(g)}$) and large (compared to the ionization potentials) ω, then, because $\alpha_d^{(g)} \sim -Z_g/\omega^2$, Eq. (7.58) takes the especially simple form

$$F_{pq}(\omega) = (4\pi(\mathbf{eq})/q^2\omega)[Z_2Q^{(1)}(q) - Z_1Q^{(2)}(q)], \tag{7.59}$$

and now describes the bremsstrahlung of the free electrons of one atom in the field of the nucleus of the other.

We note that, according to Eqs. (7.58) and (7.59), we have $F_{pq}(\omega) = 0$ for identical particles, although they may have internal structure, since they cannot have a dipole moment as a whole. The contributions of both of the particles to the bremsstrahlung amplitude (7.58) are equal in magnitude and opposite in sign. In this case, the bremsstrahlung is determined by the relativistic corrections to Eq. (7.58). Crudely speaking, these corrections reduce to having each of the particles generate its own emission spectrum with the spectrum of the incident particle shifted in frequency by an amount determined by the Doppler effect. Thus, if $\alpha_d(\omega)$ has a narrow maximum, then the bremsstrahlung spectrum will have two peaks, of which the one corresponding to the incident particle will be shifted and broadened because of the Doppler effect.

The differential bremsstrahlung cross section with respect to q and ω is obtained by substituting Eq. (7.58) in Eq. (7.14) and replacing p^2 by v_1^2. The total spectrum is obtained by integrating this cross section with respect to q from $q_{\min} \sim \omega/v_1$ to $q_{\max} = 2m_1m_2v_1/(m_1 + m_2) \equiv 2\mu v_1$. We note that for a light particle ($m_1 \ll m_2$ and $q_{\max} \sim 2p_1$) [see Eq. (7.23)], we naturally obtain

$$d\sigma(\omega) = \frac{16}{3}\frac{d\omega}{\omega c^3 v_1^2}\int\limits_{\omega/v_1}^{2\mu v_1}\frac{dq}{q}\left|\left\{(Z_1 - Q^{(1)}(q))(Z_2 - Q^{(2)}(q)) \times \right.\right.$$
$$\times\left(\frac{Z_2 - N_2}{m_2} - \frac{Z_1 - N_1}{m_1}\right) + \omega^2\left[(Z_2 - Q^{(2)}(q))\,\alpha_d^{(1)}(\omega, q) - \right.$$
$$\left.\left.\left. - (Z_1 - Q^{(1)}(q))\,\alpha_d^{(2)}(\omega, q)\right]\right\}\right|^2. \tag{7.60}$$

The bremsstrahlung spectrum during collisions of ions and atoms is determined by different ranges of the momentum q. For neutral particles there is no first term in Eq. (7.60) and the main contribution is from the region with $q \sim R_a^{-1}$ since when $q \ll R_a^{-1}$ the distributed charges $[Z_g - Q^{(g)}(q)]$ are small and when $q \gg R_a^{-1}$ the generalized polarizabilities $\alpha_d^{(g)}(\omega, q)$ are small.

For an ion, on the other hand, the main contribution is from small q, beginning with $q_{\min} = \omega v_1$ (if, of course, $q_{\min} \ll R_a^{-1}$).

The spectrum from bremsstrahlung of an ion on an atom given by Eq. (7.60) transforms to the bremsstrahlung of a heavy charge on an atom if we set $Z_1 = N_1$ and $m_2 = m_0$ and treat the charge as unstructured, i.e., replace $Q^{(2)}(q)$ by N_2 and

$-Z_2 + N_2$ by e_0 (the electronic charge equals unity). The spectrum from bremsstrahlung of an ion on an atom can be found in the logarithmic approximation and it is given by Eq. (7.44) with the above substitution.

The bremsstrahlung spectrum from atom–atom collisions must be calculated with the aid of Eq. (7.60). This formula becomes simpler for $\omega \gg I$, when, according to Eq. (7.59), we have

$$d\sigma(\omega) = \frac{16}{3}\frac{d\omega}{\omega}\frac{1}{c^3 v_1^2}\int_{\omega/v_1}^{2\mu v_1}\frac{dq}{q}\,|Z_2 Q^{(1)}(q) - Z_1 Q^{(2)}(q)|^2 \approx$$

$$\approx \frac{16}{3}\frac{d\omega}{\omega}\frac{1}{c^3 v^2}(Z_1 Z_2)^2 (I_1 - I_2)^2 \int_0^{\infty}\frac{t\,dt}{(t + I_1)^2 (t + I_2)^2}. \tag{7.61}$$

If $q_{\min} \geq R_a^{-1}$, then the spectrum of bremsstrahlung from ion–atom and atom–atom collisions is determined by the same range of values of q, so that going from an atom to an ion does not change the total bremsstrahlung spectrum very strongly.

As with the polarization bremsstrahlung from collisions of electrons with atoms, the angular distribution of the photons is characterized by a factor $(1 + \cos^2\theta)$, where θ is the angle between the momenta of the incident electron and the photon. As noted in connection with Eq. (7.36), the dependence $(1 + \cos^2\theta)$ is typical of the emission from a uniformly rotating dipole. Thus, the angular distribution of the photons confirms the qualitative picture of polarization bremsstrahlung which determines the total emission spectrum from collisions of atomic particles: each of the particles polarizes its partner and the polarization vectors rotate in space as they "follow" the particles which induce them.

7.9. Radiation Produced by Scattering of Helium and α-Particles on Xenon

In this section we examine the spectra of bremsstrahlung produced in collisions with xenon atoms of helium atoms, α-particles, and electrons moving at equal velocities [21]. We shall show that the bremsstrahlung intensities at frequencies on the order of the ionization potential of the intermediate, easily polarized, many-electron $4d^{10}$-subshell of xenon are of the same order of magnitude for electrons, α-particles, and helium atoms [20].

The bremsstrahlung spectrum of helium and α-particles is given by Eq. (7.60), while that of electrons is given by Eqs. (7.23) and (7.24). The form factor for He was taken from [33] and the form factor and generalized polarizability $\alpha_d^{(\mathrm{Xe})}(\omega, q)$ for xenon were taken from [11, 12], in which the form factor was calculated in the Hartree–Fock approximation, while $\alpha_d^{(\mathrm{Xe})}(\omega, q)$ was calculated in the random phase approximation with exchange using a special program [34].

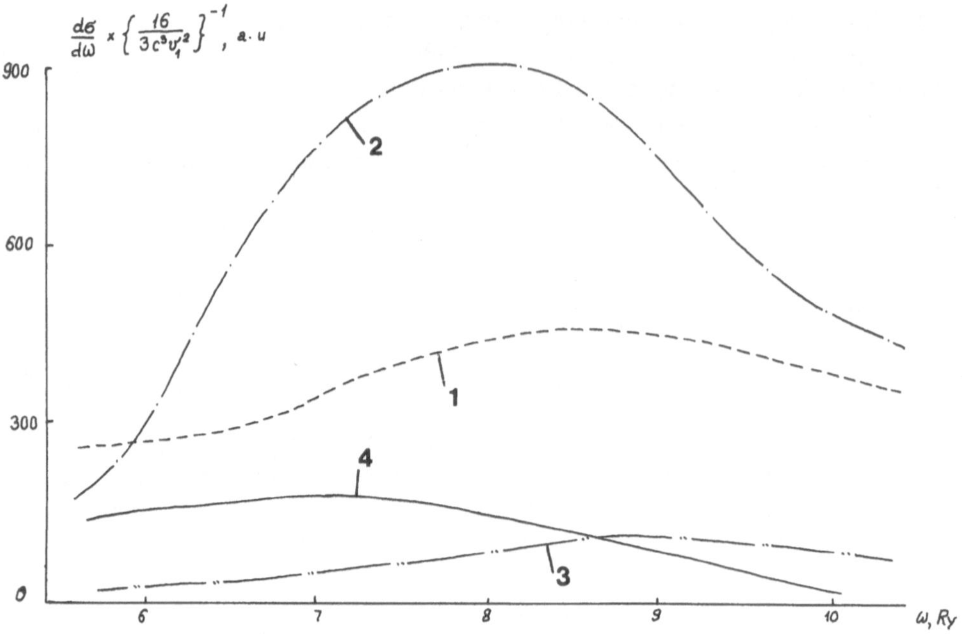

Fig. 7.25. The bremsstrahlung spectrum of fast particles on xenon [20] ($v_1 = 5$): (1) electron, (2) α-particle, (3) He (neglecting its polarizability), and (4) He (including the polarizability of both atoms).

We neglect the terms containing the mass m_2 (and, of course, m_1), even for α-particles. Let us suppose that the velocity $v_1 = 5$. This corresponds to an energy of 2.5 MeV for α-particles and He and 350 eV for electrons. This energy is somewhat low for high accuracy with the Born approximation. However, a large increase in v_1 would lead to a reduction in the bremsstrahlung intensity and, thus, be inconvenient for doing an experiment to observe polarization (atomic) bremsstrahlung. Since $q_{\min}$ and $R_a^{-1} \sim 1$ in this range of frequencies, we should expect (as noted in Sec. 7.8) that bremsstrahlung for He, α-particles, and electrons on Xe would be comparable. This is confirmed by experiment.

At first glance, it seems that one can neglect the generalized polarizability of helium, which for these values of ω is equal to $-Q^{(\mathrm{He})}(q)/\omega^2$ and is much less than $\alpha_d^{(\mathrm{Xe})}(\omega, q)$. However, for momentum transfers $qR_{id} \sim 1$, which lie within the range that controls the magnitude of the total bremsstrahlung spectrum, the contributions of both terms, $\mathrm{Re}\,\alpha_d^{(\mathrm{Xe})}(\omega, q)(Z^{(\mathrm{He})} - Q^{(\mathrm{He})}(q))$ and $\mathrm{Re}\,\alpha_d^{(\mathrm{He})}(\omega, q)(Z^{(\mathrm{Xe})} - Q^{(\mathrm{Xe})}(q))$, are of the same order, so that their interference is significant. The influence of the term proportional to $\alpha_d^{(\mathrm{He})}(\omega, q)$ is enhanced because $(Z^{(\mathrm{Xe})} - Q^{(\mathrm{Xe})}(q)) \gg 1$. Since $\mathrm{Re}\,\alpha_d^{(\mathrm{He})}(\omega, q) < 0$ in this range of values of ω, when $\mathrm{Re}\,\alpha_d^{(\mathrm{Xe})}(\omega, q) < 0$ the interference is destructive and when $\alpha_d^{(\mathrm{Xe})}(\omega, q) > 0$ it is constructive. There is actually no interference of the imaginary parts of the polarizability, since in this range of frequencies we have $\text{Im}\,\alpha_d^{(\text{He})}(\omega, q) \approx 0$, while $\text{Im}\,\alpha_d^{(\text{Xe})}(\omega, q)$ is large for $\omega \geq I_{4d}$.

Some calculated bremsstrahlung spectra of He [with and without taking $\alpha_d^{(\text{He})}(\omega, q)$ into account], α-particles, and electrons are shown in Fig. 7.25. These curves all have a maximum which is clearly related to the contribution of polarization (atomic) bremsstrahlung. It is four times greater for an α-particle than for an electron, since $Z_\alpha = 2$. The maximum of $d\sigma(\omega)/d\omega$ for an α-particle is shifted toward higher energies by 0.75 Ry relative to the location of the peak in the photoionization cross section for the $4d^{10}$-subshell of Xe. This is explained by the fact that with increasing q the proportionality between $\text{Im}\,\alpha_d(\omega, q)$ and $\sigma_\gamma(\omega)$ is disrupted. For q on the order of R_a^{-1} (which are only important in this range of ω for the chosen velocity, $v_1 = 5$), the deviation from proportionality between $\text{Im}\,\alpha_d(\omega, q)$ and $\sigma_\gamma(\omega)$ is already quite noticeable and the peak in $\text{Im}\,\alpha_d(\omega, R_a^{-1})$ is shifted toward higher ω compared to the maximum of $\sigma_\gamma(\omega)$.

The still greater shift in the peak in the (atomic) bremsstrahlung for an electron is related to the important contribution of polarization and ordinary bremsstrahlung to the bremsstrahlung spectrum, which is quite noticeable, since $q_{\min} = \omega/v_1 \sim R_a^{-1}$. The important contribution from interference of the bremsstrahlung amplitudes for helium on xenon and xenon on helium which enter Eq. (7.60) shows up in the significant difference (see Fig. 7.25) between the curves in which $\alpha_d^{(\text{He})}(\omega, q)$ is taken into account and those in which it is not. The increase in $d\sigma(\omega)/d\omega$ when $\omega < 7.8$ Ry is explained by constructive interference, and the decrease in $d\sigma(\omega)/d\omega$ when $\omega > 7.8$ Ry, by destructive interference, of the two amplitudes (helium on xenon and xenon on helium).

Beyond the ionization threshold the real parts of the amplitudes compensate one another to a significant degree, so that $d\sigma(\omega)/d\omega$ is determined by the contribution from $\text{Im}\,\alpha_d^{(\text{Xe})}(\omega, q)$.

The noteworthy difference in $d\sigma(\omega)/d\omega$ at the lowest and highest values of ω in Fig. 7.25 for electrons and α-particles is a manifestation of the "stripping" of electron shells during bremsstrahlung discussed above (in Secs. 7.5 and 7.7).

7.10. Total Bremsstrahlung Spectrum: Elastic and Inelastic

Thus far, we have only considered elastic bremsstrahlung, in which all the change in the kinetic energy of the colliding particles is carried away by the emitted photon. As noted in Chapter 1 and Sec. 7.1, however, inelastic bremsstrahlung also occurs. Examples of diagrams that describe inelastic bremsstrahlung are given in Figs. 7.11 and 7.12.

From the standpoint of the feasibility of experimental observation of the contribution of polarization bremsstrahlung, it is important to clarify the role of inelastic processes. If this role is large, then the observation of polarization bremsstrahlung requires a coincidence experiment in which the kinetic energy of the particles after the collision and the photon must be recorded simultaneously. At the same time, the possibility arises of separating elastic and inelastic bremsstrahlung and, by studying the contribution of the former, to evaluate the role of polarization bremsstrahlung. If a light (low-mass) particle is incident on the target, then this sort of program seems to be technically possible, although difficult: comparable quantities must be measured in coincidence. For example, at electron energies on the order of a few keV, one must detect photons with energies of tens and hundreds of eV. Present-day experimental studies of bremsstrahlung [10, 13] are limited just to measurements of the total spectrum, to which both elastic and inelastic bremsstrahlung contribute. It may be hoped that technical improvements in the experiments will make it possible to isolate the contribution of elastic bremsstrahlung to the scattering of electrons on atoms (or ions) and confirm the important role of the polarization of target atoms in the formation of the total bremsstrahlung spectrum.

The situation is much more complicated in heavy particle collisions. In order for the particles to be regarded as fast, their energies must reach tens or hundreds of MeV. It is extremely doubtful whether it will be possible to detect, in coincidence, energy losses of several tens or even hundreds of eV against a background of such high energies. Thus, although polarization bremsstrahlung increases the total bremsstrahlung spectrum by six to eight orders of magnitude, it is extremely important to understand that here the main contribution is from polarization bremsstrahlung or the emission by δ (secondary)-electrons that are removed from atomic particles during collisions. The δ-electrons may radiate in the static field of one of the colliding particles, and not necessarily because of their polarization. Since the probability of ionization in a collision is high, it seems natural that the main contribution to the total bremsstrahlung spectrum should be from emission by δ-electrons, which may be due mainly to purely traditional, rather than polarization, bremsstrahlung.

For heavy particles, therefore, it is especially important to calculate the total bremsstrahlung spectrum and to clarify the relative contributions of elastic and inelastic bremsstrahlung. We now show [23] that elastic bremsstrahlung predominates over a wide range of emission frequencies ω. The reason for its predominance in the total emission spectrum is that the contributions from virtual excitations of the atomic electrons combine coherently, similarly to the way this happens in Rayleigh scattering, while in the cross section for emission accompanied by ionization or excitation of the colliding particles, the contributions of the atomic electrons are incoherent. These analogies are determined by the fact that the colliding particles interact and exchange a photon (true, at nonrelativistic velocities this is a longitudinal photon, which leads to a Coulomb interaction, rather than a transverse photon).

We now consider the total emission spectrum, including elastic and inelastic bremsstrahlung by a fast, heavy charged particle with mass m_0 and charge e_0 incident on an atom. In this case, the contribution of the diagrams in Fig. 7.11 should be neglected, since their contribution to the bremsstrahlung amplitude contains a factor of $1/m_0$. Because of the high velocity of the incident particle, we must include (as a correction in the second-order Born approximation) the contribution of the diagrams which describe the emission of a photon by an electron in the target atom and the ionization of the target atom by the incident particle. The diagram of Fig. 7.26 is an example of such a process. Thus, inelastic bremsstrahlung is determined by a set of diagrams, of which the simplest are shown in Fig. 7.27. The contribution of more complicated diagrams must also be taken into account.

Experience with studies of photoionization and inelastic scattering [25] shows that the most important corrections to the diagrams of Fig. 7.27 come from diagrams which include the interaction of an electron $\varepsilon^{(j)}$ or ε'' with holes j or k, as shown in the diagrams of Fig. 7.4, and also from diagrams which take into account the readjustment (see Fig. 7.7) of the outer (intermediate) shells of a target atom during ionization of the inner j-shell.

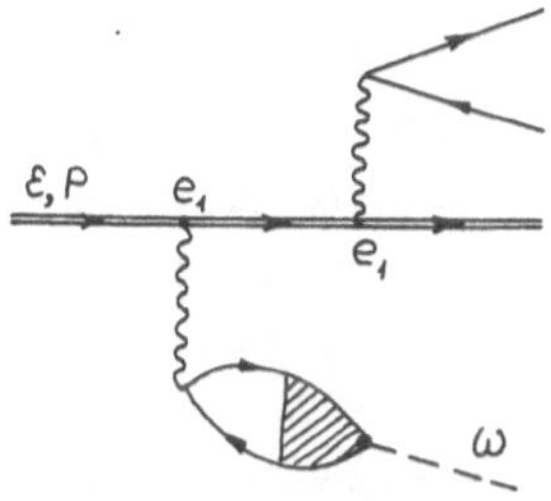

Fig. 7.26. An example of a diagram describing an inelastic process accompanying polarization bremsstrahlung.

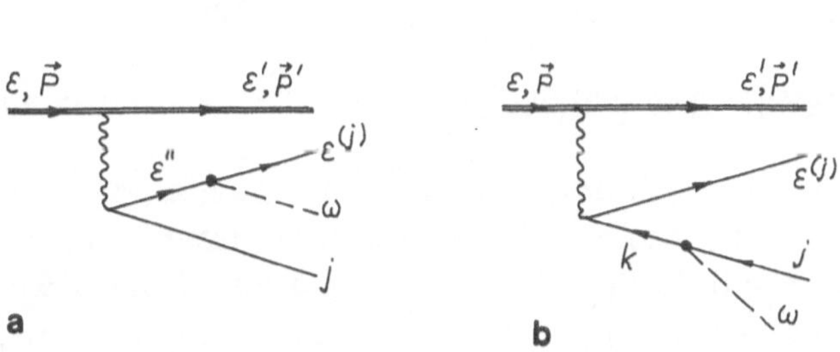

Fig. 7.27. Examples of diagrams describing inelastic bremsstrahlung of a fast particle.

The corresponding analytic expressions for the amplitude $F_{pq}(\omega, q, \varepsilon^{(j)})$ are considerably more complicated than Eq. (7.40), since, instead of the generalized polarizability $\alpha_d(\omega, q)$, it contains an expression which differs from Eq. (7.12) in having a different denominator where ω is replaced by the sum of the energies $[\omega + \varepsilon^{(j)} + I_j)]$ and the numerator contains the expression (for the diagram of Fig. 7.27a)

$$\langle j | e^{-i\mathbf{qr}} | \varepsilon'' \rangle \langle \varepsilon'' | (\mathbf{ed}) | \varepsilon^{(j)} \rangle,$$

which differs from that in Eq. (7.12) in having j replaced by $\varepsilon^{(j)}$ and $D(\omega)$ replaced by **d**. The last change is made because the correlations of the atomic electrons described by the sequence of diagrams in Fig. 7.4 has been neglected in the diagram of Fig. 7.27a.

When $\omega \ll I$ the main contribution to the total bremsstrahlung spectrum of a heavy particle is from inelastic bremsstrahlung, or a diagram similar to that of Fig. 7.27a, where a low-energy photon is emitted by an atom that has been removed from the atom. It is easy to confirm that the corresponding inelastic bremsstrahlung amplitude will be proportional to the inelastic scattering amplitude and contains a factor of $1/\omega$, while the elastic bremsstrahlung amplitude is proportional to ω. The amplitudes of the diagrams of Figs. 7.11a and 7.12a contain a factor of $1/\omega$, but along with $1/m_0$. The amplitudes of the diagrams of Figs. 7.11b and 7.12c contain a factor of $1/\omega_1$ for frequencies ω_2 that are not low. A term proportional to $1/\omega$ also exists in the elastic scattering amplitude, but it appears along with a factor of $1/m_0 \ll 1$. With increasing ω the contribution of elastic bremsstrahlung increases. Since it is of interest to find whether, and if so at what ω, elastic bremsstrahlung can be dominant, we shall immediately consider frequencies $\omega \gg I$. This condition, as we have demonstrated repeatedly above, makes it possible to neglect all the energies in the denominator except ω, i.e., to neglect the excitation energies of all the atomic shells. The sum over ε'' is easily taken in this case with the aid of the completeness theorem for the states ε'' in the diagram of Fig. 7.27a. As a result, we obtain

$$d\sigma^{\mathrm{tot}}(\omega) = \frac{16 e_0^2 d\omega}{3 c^3 v_0^2 \omega} \sum_{\omega_{n_0}=0}^{\infty} \int_{(\omega+\omega_{n_0})/v_0}^{\infty} \frac{dq}{q} \left| \left\langle n \left| \sum_{j=1}^{N} e^{-i\mathbf{qr}} \right| 0 \right\rangle \right|^2 \tag{7.62}$$

for the total emission spectrum of dipole photons, summed over all final states of the target atom.

Here ω_{n0} is the energy of atomic excitation or ionization. In the diagrams of Fig. 7.27, $\omega_{n0} = \varepsilon^{(j)} + I_j$ and N is the number of electrons in the target atom. We note that when $n = 0$, Eq. (7.43) transforms to Eq. (7.62) if we neglect terms $\sim 1/m_0$ in it and use the fact that when $\omega \gg I$, the formula $\alpha_d(\omega, q) = -Q(q)/\omega^2$ for the generalized polarizability (which we have used repeatedly above) is valid. The upper limit in the sum and integral is set equal to ∞, since the main contribution to Eq. (7.62) is from low momenta q and energies ω_{n0} (which allows us to replace the actual upper limits by ∞, at least as an estimate).

Since we are considering $\omega \gg I$, and the main contribution to Eq. (7.62) is from small ω_{n0}, for all the terms in the sum over ω_{n0}, we must neglect ω_{n0} compared to ω. For the sum over the excited states with $\omega_{n0} \gg I$, only large q, which are almost entirely transferred to an atomic electron, are significant. The restrictions on q can be obtained from the conditions $q \geq (\omega + \omega_{n0})/v_0$ and $\omega_{n0} \sim q^2/2$. The following restrictions on q follow from this for $\omega \ll v^2$: $\omega/v_0 < q < 2v_0$. As a result, on taking the sum over ω_{n0}, we obtain

$$d\sigma^{(p)}(\omega) \approx \frac{16e_0^2}{3c^3v_0^2}\frac{d\omega}{\omega}\left\{\int_{\omega/v_0}^{2v_0}\frac{dq}{q}|Q(q)|^2 + \int_{\omega/v_0}^{2v_0}\frac{dq}{q}[N - |Q(q)|^2 + \tilde{W}(q)]\right\}, \quad (7.63)$$

from Eq. (7.62), where $\tilde{W}(q) = \langle 0|\Sigma_{i\neq j}^{N} \exp\{-i\mathbf{q}(\mathbf{r}_i - \mathbf{r}_j)\}|0\rangle$, while $\mathbf{r}_i$ and $\mathbf{r}_j$ are the coordinates of the ith and jth electrons in the atom. For small q, we have $\tilde{W}(0) = N(N-1)$, while $\tilde{W}(q)$ falls off rapidly with increasing q.

The first term in Eq. (7.63) describes the contribution of elastic bremsstrahlung to the total spectrum, while the second corresponds to inelastic bremsstrahlung. Inelastic bremsstrahlung processes occur mainly in the region $\omega \leq v_0^2/2$, i.e., are determined by the translational energy and momentum of the atomic electron relative to the incident particle. The bonding of the electrons to the nucleus of the atom means that, although inelastic bremsstrahlung is possible for $\omega > v_0^2/2$, it is improbable in this range of ω.

As noted in Sec. 7.7, the main contribution to the first term of Eq. (7.63) is from momenta q ranging between ω/v_0 and R_a^{-1} (provided, of course, that $\omega R_a/v_0 \leq 1$). Then the first term of Eq. (7.63) coincides with Eq. (7.47) in the logarithmic approximation if we replace Z by N.

The integrand in the second term is close to zero for $q < R_a^{-1}$ and increases with rising q. Thus, in the logarithmic approximation, on integrating with respect to q from R_a^{-1} to $2v$, we obtain

$$d\sigma^{\text{inel}}(\omega) \approx \frac{16e_0^2}{3c^3v_0^2}\frac{d\omega}{\omega} N \ln(2v_0 R_a). \quad (7.64)$$

The ratio of the contributions of elastic and inelastic bremsstrahlung to the total spectrum is

$$N\frac{\ln(v_0/\omega R_a)}{\ln 2v_0 R_a} \gg 1. \quad (7.65)$$

Therefore, for $\omega/v_0 < R_a^{-1}$ or $I \gg \omega < v_0 R_a^{-1}$, elastic bremsstrahlung predominates. As ω increases within the interval $v_0 R_a^{-1} \ll \omega \leq v_0^2/2$, the lower limit of integration in Eq. (7.63) increases and this leads to a reduction in the contribution from elastic bremsstrahlung. The region of very high ω requires special study, since the restrictions employed in deriving Eq. (7.63) are not applicable there. A numerical calculation [35] has shown that elastic bremsstrahlung during scattering of protons with energies of several MeV on Al and C atoms exceeds the inelastic bremsstrahlung by an order of magnitude, in agreement with the estimate of Eq. (7.65).

Equations (7.43) and (7.60) for the elastic bremsstrahlung spectrum in atom–atom and charged particle–atom processes are extremely similar. The situation is analogous for inelastic bremsstrahlung. The destructive interference of the two atoms reduces the intensity of elastic bremsstrahlung. However, if the colliding particles differ noticeably in their nuclear charge, then elastic bremsstrahlung predominates for frequencies $\omega \ll v_0 R_a^{-1}$.

In fact, Eqs. (7.62) and (7.63) can be applied separately to each atomic shell. With increasing ω the inequality $\omega \ll vR_{\text{ext}}^{-1}$ begins to fail for the outer shells and their contribution becomes negligible. The main contribution to elastic bremsstrahlung is from an intermediate shell for which $\omega > I_{\text{med}}$ and $\omega \leq v_0 R_{\text{med}}^{-1}$, and that to inelastic bremsstrahlung is from all the outer shells, so that the ratio of the contributions of "elastic" and "inelastic" bremsstrahlung to the total spectrum is given by

$$\frac{N_{\text{med}}^2 \ln (v_0/\omega R_{\text{med}})}{(N_{\text{med}} + N_{\text{ext}}) \ln 2v_0 R_{\text{med}}} \tag{7.66}$$

rather than Eq. (7.65). Here N_{ext} is the number of electrons in the shells lying outward of the intermediate shell under consideration. The sum $(N_{\text{ext}} + N_{\text{med}}) < N$ appears in place of the total number N of electrons in Eq. (7.66), since those shells for which the ionization potential exceeds ω do not contribute to Eq. (7.62).

Although Eqs. (7.62)–(7.66) have been derived under the assumption that $\omega \gg I$, they can be used for estimates in the region of $\omega \geq I$. This means that the contribution of elastic bremsstrahlung to the total spectrum will still be predominant when $\omega \geq I$. Thus, the elastic bremsstrahlung spectrum calculated with the aid of Eq. (7.43) is, at least in terms of the order of magnitude, equivalent to the total spectrum, which can be measured directly in an experiment. For example, for α-particles in xenon this spectrum has a strong maximum beyond the threshold for ionization of the $4d^{10}$-subshell (see Fig. 7.25, curve 2).

Chapter 8

POLARIZATION BREMSSTRAHLUNG INVOLVING POSITRONS, μ-MESONS, AND NUCLEAR PARTICLES

M. Ya. Amus'ya

8.1. Introductory Comments

Here we shall discuss bremsstrahlung processes involving positrons, such as bremsstrahlung of electrons (and positrons) on positronium [1], as well as on mesic hydrogen. Bremsstrahlung of electrons (and positrons) on positronium is entirely polarization bremsstrahlung, since at high frequencies the inelastic and elastic processes are of the same order. Bremsstrahlung on mesic hydrogen [2] is also interesting in that it represents an important inelastic loss mechanism over an extremely wide range of incident particle energies.

Bremsstrahlung also occurs when the interaction between the colliding particles is not of an electromagnetic nature. For example, the scattering of neutrinos on atoms, as well as on atomic nuclei, is accompanied by the emission of photons [3, 4]. Of course, the intensity of this radiation is very low, but it is one of the manifestations of the passage of neutrinos through matter and can, therefore, serve as a method of observing them. Naturally, the intensity of resonant bremsstrahlung (the emission of photons with frequencies very close to or coinciding with the excitation energy of atomic levels) may be many orders of magnitude greater than nonresonant bremsstrahlung.

The polarization of a nucleus as it collides with nucleons and other nuclei plays an important role in the formation of the γ-ray bremsstrahlung spectrum at energies on the order of the binding energy of nucleons in nuclei [5]. With respect to nuclear collisions, polarization bremsstrahlung refers to the emission of γ-ray photons owing to the polarization of either the target or the incident particle. Resonant polarization bremsstrahlung, on the other hand, refers to the so-called giant nuclear resonances. Near a resonance, polarization bremsstrahlung is almost an order of magnitude greater than traditional bremsstrahlung. Polarization bremsstrahlung probably explains the broad maximum in the γ-ray spectrum corresponding to the giant quadrupole resonance of C^{12} carbon nuclei [5–7].

The contribution of polarization bremsstrahlung to the total emission spectrum resulting from collisions of μ-mesons with nuclei should be extremely important. Charged mesons can efficiently polarize a target nucleus, especially when the energy transfer to the latter lies in the region of a giant dipole resonance, so that the total emission spectrum acquires a substantial maximum at that energy. The contribution of polarization bremsstrahlung during the scattering of π-mesons on nuclei should be large. This includes π^0-mesons, for which one must take into account both the polarization of the nucleus by the meson and the polarization of the π^0-meson by the nucleus.

Polarization bremsstrahlung should also be significant during collisions of other structured particles (nucleons and mesons) among themselves if we treat them as consisting of quarks and undergoing polarization during the collision process [8].

The creation of gluons (the quanta of the chromodynamic field in the physics of strong interactions) is essentially similar to polarization bremsstrahlung of neutral particles. The similarity is, to a great extent, determined by the fact that, as with photons in neutral atom collisions, gluons are only emitted by a component of the elementary particle itself, i.e., a quark (or another gluon).

8.2. Emission during Collisions of Positrons with Atoms

Polarization bremsstrahlung is extremely important if the incident particle is a positron. The bremsstrahlung amplitude for a positron, as for an electron, is determined by the diagrams of Fig. 7.15, where, obviously, the straight line now represents the positron. The original formulas (7.1) and (7.2) and the approximations (7.3) and (7.4) are still valid for slow positrons. The opposite signs of the positron and electron do not appear explicitly, since the charge of the incident particle shows up the same way in the first order (see the bold point of Fig. 7.15) in the ordinary (electron) and polarization (atomic) amplitudes. The positron is repelled by the atom and does not penetrate (if it is slow) close to the nucleus or undergo exchange with the electrons in the target, so that replacing the wave function of the positron by a plane wave is even more justified than for an electron when calculating the polarization bremsstrahlung amplitude in Eq. (7.4). As a result, Eq. (7.6) for the bremsstrahlung amplitude is valid for a slow positron, where a now represents the scattering length of the positron on the atom.

The ratio $\zeta(\omega)$ of the total bremsstrahlung spectrum to the ordinary (electron) spectrum is given by Eq. (7.8) as

$$\zeta(\omega)=[1+(\omega/a)\alpha_d(\omega)]^2. \tag{8.1}$$

The presence of polarization bremsstrahlung means that if $a < 0$, then the cross section for positron bremsstrahlung can go to zero before the threshold for ionization of a target atom.

As an example, for He, we have $\alpha_d(0) = 1.3$ [9], while $a = -0.43$ [10], so that before the ionization threshold ($\omega < I = 1.4$ Ry) we have $\zeta(\omega) \sim (1 - 3\omega)^2$ and the bremsstrahlung cross section goes to zero for $\omega = 1/3 < I$ and increases on approaching the threshold, exceeding the "traditional" cross section by an order of magnitude or more.

Radiation may be produced during scattering of slow positrons as a result of the transfer of an electron from the target atom to the positron with the formation of positronium Ps. The binding energy of electrons in negative ions (and in alkali metal atoms) is less than in positronium. Thus, the transfer leading to positronium formation is energetically favorable and at low collision speeds all the released energy is emitted as a photon. The cross section for such a process is large and after a charge exchange event of this type to an excited level of Ps, a transition to the ground state proceeds with a probability of unity.

Whereas at low energies the difference between electron and positron scattering on an atom shows up indirectly through the phase δ_0, for fast electrons and positrons it shows up explicitly. The bremsstrahlung amplitude for a fast positron is determined by the diagrams of Fig. 7.17, and the analytic expression differs from Eq. (7.11) in that the "plus" sign in front of the contribution from polarization bremsstrahlung is replaced by a "minus" sign. This replacement is made because, as can be seen from Fig. 7.17, the amplitude of traditional bremsstrahlung is proportional to the square of the charge of the incident particle, while the amplitude of polarization bremsstrahlung is proportional to the first power of the charge. The bremsstrahlung amplitude of a fast positron is obtained if we change the sign in front of the polarizability in Eqs. (7.4)–(7.6). Of course, changing the sign leads to quantitative and qualitative changes in the bremsstrahlung spectrum of a positron compared to that of an electron.

The basic qualitative difference between positron and electron bremsstrahlung is that stripping of the electron shells does not occur with positrons. As a result, at high frequencies $\omega \gg I$ the amplitude of positron bremsstrahlung on an atom does not reduce to the bremsstrahlung amplitude of an electron on a nucleus. In fact, since an incident positron and the atomic electrons (even if we regard them as free, which is correct for high frequencies ω) have a nonzero dipole moment, they can emit photons. As a result, for example, instead of Eq. (7.17) for the amplitude we find $F_{pq}^{(+)}(\omega)$ to be

$$F_{pq}^{(+)}(\omega)=\frac{4\pi\,(\mathbf{eq})}{\omega q^2}(Z-2Q(q)). \tag{8.2}$$

Thus, we note that bremsstrahlung on an atom does not reduce to bremsstrahlung in the static field of a shielded nucleus for either electrons or positrons. The expression $Z - Q(q)$ corresponds to the static shielding model, while the additional term $[-Q(q)]$ is the contribution of polarization bremsstrahlung (we note that the bremsstrahlung of fast positrons on hydrogen including the polarizability of the target atom has been examined in [11, 12]). The expressions which take the shell structure of the target atom into account explicitly are also modified. Thus, they retain the terms for bremsstrahlung on electrons in the outer shells of the target atom, and the bremsstrahlung amplitude for a fast positron will be given by

$$F_{pq}^{(+)}(\omega) = -\frac{4\pi\,(eq)}{\omega q^2}\left[-Z + 2Q^{\text{ext}}(q) + Q^{\text{int}}(q) - \omega^2\alpha_d^{\text{int}}(\omega, q)\right] \tag{8.3}$$

rather than by Eq. (7.18), where $Q^{\text{ext}}(q)$ is the form factor for all the outer atoms of the shells and $Q^{\text{int}}(q)$ and $\alpha_d^{\text{int}}(\omega, q)$ are the form factor and polarizability, respectively, of the shells whose ionization potential exceeds ω. As ω is increased from values well below the ionization potential I_n of a given shell to $\omega > I_n$, the contribution of the term $-\omega^2\alpha_d^n(\omega, q)$ goes from being negligibly small and negative to positive and equal to $Q^n(q)$. Unlike in the case of electron bremsstrahlung, this term does not compensate the analogous term in $Q^n(q)$, but adds to it. In this sense, when speaking of positron bremsstrahlung it is appropriate to speak of "dressing" the electron shells, rather than "stripping" them.

It is clear that $F_{pq}^{(+)}(\omega)$, as well as the differential cross section for bremsstrahlung with respect to q, can also go to zero when $\omega \gg I$, as opposed to the case for $F_{pq}(\omega)$ [see Eq. (7.18)].

We note, however, that to within logarithmic accuracy the bremsstrahlung cross sections for electrons and positrons are the same at high ω. This is easily confirmed by using Eq. (8.2) for $F_{pq}^{(+)}(\omega)$ and keeping in mind that for both small q $(\omega/p < q \ll R_a^{-1})$ and large q $(R_a^{-1} \ll q < 2p)$, the expression $[Z - 2Q(q)]^2$ is the same and roughly equal to Z^2. The situation is similar with regard to the discontinuity in the cross section $\omega d\sigma(\omega)/d\omega$ owing to "dressing" of the electron shells during positron bremsstrahlung on an atom. The magnitude of the discontinuity is given by

$$\Delta\,(\omega d\sigma\,(\omega)/d\omega)^{(+)} = -2\int_{I_n/p}^{2p}\left[Z - Q(q) - Q^{\text{ext}}(q)\right]Q^n(q)\,\frac{dq}{q}, \tag{8.4}$$

where I_n and $Q^n(q)$ are the ionization potential and form factor of the "dressed" shell. At first glance this expression differs from the analogous expression for the bremsstrahlung cross section of an electron [see Eq. (7.25)]

$$\Delta\,(\omega d\sigma\,(\omega)/d\omega) = 2\int_{I_n/p}^{2p}\left[Z - Q(q) + Q^{\text{ext}}(q)\right]Q^n(q)\,\frac{dq}{q}.$$

For both positrons and electrons, however, the major contribution to $\Delta(\omega d\sigma(\omega)/d\omega)$ is from small q, where the two expressions under the integrand differ only in their sign, being equal to $-N^{\text{ext}}N^n/q$ and $N^{\text{ext}}N^n/q$, respectively. Given the negative sign in front of the integral of Eq. (8.4), it is clear that to within logarithmic accuracy, we have

$$\Delta(\omega d\sigma(\omega)/d\omega)^{(+)} \approx \Delta(\omega d\sigma(\omega)/d\omega).$$

The difference between electron and positron bremsstrahlung is important in the differential cross sections in the region $q \sim R_a^{-1}$, which makes a small contribution to $d\sigma(\omega)/d\omega$. Transition polarization scattering [12] of virtual photons lies at the foundation of the physics of polarization bremsstrahlung. The difference between the scattering of positrons and electrons when polarization effects are included has been pointed out elsewhere [12].

8.3. Scattering of Positrons or Electrons on Positronium

Let us consider the elastic and inelastic bremsstrahlung which arises during collisions of fast electrons or positrons with positronium (Ps) [1]. Despite the somewhat exotic character of the target, this kind of collision is of considerable fundamental interest, since the analog of the quantity $Z - Q(q)$ in Eq. (7.15) goes to zero for positronium. Thus, all the bremsstrahlung originates in the polarization of the positronium or its separation into an electron and positron, as well as in its excitation. In this way bremsstrahlung on positronium is fundamentally different from bremsstrahlung on hydrogen and other atoms.

A positronium atom has no nucleus, so that in determining the bremsstrahlung amplitude one must include the motion of the electron and positron that form it relative to the common center of mass.

The amplitude of a bremsstrahlung process in which a photon of frequency ω is emitted during the scattering of a fast, but nonrelativistic, positron (or electron) on positronium, while the positronium undergoes a transition from the ground state $|0\rangle$ into an excited state $|m\rangle$, can be written in the form [1]

$$F_{pq}^{mo}(\omega) = \frac{4\pi e_0\,(\mathbf{eq})}{\omega q^2}\,[e_0 Q_{mo}(\mathbf{q}) - \omega^2 A_{mo}(\omega, \mathbf{q})] \equiv F_{pq}^{mo(1)}(\omega) + F_{pq}^{mo(2)}(\omega), \qquad (8.5)$$

where

$$Q_{mo}(\mathbf{q}) = \langle m | (e^{-i\mathbf{q}\mathbf{r}_+} - e^{i\mathbf{q}\mathbf{r}_-}) | 0 \rangle. \qquad (8.6)$$

Here $r_\pm$ denotes the coordinates of the electron and positron in the Ps relative to its center of mass and e_0 is the charge of the scattering particle, which equals -1 for an electron and 1 for a positron in the system of units we have chosen.

The first term in Eq. (8.5) is the amplitude of traditional bremsstrahlung and the second is the amplitude of elastic and inelastic bremsstrahlung in the language of the diagrams shown in Figs. 7.11a and 7.12a, b, respectively. The expression $A_{m0}(\omega, \mathbf{q})$ in Eq. (8.5) is the analog of the generalized polarizability $\alpha_d(\omega, \mathbf{q})$ [see Eq. (7.12)] for polarization inelastic bremsstrahlung processes and is given by

$$A_{m0} = \frac{i}{(eq)\,\omega} \sum_n \left\{ \frac{\omega_{mn}\,(\mathbf{ed})_{m0} Q_{n0}(\mathbf{q})}{\omega_{nm} - \omega} + \frac{\omega_{n0} Q_{mn}(\mathbf{q})\,(\mathbf{ed})_{n0}}{\omega_{n0} + \omega} \right\}, \tag{8.7}$$

where $\mathbf{d}$ is the dipole moment operator for positronium and $\omega_{mn} = E_m - E_n$ is the transition energy. We note that for $\langle m| = \langle 0|$ and small momentum transfers $q \ll 1$, we have $A_{00}(\omega, q) = \alpha_d^{\mathrm{Ps}}(\omega)$, where $\alpha_d^{\mathrm{Ps}}(\omega)$ is the dipole dynamic polarizability of the positronium. Equation (8.5) coincides with the amplitude of bremsstrahlung on a hydrogen atom if $\mathbf{r}_+$ and $\mathbf{r}_-$ in Eq. (8.6) are taken to be the coordinates of the proton ($r_p \approx 0$) and electron. This agreement is possible because of the fact that positronium, like hydrogen, can be considered to be at rest during the collision process [including the recoil of the positronium introduces a correction of order $\delta v/c \ll 1$ in Eq. (8.4), where δv is the velocity acquired by the positronium because of the recoil]. We point out that Ps is a truly neutral particle, since $Q_{00}(q) = 0$ for arbitrary q. Thus, in Eq. (8.5) there is no traditional part of the amplitude for elastic bremsstrahlung, which contains $Q_{00}(q) = 0$. Since $Q_{00}(q) = 0$, elastic scattering of a fast charged particle on positronium does not occur in the first Born approximation.

The differential cross section for inelastic bremsstrahlung with respect to ω, q, and the direction of emission of the photon, $\Omega_{\mathbf{k}}$, is obtained using Eq. (8.5):

$$d\sigma^{m0}(\omega, q, \Omega_{\mathbf{k}}) = \frac{\omega q\, d\omega\, dq\, d\Omega_{\mathbf{k}}}{(2\pi)^3 c^3 v^2} \{ |F_{pq}^{m0(1)}(\omega)|^2 + |F_{pq}^{m0(2)}(\omega)|^2 \}. \tag{8.8a}$$

It is understood that in Eq. (8.8a) the sum has been taken over the fine structure levels of a given state $|m\rangle$, so that the different $|m\rangle$ are distinguished only by their principal quantum numbers.

It is interesting that there is no interference term in Eq. (8.8a). This is related to the fact that, according to the definition (8.6), the ordinary part of the amplitude is nonzero only for odd transitions, while for even transitions we have $Q_{m0}(q) = 0$ because the expression $(e^{-i\mathbf{qr}_+} - e^{-i\mathbf{qr}_-})$ is odd. On the other hand, $F_{pq}^{m0(2)}(\omega) \neq 0$ only for even transitions, as can be seen from Eq. (8.7). The parity of the states $|0\rangle$ and $|n\rangle$, and $|m\rangle$ and $|n\rangle$ must be opposite, so that the states $|0\rangle$ and $|m\rangle$ must, therefore, have the same parity.

The absence of an interference term also follows from the evident fact that the emission during $(e^+ + \mathrm{Ps})$ and $(e^- + \mathrm{Ps})$ collisions is the same because of the complete symmetry with respect to the substitution $e^+ \rightleftharpoons e^-$, while there is a sign change in front of only one term in the amplitude (8.5) on going from an incident e^- to an incident e^+.

In the case $|m\rangle = |0\rangle$, we have

$$d\sigma(\omega, q) = \frac{16}{3}\frac{d\omega}{\omega c^3 v_0^2}|\omega^2 \alpha_d^{Ps}(\omega, q)|^2 \frac{dq}{q}. \tag{8.8b}$$

from Eq. (8.5). It is clear that, unlike bremsstrahlung on any atom, bremsstrahlung on positronium does not contain an ordinary component and is entirely produced by virtual excitation of the positronium in the course of the collision.

The differential cross section with respect to ω and the momentum q for bremsstrahlung of an electron or positron on positronium is greater than on a hydrogen atom for arbitrary (but not too low) ω and q. The ratio χ of these cross sections can be obtained from Eqs. (8.8b), (7.14), and (7.15) for an electron and, with a negative sign in front of $\alpha_d(\omega, q)$, for a positron [on setting $Z = 1$ and $Q(q)$ equal to $(1 + 1/4q^2)^{-2}$, the form factor for the hydrogen atom, in Eq. (7.15)]. For electron scattering and $\omega \gg I$, we have $\chi = 4$. A value of χ in this range can be explained qualitatively as follows: for $\omega \gg I$ the target atom can be regarded as a set of free particles. Since free electrons do not radiate as dipoles, the bremsstrahlung of an electron on positronium reduces to the emission of a photon by an "incident electron + positron" system, while the bremsstrahlung of an electron on hydrogen reduces to the emission of an electron on a proton. In the first case, the reduced mass of the colliding particles is 1/2, while the bremsstrahlung cross section, which is inversely proportional to the square of the mass, is four times greater in the first case than in the second. For low ω ($\omega \lesssim I$) and $q \leq 1$, the quantity χ reaches 64.

It is extremely easy to find the total cross section for emission by an electron (positron) on positronium, $d\sigma^{tot}(\omega, q)$, which is obtained from Eq. (8.7) by integrating over $d\Omega_k$ and summing over m. From Eq. (8.7) we obtain

$$d\sigma^{tot}(\omega, q) = \frac{16}{3}\frac{d\omega dq}{\omega q}\frac{1}{c^3 v_0^2}\sum_m \{|Q_{mo}(\mathbf{q})|^2 + \omega^4 |A_{mo}(\omega, \mathbf{q})|^2\}. \tag{8.9}$$

For frequencies $\omega \gg I$, the second term in Eq. (8.8a) becomes much simpler:

$$\sum_m \omega^4 |A_{mo}(\omega, \mathbf{q})|^2 = 4\langle 0|\cos^2(\mathbf{qr}/2)|0\rangle.$$

Given also that $\Sigma_m |Q_{m0}(\mathbf{q})|^2 = 4\langle 0|\sin^2(\mathbf{qr}/2)|0\rangle$, we find the differential cross section with respect to ω and q for elastic and inelastic bremsstrahlung at high frequencies $\omega \gg I$:

$$d\sigma^{tot}(\omega, q) = \frac{64}{3}\frac{d\omega dq}{\omega q}\frac{1}{c^3 v_0^2}\left\langle 0\left|\left(\sin^2\left(\frac{\mathbf{qr}}{2}\right) + \cos^2\left(\frac{\mathbf{qr}}{2}\right)\right)\right|0\right\rangle = \frac{64}{3}\frac{d\omega dq}{\omega q}\frac{1}{c^3 v_0^2}. \tag{8.10}$$

The total emission spectrum is obtained from Eq. (8.10) by integrating with respect to q from $q_{min} = \omega/v_0$ to $q_{max} = 2\mu v_0$, where μ is the reduced mass of the electron (positron) in the collision with positronium ($\mu = 2/3$):

$$d\sigma^{tot}(\omega) = \frac{64}{3}\frac{d\omega}{\omega c^3 v_0^2}\ln\frac{4v_0^2}{3\omega}. \tag{8.11}$$

The contribution to the total spectrum from elastic bremsstrahlung can be found separately by integrating Eq. (8.8a) with respect to q from q_{min} to q_{max}. Since $\alpha_d^{Ps}(\omega, q)$ falls off rapidly with increasing q, we find that the elastic bremsstrahlung spectrum for $q > 1$ is given to within logarithmic accuracy by

$$d\sigma^{el}(\omega) = \frac{64}{3}\frac{d\omega}{\omega c^3 v_0^2}|\alpha_d^{Ps}(\omega)\,\omega^2|^2\ln\frac{v_0}{\omega}. \tag{8.12}$$

For high ω the polarizability of positronium is twice that of hydrogen and equals $-2/\omega^2$; hence, for $\omega \gg I$ Eq. (8.12) yields

$$d\sigma^{el}(\omega) = \frac{64}{3}\frac{d\omega}{\omega c^3 v_0^2}\ln\frac{v_0}{\omega}. \tag{8.13}$$

The difference between the total and elastic bremsstrahlung spectra is simply the contribution of the inelastic bremsstrahlung,

$$d\sigma^{inel}(\omega) = \frac{64}{3}\frac{d\omega}{\omega c^3 v_0^2}\ln\frac{4v_0}{3}. \tag{8.14}$$

Therefore, inelastic bremsstrahlung makes a logarithmically large contribution to the emission spectrum. Note that this is a consequence of the relatively low mass of positronium. In collisions of electrons with hydrogen [12] or any other atom, the contribution of inelastic bremsstrahlung is small for $\omega \gg I$. In this case, the atomic electrons can be regarded as free. A system of free electrons, including the incident electron, does not have a dipole moment and cannot, therefore, emit dipole photons. Only the incident electron changes its state, while the atom is not excited during the process of scattering with radiation.

If a positron is incident on an atom, then polarization bremsstrahlung, both elastic and inelastic (i.e., bremsstrahlung on electrons of the target atom), takes place with a high probability. Analogously, for positronium as well, inelastic bremsstrahlung is an extremely likely process, in which the "incident electron (positron) + positron (electron)" emits at $\omega \gg I$ and either of the colliding particles can change its state. This also leads to a large contribution from inelastic bremsstrahlung (8.14). We note that the above Eqs. (8.8a) and (8.12) need to be made more accurate for $\omega \ll I$, since they do not contain a singularity of the type $1/\omega$ or a term independent of ω, which are characteristic of the bremsstrahlung spectrum at low frequencies for arbitrary colliding and emitting particles [13]. This defect of Eqs. (8.8a) and (8.12) is related to the fact that the cross section for elastic scattering on positronium is equal to zero in the first Born approximation. A term proportional to $1/\omega$ and one that is independent of ω appear if we take the polarization potential into account for electron (positron) scattering on positronium.

8.4. Inelastic Scattering on Mesic Hydrogen

We now consider electron bremsstrahlung on μ-hydrogen (H_μ atoms), consisting of a proton and a μ-meson. In this system bremsstrahlung is the only mechanism for inelastic losses over an extremely wide range of incident particle energies up to the excitation (ionization) threshold for H_μ, 2.1 (2.8) keV. It is also important that we can limit ourselves to the first Born approximation for describing the scattering of an electron (or even a positron) on H_μ for arbitrary incident particle energies [14]. The smallness of the elastic scattering cross section means, of course, that the contribution of polarization (atomic) bremsstrahlung is relatively large.

The elastic scattering cross section for electrons (positrons) on H_μ at very low energies is not equal to the geometric cross section, which is of order $R_\mu{}^2$ (where $R_\mu = 1/m_\mu$ is the Bohr radius for μ-hydrogen and $m_\mu \sim 208$ is the mass of the μ-meson), but is several orders of magnitude smaller: $\sigma_{el} = 4\pi R_\mu{}^2/m_\mu{}^2$, where R_μ is measured in Bohr radii and m_μ, in electron masses.

As in the elastic scattering problem for electrons (positrons) on H_μ, the inelastic scattering cross section can be calculated by using perturbation theory with respect to the interaction between the colliding particles at arbitrary collision energies. For just this reason, we can use Eq. (7.13) which is applicable for arbitrary collision energies, including low energies (unlike in the case of electron–atom scattering). Thus, the amplitude of elastic (positron) electron bremsstrahlung on H_μ is given by the following formula [2] ($\omega < p^2/2$ and $q < 2p$):

$$F_{pq}^{H_\mu}(\omega) = \frac{4\pi\,(\mathbf{eq})}{\omega q^2}\left(1 - Q^{H_\mu}(q)\right) + \frac{4\pi e_0 \omega}{q^2}(\mathbf{eq})\,\alpha_d^{H_\mu}(\omega, q). \tag{8.15}$$

Here $Q^{H_\mu}(q)$ and $\alpha_d{}^{H_\mu}(\omega, q)$ are the form factor and generalized polarizability of the μ-hydrogen, respectively, given by Eqs. (7.15) and (7.12), where the density distribution $\rho(r)$ and the wave function refer to the H_μ. The charge $e_0 = \pm 1$ in Eq. (8.15) changes sign on going from an electron to a positron.

We restrict ourselves to nonrelativistic energies of the incident electron, where $\rho \ll c$ and, therefore, $q_{max} R_\mu \sim p/m_\mu \ll 1$. This means that bremsstrahlung is produced mostly at large distances from the μ-hydrogen, which is explained by its small radius. Since $qR_\mu \ll 1$, it is sufficient to limit ourselves to the lowest-order term in the expansion of $F_{pq}{}^{H_\mu}(\omega)$ in powers of (qR_μ). On making this expansion, from Eq. (8.15) we find that

$$F_{pq}^{H_\mu}(\omega) = \frac{4\pi\,(\mathbf{eq})}{\omega q^2}\left[\frac{q^2}{2m_\mu^2} + e_0\omega^2\alpha_d^{H_\mu}(\omega)\right] \equiv F_1(\omega, q)\,\frac{(\mathbf{eq})}{q}. \tag{8.16}$$

The differential cross section with respect to q and ω for bremsstrahlung is obtained from Eq. (8.15) with the aid of Eq. (7.14), from which we obtain

$$d\sigma(\omega, q) = \frac{\omega q\, d\omega\, dq}{3\pi^2 c^3 v_0^2}\,|F_1(\omega, q)|^2. \tag{8.17}$$

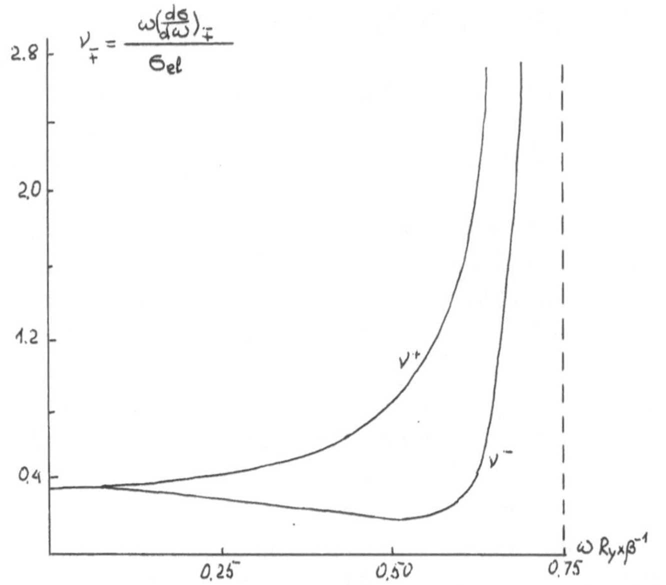

Fig. 8.1. The ratio of the cross section for bremsstrahlung of an electron on H_μ to the corresponding cross section for elastic scattering σ_{el} [2]. (The collision energy is 225 Ry.)

on summing over the polarizations of the photon and integrating over the angles of emission. The ratio η of the amplitudes of the polarization and traditional terms [the first and second terms in Eq. (8.16)] can be estimated by replacing the polarizability $\alpha_d^{H_\mu}(\omega)$ by its static value $9/2R_\mu{}^3$ ($\alpha_d = 9/2$ for hydrogen):

$$\eta > 9\omega^2/q^2 m_\mu. \tag{8.18}$$

This inequality shows that for low momentum transfers $q \sim q_{\min} = \omega/v_0$ with $p \geq (m_\mu)^{1/2}$, the ratio $\eta > 9$. Thus, in the differential cross section with respect to q for bremsstrahlung, there is a region of q and p in which the contribution of polarization bremsstrahlung will be predominant for arbitrary $\omega < (m_\mu)^{1/2}$.

The bremsstrahlung spectrum $d\sigma(\omega)/d\omega$ is obtained by integrating Eq. (8.17) with respect to the variable q:

$$d\sigma(\omega) = \frac{16}{3}\frac{d\omega}{c^3}\left\{\frac{p^2R_\mu^4}{\omega}\left[1-\left(\frac{\omega}{2p^2}\right)^4\right] + e_0\omega R_\mu^2 \times \right.$$

$$\left. \times\, [2\operatorname{Re}\alpha_d^{H_\mu}(\omega)]\left[1-\left(\frac{\omega}{2p^2}\right)^2\right] + \frac{\omega^3}{p^2}|\alpha_d^{H_\mu}(\omega)|^2 \ln\left(\frac{2p^2}{\omega}\right)\right\}. \tag{8.19}$$

We note that, since $\omega < p^2/2$, the terms in $\omega/2p^2$ are small compared to unity. We can now estimate the ratio η of the contributions of polarization bremsstrahlung and the interference terms [the second and third terms in Eq. (8.19)] to the first term, which represents the contribution of traditional bremsstrahlung to the bremsstrahlung spectrum. Replacing $\alpha_d^{H_\mu}(\omega)$ by $9/2m_\mu^3$, which is legitimate for $\omega \leq m_\mu$, we obtain, respectively,

$$\eta_1(\omega) \gtrsim 9\left(\frac{\omega}{p}\right)^2 \frac{1}{m_\mu},$$
$$\eta_2(\omega) \gtrsim \left(\frac{9\omega^2}{2p^2 m_\mu}\right)^2 \ln\frac{2p^2}{\omega}. \quad (8.20)$$

These inequalities show that in the region of $p \sim m_\mu^{1/2}$ and $\omega \sim m_\mu$, the bremsstrahlung spectrum is mainly determined by the polarization term in Eq. (8.19). With decreasing frequency the contribution of ordinary bremsstrahlung increases, and is predominant for low ω, as is to be expected.

We note that bremsstrahlung in the static field of a target atom usually predominates for a light particle. In the situation discussed above, the extremely important role of polarization bremsstrahlung is a consequence of the electrical neutrality and compactness of the target atom.

A bremsstrahlung spectrum calculated according to Eq. (8.19) is shown in Fig. 8.1 in the form of the ratio $\nu_\pm = \omega(d\sigma/d\omega)_\pm/\sigma_{el}$ for electrons and positrons, respectively. Figure 8.1 demonstrates the importance of the deformability of the target atom in the bremsstrahlung process, which leads to a complicated dependence of the total bremsstrahlung cross section on ω. The formula for the dynamic polarizability $\alpha_d^{H_\mu}(\omega)$ needed for calculating these spectra is taken from [15]. For making parametric comparisons of the cross section for elastic scattering on H_μ ($\sigma_{el} = 4\pi R_\mu^2/m_\mu^2$ [14]) with the bremsstrahlung cross section $\omega(d\sigma(\omega)/d\omega)_\pm$ at equal energies of the incident electron (positron), we can write the ratios explicitly as

$$\nu_\mp = \frac{\omega\,(d\sigma(\omega)/d\omega)_\mp}{\sigma_{el}} = \frac{4}{3\pi}\frac{m_\mu^2 p^2 R_\mu^2}{c^3}\,[1 \mp \delta_1(\omega) + \delta_2)\omega)]. \quad (8.21)$$

Equations (8.20) and (8.21) show that in the region $p \sim \sqrt{m_\mu}$, $\omega \ll m_\mu$, the bremsstrahlung cross section is of order m_μ/c^3 times smaller than the cross section for elastic scattering of an electron on H_μ. This is illustrated clearly in Fig. 8.1. Because of the substantial growth in the polarizability and, therefore, in $\delta_1(\omega)$ and $\delta_2(\omega)$ in the region of the threshold for excitation of the $2s$ and $2p$ states of H_μ, the cross section ratio $\nu_\pm$ increases significantly, reaching values of order $1/m_\mu$ several Rydbergs before the threshold and continuing to rise further. In this regard, there is some interest in determining the upper limit to the growth in the ratios $\nu_\mp$. Introducing the finite width Γ of the $1s \to 2p$ line of H_μ, we estimate the maximum polarizability to be $\alpha_d^{max} \sim i/\Gamma$. This means that at the resonance $\nu_\mp$ increases to the level of 10^6–10^7.

The simplicity of the formulas for the bremsstrahlung spectrum resulting from electron collisions with H_μ atoms allows us to write down a formula for the bremsstrahlung spectrum in a frequency range $\varepsilon \sim \omega \sim p^2/2$ that we have not discussed much before. Here the energy dependence of $d\sigma/d\omega$ is determined by integrating over the statistical weight of the final states of the incident electron while taking the conservation of energy, $\int d^4p'\delta(\varepsilon - \omega - p'^2/2) \sim (\varepsilon - \omega)^{1/2}$, into account. On noting the dependence of q_{min} and q_{max} on p and ω, with the aid of Eqs. (8.16) and (8.17) we arrive at the expression

$$\frac{d\sigma^{\pm}(\omega)}{d\omega} = \frac{16}{3c^3}\left[\left(\frac{1}{m_\mu}\right)^4 - \frac{p^2}{m_\mu^2} e_0 \operatorname{Re} \alpha_d^{H_\mu}\left(\frac{p^2}{2}\right) + \frac{p^4}{8}\left|\alpha_d^{H_\mu}\left(\frac{p^2}{2}\right)\right|^2\right]\sqrt{\frac{\varepsilon - \omega}{\varepsilon}}, \quad (8.22)$$

which is valid for $(\varepsilon - \omega)/\varepsilon \ll 1$. Equation (8.22) is approximately satisfactory in the region $(\varepsilon - \omega)/\varepsilon \lesssim 1$. Naturally, the energy dependence near the edge is of the form $d\sigma/d\omega \propto (\varepsilon - \omega)^{1/2}$.

As can be seen from Eq. (8.22), the contributions of polarization bremsstrahlung and the interference term increase with increasing p, and in the region $p^2 \sim m_\mu$, they become comparable to the contribution from traditional bremsstrahlung. When $p^2/2 < I_{H_\mu}$ ($p^2 \simeq m_\mu$) the interference is destructive for an electron ($e_0 = -1$) and constructive for a positron, so that the $d\sigma^{\pm}(\omega)/d\omega$ are equal, respectively, to 0.5 and 5 times the contribution from the traditional term when $\omega = p^2/2 = m_\mu/2$ if we set $\alpha_d^{H_\mu}(m_\mu) = 9/2m_\mu^3$. With further increases in p in the region $m_\mu \ll p^2 < m_\mu^2$, polarization bremsstrahlung continues to be important. For $\omega \gg I_{H_\mu}$, the expression $\alpha = -1/\omega^2$ for the polarizability at $p = m_\mu$ and Eq. (8.22) yield a ratio of $d\sigma(\omega)/d\omega$ to the ordinary (traditional) contribution of 4.4 for a positron and 0.4 for an electron. We can see that this ratio is the reciprocal of that for $\omega < I_{H_\mu}$. For large p and arbitrary ω, we must account for the relativistic character of the incident particle, which leads, as we saw in Sec. 7.6, to an increased contribution from polarization bremsstrahlung. If the target is a $\mu^+\mu^-$ atom, then the contribution of polarization bremsstrahlung during electron (positron) scattering increases, since the static potential created by a $\mu^+\mu^-$ atom goes to zero. Therefore, ordinary bremsstrahlung is determined by the elastic scattering amplitude at the polarization potential, which, as shown in [14], is smaller than the amplitude for scattering on the static potential by a factor of $m_\mu^{1/2}$. This implies that here m_μ^{-1} will be replaced by $m_\mu^{-1/2}$ (which is an order of magnitude greater) in the estimates (8.18) and (8.20) of the relative contribution of polarization bremsstrahlung. The first term in Eq. (8.22) is also smaller by a factor of m_μ .

8.5. Collisions of Mesons and Electrons with Atomic Nuclei

Bremsstrahlung is also produced during scattering of electrons, mesons, and nucleons on atomic nuclei. It can be caused either by slowing down of the incident particle in the static field of the nucleus or by changes in the dipole-polarized target nucleus with time. As in the three previous sections, we shall refer to the first mechanism as ordinary (traditional) bremsstrahlung and the second as polarization bremsstrahlung.

It is simplest to treat the bremsstrahlung of a μ-meson (a comparatively heavy particle) interacting with the nucleons of a nucleus in the same way as that of an electron. By analogy with Eq. (7.15), the amplitude of the bremsstrahlung from a $\mu^{\pm}$-meson is given by

$$F_{pq}(\omega) = \frac{4\pi\,(\mathrm{eq})\,Q^n(q)}{\omega q^2 m_\mu} + (\mathrm{eq})\frac{4\pi e_\mu \omega}{q^2}\,\alpha_d^n(\omega, q), \tag{8.23}$$

where m_μ is the mass of a μ-meson; e_μ is its charge, equal to ±1; $Q^n(q)$ is the form factor of the nucleus, with $Q^n(0) = Z$; Z is its charge; and, $\alpha_d^n(\omega, q)$ is the generalized dipole polarizability of the nucleus given by Eq. (7.12), where the wave functions and energies ω_{nk} describe the ground and excited states of the nucleus and their energies. In the following we shall use an analog of the atomic system of units for the nucleus (a nuclear system), i.e., we set $e = \hbar = m_n = 1$, where m_n is the mass of a nucleon. In this system of units $m_e \approx 1/1835$ and $m_\mu = 0.11$. The characteristic dimension of a nucleus is 1 Fermi, or 10^{-13} cm, which equals 0.035 nuclear length units in this system, while an energy of 1 MeV is equal to 20 nuclear units of energy. Equation (8.23) only includes dipole photons, i.e., energies up to a few tens or hundreds of MeV, and is valid only for fast incident particles, for which we can limit ourselves to the first Born approximation. It is valid if $Z/v_0 \ll 1$. In calculating the generalized dipole polarizability of the nucleus, $\alpha_d^n(\omega, q)$, we should note that both nuclear particles (protons and neutrons) have effective charges that have opposite signs and are equal in magnitude to half the electronic charge [16].

The ratio of the total bremsstrahlung spectrum to the contribution from traditional bremsstrahlung for small q is given by

$$\xi_\mu(\omega, q) = \left|\left[1 + \frac{m_\mu e_\mu \omega^2 \alpha_d^n(\omega)}{Q^n(q)}\right]\right|^2. \tag{8.24}$$

The photoabsorption cross-section curve $\sigma_\gamma(\omega)$ for nuclei has one strong, comparatively narrow peak at an energy of $\omega \approx \Omega_d$, the giant resonance. The cross section $\sigma_\gamma(\omega)$ is symmetric with respect to Ω_d; hence, $\alpha_d^n(\Omega_d) = ic\sigma_\gamma(\Omega_d)/4\pi\Omega_d$ within the region of the giant resonance [see Eq. (7.7)]. As a result, according to Eq. (8.24), the ratio of the amplitudes of polarization and traditional bremsstrahlung for $\omega \sim \Omega_d$ is

$$\eta_\mu = (m_\mu c\sigma_\gamma(\Omega_d)\,\Omega_d)/4\pi Q^n(q) > (m_\mu c\sigma_\gamma(\Omega_d)\,\Omega_d/4\pi Z) \equiv \eta_{0\mu}. \tag{8.25}$$

The giant dipole resonance [16] basically saturates the dipole sum rule, so that in our system of units we have

$$\int_{I}^{\infty} \sigma_\gamma(\omega)\, d\omega = 0.147 \frac{NZ}{A}. \tag{8.26}$$

In the conventional system of units for nuclear physics, this integral is $6NZ/A$ (MeV/Fermi^2), where the number A of nucleons in the nucleus is equal to $N + Z$. The total width of the giant resonance peak is 100 nuclear units of energy (5 MeV) [16], so that the peak cross section $\sigma_\gamma(\omega)$ is given roughly by

$$\sigma_\gamma(\Omega_d) \approx 0.147NZ/A\Gamma \approx 0.0007Z. \tag{8.27}$$

In Eq. (8.27) we have used the fact that for light and medium nuclei $N \sim Z \sim A/2$. The frequency of the giant resonance is given by $\Omega_d \sim 1500\, A^{-1/3} \sim 1200\, Z^{-1/3}$. As a result, for $\eta_{0\mu}$ we obtain the very simple estimate

$$\eta_{0\mu} \approx Z^{-1/3}, \tag{8.28}$$

which is much less than unity. With the aid of Eq. (8.24) this yields an expression for the ratio $\xi_\mu(\omega)$:

$$\xi_\mu(\omega) \approx 1 - 2Z^{-1/3}, \quad Z \gg 1, \tag{8.29}$$

which determines the contribution of polarization bremsstrahlung to the total bremsstrahlung spectrum generated in a collisions of μ-mesons with nuclei.

According to Eq. (8.23), the angular distribution of the bremsstrahlung photons (for fixed ω and q) is the same as when polarization bremsstrahlung is neglected and is determined by the factor (**eq**) in the amplitude. Thus, it coincides with $d\sigma(\omega, \Omega_{\mathbf{k}})/d\omega d\Omega_{\mathbf{k}}$ for bremsstrahlung during scattering of a fast electron [see Eq. (7.27)].

For high ω we shall use the expression $\alpha_d^n(\omega) = -NZ/A\omega^2$ for the polarizability of a nucleus. On substituting it into Eq. (8.24), we obtain

$$\xi_\mu = 1 + 2e_\mu m_\mu N/A \approx 1 + 0.1e_\mu. \tag{8.30}$$

In making the above estimates, we assumed that $Q(q) \sim Z$ and neglected the dependence of $\alpha_d^n(\omega, q)$ on q. This is correct if $q < R_n^{-1} \sim 30A^{-1/3}$. We note that the minimum momentum transfer $q_{\min}$ consistent with the conservation of energy is $q_{\min} = \Omega_d m_\mu/p \ll (m_\mu\Omega_d/2)^{1/2} \sim 9A^{-1/6}$, since the energy of the μ-meson obeys $p^2/2m_\mu \gg \Omega_d$. Thus, we have $q_{\min} < R_n^{-1}$ if $(m_\mu\Omega_d/2)^{1/2} < 30A^{-1/3}$ or, in fact, for all $A < 1700$.

With increasing q in the region $q > R_n^{-1}$, the contribution of polarization bremsstrahlung does not become significant since both the form factor $Q^n(q)$ and the generalized polarizability $\alpha_d^n(\omega, q)$ fall off rapidly. The following simple relation between them holds for high ω:

$$\alpha_d^n(\omega, q) \approx -Q^n(q)N/A/\omega^2.$$

All the results given above are valid when the photon momentum is neglected. This is legitimate for estimates, but not for accurate calculations. In fact, the photon momentum corresponding to an energy of Ω_d is $k \sim 11A^{-1/3}$, whereas $R_n^{-1} \sim 30A^{-1/3}$, i.e., although k is also less than R_n^{-1}, but only a little. We note that k is less than the upper limit $q_{\min} = (m_\mu \Omega_d/2)^{1/2}$.

The contribution of polarization bremsstrahlung becomes more noticeable if the meson moves at a relativistic velocity. As noted in Sec. 7.6, in this case the contribution of polarization bremsstrahlung increases. The angular distributions also become distinctive: close to isotropic for polarization bremsstrahlung and concentrated in a narrow cone with an aperture angle of $\theta < (1 - v_0^2/c^2)^{1/2}$ for ordinary bremsstrahlung. Outside this cone the ordinary bremsstrahlung can be neglected and polarization bremsstrahlung predominates. This should be studied experimentally to yield, as Eq. (8.23) shows, information on the polarizability of the atomic nucleus, a characteristic which is difficult to measure directly by other methods.

Since an electron is two hundred times lighter than a μ-meson, the contribution of ordinary bremsstrahlung should be still smaller for it. However, in order for emission of a photon with an energy on the order of the characteristic nuclear frequencies ($\omega = \Omega_d$) to be possible, the incident electron would actually have to be ultrarelativistic. Thus, the angular distribution of the electron bremsstrahlung is extremely anisotropic and concentrated in a cone with an aperture angle of $\theta < 0.05$ rad. Outside that cone, the emission is determined by the dynamic polarizability of the target nucleus and the differential cross section for bremsstrahlung originating in the collision of an electron with a nucleus at angles greater than a few degrees is determined by the dynamic polarizability of the target nucleus. As opposed to an atom, where $\alpha_d(\omega, q)$ can be calculated with fair accuracy from first principles, a reliable calculation of $\alpha_d{}^n(\omega, q)$ requires very much more effort, so that it is much less well known. Studies of bremsstrahlung from electron–nuclear collisions make it possible to determine $\alpha_d{}^n(\omega, q)$ almost directly from experiment.

8.6. Nucleon–Nucleus Collisions

On going to particles with larger masses, the relative contribution of polarization bremsstrahlung increases. Far from a nucleus (for small q), all interactions of a proton except its purely Coulomb interaction with the nucleus and nucleons in the nucleus can be neglected and Eq. (8.24) can be used to estimate the relative contribution of polarization bremsstrahlung on replacing the mass of the μ-meson by the nine-times larger mass of a proton. The ratio of polarization and ordinary bremsstrahlung changes similarly for a proton, so that instead of Eq. (8.28) we obtain the estimate

$$\eta_{0p} \approx 9Z^{-1/3} > 1, \tag{8.31}$$

which is correct for the entire periodic table. For small q the contribution of polarization bremsstrahlung is significant at large $\omega \gg \Omega_d$, as well as for $\omega \sim \Omega_d$. Instead of Eq. (8.30) we obtain

$$\xi_p \approx (1 - N/A)^2 \leqslant 0.25. \tag{8.32}$$

All of the above results have been obtained neglecting the momentum of the photon and the emission of multipoles higher than dipole. This approximation is entirely legitimate for scattering on atoms. As noted in the preceding section, the situation is different for bremsstrahlung on a nucleus: taking the momentum of the photon to be roughly $k = \Omega_d/c \sim 11A^{-1/3}$, we find that kR_n is not too small at $kR_n \sim 0.3$, so that $R_n \sim 0.03A^{1/3}$. Thus, an accurate calculation must be done without neglecting the momentum $\mathbf{k}$. As q of the incident proton increases (and for arbitrary q in the case of a neutron), in addition to the Coulomb forces we must include the nuclear interaction which is considerably stronger than the Coulomb interaction when the incident particle and the nucleus come closer than distances on the order of the nuclear radius. Diagrams describing the bremsstrahlung of a nucleon on a nucleus are shown in Fig. 7.2 and the corrections which take the excitation interaction between the excited nucleon and a vacancy into account are shown in the sequence of diagrams of Fig. 7.4. As in an atom, they are also more important in a nucleus. When they are included, it is possible to obtain a satisfactory description of the giant nuclear resonances if the internucleon interaction in the nucleus is chosen phenomenologically, so that even in the Hartree–Fock approximation one can describe with fair accuracy the excitation spectrum and the ground states of the nuclei, their energy, density of nucleons, and average radius [17].

The bremsstrahlung amplitude of a nucleon on a nucleus corresponding to Fig. 7.2 has the form

$$F_{if}(\omega, k) = \langle i | (\mathbf{e}\nabla)\, e^{i\mathbf{kr}} e_0 | f \rangle + i \sum_{n \neq 0} \frac{2\omega \langle i, 0 | \hat{u} | f, n \rangle \langle n | (\mathbf{e}\nabla)\, e^{i\mathbf{kr}} e_{\text{ef}} | 0 \rangle}{\omega^2 - \omega_{n0}^2 + i\omega\delta}. \tag{8.33}$$

Here $|i\rangle$ and $|f\rangle$ are the wave functions of the initial and final states of the incident nucleon moving in the field of the nucleus; $|0\rangle$ and $|n\rangle$ denote nuclear states; the excitation energy of state $|n\rangle$ is $\omega_{n0} = \varepsilon_n - \varepsilon_0$; e_0 is the charge of the incident nucleon, which equals zero for a neutron and unity for a proton (in our system of units); and $\hat{u}$ is the interaction potential of the incident nucleon with the nucleus. In Eq. (8.33) we have neglected the small mass difference between the proton and neutron. There e_{ef} denotes the effective charge of a nuclear nucleon, i.e., of a proton e_p or a neutron e_n. The appearance of e_{ef} unequal to 1 or 0 in the nucleus is associated with the fact that the nucleus does not have a heavy "core" (as in an atom) and the displacement of a nucleon relative to the nucleus is actually an excitation of the nucleus. On the contrary, a nuclear excitation is the displacement of one or several of the nucleons that make up the nucleus relative to the center of gravity of the nucleus. A displacement of the center of gravity itself represents scattering of the nucleus and as such does not involve any excitation of its internal degrees of freedom.

The requirement that the center of gravity be motionless during excitation makes the effective electrical charge of a nucleon differ from the actual charge and depend on the multipole order of the excitation [16]. Thus, for dipole excitation we can assume with fair accuracy that the charge on a proton is $e_p = N/A$ and the charge on a neutron is $e_n = -Z/A$. For excitation of higher multipole orders, the charge on a proton is 1 and that on a neutron is 0.

The formula for $F_{if}(\omega, k)$ and the estimates of the relative contributions of the first and second terms in Eq. (8.33) are considerably simpler for a fast incident nucleon. In this case, the bremsstrahlung amplitude is given by

$$F_{pq}(\omega) \approx -U(q)\frac{(\mathbf{eq})}{\omega} e_0 + 2iu(q)\,\omega \sum_{n\neq 0} \frac{\langle 0|\exp(-i\mathbf{qr})|n\rangle\langle n|(\mathbf{e}\nabla)\exp(i\mathbf{kr})\, e_{\mathrm{ef}}|0\rangle}{\omega^2 - \omega_{n0}^2 + i\omega\delta}. \quad (8.34)$$

In the first term of (8.34) we have neglected the photon momentum $\mathbf{k}$ compared to the momentum $\mathbf{q} = \mathbf{p}' - \mathbf{p} + \mathbf{k}$ transferred to the nucleus during the collision. As noted above, in the frequency range $\omega \sim \Omega_d$ of interest to us, $kR_n \sim 1$, while since $\omega = kc \sim pq$, where p is the momentum of the incident nucleon ($p < c$), we have $q > k \sim R_n^{-1}$ and the momentum k can be neglected compared to q.

In Eq. (8.34) both $U(q)$ and $u(q)$ are the Fourier transforms of the potentials of the interactions of the incident nucleon with the nucleus and with a nucleon inside the nucleus, respectively. Both $U(q)$ and $u(q)$ contain real and imaginary parts, as well as a Coulomb interaction (if the incident particle is a proton). The interaction among the nucleons in the nucleus also enters Eq. (8.34) implicitly through the state $|n\rangle$ and is generally different from $u(q)$. The characteristic region of the interaction $U(q)$ is given by the nuclear radius R_n and that of $u(q)$ is given by the average internucleon separation r_0. Since $r_0 \ll R_n$ and $q > R_n^{-1}$, for estimating the bremsstrahlung amplitude we can assume that $r_0^{-1} > q \gg R_n^{-1}$ in Eq. (8.34). We note that if the momentum k of the photon is neglected in the second term of Eq. (8.34), then the remaining expression will be proportional to the generalized dipole polarizability of the nucleus.

For substantial energies of the incident nucleon, the interactions $U(q)$ and $u(q)$ are related: $u(q)$ is expressed in terms of the amplitude of nucleon–nucleon scattering, while $U(q)$ is the Hartree potential created by the nucleus [by a distribution of nucleons with density $\rho(r)$ interacting with the incident nucleon through the potential $u(r)$]. With increasing q the nuclear interaction of the proton begins to dominate the pure Coulomb interaction, initially in the first and then in the second term of the amplitude (8.34). For substantial (of order 1 GeV) energies of the incident nucleon and small momenta q, the imaginary part of $U(q)$ is much greater than the real part and the latter can be neglected.

Nuclear forces are such that a fast incident nucleon (with energies of hundreds of MeV to 1 GeV) interacts in almost the same way with the protons and neutrons in a nucleus. Dipole excitation, however, shows up when particles with different charges (protons and neutrons) are displaced relative to one another. Thus, the probability of dipole excitation by fast incident nucleons is low and quadrupole excitation becomes more important. Accordingly the dipole component of the bremsstrahlung is also suppressed with increasing q. The angular distribution differs substantially from that given by Eq. (7.27). Besides a term with $P_2(\cos\theta)$, $d\sigma(\omega, \Omega_k)/d\omega d\Omega_k$ will also contain a term with $P_4(\cos\theta)$.

The main contribution to the first term of Eq. (8.34) is from dipole emission, while that in the second is from quadrupole emission. Because of the different multipole order of the photons whose emission is described by the first and second terms of Eq. (8.34), they do not interfere in the total bremsstrahlung spectrum.

The ratio of the contributions of polarization and traditional bremsstrahlung in the total bremsstrahlung spectrum is given roughly by

$$\xi(\omega) = 1 + |kR_n u(q)\,\omega^2 \alpha_d^n(\omega)|^2 / |U(q)|^2. \tag{8.35}$$

For purposes of estimates, the quadrupole polarizability $\alpha_q^n(\omega)$ in Eq. (8.35) is expressed in terms of the dipole polarizability with the aid of the simple formula $\alpha_q^n(\omega) \sim \alpha_d^n(\omega)kR_n = \omega\alpha_d^n(\omega)R_n/c$. The ratio $u(q)/U(q)$ for $q \leq R_n^{-1}$ is given approximately by A^{-1}, so that Eq. (8.35) simplifies to

$$\xi(\omega) = 1 + (kR_n\omega^2\alpha_d^n(\omega)/A)^2. \tag{8.36}$$

It follows from this equation and the approximation (8.25) that, as q increases, the total bremsstrahlung spectrum continues to be slightly higher than the contribution from the first term in the amplitude (8.34). In fact, it is difficult to make an estimate near $\omega \sim \Omega_d$; however, it is clear that the second term in Eq. (8.36) is actually greater because $U(q)$ is considerably smaller than $U(0)$ for $q \sim R_n^{-1}$.

The second term in Eq. (8.34) is also nonzero for neutrons, for which the first term is absent and there is no Coulomb interaction in $u(q)$ when magnetic multipole radiation is neglected. For scattering angles such that the quadrupole contribution predominates in the second term, proton and neutron bremsstrahlung are similar in magnitude [5]. Since the frequencies of the dipole and quadrupole giant resonances Ω_d and Ω_q are different, the bremsstrahlung spectrum may contain two maxima at $\omega \sim \Omega_d$ and $\omega \sim \Omega_q$.

The results of a calculation of the differential bremsstrahlung cross section with respect to the scattering angle for an incident nucleon with an energy of 1 GeV on a Ca^{40} nucleus using Eqs. (8.34) and (7.14) [5] are shown in Fig. 8.2. The contribution from the polarization of the nucleus [i.e., the second term in Eq. (8.34)] was calculated in the random phase approximation (described by the diagrams of Fig. 7.4). For the internucleon interaction we have used the so-called

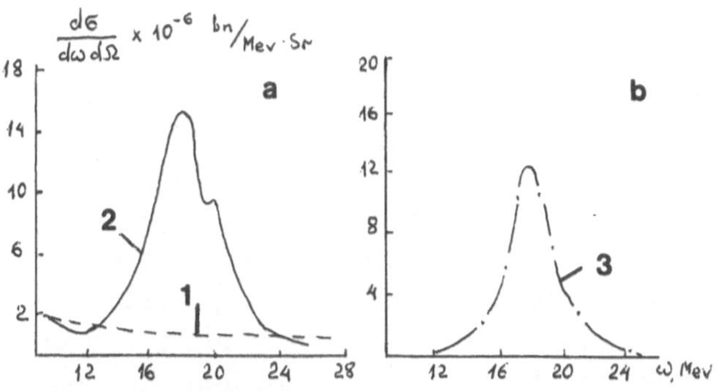

Fig. 8.2. The differential cross section for bremsstrahlung of a nucleon with an energy of 1 GeV on a Ca^{40} nucleus at a scattering angle of 2.5° [5]: 1) ordinary proton bremsstrahlung; 2) total proton bremsstrahlung; 3) neutron bremsstrahlung.

Skyrm forces, and the interaction of the incident nucleon with the nucleus was taken to be a point interaction of a type such that the optical potential for elastic scattering of a nucleon on a nucleus was recovered in the eikonal approximation [17]. The effect of the field of the optical potential on the incident nucleon was taken into account and at this energy it reduced mainly to absorption of the nucleon wave function. This reduced the bremsstrahlung cross section, but, of course, did not change the relative contributions of the first and second terms to the amplitude. It is clear from Fig. 8.2 that the second term is dominant for $\omega \sim \Omega_d, \Omega_q$. In the calculations [5], the width in the random phase equation was derived phenomenologically so as to ensure agreement with experiment in the description of the giant nuclear resonances (both dipole and quadrupole).

The additional maximum in the proton bremsstrahlung spectrum is the result of a significant probability of emission of dipole photons owing to the Coulomb interaction with the nucleons of the nucleus, which is significant even for such a large angle as 2.5°.

For incident nucleon velocities close to that of light, as well as for a relativistic electron, traditional and polarization bremsstrahlung differ significantly in the angular distribution of the bremsstrahlung photons. The first mechanism leads to predominantly forward emission and the second, to lateral emission.

The cross section for bremsstrahlung of a nucleon on a nucleus is of an experimentally observable magnitude and requires coincidence detection of the fast nucleon and a photon. If the incident particle is not an isolated nucleon but one contained in a nucleus, then the contribution of polarization bremsstrahlung is greater, since the first term in Eq. (8.34) is inversely proportional to the mass m_0 of the incident particle.

In collisions of complex particles [two nuclei, as well as two atoms; see Eqs. (7.58)–(7.61)] the radiation source may be either the target or the projectile, but the emission spectrum of the projectile is shifted and broadened as a result of the

Doppler effect. If the ratio Z/A is the same in the target and projectile, then dipole emission is forbidden, at least for small q, since the "projectile + target" system does not have a dipole moment. Thus, the bremsstrahlung spectrum for collisions of identical nuclei will have a peak only at the energy of the quadrupole giant resonance, Ω_q. Experimental observation of such a maximum would serve as convincing proof of the existence of this mechanism for bremsstrahlung.

Polarization bremsstrahlung is also undoubtedly important in the case where the projectile or target are not atoms, but molecules, or not nuclei, but elementary particles. In all cases, when one or both of the colliding particles have an internal structure (whether the constituents are atoms, electrons, nucleons, or quarks), virtual or real excitation shows up in the bremsstrahlung that involves this structured particle. Here an investigation of the polarization bremsstrahlung is extremely important and sometimes is of fundamental importance. In addition, the emission spectrum is determined by the internal structure of the colliding particles and studies of this spectrum can provide additional and sometimes unique information on this structure. It is also important and appropriate to study the effect (and manifestations) of the internal structure of atoms on the spectra produced during the interaction of incident particles with complex polyatomic formations, such as clusters, mixtures, and solids.

Chapter 9

POLARIZATION BREMSSTRAHLUNG IN COLLISIONS OF RELATIVISTIC ATOMS AND RELATIVISTIC PARTICLES WITH ATOMS

M. Ya. Amus'ya, A. V. Korol',
and A. V. Solov'ev

9.1. Emission during Collisions of Structureless Charged Particles with Atoms

We now consider bremsstrahlung produced in atomic collisions when the colliding particles have relativistic energies. Besides bremsstrahlung of charged structureless particles (electrons, positrons, muons, protons) on atoms (see Chapters 5, 7, and 8 as well), here we shall consider emission from atom–atom and ion–atom collisions when both the target and the incident particle (projectile) have internal structure. Using the neutrino as an example, we then examine bremsstrahlung from particles whose interactions with an atom are not electromagnetic in character. In this case, along with the characteristic features of polarization bremsstrahlung (and for neutrinos, there is no other radiation but polarization bremsstrahlung), certain features owing to weak interactions will show up.

The bremsstrahlung of relativistic structureless charged particles with the polarization of the target atom taken into account has been examined by several authors [1–4] (also, see Chapters 5 and 7). For the sake of coherence in the discussion, in the following we shall present the basic results on bremsstrahlung of relativistic particles on atoms, dwelling in a bit more detail on some points that have been discussed less in the past, but are nevertheless important.

The relativistic treatment of the influence of the polarization of a target atom on charged particle bremsstrahlung has led to some fundamentally new results compared to the nonrelativistic case. Thus, it turns out that the spectrum of the polarization part of bremsstrahlung (and the total emission spectrum along with it) increases logarithmically with the energy ε of the incident particle. The contributions of the polarization and traditional bremsstrahlung have different angular distributions, since the bremsstrahlung of the incident particle is concentrated mainly in a narrow cone along the direction of motion and the aperture angle of the cone becomes smaller with increasing ε. At the same time, polarization bremsstrahlung is weakly anisotropic and its form is independent of the velocity of the particle (this has been emphasized repeatedly above, in Chapters 1–7).

A coherence effect occurs in the total elastic bremsstrahlung spectrum $d\sigma^{el}(\omega)$ of a relativistic particle, according to which, in some region of ω, $d\sigma^{el}(\omega)$ is proportional to the square of the number of electrons in the atom N^2. In this region, which is referred to as the coherence region, the cross section for elastic bremsstrahlung exceeds that for inelastic bremsstrahlung, which is proportional to N.

Our discussion is based in many respects on [2, 4–6a]. Some of the problems analyzed below have been discussed in [3]. As in the preceding Chapters 7 and 8, we use atomic units, where $\hbar = m_e = e = 1$ and the charge on an electron is taken to be –1.

We begin with the amplitude of the emission process during a collision of a charged particle of mass m_0 and energy $\varepsilon = c(p^2 + m_0{}^2c^2)^{1/2}$ with a many-electron atom in its ground state $|0\rangle$. As a result of the interaction, the energy of the particle is equal to $\varepsilon' = c(p'^2 + m_0{}^2c^2)^{1/2}$ and a photon of frequency ω is emitted. The atom may remain in the ground state (elastic bremsstrahlung) or enter an excited state or be ionized (inelastic bremsstrahlung). Let the energy of the photon and the particle satisfy the inequality $\omega \ll \varepsilon, \varepsilon'$. In addition, we shall assume that the Born condition is satisfied for the incident and scattered particles; i.e., that $|e_0Z| \ll v = v_0, v'$, where Z is the nuclear charge, e_0 is the charge on the particle, and v and v' are the velocities of the projectile before and after the collision. In this case, when analyzing the amplitude of this process it is sufficient to limit ourselves to the lowest order of perturbation theory with respect to the interaction of the incident particle with the atom, as well as to the interaction of the particle and the atomic electrons with the electromagnetic field.

The amplitude is represented graphically in the diagrams of Fig. 9.1. There a thin line represents the incident particle; a double line corresponds to an atom in some state ($|0\rangle$, $|n\rangle$, $|m\rangle$); a dashed line connecting the lines of the incident particle and the atom is a photon propagator for which we use a Feynman gauge; an escaping dashed line represents an emitted photon; and **k** is the wave vector of the latter.

For $|m\rangle = |0\rangle$, the first pair of diagrams (Fig. 9.1a, b) determines the well known amplitude of traditional elastic bremsstrahlung (see [7] for example). In the general case of $|0\rangle \neq |m\rangle$, on using the conditions $\varepsilon, \varepsilon' \gg \omega$ and $p, p' \gg k$ and neglecting the recoil of the atom, we obtain the following expression for this amplitude:

$$F^t_{m0} = \frac{4\pi e_0^2}{\mathbf{q}^2 - \omega_{m0}^2/c^2}\left[\frac{\mathbf{e}\mathbf{p}'}{(kp')} - \frac{\mathbf{e}\mathbf{p}}{(kp)}\right]\{b^0\,(Z\delta_{m0} - W^{(p)}_{m0}(\mathbf{q})) - \mathbf{b}\mathbf{J}^{(p)}_{m0}(\mathbf{q})\}, \tag{9.1}$$

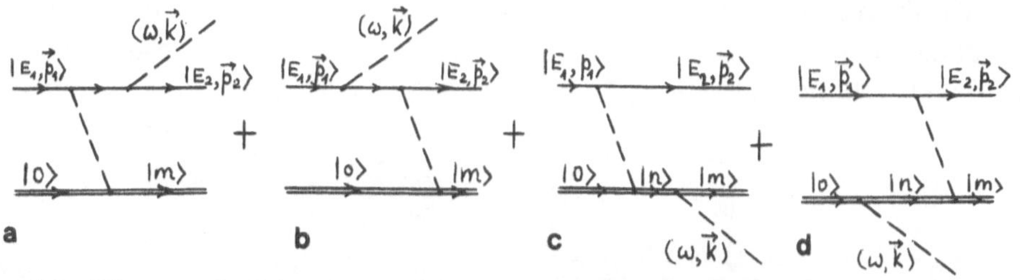

Fig. 9.1. Diagrams depicting the amplitudes of ordinary and polarization bremsstrahlung of a relativistic charged particle.

where **e** is the polarization vector of the photon; $b^\mu = \bar{u}(p', s)\gamma^\mu u(p, s)$ is the current 4-vector of the incident particle; $u(p, s)$ is the bispinor amplitude; $\mathbf{q} = \mathbf{p}' - \mathbf{p} + \mathbf{k}$ is the momentum transferred to the atom in this process; $(pk) = p_\mu k^\mu = (\varepsilon\omega/c^2)(1 - \cos\theta \cdot v/c)$; $(p'k) = (\varepsilon'\omega/c^2)(1 - \cos\theta' \cdot v'/c)$; and θ and θ' are the emission angles of the photon relative to the vectors **p** and **p'**. The functions $W_{m0}^{(p)}(\mathbf{q})$ and $J_{m0}^{(p)}(\mathbf{q})$ represent the following matrix elements for the exact relativistic wave functions of the atom:

$$W_{mo}^{(p)}(\mathbf{q}) = \left\langle m \left| \sum_{j=1}^{N} e^{-i\mathbf{q}\mathbf{r}_j}\gamma^0 \right| 0 \right\rangle,$$
$$\mathbf{J}_{mo}^{(p)}(\mathbf{q}) = -\left\langle m \left| \sum_{j=1}^{N} e^{-i\mathbf{q}\mathbf{r}_j}\boldsymbol{\gamma} \right| 0 \right\rangle, \quad N = Z, \tag{9.2}$$

where the sums are taken over all the atomic electrons.

As we are not considering heavy atoms with high Z, we can assume that the relativistic corrections to the wave functions of the atom are small and proceed to the nonrelativistic limit for the motion of the atomic electrons. Then the matrix elements (9.2) are replaced, respectively, by the nondiagonal form factor of the atom and the Fourier component of the current of the atomic electrons [7]:

$$W_{mo}^{(p)}(\mathbf{q}) \rightarrow W_{mo}(\mathbf{q}) = \left\langle m \left| \sum_{j=1}^{Z} e^{-i\mathbf{q}\mathbf{r}_j} \right| 0 \right\rangle,$$
$$\mathbf{J}_{mo}^{(p)}(\mathbf{q}) \rightarrow J_{mo}(\mathbf{q}) = -\left\langle m \left| \frac{1}{2c} \sum_{j=1}^{Z} \left(e^{-i\mathbf{q}\mathbf{r}_j}\hat{\mathbf{p}}_j + \hat{\mathbf{p}}_j e^{-i\mathbf{q}\mathbf{r}_j}\right) \right| 0 \right\rangle, \tag{9.3}$$

where $\hat{\mathbf{p}}_j$ is the momentum operator for an atomic electron. On the right-hand sides of Eqs. (9.3) the states $|0\rangle$ and $|m\rangle$ are described by nonrelativistic wave functions, the solutions of the Schrödinger equation for the atom.

The diagrams of Fig. 9.1c, d correspond to polarization bremsstrahlung. We can obtain an analytic expression for the amplitude with the aid of the standard correspondence rules [7]. Using the definitions (9.2), we have

$$F^{p}_{m0} = \frac{4\pi e_0}{q_1^2 - (\omega + \omega_{m0})^2/c^2} \sum_n \left\{ \frac{[b^0 W^{(p)}_{mn}(\mathbf{q}_1) - \mathbf{b}\mathbf{J}^{(p)}_{mn}(\mathbf{q}_1)]\,(\mathbf{e}\mathbf{J}^{(p)}_{m0}(\mathbf{k}))}{E_n - E_0 + \omega} - \frac{\mathbf{e}\mathbf{J}^{(p)}_{mn}(\mathbf{k})\,[b^0 W^{(p)}_{n0}(\mathbf{q}_1) - \mathbf{b}\mathbf{J}^{(p)}_{n0}(\mathbf{q}_1)]}{\omega - E_n + E_m} \right\}, \tag{9.4}$$

where $\mathbf{q}_1 = \mathbf{p}' - \mathbf{p} = \mathbf{q} - \mathbf{k}$ and $\omega_{m0} = E_m - E_0$.

On transforming to the nonrelativistic limit for the motion of the atomic electrons in Eq. (9.4), one must note that the sum over the intermediate states $|n\rangle$ is taken over states with both positive and negative E_n (here E_n is the total energy of an atomic electron in state $|n\rangle$ including the rest mass). The transition to the nonrelativistic limit in the negative-frequency part of the sum has been examined in detail in [8]. For states with $E_n > 0$, the transition in Eq. (9.4) is made with the aid of Eq. (9.3). As a result, for dipole frequencies of the emitted photon, i.e., when $(\omega/c)R_a \ll 1$ (where R_a is the size of the atom), we arrive at the following expression for the amplitude of polarization bremsstrahlung including both the elastic and inelastic channels [3]:

$$F^{p}_{m0} = \frac{-4\pi e_0}{q_1^2 - (\omega + \omega_{m0})^2/c^2} \left\{ (\mathbf{e}\mathbf{q}_1)\, b^0 \alpha_{m0}(\omega, \mathbf{q}_1) + \frac{\omega}{c} e^i b^j \beta^{ij}_{m0}(\omega, \mathbf{q}_1) \right\} \tag{9.5}$$

Here $\alpha_{m0}(\omega, \mathbf{q})$ is the generalized dynamic polarizability of the atom given by

$$(\mathbf{e}\mathbf{q}_1)\, \alpha_{m0}(\omega, \mathbf{q}_1) = -i \sum_n \left\{ \frac{W_{mn}(\mathbf{q}_1)\,(\mathbf{e}\mathbf{D})_{n0}}{\omega + \omega_{n0}} + \frac{(\mathbf{e}\mathbf{D})_{mn} W_{n0}(\mathbf{q}_1)}{\omega - \omega_{n0}} \right\}, \tag{9.6a}$$

and the three-dimensional tensor $\beta_{m0}{}^{ij}(\omega, q)$ is

$$\beta^{ij}_{m0}(\omega, \mathbf{q}_1) = \frac{ci}{\omega} \sum_n \left\{ \frac{\mathbf{D}^i_{mn} \mathbf{J}^j_{n0}(\mathbf{q}_1)}{\omega - \omega_{n0}} - \frac{(\mathbf{J}^i_{mn}(\mathbf{q}_1))\,\mathbf{D}^j_{n0}}{\omega + \omega_{n0}} \right\}, \tag{9.6b}$$

where $\mathbf{D} = \sum_{j=1}^{Z} \mathbf{r}_j$ is the operator for the dipole moment of the atom. We note that Eq. (9.5) is valid for $\omega \ll c^2$ and $q \ll c$ [8].

The expression for the amplitude $F_{m0}{}^p$ is considerably simpler in two limiting cases.

When $|m\rangle = |0\rangle$, the elastic polarization bremsstrahlung amplitude (9.5) is given in terms of the dipole polarizability of the atom, $\alpha_d(\omega)$, through

$$\alpha_{00}(\omega, \mathbf{q}_1) \approx \omega \alpha_d(\omega), \qquad \beta^{ij}_{00}(\omega, \mathbf{q}_1) \approx \delta^{ij} \alpha_d(\omega).$$

We find that $F_{el}^p = F_{00}^p$ for $qR_a \ll 1$:

$$F_{el}^p = -\frac{4\pi e_0 \omega \alpha_d(\omega)}{\mathbf{q}_1^2 - \omega^2/c^2}\left[(\mathbf{e}\mathbf{q}_1)\, b^0 + \frac{\omega}{c}(\mathbf{e}\mathbf{b})\right], \tag{9.7}$$

and in the nonrelativistic limit we have

$$F_{el}^p = -\frac{4\pi e_0 \omega \alpha_d(\omega)}{\mathbf{q}_1^2}(\mathbf{e}\mathbf{q}_1).$$

The other limit in which Eq. (9.5) becomes considerably simpler is the region of high frequencies (compared to atomic frequencies) ($\omega \gg 1$).

Then we have the following equations:

$$\beta_{m0}^{ij}(\omega, \mathbf{q}_1) \approx \frac{\delta^{ij}}{\omega}\alpha_{m0}(\omega, \mathbf{q}_1) \approx -\frac{\delta^{ij}}{\omega^2}W_{m0}(\mathbf{q}),$$

and, therefore,

$$F_{m0}^p = \frac{4\pi e_0}{q_1^2 - (\omega + \omega_{m0})^2/c^2}\,\frac{W_{m0}(q_1)}{\omega}\left[(\mathbf{e}\mathbf{q}_1)\, b^0 + \frac{\omega}{c}(\mathbf{e}\mathbf{b})\right]. \tag{9.8}$$

The total amplitude of the bremsstrahlung process is the sum of Eqs. (9.5) and (9.1).

The general formula for the differential cross section for bremsstrahlung with a simultaneous transition of the atom from state $|0\rangle$ into state $|m\rangle$ is

$$d\sigma_{m0}^{tot} = d\sigma_{m0}^t + d\sigma_{m0}^p + d\sigma_{m0}^{int} = \frac{q\omega\varepsilon' dq d\omega d\Omega_k d\varphi_{p1}}{2v_0 pc^3 (2\pi)^4};\sum_{\lambda,s,s'} |F^t + F_{m0}^p|^2, \tag{9.9}$$

where $d\Omega_\mathbf{k}$ is the solid angle at which the photon leaves; $\varphi_{\mathbf{p}1}$ is the azimuthal angle of the vector $\mathbf{p}'$; and, the z axis is chosen to be along the direction of $\mathbf{p}$. The sum in Eq. (9.9) is taken over the polarizations λ of the photons and over the polarizations of the projectile before and after the collision, s and s'. The energy ε' of the particle after the collision is found from the conservation of energy to be

$$\varepsilon' = \varepsilon - \omega - \omega_{m0}.$$

Let us examine the elastic (with $|m\rangle = |0\rangle$) bremsstrahlung further.

Substituting Eq. (9.7) for the amplitude F_{el}^p in Eq. (9.9) and then taking the sum over the photon polarization and spins of the incident particle and integrating over the angle $\varphi_{\mathbf{p}1}$ and over the momentum transfer q (from $q_{min} = \omega/v_0$ to some $q_{max} \sim 1/R_a$), we find to within logarithmic accuracy that

$$d\sigma_{el}^p(\omega, \Omega_\mathbf{k}) = \frac{e_0^2 d\omega d\Omega_\mathbf{k}}{c^3\omega}(1 + \cos^2\theta)\,|\omega^2\alpha_d(\omega)|^2 \ln\frac{v_0\gamma}{\omega R_a},$$
$$\gamma = \frac{\varepsilon}{mc^2}. \tag{9.10}$$

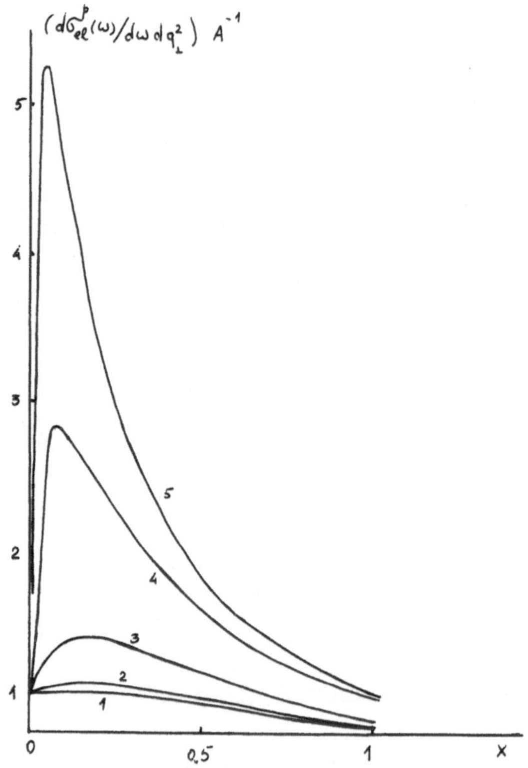

Fig. 9.2. Dependence of the cross section $d\sigma_{el}^{P}(\omega)/(d\omega dq_{\perp}{}^{2}A)$ on the dimensionless variable $x = q_{\perp 2}c^2/\omega^2$ for different velocities v_0 (ω = 13.7 Ry, m = 1): 1) $v_0{}^2 = 0.5c^2$; 2) 0.6 c^2; 3) 0.75 c^2; 4) 0.9 c^2; 5) 0.95 c^2. The angles θ corresponding to the peak in the cross section are: 1) θ = 0; 2) $1.49 \cdot 10^{-2}$; 3) $1.17 \cdot 10^{-2}$; 4) $5 \cdot 10^{-3}$; 5) $2.6 \cdot 10^{-3}$ rad.

As the velocity of the incident particle increases, a singularity will appear in its angular distribution in polarization bremsstrahlung. Let us write the differential cross section for elastic polarization bremsstrahlung as a function of the momentum transfer q_1:

$$d\sigma_{\text{el}}^{P}(\omega, q_1^2) = \frac{A d\omega dq_1^2}{v_0^2 (q_1^2 - \omega^2/c^2)} \left[1 - \frac{\omega^2}{c^2\gamma^2 (q_1^2 - \omega^2/c^2)} \right], \tag{9.11}$$

where $A = (8/3\omega c^3)|\omega^2 \alpha_d(\omega)|^2$.

The angular variation of Eq. (9.11) is most conveniently analyzed by taking as a variable $\mathbf{q}_{\perp}$, the component of $\mathbf{q}_1$ perpendicular to $\mathbf{p}$, whose magnitude is related to the scattering angle of the particle θ by the simple formula $q_{\perp} = p\theta_p(p, p' \gg q_1)$. Transforming to the variable $q_{\perp}{}^2$ in Eq. (9.11), we obtain

$$d\sigma_{el}^{p}(\omega, q_{\perp}^{2}) = \frac{A d\omega dq_{\perp}^{2}}{v_{0}^{2}(q_{\perp}^{2} + \omega^{2}/v_{0}^{2}\gamma^{2})}\left[1 - \frac{\omega^{2}}{c^{2}\gamma^{2}(q_{\perp}^{2} + \omega^{2}/v_{0}^{2}\gamma^{2})}\right]. \tag{9.12}$$

The cross section (9.12) is plotted as a function of the dimensionless variable $x = q_{\perp}^2c^2/\omega^2$ in Fig. 9.2, which reflects the following basic features:

a) a monotonic decrease in $d\sigma_{el}^p(\omega, q_{\perp}^2)$ with increasing $q_{\perp}^2$ for velocities $v_0^2 \leq 0.5c^2$. Then the cross section reaches a peak at the point $q_{\perp}^2 = 0$: $d\sigma_{el}^p(q_{\perp}^2 = 0) = A/\omega^2$; and

b) the appearance of a maximum for $v_0^2 > 0.5c^2$ at $q_{\perp}^2 = q_{\perp 0}^2 = (\omega^2/\gamma^2 v_0^2)(2v_0^2/c^2 - 1)$; $d\sigma_{el}^p(\omega, q_{\perp 0}^2) = A d\omega dq_{\perp}^2\gamma^2c^2/4v_0^2\omega^2$.

It is clear that, depending on the scattering angle θ, there is a "dip" in the cross section (9.12) for $\theta < q_{\perp 0}/p$. It is, in fact, possible that going beyond the framework of the Born approximation used here will remove the "dip" in the angular distribution (9.12).

With increasing v_0, the value of $q_{\perp 0}^2$ drops and the cross section (9.12) becomes steeper with a sharp maximum at $q_{\perp} \sim q_{\perp 0}$. Then the absolute magnitude of $d\sigma_{el}^p(\omega, q_{\perp 0}^2)$ at its peak increases without bound in the limit $v_0 \to c$. Qualitatively this can be explained by the fact that when $v_0 \sim c$, exchange of transverse virtual photons plays an ever increasing role in the interaction between the incident particle and an atomic electron. Here the characteristic distances between the atom and projectile increase with energy in proportion to $q_{\perp 0}^{-1} \sim \gamma v_0/\omega$, which leads to an increase in the cross section.

The sharp dependence of Eq. (9.12) on $q_{\perp}^2$ for low momentum transfers makes it possible to isolate the polarization and traditional parts of the total differential cross section for elastic bremsstrahlung with respect to q. In fact, momentum transfers $q \gg R_a^{-1}$, at which there is little shielding of the nucleus by the atomic electrons, are important for traditional bremsstrahlung. Thus, during traditional bremsstrahlung an incident particle is scattered by an angle θ_p, which is significantly greater than the scattering angle for polarization bremsstrahlung, $\theta_{p0} = q_{\perp 0}/p \sim \omega/v_0\gamma p$. These considerations are valid for a neutral atom, but are not applicable to bremsstrahlung on ions since the emission of a photon by the incident particle in the Coulomb field of an ion involves its being scattered at small angles. In this case it is possible to isolate the polarization bremsstrahlung from the traditional bremsstrahlung by examining the angular distribution of the emission, which is different for the two bremsstrahlung mechanisms.

Integrating Eq. (9.12) with respect to the variable from q_1 to some $q_{max} \sim R_a^{-1}$, we obtain, to within logarithmic accuracy, the well known expression for the elastic polarization bremsstrahlung spectrum,

$$\frac{d\sigma_{el}(\omega)}{d\omega} = \frac{16}{3} e_0^2 \frac{1}{v_0^2 c^3 \omega} |\omega^2 \alpha_d(\omega)|^2 \ln \frac{v_0\gamma}{\omega R_a}. \tag{9.13}$$

One important feature of the cross section (9.13) is that it increases with the energy of the incident particle. As shown in [2], this happens because of an increasing contribution from transverse virtual photons to the interaction of the projectile with an atomic electron, which is the source of the factor ε in the logarithm. The Coulomb part of the interaction does not, however, depend on ε, and its contribution, which is proportional to $\ln v_0/\omega R_a$ in Eq. (9.13), is determined solely by the velocity of the particle. This is easily shown by isolating the contribution of the Coulomb interaction, which depends on $1/q_1^2$, in Eq. (9.12). The remaining terms will correspond to the contributions from transverse virtual photons. This partition of the cross section (9.12) and, with it, Eq. (9.13), arises naturally when a Coulomb gauge is used for the photon propagator, as in [2].

The traditional part of the elastic bremsstrahlung spectrum for ultrarelativistic particles is independent of ε to within logarithmic accuracy and is given by [7]

$$\frac{d\sigma_{\text{el}}^{t}(\omega)}{d\omega}=\frac{16}{3}\frac{e_0^4}{m_0^2c^3}\frac{1}{\omega}\ln(cR_a). \tag{9.14}$$

If follows from a comparison of Eqs. (9.13) and (9.14) that when $\gamma = \varepsilon/mc^2 \gg 1$ the total cross section for elastic bremsstrahlung increases with the energy of the incident particle and that the major contribution to it is from $d\sigma_{\text{el}}{}^{p}$. In addition, traditional bremsstrahlung can generally be neglected for scattering of particles heavier than an electron, since Eq. (9.14) is then suppressed by a factor of $m_0{}^2$ compared to $d\sigma_{\text{el}}{}^{p}$. The region of very low frequencies $\omega \ll (m_0\alpha_{\text{st}})^{-1/2}$, where the traditional bremsstrahlung of a heavy particle predominates over polarization bremsstrahlung, is an exception.

It is easy to show that the contribution of the interference term to the bremsstrahlung spectrum can be neglected for all incident particle velocities and over the entire frequency range.

The total emission spectrum $d\sigma^{p}(\omega)$ from the collision of charged relativistic particles with an atom is the sum of the elastic and inelastic bremsstrahlung spectra, which may originate through either the traditional or the polarization mechanisms. We first consider the total polarization bremsstrahlung spectrum $d\sigma^{\text{tot},p}(\omega)$ which basically determines the emission of heavy charged particles, since the probability of emission of a photon by the particle itself is small owing to its large mass $m_0 \gg 1$. Following [4], for concreteness we shall speak of a proton in the following discussion (see [5] also).

We now obtain an expression for $d\sigma^{\text{tot},p}(\omega)$ in the region $\omega \gg 1$, where the differential cross section with respect to the photon frequency integrated over the momentum transfer $\mathbf{q}_1$ for polarization bremsstrahlung with a simultaneous transition of the atom from the ground state $|0\rangle$ to all final states $|m\rangle$ is given by

$$d\sigma^{p,\text{tot}}(\omega)=d\omega\sum_m\int_{q_{\min}^m}^{q_{\max}^m}d\sigma_{m0}^{p,\text{tot}}(\omega,q_1)\,dq_1,\ \mathbf{q}\approx\mathbf{q}_1, \tag{9.15}$$

where

$$d\sigma_{mo}^{p,\mathrm{tot}}(\omega, q_1) = \frac{16}{3}\frac{e_0^2}{c^3}\frac{d\omega}{\omega}\frac{|W_{mo}(q_1)|^2}{v_0^2\omega}\frac{q_1 dq_1}{q_1^2 - (\omega + \omega_{mo})^2/c^2}\left[1 - \frac{\omega^2/\gamma^2c^2}{q_1^2 - (\omega + \omega_{mo})^2/c^2}\right];$$

$$W_{mo}(\mathbf{q}_1) = \left\langle m \left| \sum_{j=1}^{N} e^{-i\mathbf{q}_1\mathbf{r}_j} \right| 0 \right\rangle .$$

The lower limit of integration in Eq. (9.15), which equals $q_{1,\min}{}^m = |\mathbf{p}' - \mathbf{p}|_{\min}$, is found from the conservation of energy $\varepsilon = \varepsilon' + \omega + \omega_{m0}$, where ω_{m0} is the frequency of the atomic transition. As a result, for $\omega \ll \varepsilon$ and $\omega_{m0} \ll \varepsilon$ we find that $q_{1,\min}{}^m = (\omega + \omega_{m0})/v$. The upper limit $q_{1,\max}{}^m = 2p - q_{1,\min}{}^m$ in Eq. (9.15) can be set to infinity, since, as will be shown below, the main contribution to the cross section is from nonrelativistic excitation energies ω_{m0} and momentum transfers q_1.

We can show that for both small $\omega_{m0} \sim 1$ and large $\omega_{m0} \gg 1$ degrees of excitation, one can neglect the dependence of $q_{\min}{}^m$ and the integrand in Eq. (9.15) on ω_{m0}. This allows us to switch the order of summation and integration in Eq. (9.15). Separating the contributions of elastic and inelastic bremsstrahlung in Eq. (9.15) and performing some simple calculations, we obtain

$$d\sigma^{p,\mathrm{tot}}(\omega) = \frac{16}{3}\frac{e_0^2 d\omega}{c^3 v_0^2 \omega}\left\{\int_{\omega/v_0}^{\infty}\frac{q_1|W(q_1)|^2 dq_1}{(q_1^2 - \omega^2/c^2)}\left[1 - \frac{\omega^2/\gamma^2c^2}{q_1^2 - \omega^2/c^2}\right] + \right.$$

$$\left. + \int_{\omega/v_0}^{2v_0}\frac{(N + F(q_1) - |W(q_1)|^2)}{q_1^2 - \omega^2/c^2} q_1 dq_1 \left[1 - \frac{\omega^2/\gamma^2c^2}{q_1^2 - \omega^2/c^2}\right]\right\}, \tag{9.16}$$

$$W(q) = W_{00}(q),$$

$$F(\mathbf{q}_1) = \sum_{j \neq j'} \langle 0 | \exp[-i\mathbf{q}_1(\mathbf{r}_j - \mathbf{r}_{j'})] | 0 \rangle .$$

The first and second terms in Eq. (9.16) describe elastic and inelastic polarization bremsstrahlung, respectively. In the nonrelativistic case we have

$$d\sigma^{p,\mathrm{tot}}(\omega) = \frac{16 e_0^2 d\omega}{3 c^3 v_0^2 \omega}\left\{\int_{\omega/v_0}^{\infty}\frac{|W(q_1)|^2 dq_1}{q_1} + \int_{\omega/v_0}^{2v_0}(N + F(q_1) - |W(q_1)|^2)\frac{dq_1}{q_1}\right\}, \tag{9.17}$$

which was first obtained in [5] [see Eq. (7.63) as well]. The upper limit of integration $q_1 = 2v_0$, in Eq. (9.17), as well as the limitation $\omega < v_0^2/2$ on the frequency are easily understood qualitatively. In fact, inelastic polarization bremsstrahlung at large q is determined by processes leading to ionization of an atom whose electrons are seemingly free.

Using Eq. (9.16) it is easy to calculate the spectra of elastic $d\sigma_{\mathrm{el}}{}^p(\omega)$ and inelastic $d\sigma_{\mathrm{inel}}{}^p(\omega)$ polarization bremsstrahlung for $\omega < v_0 R_a^{-1}$ $(q_{\min} R_a < 1)$.

Performing the required calculations, we find that, to within logarithmic accuracy,

$$d\sigma_{el}^{p}(\omega)=\frac{16e_0^2N^2d\omega}{3c^3v_0^2\omega}\ln\frac{\gamma v_0}{\omega R_a}, \tag{9.18}$$

and

$$d\sigma_{inel}^{p}(\omega)=\frac{16e_0^2Nd\omega}{3c^3v_0^2\omega}\ln(cR_a). \tag{9.19}$$

Equations (9.18) and (9.19) coincide with the first term of Eqs. (5.55). We can confirm that for $N \gg 1$, elastic bremsstrahlung predominates in $d\sigma^{p,\text{tot}}(\omega)$ and that this is enhanced by its logarithmic growth with the energy of the incident particle. The factor N^2 appears in Eq. (9.19) because the emission of the atomic electrons during elastic bremsstrahlung adds coherently, similarly to the way this happens in Rayleigh scattering of light. On the other hand, in inelastic bremsstrahlung processes, the contribution of the individual atomic electrons to the spectrum are incoherent, similarly to the case of Raman scattering of light. The coherent emission by the electrons of a target atom during elastic bremsstrahlung leads to a dominant role for $d\sigma_{el}{}^{p}(\omega)$ at nonrelativistic collision velocities [5], as well, as can be confirmed easily with the aid of Eq. (9.17).

When $\omega \gg vR_a{}^{-1}$, inelastic bremsstrahlung processes predominate in $d\sigma^{p,\text{tot}}(\omega)$. At these frequencies, the relation $q_{1,\text{min}}R_a \gg 1$ holds, so that for all q_1, we have $W(q) \sim 0$ and $[N + F(q) - |W(q)|^2] \sim N$. Formulas for $d\sigma_{el}{}^{p}(\omega)$ can be obtained with logarithmic accuracy from Eq. (9.16) [see also Eqs. (5.46), (5.47), (5.55), and (5.57a)]:

$$d\sigma_{el}^{p}=\frac{16e_0^2Nd\omega}{3c^3v_0^2\omega}\ln\frac{\varepsilon}{\omega},\qquad d\sigma_{el}^{p}\ll d\sigma_{inel}^{p},\quad v_0R_a^{-1}\ll\omega\ll c^2. \tag{9.20}$$

According to Eq. (9.20), the cross section is given to within logarithmic accuracy by

$$d\sigma_{inel}^{p}(\omega)=Nd\sigma_0(\omega), \tag{9.21}$$

where $d\sigma_0(\omega)$ is the cross section for bremsstrahlung of a recoil electron during scattering of an ultrarelativistic particle with charge e_0 on a slow free electron. This relationship among the cross sections is not coincidental, and the atomic electrons can be regarded as free for sufficiently high frequencies ω. Equation (9.21) allows us to reach the conclusion, used above, to the effect that the region of nonrelativistic momentum transfers q_1 gives a logarithmic contribution to the cross section. Including the relativistic motion of an ionized electron leads to improved accuracy (see [7]), but does not change Eq. (9.21) for $\omega \gg v_0R_a{}^{-1}$ when the details of the atomic structure are unimportant.

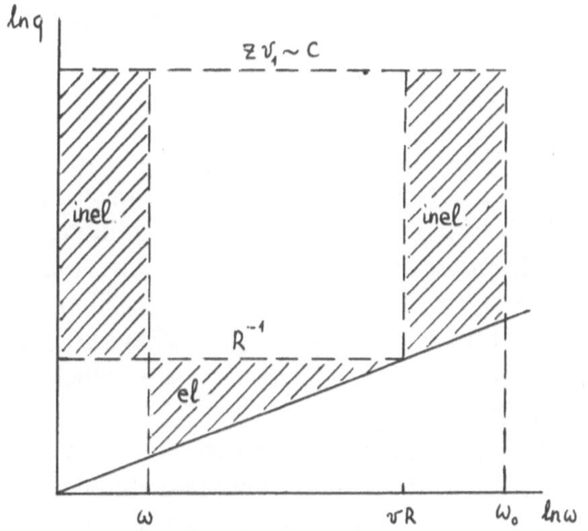

Fig. 9.3. The principal contributions to the spectrum $d\sigma^p(\omega)/d\omega$ for different ranges of ω and q_1. el) denotes elastic bremsstrahlung; inel) inelastic bremsstrahlung.

Equations (9.19)–(9.21) have been obtained for $\omega \gg 1$, but the conclusion that elastic bremsstrahlung predominates in $d\sigma^{p,\mathrm{tot}}(\omega)$ owing to the coherence effect can also be extended to the region $\omega \sim 1$, where $d\sigma_{\mathrm{el}}{}^p(\omega)$ is given by Eq. (9.13). When $\omega \ll 1$, $d\sigma_{\mathrm{el}}{}^p(\omega)$ falls off as ω^3. The spectrum $d\sigma_{\mathrm{inel}}{}^p(\omega)$ also changes in the region $\omega \ll 1$. However, the contribution to it from bremsstrahlung processes involving ionization of the atom during large momentum transfers $q_1R_a \gg 1$ will be given by Eq. (9.19), as before (this is easily verified). It follows from a comparison of Eqs. (9.23) and (9.29) that for $\omega \ll 1$, when $N \ll \omega^4|\alpha_d(\omega)|^2$, inelastic bremsstrahlung processes predominate over elastic ones.

Figure 9.3 is a schematic representation of the region of the most important momenta q_1 as a function of ω (see the shaded parts of the figure.)

We now proceed to a discussion of the total spectrum from collisions of light charged particles (electrons, positrons) with atoms. In this case, the contributions $d\sigma^{p,\mathrm{tot}}$ from polarization and traditional bremsstrahlung are comparable. The spectra $d\sigma_{\mathrm{el}}{}^p$ and $d\sigma_{\mathrm{inel}}{}^p$ are given by the same Eqs. (9.18) and (9.19) as for protons. The formulas for the spectra of elastic and inelastic traditional bremsstrahlung are known [7]. For arbitrary v_0 they are quite complicated, while in the ultrarelativistic limit they take the following extremely simple forms:

$$d\sigma_{\mathrm{el}}^{t}(\omega) = \frac{16Z^2 d\omega}{3c^5\omega}\ln(cR_a), \qquad \omega \ll cR_a^{-1}\frac{\varepsilon^2}{c^4}; \tag{9.22a}$$

$$d\sigma_{\mathrm{el}}^{t}(\omega) = \frac{16Z^2 d\omega}{3c^5\omega}\ln\frac{2\varepsilon^2}{c^2\omega}, \qquad \varepsilon \gg \omega \gg cR_a^{-1}\frac{\varepsilon^2}{c^4}; \tag{9.22b}$$

$$d\sigma^{t}_{\text{inel}}(\omega) = \frac{16N}{3c^{5}\omega}\ln cR_a, \qquad \omega \ll cR_a^{-1}\frac{\varepsilon^2}{c^4}; \tag{9.22c}$$

$$d\sigma^{t}_{\text{inel}}(\omega) = \frac{16N}{3c^{5}\omega}\ln\frac{2\varepsilon^2}{c^3\omega}, \qquad \varepsilon \gg \omega \gg cR_a^{-1}\frac{\varepsilon^2}{c^4}. \tag{9.22d}$$

The momentum transfer $(\mathbf{q})_{\min} = (\mathbf{q}_1 - \mathbf{k})_{\min} = \omega c^3/2\varepsilon^2$ in traditional bremsstrahlung determines the degree of shielding of the nucleus by the electrons [7]. Shielding becomes important in the traditional bremsstrahlung spectra when $(q)_{\min} < R_a^{-1}$, as can be seen from Eqs. (9.22a) and (9.22c). In the opposite case of $(q)_{\min} \gg R_a^{-1}$, the spectrum $d\sigma_{\text{el}}^{t}(\omega)$ reduces to the bremsstrahlung of an electron in the field of the nucleus, while $d\sigma_{\text{inel}}^{t}(\omega) = Nd\sigma_e(\omega)$, where $d\sigma_e(\omega)$ is the bremsstrahlung cross section for an ultrarelativistic electron in a collision with a free electron neglecting the recoil of the latter.

The interference terms in the total emission spectrum can be neglected in the ultrarelativistic limit, since all the traditional bremsstrahlung is concentrated in a narrow range of angles $\theta < \gamma^{-1}$, while the polarization bremsstrahlung and inelastic bremsstrahlung are only weakly anisotropic. Thus, $d\sigma^{\text{tot}}(\omega)$ is given by the sum

$$d\sigma^{\text{tot}}(\omega) = d\sigma^{t}_{\text{el}}(\omega) + d\sigma^{p}_{\text{el}}(\omega) + d\sigma^{t}_{\text{inel}}(\omega) + d\sigma^{p}_{\text{inel}}(\omega), \tag{9.23}$$

where all the terms have been defined above.

Substituting the spectra (9.18), (9.19), (9.22a), and (9.22c) in Eq. (9.23) and assuming for simplicity that the target atom is neutral ($N = Z$), we find the total emission spectrum in the coherence region of the polarization bremsstrahlung, i.e., for $1 \ll \omega < cR_a^{-1}$ ($v_0 \to c$), to be

$$d\sigma^{\text{tot}}(\omega) = \frac{16Z^2 d\omega}{3c^5\omega}\ln\frac{\varepsilon}{\omega} + \frac{32Z}{3\omega c^5}\ln(cR_a). \tag{9.24}$$

The first term in this expression is the sum of the elastic polarization and traditional spectra, while the second describes the contribution of inelastic bremsstrahlung processes to $d\sigma^{\text{tot}}(\omega)$ and is negligibly small compared to the first for many-electron atoms ($Z \gg 1$). Equation (9.24) shows that including elastic polarization bremsstrahlung leads to a logarithmic growth in $d\sigma^{\text{tot}}(\omega)$ with ε.

Beyond the limits of the coherence region examined above, the inelastic processes, whose cross section is proportional to Z, predominate in the polarization bremsstrahlung spectrum. Thus, in these regions of ω the main contribution to $d\sigma^{\text{tot}}(\omega)$ for a many-electron atom is $d\sigma_{\text{el}}^{t} \sim Z^2 \gg d\sigma_{\text{inel}}^{t} \sim d\sigma_{\text{inel}}^{p} \sim Z$. For atoms with small Z, the spectra $d\sigma_{\text{el}}^{t}$, $d\sigma_{\text{inel}}^{t}$, and $d\sigma_{\text{inel}}^{p}$ must be taken into account simultaneously.

We have examined $d\sigma^{\text{tot}}(\omega)$ for ultrarelativistic electrons. The results given above are equally applicable to an ultrarelativistic positron, since according to Eq. (9.23), $d\sigma^{\text{tot}}(\omega)$ does not contain interference terms that depend on the sign of the charge of the incident particle. When $v_0 \ll c$, however, the total emission spectrum for a positron differs substantially from the nonrelativistic spectrum [10] because of the possibility of dipole radiation during scattering of a positron on the electrons of an atom at $\omega \gg 1$ (see Chapter 4 as well).

In sum, we have found that in a whole range of cases one can evaluate the properties of individual processes in terms of the total emission spectrum. Thus, the predominance of polarization bremsstrahlung in $d\sigma^{tot}$ for $1 \leq \omega \leq v_0 R_a^{-1}$ because of the coherence of the emission from the electrons in the target atom is of a universal character that shows up for both relativistic and nonrelativistic velocities of the scattered particle, which may be either heavy or light. The dominance of elastic over inelastic bremsstrahlung makes it possible to observe the elastic polarization bremsstrahlung of heavy charged particles (such as protons) by choosing a many-electron system as a target and measuring only the total emission spectrum [4, 5].

The coherence effect in bremsstrahlung does not depend on the type of interaction between the incident particle and an atomic electron. It shows up, in particular, during scattering of a neutrino on an atom. In this case, the atomic shell is polarized because of the weak interaction. The emission produced in collisions of neutrinos with atoms also has some specific features which are characteristic of weak interactions. It turns out that the spectrum and angular distribution of the bremsstrahlung from neutrinos depend significantly on the mutual orientation of the momentum p of the neutrino before scattering and the spin s of the atomic electron. Thus, the intensity of the radiation reaches its maximum and minimum values for co- and counter-orientations of these vectors, respectively, which is an indication that parity is not conserved in weak interactions. This difference in the intensities of bremsstrahlung shows up especially strongly in the scattering of neutrinos on hydrogen or hydrogenlike ions. In this way, by studying the spectrum of neutrino bremsstrahlung on a polarized target (we mean electron polarization), one can determine the direction of motion of a beam of incident neutrinos. The absolute intensity of the bremsstrahlung is especially high if the frequencies of the radiation are close to atomic resonance frequencies.

9.2. Amplitude of Bremsstrahlung in Atom–Atom Collisions at Relativistic Velocities

The bremsstrahlung produced by collisions of fast, but nonrelativistic, atomic particles [6] is very different from the case of relativistic particles. The reason for the variation in the bremsstrahlung cross section with increasing collision velocities is that the frequency and angle of emission of the photon emitted by an atomic projectile change, because of the Doppler effect and the aberration of light [1, 11]. These effects show up especially strongly when the relative velocities of the atoms $v \sim c$ (where c is the speed of light), but it is also important to take them into account for collisions of fast, but nonrelativistic, atoms, i.e., when $1 \ll v \ll c$ (in atomic units $c = 137$). In particular, the Doppler effect and the aberration of light remove the prohibition on dipole emission during collisions of identical atoms [6], and near the characteristic atomic frequencies the intensity of bremsstrahlung owing to these relativistic effects is parametrically dominant over the quadrupole bremsstrahlung of the system. Let us consider an atom–atom collision in which a photon with momentum **k** and polarization **e** is emitted, while the atoms are in the ground states at both the beginning and end of the process. In order to describe bremsstrahlung at relativistic velocities we shall use the lowest orders of perturbation theory with respect to the interaction among the atoms, which reduces to one-photon exchange, and with respect to the interaction between the atoms and the radiation field.

The Feynman diagrams for the bremsstrahlung amplitude have the form shown in Fig. 9.4.

Here a double line describes an atom and a dashed line, a photon. The states of the particles are indicated by a pair of letters, of which the first is the total energy of the particle and the second, its momentum. The total energy of a relativistic atom includes its intrinsic and kinetic energies. The subscripts 1 and 2 on the diagrams and in the following discussion refer to the projectile atom and target atom, respectively.

We now write down a formula for the amplitude of bremsstrahlung in atom–atom collisions in the form of a sum of two terms,

$$F = F_1 + F_2, \tag{9.25}$$

where F_1 describes emission of a photon by the second atom (the first two diagrams of Fig. 9.4) and F_2 describes emission by the first atom (the third and fourth diagrams of Fig. 9.4). The analytic expression for the amplitude, written in accordance with the general rules for constructing Feynman diagrams [7] taking the many-electron character of the colliding atoms into account, has the form

$$\left.\begin{aligned} \langle f|\hat{F}_1|i\rangle &= i(2\pi)^4\,\delta(E_1+E_2-E_1'-E_2'-\omega)\,\delta(\mathbf{p}_1+ \\ &+\mathbf{p}_2-\mathbf{p}_1'-\mathbf{p}_2'-\mathbf{k})\sqrt{\frac{2\pi c^2}{\omega L^3}\frac{m_1c^2}{E_1}\left(\frac{m_2c^2}{E_2}\right)}\frac{1}{L^6}F_1, \end{aligned}\right\} \tag{9.26}$$

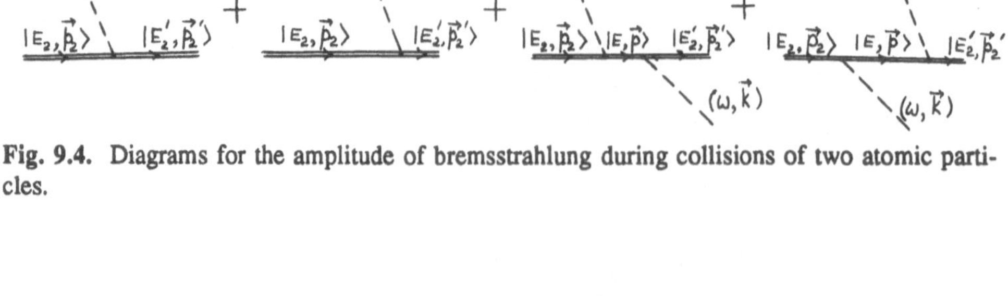

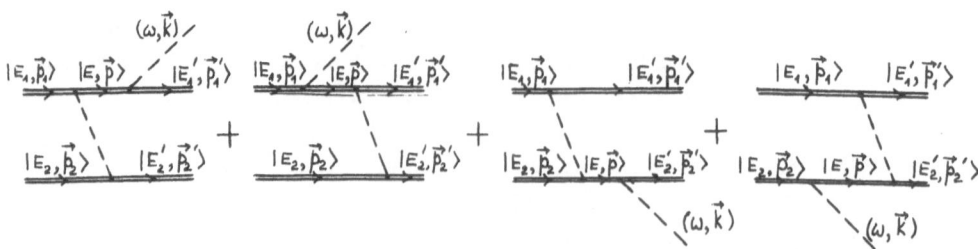

Fig. 9.4. Diagrams for the amplitude of bremsstrahlung during collisions of two atomic particles.

$$
\left.\begin{aligned}
F_1 = &\left\langle E_1' \left| \sum_{j=1}^{N_1+1} e_j \gamma_\mu e^{i\mathbf{q}_1'\mathbf{r}_j} \right| E_1 \right\rangle D^{\mu\nu}(E_1 - E_1', -\mathbf{q}_1') \times \\
&\times \sum_E \left\{ \frac{\left\langle E_2' \left| \sum_{j=1}^{N_2+1} e_j \hat{e} e^{-i\mathbf{k}\mathbf{r}_j} \right| E \right\rangle \left\langle E \left| \sum_{j=1}^{N_2+1} \gamma_\nu e_j e^{i\mathbf{q}_2\mathbf{r}_j} \right| E_2 \right\rangle}{\omega - (E - E_2')} - \right. \\
&\left. - \frac{\left\langle E_2' \left| \sum_{j=1}^{N_2+1} e_j \gamma_\nu e^{i\mathbf{q}_2\mathbf{r}_j} \right| E \right\rangle \left\langle E \left| \sum_{j=1}^{N_2+1} e_j \hat{e} e^{-i\mathbf{k}\mathbf{r}_j} \right| E_2 \right\rangle}{\omega + (E - E_2)} \right\},
\end{aligned}\right\} \quad (9.26)
$$

where

$$
\mathbf{q}_{1,2} = \mathbf{p}_{1,2}' - \mathbf{p}_{1,2} + \mathbf{k}, \quad \mathbf{q}_{1,2}' = \mathbf{q}_{1,2} - \mathbf{k}, \quad \mathbf{q}_1 + \mathbf{q}_2^{*} = \mathbf{q}_2 + \mathbf{q}_1^{*} = 0.
$$

In Eq. (9.26) we have integrated over the coordinates of the centers of inertia of each of the atoms. The following notation has been used: m_1 is the mass of the projectile atom (the first atom); $\{\mathbf{r}_j\}, j = 1, ..., N_1,$ is the set of relative coordinates of the electrons in it; $\mathbf{r}_{N_1+1} \equiv \mathbf{r}_{n,1} \sim 0$ is the relative coordinate of the nucleus of the atom neglecting its recoil during the collision process; $e_j = -1$ with $j = 1, ..., N_1$ and $e_{N_1+1} = Z_1$ are the charges of the electrons and nucleus of the first atom; $\mathbf{q}_1' = \mathbf{p}_1' - \mathbf{p}_1$; $\mathbf{q}_2 = \mathbf{p}_2' - \mathbf{p}_2 + \mathbf{k}$; and L^3 is the normalization volume. Analogous notation has been used for the target atom (second atom). The symbol $D^{\mu\nu}(E_1 - E_1', -\mathbf{q}_1')$ denotes the Green function of the photon and $\hat{e} = e_\mu \gamma^\mu$ ($\mu = 0, 1, 2, 3$), where γ^μ are the Dirac matrices and e^μ is the polarization vector of the photon. The sum in Eq. (9.26) is taken over the entire spectrum of positive and negative energies of the target atom in the intermediate state. For convenience in writing down the following formulas we have taken the normalizing factors in the wave functions of the atoms $(m_i c^2/E_i L^3)^{1/2}$, $i = 1, 2$ and emitted photon $(2\pi c^2/\omega L^3)^{1/2}$ out of the definition of F_1, along with the δ-functions which ensure that the momentum and energy are conserved. Here we have neglected the changes in the factors $m_i c^2/E_i = \gamma_i^{-1}$ and $m_i c^2/E_i' = (\gamma_i')^{-1}$, $i = 1, 2$, during the bremsstrahlung process. These factors take into account the Lorentz contraction of the normalization volume for the wave functions of the translational motion of the atoms. This is correct, since in the kinematic region of subsequent interest to us, the atoms move almost rectilinearly during the bremsstrahlung process, i.e., $E_i - E_i' \ll E_i$ and $\mathbf{p}_i - \mathbf{p}_i' \ll \mathbf{p}_i$, where $i = 1, 2$.

Let us write the amplitude F_1 in the symbolic form

$$
F_1 = J_\mu^{(1)}(\mathbf{q}_1') D^{\mu\nu}(E_1 - E_1', -\mathbf{q}_1') T_{\lambda\nu}^{(2)}(\omega, k, \mathbf{q}_2) e^\lambda, \quad (9.27)
$$

where it is easy to find the explicit forms of $J_\mu^{(1)}(\mathbf{q})$ and $T_{\lambda\nu}^{(2)}(\omega, \mathbf{k}, \mathbf{q})$ by comparing Eqs. (9.27) and (9.26). Here the 4-vector $J_\mu^{(1)}(\mathbf{q}_1')$ describes the current of the first atom and the 4-tensor $T_{\lambda\nu}^{(2)}(\omega, \mathbf{k}, \mathbf{q}_2)$, the dynamic response of its partner, which leads to the emission of a photon. The interaction between the atoms takes place through the exchange of a photon, whose Green function $D^{\mu\nu}(E_1 - E_1',$

$-\mathbf{q}_1'$) relates $J_\mu^{(1)}(\mathbf{q}_1')$ and $T_{\lambda\nu}^{(2)}(\omega, \mathbf{k}, \mathbf{q}_2)$. Therefore, we see that the structure of the amplitude F_1 corresponds correctly to the qualitative picture of the process.

The quantities $J_\mu^{(1)}(\mathbf{q}_1')$ and $T_{\lambda\nu}^{(2)}(\omega, \mathbf{k}, \mathbf{q}_2)$ have their simplest forms in a reference frame where the first or second atom is at rest as a whole, i.e., in the rest frame. In this process a reference system of this sort can be introduced for each of the atoms, since at the small momentum transfers $\mathbf{q}_1'$ and $\mathbf{q}_2'$ of interest to us, the recoil of the atoms is negligible. The time and space components of $J_\mu^{(1)}(\mathbf{q}_1')$ in the rest frame of the first atom are given, respectively, by

$$J_0^{(1)}(\mathbf{q}_1'^s) \equiv Z_1\delta_{n_1',n_1} - \left\langle n_1' \left| \sum_{j=1}^{N_1} e^{i\mathbf{q}_1'^s\mathbf{r}_j} \right| n_1 \right\rangle = Z_1\delta_{n_1',n_1} - W^{(1)}_{n_1',n_1}(\mathbf{q}_1'^s),$$
$$\mathbf{J}^{(1)}(\mathbf{q}_1'^s) \equiv \mathbf{J}^{(1)}_{n_1',n_1}(\mathbf{q}_1'^s) = -\left\langle n_1' \left| \sum_{j=1}^{N_1} [\exp(i\mathbf{q}_1'^s\mathbf{r}_j)]\,\hat{\mathbf{p}}_j/c \right| n_1 \right\rangle. \tag{9.28}$$

In Eq. (9.28) $\hat{\mathbf{p}}_j$ is the momentum operator of the jth atomic particle; $|n_1\rangle$ and $\langle n_1'|$ are the states of the atom in its rest frame; and $\mathbf{q}_1'^s$ is the momentum transfer, which is the spatial part of the 4-vector

$$-q_1'^s = \left\{ \frac{\omega^s_{n_1,n_1'}}{c};\ -\mathbf{q}_1'^s \right\}, \tag{9.29}$$

where $\omega_{n_1,n_1'}{}^s = E_{n_1}{}^s - E_{n_1'}{}^s$ is the frequency of the transition between states $|n_1\rangle$ and $|n_1'\rangle$, and the superscript s means that a quantity belongs to the rest frame. In deriving Eqs. (9.28) and (9.29) we have omitted relativistic corrections of order $q_1'^s/c \ll 1$ and $1/m_1 \ll 1$.

Once $J_\mu^{(1)}(\mathbf{q}_1'^s)$ is known in the rest frame, it is simple to obtain this 4-vector in the laboratory coordinate system, where the atom moves at velocity v_1 during the collision process, neglecting the recoil of its nucleus:

$$J_0^{(1)L}(\mathbf{q}_1'^{\parallel}, \mathbf{q}_s'^{\perp}) = \gamma\left\{ Z_1\delta_{n_1',n_1} - W^{(1)}_{n_1',n_1}\left(\mathbf{q}_1'^{\parallel}\gamma^{-1} + \frac{\mathbf{v}_1}{c}\frac{\omega^s_{n_1,n_1'}}{c} + \mathbf{q}_1'^{\perp}\right) + \frac{v_1}{c}\mathbf{n}_1\mathbf{J}^{(1)}_{n_1',n_1}\left(\mathbf{q}_1'^{\parallel}\gamma^{-1} + \frac{\mathbf{v}_1}{c}\frac{\omega^s_{n_1,n_1'}}{c} + \mathbf{q}_1'^{\perp}\right)\right\},$$
$$\mathbf{J}^{(1)L}(\mathbf{q}_1'^{\parallel}, \mathbf{q}_1'^{\perp}) = \gamma\mathbf{n}_1\left\{\mathbf{n}_1\mathbf{J}^{(1)}_{n_1',n_1}\left(\mathbf{q}_1'^{\parallel}\gamma^{-1} + \frac{\mathbf{v}_1}{c}\frac{\omega^s_{n_1,n_1'}}{c} + \mathbf{q}_1'^{\perp}\right) + \frac{v_1}{c}\left(Z_1\delta_{n_1',n_1} - W^{(1)}_{n_1',n_1}\left(\mathbf{q}_1'^{\parallel}\gamma^{-1} + \frac{\mathbf{v}_1}{c}\frac{\omega^s_{n_1,n_1'}}{c} + \mathbf{q}_1'^{\perp}\right)\right)\right\} + \mathbf{J}^{(1)\perp}_{n_1',n_1}\left(\mathbf{q}_1'^{\parallel}\gamma^{-1} + \frac{\mathbf{v}_1}{c}\frac{\omega^s_{n_1,n_1'}}{c} + \mathbf{q}_1'^{\perp}\right). \tag{9.30}$$

In Eq. (9.30) and subsequently, the indices $\parallel$ and $\perp$ denote the parallel and perpendicular components of the velocity vectors of the atoms in the laboratory system $\mathbf{n}_1 = \mathbf{v}_1/v_1$, and $\gamma = (1 - v_1^2/c^2)^{-1/2}$.

It is easy to verify that for diagonal transitions,

$$|\mathbf{J}^{(1)}_{n_1,n_1}(+\mathbf{q}_1')| = \frac{1}{2c} q_1' W^{(1)}_{n_1,n_1}(\mathbf{q}_1') \ll W^{(1)}_{n_1,n_1}(\mathbf{q}_1')$$

in the region $q_1' \ll c$ of interest to us. Thus, $J_0^{(1)L}$ and $\mathbf{J}^{(1)L}$ can be simplified greatly by neglecting the terms in Eq. (9.30) that are proportional to $J_{n1,n1}^{(1)}(\mathbf{q}_1'^{\|}\gamma^{-1} + \mathbf{q}_1'^{\perp})$. As a result, we obtain

$$J_0^{(1)L}(\mathbf{q}_1'^{\|}, \mathbf{q}_1'^{\perp}) = \gamma(Z_1 - W^{(1)}_{n,n}(\mathbf{q}_1'^{\|}\gamma^{-1} + \mathbf{q}_1'^{\perp})),$$
$$J^{(1)L}(\mathbf{q}_1'^{\|}, \mathbf{q}_1'^{\perp}) = \gamma\frac{\mathbf{v}_1}{c}(Z_1 - W^{(1)}_{n,n}(\mathbf{q}_1'^{\|}\gamma^{-1} + \mathbf{q}_1'^{\perp})). \tag{9.31}$$

These formulas have a simple meaning. The time component $J_0^{(1)L}(\mathbf{q}_1'^{\|}, \mathbf{q}_1'^{\perp})$ describes the form factor of the electron and nuclear charge of the atom in the $|n\rangle$ state that has been deformed as a result of Lorentz contraction, while the spatial component $J^{(1)L}(\mathbf{q}_1'^{\|}, \mathbf{q}_1'^{\perp})$ describes the current of the atomic charges associated with their translational motion at velocity v_1.

The 4-vector $T_{\lambda\nu}^{(2)}(\omega, \mathbf{k}, \mathbf{q}_2)$ is calculated analogously. In the dipole approximation $(\omega/c)R_a \ll 1$, this tensor has the following form in the rest frame of the second atom:

$$T^{(2)}_{\lambda\nu} = \left\| \begin{matrix} T^{(2)}_{00} & \mathbf{T}^{(2)}_{0j} \\ \mathbf{T}^{(2)}_{i0} & T^{(2)}_{ij} \end{matrix} \right\| \equiv \left\| \begin{matrix} -\mathbf{k}^s\mathbf{q}_2^s\alpha^{(2)}(\omega^s, \mathbf{q}_2^s) & -\dfrac{\omega^s}{c} k_i^s\beta_{ij}^{(r)}(\omega^s, \mathbf{q}_2^s) \\ -\dfrac{\omega^s}{c} q_{2i}^s\alpha^{(2)}(\omega^s, \mathbf{q}_2^s) & -\left(\dfrac{\omega^s}{c}\right)^2 \beta_{ij}^{(2)}(\omega^s, \mathbf{q}_2^s) \end{matrix} \right\|. \tag{9.32}$$

In Eq. (9.32), the indices i and j denote the spatial components of the tensor. The generalized polarizability $\alpha^{(2)}(\omega^s, \mathbf{q}^s)$ is introduced in accordance with the definition

$$\mathbf{q}^s\alpha^{(2)}(\omega^s, \mathbf{q}^s) = -i\sum_n \frac{2\omega^s_{no}}{(\omega^s)^2 - (\omega^s_{no})^2}\mathbf{D}^{(2)}_{on}W^{(2)}_{no}(\mathbf{q}^s), \tag{9.33}$$

and the polarization tensor $\beta_{ij}^{(2)}(\omega^s, \mathbf{q}^s)$ is given by

$$\frac{\omega^s}{c}\beta_{ij}^{(2)}(\omega^s, \mathbf{q}^s) = i\sum_n \left\{ \frac{(\mathbf{D}^{(2)}_{on})_i(\mathbf{J}^{(2)}_{no}(\mathbf{q}^s))_j}{\omega^s - \omega^s_{no}} - \frac{(\mathbf{J}^{(2)}_{on}(\mathbf{q}^s))_i(\mathbf{D}^{(2)}_{no})_j}{\omega^s + \omega^s_{no}} \right\}. \tag{9.34}$$

In deriving Eqs. (9.32)–(9.34) we have chosen the ground state $|0\rangle$ of the atom as its initial and final states. For small q_2^s $(q_2^sR_a \ll 1)$, we have $\alpha^{(2)}(\omega^s, \mathbf{q}^s) \sim \alpha_d^{(2)}(\omega^s)$ and $\beta_{ij}^{(2)} \sim \delta_{ij}\alpha_d^{(2)}(\omega^s)$, where $\alpha_d^{(2)}(\omega^s)$ is the dipole dynamic polarizability of the atom.

In the general case, the transformation of the components of the tensor $T_{\lambda\nu}^{(2)}$ (9.32) to the laboratory system for the second atom is quite complicated. However, it turns out that the ranges of q_2^s which are important for calculating the bremsstrahlung cross sections are those where $T_{\lambda\nu}^{(2)}$ has a very simple form. As we shall show, during a collision of a pair of neutral atoms the main contribution to the bremsstrahlung spectrum is from the region $q_2^s R_a \sim 1$; hence, $q_2^s \sim R_a^{-1} \gg \omega^s/c$ or, in the dipole approximation, $(\omega_s/c)R_a \ll 1$. In this region the tensor $T_{\lambda\nu}^{(2)}$ is approximately equal to

$$T_{\lambda\nu}^{(2)} = \alpha^{(2)}(\omega^s, \mathbf{q}_2^s) \begin{Vmatrix} -\mathbf{k}^{(s)}\mathbf{q}_2^s & 0_j \\ -\dfrac{\omega^s}{c}\mathbf{q}_{2,i}^s & 0_{ij} \end{Vmatrix}. \tag{9.35}$$

The symbols $\mathbf{0}_j$ and $\mathbf{0}_{ij}$ denote the components of the null vector and tensor, respectively.

When an atom is polarized in the field of a charged particle, such as an ion, the main contribution to the calculated spectrum is from small q_2^s ($q_2^s R_a \ll 1$), where the tensor $T_{\lambda\nu}^{(2)}$ is approximately equal to

$$T_{\lambda\nu}^{(2)} \approx \alpha_d^{(2)}(\omega^s) \begin{Vmatrix} -k^s\mathbf{q}_2^s & -\dfrac{\omega^s}{c}\mathbf{k}_j^s \\ -\dfrac{\omega^s}{c}\mathbf{q}_{2,i}^s & -\left(\dfrac{\omega^s}{c}\right)^2 \delta_{ij} \end{Vmatrix}. \tag{9.36}$$

Here $\alpha_d^{(2)}(\omega^s)$ is the dipole dynamic polarizability of the atom. Then it is easy to write down the components of $T_{\lambda\nu}^{(2)}$ [see Eqs. (9.35) and (9.36)] in the laboratory system with the aid of the Lorentz transformation for 4-vectors, similarly to the way it was done for $J_\mu^{(1)}$ in deriving Eq. (9.30).

Having constructed $J_\mu^{(1)}(\mathbf{q}_1')$ and $T_{\lambda\nu}^{(2)}(\omega, \mathbf{k}, \mathbf{q}_2)$, and, analogously, $J_\mu^{(2)}(\mathbf{q}_2')$ and $T_{\lambda\nu}^{(1)}(\omega, \mathbf{k}, \mathbf{q}_1)$ in the rest and laboratory coordinate systems, we now write the amplitudes F_1 and F_2 [see Eq. (9.25)] in terms of these quantities. Since we can assume that the motion of the atoms is almost rectilinear in this kinematic region, it is always possible to choose a coordinate system in which one of the atoms, for example the first, is at rest during the bremsstrahlung process and its partner moves at velocity $\mathbf{v}_1$. Then the amplitude F_1 is given in terms of the 4-vector $J_\mu^{(1)}(+\mathbf{q}_1')$, calculated in the laboratory reference frame of the projectile, and $T_{\lambda\nu}^{(2)}(\omega, \mathbf{k}^s, \mathbf{q}_2^s)$, calculated in the rest frame of the target. F_2, on the other hand, includes $J_\mu^{(2)}(+\mathbf{q}_2')$ in the rest frame of the target atom and $T_{\lambda\nu}^{(1)}(\omega, \mathbf{k}, +\mathbf{q}_1^s)$ in the laboratory system of the projectile. We now write down the amplitudes F_1 and F_2 without performing the intermediate calculations.

The amplitudes F_1 and F_2 for two colliding neutral atoms are given by

$$F_1 = -\frac{4\pi\gamma\,(Z_1 - W^{(1)}(\mathbf{q}_1^\perp))}{(q_1^\perp)^2}(\mathbf{e}\mathbf{q}_1^\perp)\frac{\omega}{c}\alpha^{(2)}(\omega, \mathbf{q}_1^\perp) \tag{9.37}$$

and

$$F_2 = + \frac{4\pi\gamma^3 (Z_2 - W^{(2)}(-\mathbf{q}_1^{\perp}))}{(\mathbf{q}_1^{\perp})^2} \left[\frac{\omega}{c} \left(1 - \frac{v_1}{c} \cos\theta\right) (\mathbf{e}\mathbf{q}_1^{\perp}) + \right.$$

$$\left. + \frac{v_1}{c} (\mathbf{e}\mathbf{n}_1)(\mathbf{k}\mathbf{q}_1^{\perp}) \right] \alpha^{(1)} \left(\omega\gamma \left(1 - \frac{v_1}{c} \cos\theta\right); -\mathbf{q}_1^{\perp} \right) \Big], \tag{9.38}$$

where $\mathbf{q}_2 = -\mathbf{q}_1$; $\mathbf{n}_1 = \mathbf{v}_1/v_1$; θ is the angle between the vectors $\mathbf{k}$ and $\mathbf{v}_1$; $\mathbf{q}_1' \approx \mathbf{q}_1$; $\mathbf{q}_2 \approx -\mathbf{q}_1 \approx \mathbf{q}_2$; $W^{(i)}(\mathbf{q}_1^{\perp}) = W_{00}^{(i)}(\mathbf{q}_{\perp})$; $q_1^{\parallel} \approx \omega/v_1 \ll q_1^{\perp} \sim R_a^{-1}$, since $(\omega/v_1)R_a \ll 1$; $\gamma = (1 - v_1^2/c^2)^{-1/2}$.

In ion–ion collisions F_1 and F_2 for small momentum transfers are equal to

$$F_1 = - \frac{4\pi (Z_1 - N_1)}{q_1'^2 - \omega^2/c^2} \left[\mathbf{e}\mathbf{q}_1' + \frac{\omega}{c^2} (\mathbf{e}\mathbf{v}_1) \right] \frac{\omega}{c} \alpha_d^{(2)}(\omega), \tag{9.39}$$

and

$$F_2 = - \frac{4\pi\gamma^3 (Z_2 - N_2)}{q_2'^2} \left[\frac{\omega}{c} \left(1 - \frac{v_1}{c} \cos\theta\right) (\mathbf{e}\mathbf{q}_2'^{\perp}) + \right.$$

$$\left. + \frac{v_1}{c} (\mathbf{e}\mathbf{n}_1)(\mathbf{k}\mathbf{q}_2'^{\perp}) - \gamma^{-2} \frac{\omega}{c} \left(1 - \frac{v_1}{c} \cos\theta\right) (\mathbf{e}\mathbf{q}_1'^{\parallel}) \right] \times$$

$$\times \alpha_d^{(1)} \left(\omega\gamma \left(1 - \frac{v_1}{c} \cos\theta\right) \right). \tag{9.40}$$

These formulas also describe bremsstrahlung in ion–atom collisions, since then the major contribution to the cross section is from bremsstrahlung of the atom in the Coulomb field of the atom. Here, depending on whether the atom is at rest or moving, the amplitude for the process is given by Eq. (9.39) or (9.40), respectively.

We now discuss the formulas given above. The amplitude F_1 (9.37) including the normalizing factors for the wave functions [Eq. (9.26)] coincides with the expression for the bremsstrahlung amplitude of a target atom obtained in [6] for nonrelativistic collision velocities. This agreement is not coincidental. The process in which the atoms collide and emit a photon occurs over distances of $R \sim q_1^{-1} \sim R_a$, where the interaction between them is described primarily by the potentials of the pairwise Coulomb interactions among the particles making up the projectile and target. The retardation of the electromagnetic interaction and the transverse component of the latter show up weakly in this range of R. The pairwise Coulomb interaction between the particles of the colliding atoms develops mainly through the transverse component of the momentum transfer $q_1^{\perp} \sim R_a^{-1}$, which is considerably greater than its longitudinal component; i.e., $q_1^{\parallel} = \omega/v \ll q_1^{\perp} \sim R_a^{-1}$, since $\omega/v_1 \ll R_a^{-1}$. The longitudinal component of the momentum transfer $q_1^{\parallel} = \omega/v_1$ is fixed by the conservation of energy. The transverse component $\mathbf{q}_1^{\perp}$ is related to the transverse polarization of a target atom during the collision process, which does not depend on the choice of reference system. Thus, the amplitude F_2 (9.38) can be obtained from F_1 (9.37) by transforming to a coordinate system moving at velocity $\mathbf{v}_1$ and then interchanging the atoms. Here only the frequency ω and polarization vector $\mathbf{e}$ of the photon are transformed, while the atomic characteristics in Eq. (9.37) are unaffected by the transformation.

The bremsstrahlung process proceeds somewhat differently in ion–ion collisions, since then the main contribution to the cross section is from small momentum transfers $qR_a \ll 1$. In this kinematic region collisions take place over large distances, so that at relativistic velocities the retardation of the interaction between the ions and the transverse nature of the interaction become important. The polarization of each ion at large distances is caused mainly by the charge $(Z_i - N_i)$ of the partner ion. The amplitude of the emission from the projectile (9.40) can be found from Eq. (9.39) by transforming to a reference system moving at velocity $\mathbf{v}_1$ followed by interchange of the ions. Here, as opposed to the analogous transformation for neutral atom collisions, we must also include the transformation of the longitudinal component $q_1^{\parallel}$ of the momentum transfer, since for $q_1 \sim q_1^{\min} = \omega/v_1 \ll R_a^{-1}$ the two components $\mathbf{q}_1^{\parallel}$ and $\mathbf{q}_1^{\perp}$ are comparable. The longitudinal polarization of the ion, whose magnitude is γ^{-2} times smaller than its transverse polarization in the laboratory frame, is related to the longitudinal momentum transfer [see the third term in the brackets of Eq. (9.40)]. This circumstance must be taken into account in going from Eq. (9.39) to Eq. (9.40).

We now expand these amplitudes in terms of the parameter v_1/c. This yields the main relativistic correction to the equations of [6]. As a result of some simple transformations we find that for a collision of a pair of neutral atoms, the amplitude $F = F_1 + F_2$ of the process has the form

$$F = -\frac{4\pi}{(\mathbf{q}_1^{\perp})^2}\left[(Z_1 - \overset{(1)}{W}(\mathbf{q}_1^{\perp}))\,\alpha^{(2)}(\omega, \mathbf{q}_1^{\perp}) - (Z_2 - W^{(2)}(\mathbf{q}_1^{\perp}))\times\right.$$
$$\left.\times\,\alpha^{(1)}(\omega, \mathbf{q}_1^{\perp})\right]\frac{\omega}{c}(\mathbf{e}\mathbf{q}_1^{\perp}) - \frac{4\pi\,(Z_2 - W_2(\mathbf{q}_1^{\perp}))}{(\mathbf{q}_1^{\perp})^2}\frac{v_1}{c}\times$$
$$\times\left[\cos\theta\,\frac{\omega}{c}(\mathbf{e}\mathbf{q}_1^{\perp})\frac{\partial}{\partial\omega}\{\omega\alpha^{(1)}(\omega, \mathbf{q}_1^{\perp})\} - (\mathbf{e}\mathbf{n}_1)(\mathbf{k}\mathbf{q}_1^{\perp})\,\alpha^{(1)}(\omega, \mathbf{q}_1^{\perp})\right], \qquad (9.41)$$

where the first term describes the bremsstrahlung amplitude with complete neglect of relativistic effects, while the second term is proportional to v_1/c and is the main relativistic correction to this amplitude.

In the case of ion–ion collisions, an analogous expansion of the amplitudes (9.39) and (9.40) yields the result

$$F = -\frac{4\pi}{q_1^2}\frac{\omega}{c}(\mathbf{e}\mathbf{q}_1)\left[(Z_1 - N_1)\,\alpha_d^{(2)}(\omega) - (Z_2 - N_2)\,\alpha_d^{(1)}(\omega)\right] +$$
$$+\frac{8\pi}{q_1^2}\frac{\omega}{c}(\mathbf{e}\mathbf{q}_1)(Z_2 - N_2)\,\alpha_d^{(1)}(\omega)\,\frac{(\mathbf{k}\mathbf{q}_1)}{q_1^2} -$$

$$-\frac{4\pi(Z_2-N_2)}{q_1^2}\frac{v_1}{c}\left[\frac{\omega}{c}\cos\theta\frac{\partial}{\partial\omega}\{\omega\alpha_d^{(1)}(\omega)\}(\mathbf{eq}_1)-(\mathbf{en}_1)(\mathbf{kq}_1)\alpha_d^{(1)}(\omega)\right]. \quad (9.42)$$

Here Z_1-N_1 and Z_2-N_2 are the charges of the ions. The second and third terms in Eq. (9.42) give the relativistic correction to the nonrelativistic result.

For collisions of pairs of identical atoms or ions, the first terms in the amplitudes (9.41) and (9.42) drop out, so that F is entirely determined by the relativistic corrections found here.

The amplitudes (9.41) and (9.42) include the lowest-order terms in an expansion in the parameter v_1/c obtained above in the dipole approximation of the general expressions (9.38) and (9.40). Along with the terms written in Eqs. (9.41) and (9.42), in general the bremsstrahlung amplitude includes terms for the emission of quadrupole photons. They provide a correction to the bremsstrahlung amplitude on the order of $(\omega/c)R_a \ll 1$. This correction, however, can be neglected for $(\omega/v_1)R_a \ll 1$, since here it is of a much smaller order of magnitude than the above relativistic correction for $v_1/c \gg (\omega/c)R_a$.

9.3. Cross Section for Bremsstrahlung in Atom–Atom Collisions at Relativistic Velocities

The bremsstrahlung cross section is the sum of three parts:

$$d\sigma=d\sigma^{\text{targ}}+d\sigma^{\text{bul}}+d\sigma^{\text{int}}, \quad (9.43)$$

where $d\sigma^{\text{targ}}$ and $d\sigma^{\text{bul}}$ are the cross sections for bremsstrahlung of the target and of the projectile (bullet), and $d\sigma^{\text{int}}$ is the interference term. Using the amplitudes F_1 and F_2 found in Sec. 9.2, we now calculate $d\sigma(\theta, \Omega_{\mathbf{k}})$ similarly to the nonrelativistic case [6] (see Chapter 8).

The cross sections for these processes in an atom–atom collision, integrated over the momenta of the atoms in the final state and summed over the polarizations of the photon, are given by

$$d\sigma^{\text{targ}}=\frac{d\omega d\Omega_{\mathbf{k}}(1+\cos^2\theta)}{\pi v_1^2c^3\omega}\int_0^\infty\frac{dq_1^\perp}{q_1^\perp}|(Z_1-W^{(1)}(q_1^\perp))\,\omega^2\alpha^{(2)}(\omega,q_1^\perp)|^2, \quad (9.44)$$

$$d\sigma^{\text{bul}}=\frac{\gamma^2(1-\cos\theta\cdot v_1/c)^2\,d\omega d\Omega_{\mathbf{k}}}{\pi v_1^2c^3\omega}\left(1+\frac{(\cos\theta-v_1/c)^2}{(1-\cos\theta\cdot v_1/c)^2}\right)\times$$

$$\times\int_0^\infty\frac{dq_1^\perp}{q_1^\perp}\left|(Z_2-W^{(2)}(q_1^\perp))\,\omega^2\alpha^{(1)}\left(\omega\gamma\left(1-\frac{v_1}{c}\cos\theta\right),q_1^\perp\right)\right|^2, \quad (9.45)$$

and

$$d\sigma^{\text{int}}=-\frac{2\omega^3\gamma(1-\cos\theta\cdot v_1/c)\,d\omega d\Omega_{\mathbf{k}}}{\pi v_1^2c^3}\left(1+\cos\theta\frac{(\cos\theta-v_1/c)}{(1-\cos\theta\cdot v_1/c)}\right)\times$$

$$\times \int_0^\infty \frac{dq_1^\perp}{q_1^\perp} \mathrm{Re}\{(Z_1 - W^{(1)}(q_1^\perp))(Z_2 - W^{(2)}(q_1^\perp)) \times$$
$$\times \alpha^{(1)}(\omega\gamma(1 - \cos\theta \cdot v_1/c), q_1^\perp)[\alpha^{(2)}(\omega, q_1^\perp)]^*\}. \quad (9.46)$$

Equations (9.44)–(9.46) show that the main contribution to the bremsstrahlung cross section for collisions of pairs of neutral atoms is from momentum transfers in the range $q_1{}^\perp R_a \sim 1$, i.e., the same region in which the F_1 and F_2 [see Eqs. (9.37) and (9.38)] used to calculate the cross section for this process were derived. In fact, outside this region, we either have $Z_i - W^{(i)}(q_1{}^\perp) \ll 1$ $(q_1{}^\perp \ll R_a{}^{-1})$ or $\alpha^{(i)}(\omega, q_1{}^\perp) \ll 1$ $(q_1{}^\perp \gg R_a{}^{-1})$.

On comparing $d\sigma^{\text{targ}}$ and $d\sigma^{\text{bul}}$, we can see that the phenomenon of aberration shows up in $d\sigma^{\text{bul}}$ through a change in the angular dependence of the cross section, $(1 + \cos^2\theta) \to [1 + (\cos\theta - v_1/c)^2/(1 - \cos\theta \cdot v/c)^2]$, and the Doppler effect shows up through a frequency shift, $\omega \to \omega\gamma(1 - \cos\theta \cdot v_1/c)$.

The character of the angular distribution of the emission from the projectile and the interference term depend strongly on the behavior of its polarizability. This follows from Eqs. (9.45) and (9.46).

In the limit of low frequencies $\omega \ll 1/\gamma$ such that the relation $\omega\gamma(1 - \cos\theta \cdot v_1/c) \ll 1$ holds for all angles $0 \le \theta \le \pi$, the polarizability $\alpha^{(1)}(\omega\gamma(1 - \cos\theta \cdot v_1/c), q_1{}^\perp)$ is actually independent of ω and θ and approaches its static value. In this case, the angular distribution of the photons in Eq. (9.45) has the form $d\sigma^{\text{bul}} \propto (1 - \cos\theta \cdot v_1/c)^2 + (\cos\theta - v_1/c)^2$ and does not contain the well-known singularity in the angular distribution of the emission from charged particles [7], where photons are predominantly emitted along the direction of motion of the particle in a cone with an aperture angle of $\theta \le \gamma^{-1}$. A comparison of the cross sections (9.44) and (9.45) shows that the ratio $\eta = d\sigma^{\text{targ}}/d\sigma^{\text{bul}}$ is on the order of $\eta \sim \gamma^2$ for angles $\theta \le \gamma^{-1}$ and of $\eta \sim \gamma^{-2}$ for $\theta \gg \gamma^{-1}$. Thus, when $\gamma \gg 1$, for small angles $\theta \le \gamma^{-1}$ the bremsstrahlung of the target atom predominates, while in the rest of the angular interval $\theta \gg \gamma^{-1}$, the bremsstrahlung of the projectile is more intense.

A singularity in the angular distribution of bremsstrahlung from the projectile atom at small angles shows up at high frequencies $\omega \gg \gamma$. In this case, for all angles θ the condition $\omega\gamma(1 - \cos\theta) \gg 1$ holds and the polarizability $\alpha^{(1)}(\omega\gamma(1 - \cos\theta \cdot v_1/c)) \sim [\gamma\omega(1 - \cos\theta \cdot v_1/c)]^{-2}$, so that a singularity $d\sigma^{\text{bul}}$ shows up in the cross section $d\sigma^{\text{bul}} \sim [\gamma(1 - \cos\theta \cdot v/c)]^{-2}$ (9.45). Taking the ratio η of the cross sections (9.44) and (9.45), it is easy to find that for $\theta \le \gamma^{-1}$, we have $\eta \sim \gamma^{-2}$, while for angles $\theta \gg \gamma^{-1}$, we have $\eta \sim \gamma^2$. Thus, for emission at high frequencies and small angles $\theta \ll \gamma^{-1}$, the bremsstrahlung of the projectile atom predominates, while for large $\theta \gg \gamma^{-1}$, the emission of the target predominates. The angular distribution of the bremsstrahlung of a relativistic electron on an atom [2] has a similar character, as shown in Sec. 9.1 (see Chapters 5 and 7). This agreement is not coincidental, since at high frequencies the electrons in a projectile atom can be regarded as free and the bremsstrahlung of the projectile can be represented as the coherent emission of these electrons in the field of the target.

In general, we can state that if the frequency and angle are chosen so that the condition

$$\gamma^2(1-\cos\theta\cdot v_1/c)^2|\alpha_d^{(1)}(\omega\gamma(1-\cos\theta\cdot v_1/c))|^2 \gg |\alpha_d^{(2)}(\omega)|^2, \qquad (9.47)$$

is satisfied, then in this range of ω and θ the bremsstrahlung of the projectile will exceed that of the target. When the direction of this inequality is reversed, the predominance of the projectile emission is replaced by that of the target. Thus, we may conclude that for sufficiently high collision velocities, there is a possibility, in principle, of separating the bremsstrahlung of the projectile and target. This result is a generalization of an analogous conclusion from an earlier study [2] (see Sec. 9.1) of the bremsstrahlung of relativistic electrons on atoms.

Integrating the cross sections (9.43)–(9.46) over $\Omega_{\mathbf{k}}$, we obtain formulas for the corresponding spectra:

$$d\sigma^{\text{targ}}(\omega) = \frac{16}{3}\frac{d\omega}{\omega v_1^2 c^3}\int_0^\infty \frac{dq_1^\perp}{q_1^\perp}|\omega^2(Z_1 - W^{(1)}(q_1^\perp))\alpha^{(2)}(\omega, q_1^\perp)|^2, \qquad (9.48)$$

$$d\sigma^{\text{bul}}(\omega) = \frac{2\omega^3 d\omega}{v_1^2 c^3 \gamma}\left(\frac{c}{v_1}\right)^3 \int_0^\infty \frac{dq_1^\perp}{q_1^\perp}\int_{x_0^{-1}}^{x_0} dx\left(\left(\frac{v_1}{c}\right)^2 x^2 + \right.$$

$$\left. + \left(\frac{1}{\gamma} - x\right)^2\right)|(Z_2 - W^{(2)}(q_1^\perp))\alpha^{(1)}(\omega x, q_1^\perp)|^2, \qquad (9.49)$$

and

$$d\sigma^{\text{int}}(\omega) = -\frac{4\omega^3 d\omega}{v_1^2 c^3 \gamma}\left(\frac{c}{v_1}\right)^3 \int_0^\infty \frac{dq_1^\perp}{q_1^\perp}\int_{x_0^{-1}}^{x_0} dx\left[\left(\frac{v_1}{c}\right)^2 x + \right.$$

$$\left. + \left(1 - \frac{x}{\gamma}\right)\left(\frac{1}{\gamma} - x\right)\right] \operatorname{Re}\{(Z_1 - W^{(1)}(q_1^\perp)) \times$$

$$\times (Z_2 - W^{(2)}(q_1^\perp))\alpha^{(1)}(\omega x, q_1^\perp)\alpha^{(2)*}(\omega, q_1^\perp)\}, \qquad (9.50)$$

where

$$x_0 = \gamma(1 + v_1/c) = \sqrt{(1 + v_1/c)/(1 - v_1/c)}.$$

One important feature of these formulas is that at some frequency ω_0, the spectra of the bremsstrahlung from the projectile and the interference term depend on the behavior of the dynamic polarizability of the projectile atom over an extremely wide range of frequencies from $\omega_0/\gamma(1+\beta)$ to $\omega_0\gamma(1+\beta)$, where $\beta = v_1/c$. In particular, this means that the narrow spectrum lines which can be observed in the bremsstrahlung spectrum of an atom at rest in the region of the characteristic atomic frequencies $\omega \sim 1$ are "smeared out" over the region $1/\gamma \lesssim \omega \lesssim \gamma$ in the emission spectrum of the projectile.

Equations (9.47)–(9.50) for the bremsstrahlung spectrum are much simpler at low frequencies $\omega \ll \gamma^{-1}$, where the polarizability $\alpha^{(1)}(\omega x, q_1^{\perp})$ is approximately equal to its static value $\alpha^{(1)}(0, q_1^{\perp})$ for all $x < \gamma(1 + v_1/c)$, so that the integral with respect to x can be evaluated explicitly. A simple calculation yields

$$d\sigma^{\mathrm{bul}}(\omega) = \frac{16}{3}\frac{\omega^3 d\omega}{c^3 v_1^2}\left(1+\frac{v_1^2}{c^2}\right)\gamma^2 \int_0^\infty \frac{dq_1^{\perp}}{q_1^{\perp}} \times$$
$$\times \left| (Z_2 - W^{(2)}(q_1^{\perp}))\, \alpha^{(1)}(0, q_1^{\perp}) \right|^2, \tag{9.51}$$

and

$$d\sigma^{\mathrm{int}}(\omega) = -\frac{32\omega^3 d\omega}{3c^3 v_1^2}\gamma \int_0^\infty \frac{dq_1^{\perp}}{q_1^{\perp}} \operatorname{Re}\{(Z_1 - W^{(1)}(q_1^{\perp})) \times$$
$$\times (Z_2 - W^{(2)}(q_1^{\perp}))\, \alpha^{(1)}(0, q_1^{\perp})\, \alpha^{(2)*}(0, q_1^{\perp}). \tag{9.52}$$

The spectrum (9.48) in this region of ω is given analogously to Eq. (9.51) in terms of the static polarizability. A comparison of Eq. (9.51) with Eqs. (9.52) and (9.48) shows that the bremsstrahlung spectrum of the projectile is a factor of γ^2 more intense than that of the target and a factor of γ times more intense than the interference term.

At high frequencies $\omega \gg \gamma$, the polarizability is roughly equal to $\alpha^{(1)}(\omega x, q_1^{\perp}) \sim -(1/\omega^2 x^2) W^{(1)}(q_1^{\perp})$ for all x in the range of integration for Eqs. (9.49) and (9.50); thus, it is possible to obtain the following simple expressions for $d\sigma^{\mathrm{bul}}$ and $d\sigma^{\mathrm{int}}$ in this region of the spectrum:

$$d\sigma^{\mathrm{bul}} = \frac{16 d\omega}{3c^3 v_1^2 \omega} \int_0^\infty \frac{dq_1^{\perp}}{q_1^{\perp}} \left| W^{(1)}(q_1^{\perp})(Z_2 - W^{(2)}(q_1^{\perp}))\right|^2, \tag{9.53}$$

and

$$d\sigma^{\mathrm{int}} = \frac{16 d\omega}{c^3 v_1^2 \omega\gamma}\left(\frac{c}{v_1}\right)^2 \left[\left(\frac{c}{v_1}\right)\frac{1}{\gamma^2}\ln\gamma\left(1+\frac{v_1}{c}\right) - 1\right] \times$$
$$\times \int_0^\infty \frac{dq_1^{\perp}}{q_1^{\perp}} \operatorname{Re}\{W^{(1)}(q_1^{\perp})\, W^{(2)*}(q_1^{\perp})\,(Z_1 - W^{(1)*}(q_1^{\perp}))\,(Z_2 - W^{(2)}(q_1^{\perp}))\}. \tag{9.54}$$

These expressions are a generalization of the corresponding formulas from the nonrelativistic theory which are obtained from Eqs. (9.51)–(9.54) in the limit $v_1/c \ll 1$. Since $\alpha^{(i)}(0, q_1{}^{\perp})$ and $W^{(i)}(q_1{}^{\perp})$ are real quantities, the symbols for the complex conjugates in Eqs. (9.51)–(9.54) can be dropped. By adding Eqs. (9.48), (9.51), and (9.52), as well as (9.48), (9.53), and (9.54), one can easily obtain expressions of this sort for the total bremsstrahlung spectrum in atom–atom collisions at high and low frequencies. They yield a nonzero result for collisions of identical atoms because of the Doppler effect and aberrations in the emission from the projectile atom. For $\omega \ll \gamma^{-1}$ and a sufficiently large $\gamma \gg 1$, the emission of the projectile atom is dominant in the total bremsstrahlung spectrum. For $\omega \gg \gamma$ ($\gamma \gg 1$), the spectrum (9.53) is similar to the bremsstrahlung spectrum of the target and is γ times greater than the interference part $d\sigma^{\text{int}}$ (9.54). The smallness of the interference term in the cross section is related to the fact that the emission from the projectile for $\omega \gg 1$ is concentrated mainly in a narrow range of angles $\theta \lesssim \gamma^{-1}$, where it is a factor of γ^2 greater than the bremsstrahlung of the target atom, which is weakly anisotropic.

These expressions can be used to find the cross section for bremsstrahlung in collisions of two identical nonrelativistic atoms when $v_1/c \ll 1$. Complete neglect of the relativistic corrections in this case yields a null result [6]. This happens because the induced dipole moments of identical atoms are equal in magnitude and are oriented in opposite directions, so that the total dipole moment of such a system is zero and, therefore, the associated variation in the emission is absent. Including the Doppler effect and the aberration of light in the emission from the projectile removes this restriction on dipole bremsstrahlung. Neglecting all terms of lower order than v_1/c in Eqs. (9.44)–(9.46), we find the following expression for the total bremsstrahlung cross section (9.43):

$$d\sigma(\omega, \Omega_{\mathbf{k}}) = \frac{\omega^3 d\omega d\Omega_{\mathbf{k}}}{\pi c^3 v_1^2} \left(\frac{v_1}{c}\right)^2 \int_0^\infty \frac{dq_1^{\perp}}{q_1^{\perp}} (Z - W(q_1^{\perp}))^2 \times$$

$$\times \left[\cos^2\theta \left|\frac{\partial}{\partial\omega}(\omega\alpha(\omega, q_1^{\perp}))\right|^2 + \left|\alpha(\omega, q_1^{\perp}) + \omega\cos^2\theta \frac{\partial}{\partial\omega}\alpha(\omega, q_1^{\perp})\right|^2\right]. \tag{9.55}$$

The cross section (9.55) is independent of the collision velocity. The above expansion shows that the correction to the nonrelativistic cross sections owing to the Doppler and aberration effects is dominant for $\omega \ll v_1 R_a{}^{-1}$, since at these frequencies the order of magnitude of this correction, $\sim(v_1/c)^2$, exceeds that of the correction $(R_a\omega/c)^2$ owing to the quadrupole emission from the system. Integrating Eq. (9.55) with respect to $d\Omega_{\mathbf{k}}$, we obtain the bremsstrahlung spectrum for collisions of identical atoms:

$$d\sigma(\omega) = \frac{16}{3}\frac{\omega^3 d\omega}{c^5}\int_0^\infty \frac{dq^\perp}{q^\perp}(Z - W(q^\perp))^2\left[|\alpha(\omega, q^\perp)|^2 + \right.$$

$$\left. + \omega\,\mathrm{Re}\left\{\alpha^*(\omega, q^\perp)\frac{\partial}{\partial\omega}\alpha(\omega, q^\perp)\right\} + \frac{2\omega^2}{5}\left|\frac{\partial}{\partial\omega}\alpha(\omega, q^\perp)\right|^2\right]. \tag{9.56}$$

It is easy to obtain the relativistic correction to the bremsstrahlung cross section for collisions between different types of atoms as well. This correction may be large in those regions of the spectrum where there is substantial destructive interference of the emission from the projectile and target. This sort of analysis is fairly simple for any specific pair of atoms, so we shall not carry it out here.

In Sec. 9.2 we found the amplitude of bremsstrahlung in ion–ion collisions at relativistic velocities [Eqs. (9.39) and (9.40)]. The cross sections for this process are calculated similarly to the case examined above of atom–atom collisions and this leads to the following result:

$$d\sigma^{\mathrm{targ}}(\omega, \Omega_{\mathbf{k}}, \mathbf{q}_1) = \frac{(Z_1 - N_1)^2\omega^3|\alpha_d^{(2)}(\omega)|^2}{\pi c^3 v_1^2(q_1^2 - \omega^2/c^2)} q_1 \times$$

$$\times\left\{(1 + \cos^2\theta) - \frac{(1 - (v_1/c)^2)}{q_1^2 - \omega^2/c^2}\frac{\omega^2}{v_1^2}(3\cos^2\theta - 1) - \right.$$

$$\left. - \frac{2\omega^2}{c^2}\frac{(1 - (v_1/c)^2)}{q_1^2 - v_1^2/c^2}\sin^2\theta\right\} d\omega\, d\Omega_{\mathbf{k}}\, dq_1 \tag{9.57}$$

and

$$d\sigma^{\mathrm{bul}}(\omega, \Omega_{\mathbf{k}}, \mathbf{q}_2) = \frac{(Z_2 - N_2)^2\omega^3\gamma^2(1 - \cos\theta\cdot v_1/c)^2}{\pi c^3 v_1^2 q_2} \times$$

$$\times |\alpha_d^{(1)}(\omega\gamma(1 - \cos\theta\cdot v_1/c))|^2\left\{\sin^2\theta_{q_2}\left[1 + \frac{(\cos\theta - v_1/c)^2}{(1 - \cos\theta\cdot v_1/c)^2}\right] - \right.$$

$$\left. - \frac{2}{\gamma^2}\cos^2\theta_{q_2}\left(\frac{\gamma^{-1}\sin^2\theta}{1 - v_1\cos\theta/c}\right)^2\right\} d\omega\, d\Omega_{\mathbf{k}}\, dq_2, \tag{9.58}$$

where θ_{q2} is the angle between the vectors $\mathbf{q}_2$ and $\mathbf{v}_1$ and $\cos^2\theta_{q2} = \omega^2(1 - v_1\cos\theta/c)^2/v_1{}^2 q_2{}^2$.

The formulas (9.57) and (9.58) for the cross sections are valid for $q_1 R_a \ll 1$ and $q_2 R_a \ll 1$. The first equation coincides with the cross section for bremsstrahlung of the target in a collision with a structureless particle with charge $Z_1 - N_1$, which is natural for small momentum transfers, i.e., large impact parameters. The structure of $d\sigma^{\mathrm{bul}}(\omega, \Omega_{\mathbf{k}}, q_1)$ (9.58) is also understandable. The first term in the curly brackets in this equation describes the bremsstrahlung of a projectile owing to its transverse polarization by the static Coulomb field of a target ion, and the second, the bremsstrahlung owing to the longitudinal polarization of the projectile. The result (9.58) could have been obtained differently by first isolating the part of $d\sigma^{\text{targ}}(\omega, \Omega_{\mathbf{k}}, q_1)$ owing to the purely Coulomb interaction of the incident ion with the target (see [2]) and then transforming this part of the cross section in the reference system where the projectile is at rest and the target moves at velocity v_1, while changing the numbers of the atoms. In the course of the transformation, one should take account of the Lorentz contraction of the longitudinal component of the dipole moment of the projectile induced during the bremsstrahlung process. The factor γ^{-2}, which then appears in the cross section, naturally arises in the subsequent derivation of $d\sigma^{\text{bul}}(\omega, \Omega_{\mathbf{k}}, q_2)$ and shows up as a factor in the second term in the curly brackets of Eq. (9.58).

Equations (9.57) and (9.58) can be used to find, to within logarithmic accuracy, the cross sections for bremsstrahlung of a target and projectile, respectively, integrated over q_1 and q_2 in a fashion analogous to that for the case of the collision of a structureless charged particle with an atom, which was examined in detail in Sec. 9.1 and Chapters 5 and 7. Integrating Eqs. (9.57) and (9.58) from $q_{1,\min} = \omega/v_1$ and $q_{2,\min} = (\omega/v_1)(1 - \cos\theta\cdot v_1/c)$, respectively, to $q_{1,2,\max} \sim R_a^{-1}$ and retaining only the logarithmically large terms, we arrive at the following result:

$$d\sigma^{\text{targ}} = \frac{(Z_1 - N_1)^2\, d\omega d\Omega_{\mathbf{k}}}{\pi c^3 v_1^2 \omega} \,|\, \omega^2 \alpha_d^{(2)}(\omega) \,|^2 \,(1 + \cos^2\theta) \ln \frac{v_1 \gamma}{\omega R_a}, \tag{9.59}$$

and

$$d\sigma^{\text{bul}} = \frac{\gamma^2 (Z_2 - N_2)^2\, d\omega d\Omega_{\mathbf{k}}}{\pi c^3 v_1^2 \omega} \left| \omega^2 \alpha_d^{(1)} \left(\omega\gamma \left(1 - \frac{v_1}{c} \cos\theta \right) \right) \right|^2 \times$$
$$\times \ln \frac{v_1}{R_a \omega\,(1 - v_1 \cos\theta/c)} \left[1 + \frac{(\cos\theta - v_1/c)^2}{(1 - v_1 \cos\theta/c)^2} \right] \left(1 - \frac{v_1}{c}\cos\theta\right)^2. \tag{9.60}$$

Once the bremsstrahlung amplitudes of the projectile and target are known, one can also find the interference term in the bremsstrahlung cross section, $d\sigma^{\text{int}}(\omega, \Omega_{\mathbf{k}})$. In the logarithmic approximation, these calculations yield

$$d\sigma^{\text{int}} = -\frac{2 d\omega d\Omega_{\mathbf{k}}\, (Z_1 - N_1)(Z_2 - N_2)\, \gamma}{\pi c^3 v_1^2 \omega} \left(1 + \cos^2\theta - \frac{v_1}{c} 2\cos\theta \right) \times$$
$$\times \operatorname{Re}\left\{ \omega^4 \alpha_d^{(1)} (\omega\gamma\,(1 - v_1 \cos\theta/c))\, \alpha_d^{*(2)}(\omega) \right\} \ln \frac{v_1}{\omega R_a\,(1 - v_1 \cos\theta/c)}. \tag{9.61}$$

The structure of Eqs. (9.59)–(9.61) is analogous to that of the bremsstrahlung cross sections (9.44)–(9.46). Thus, the analysis of the behavior of the frequency and angular dependences of this cross section is completely analogous to that done above for bremsstrahlung in atom–atom collisions. It shows that the general picture of bremsstrahlung in these two cases is qualitatively similar.

The existence of long-range interaction forces in the system consisting of a pair of ions only leads to a logarithmic growth in the cross sections, which does not change the general character of the bremsstrahlung from a system of two relativistic colliding particles.

The above formulas applied to an ion–atom collision show that the bremsstrahlung of the atom in the field of the ion yields a logarithmically large contribution to the cross section for this process. Then the bremsstrahlung cross section of an atom at rest is described by Eq. (9.59) and that of a moving atom, by Eq. (9.60).

To conclude our discussion of bremsstrahlung in ion–ion collisions, we now examine the emission from a system of identical ions at velocities $v_1 \ll c$. The cross section for the process in this case can be obtained from the general equations (9.59)–(9.61) on going to the region with $v_1/c \ll 1$ or by direct calculation with the aid of the amplitude (9.42). As a result, in the logarithmic approximation we find

$$d\sigma(\omega, \Omega_k) = \frac{(Z-N)^2\omega^3}{\pi c^5}\left\{\cos^2\theta\left|\frac{\partial}{\partial\omega}(\omega\alpha_d(\omega))\right|^2 + \right.$$

$$\left. + \left|\sin^2\theta\alpha_d(\omega) + \cos^2\theta\frac{\partial}{\partial\omega}(\omega\alpha_d(\omega))\right|^2\right\}\ln\frac{v_1}{\omega R_a}\,d\omega d\Omega_k \tag{9.62}$$

and

$$d\sigma(\omega) = \frac{16\omega^3(Z-N)^2 d\omega}{3c^5}\left\{|\alpha_d(\omega)|^2 + \frac{2\omega^2}{5}\left|\frac{\partial}{\partial\omega}\alpha_d(\omega)\right|^2 + \right.$$

$$\left. + \omega\,\mathrm{Re}\left(\alpha_d^*(\omega)\frac{\partial}{\partial\omega}\alpha_d(\omega)\right)\right\}\ln\frac{v_1}{\omega R_a}. \tag{9.63}$$

As opposed to the analogous formulas (9.55) and (9.56) for atom–atom collisions, Eqs. (9.62) and (9.63) contain a logarithmic dependence on the collision velocity. This dependence is caused by the presence of long-range Coulomb forces in this system.

At relativistic collision velocities in a system consisting of two atomic particles, therefore, the specific features of the frequency and angular distributions of the bremsstrahlung are caused mainly by the Doppler effect and the aberration of light in the emission from the projectile atom.

9.4. Neutrino Bremsstrahlung

We now consider the bremsstrahlung of particles that interact with an atom in a nonelectromagnetic fashion, such as a neutrino as it is scattered on a hydrogen atom or a hydrogenlike ion. The mechanism for production of the radiation is analogous to the polarization bremsstrahlung of a charged particle. The sole difference is that the electron shell is polarized owing to the weak interaction with the incident neutrino. Limiting ourselves to the low-energy region of weak interactions, i.e., assuming that the energy of the neutrino is small compared to the mass of the intermediate boson, we can treat the interaction of the neutrino with an atomic electron in terms of the standard *V–A* theory. The operator for the e^- and ν interaction in this case is [12]

$$\hat{V}_{e,\nu} = -\frac{G}{\sqrt{2}} \int d\mathbf{r} \hat{J}_\alpha^+(\mathbf{r}) \hat{J}^\alpha(\mathbf{r}), \tag{9.64}$$

where $\hat{J}_\alpha(\mathbf{r}) = \Psi_e(r)\gamma_\alpha(1+\gamma^5)\Psi_\nu(\mathbf{r})$ is the weak current operator, Ψ_ν and Ψ_e are the bispinor wave functions (operators) of the electron and neutrino, and G is the weak interaction constant.

In the lowest order of perturbation theory with respect to the weak and electromagnetic interactions, the amplitude of elastic neutrino bremsstrahlung is determined by the two diagrams of Fig. 9.5. Here a thin line corresponds to the neutrino. The remaining notation is as before.

Restricting ourselves, as in Sec. 9.1, to the region of the dipole photon frequencies, we obtain the following analytic expression for F_{el} from the diagrams of Fig. 9.5:

$$F_{\text{el}} = -\frac{G}{\sqrt{2}} b_\alpha \sum_n \left\{ \frac{\langle 0|\mathbf{e}\gamma|n\rangle \langle n|e^{-i\mathbf{qr}}\mathrm{O}^\alpha|0\rangle}{E_n - E_0 - \omega} + \frac{\langle 0|e^{-i\mathbf{qr}}\mathrm{O}^\alpha n\rangle \langle n|\mathbf{e}\gamma|0\rangle}{E_n - E_0 + \omega} \right\}, \tag{9.65}$$

where $\mathrm{O}^\alpha = \gamma^\alpha(1+\gamma^5)$; b_α denotes the 4-vector neutrino current, $b_\alpha = \overline{u}(p', s)$, $\mathrm{O}_\alpha u(p, s)$; and $\mathbf{q} = \mathbf{p}' - \mathbf{p}$.

In Eq. (9.65) the state of an atomic electron is described by relativistic wave functions. The nonrelativistic limit for the electron's motion in Eq. (9.65) is obtained similarly to the way it was obtained in our analysis of the bremsstrahlung of a charged particle. In addition, assuming that the energy of the neutrino obeys the inequality $E/c \leq R_a^{-1}$, we now expand $e^{-i\mathbf{qr}}$ in the matrix elements of Eq. (9.65). (We shall give the generalization to the case $E/c > R_a^{-1}$ below when we consider the bremsstrahlung spectrum.) As a result, we obtain

$$F_{\text{el}} = \frac{G}{\sqrt{2}} \left\{ \omega\alpha_d(\omega) \left[b^0(\mathbf{eq}) + \frac{\omega}{c}(\mathbf{eb}) \right] + \omega \left[\omega b^0 (\mathbf{e}\beta_d(\omega)) \frac{1}{c} + (\mathbf{eq})(\beta_d(\omega)\mathbf{b}) \right] \right\}. \tag{9.66}$$

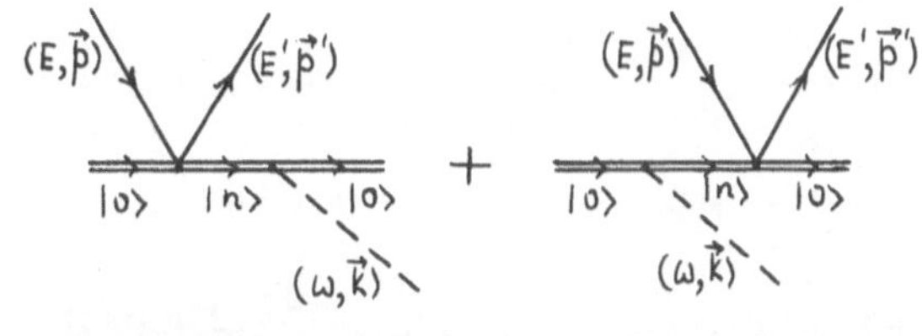

Fig. 9.5. Diagrams describing the amplitude of the polarization bremsstrahlung of a neutrino on an atom.

for the elastic bremsstrahlung amplitude. In Eq. (9.66) the term containing the dynamic dipole polarizability $\alpha_d(\omega)$ is the contribution from the vector part of the interaction operator (9.64). The contribution of the pseudovector part of Eq. (9.64) depends on the spin of the electron in the initial and final states and is described by the term proportional to the vector

$$\boldsymbol{\beta}_d(\omega) = \sum_n \frac{2\omega_{n0} |(D_z)_{n0}|^2}{\omega_{n0}^2 - \omega^2} \mathbf{s}_{fi} \tag{9.67}$$

in Eq. (9.66), where $\mathbf{s}_{fi} = 1/2\langle\bar{\eta}_f \boldsymbol{\sigma} \eta_i\rangle$ is the matrix element of the spin operator taken over the spin parts of the wave function of the atomic electron before and after the collision, η_i and η_f. Since the energy of an atomic electron is independent of its spin in the nonrelativistic limit, the spin state of the e^- may change during elastic bremsstrahlung. Thus, we have introduced the subscripts i and f for denoting η_i and η_f in Eq. (9.67), as opposed to 0 and n, which refer to the position dependence of the electron wave function. From the definition of $\alpha_d(\omega)$ and Eq. (9.67) it follows that

$$\boldsymbol{\beta}_d(\omega) = \mathbf{s}_{fi}\alpha_d(\omega) \tag{9.68}$$

for hydrogen.

Using Eq. (9.66) we find the spectrum of the photons and the angular distribution of the emission as functions of the mutual orientation of the vector $\mathbf{p}$ and the spin of an atomic electron prior to the collision. In accordance with the general rules [7], the differential cross section for neutrino bremsstrahlung can be written in the form

$$d\sigma_{el}(\omega, \Omega_{\mathbf{k}}, q, \varphi_{p'}) = \frac{\omega q\, d\omega\, d\Omega_{\mathbf{k}}\, dq\, d\varphi_{p'}}{2(2\pi)^4 c^5} \frac{1}{4} \sum_{\lambda, s, s', s} |F_{el}|^2, \tag{9.69}$$

where the sum is taken over the polarizations of the neutrino in the initial and final states and over the polarizations of the photon and atomic electron in the final state. Performing these calculations and taking the integral over the momentum transfer q from $q_{min} = \omega/c$ to $\mathbf{q}_{max} = \mathbf{p} - \mathbf{p}'$, we obtain a fairly complicated expression for the angular distribution of the bremsstrahlung [13]:

$$\begin{aligned} d\sigma(\omega, \Omega_{\mathbf{k}}) = {} & \frac{G^2\omega^3 p'^2 d\omega\, d\Omega_{\mathbf{k}}}{\pi^3 c^5} |\alpha_d(\omega)|^2 \Big\{ p^2 (1 - (\mathbf{n}\mathbf{n}_\nu)^2)(1 + \boldsymbol{\xi}\mathbf{n}_\nu) + \\ & + \frac{2}{3} p'^2 (1 + \boldsymbol{\xi}\mathbf{n}_\nu) - \frac{1}{c}\omega p\,[1 - (\mathbf{n}\mathbf{n}_\nu)^2 + (\mathbf{n}_\nu\boldsymbol{\xi}) - \\ & - (\mathbf{n}\mathbf{n}_\nu)(\boldsymbol{\xi}\mathbf{n})] + \frac{2}{3}\frac{\omega p'}{c}(1 + \boldsymbol{\xi}\mathbf{n}_\nu) + \frac{\omega^2}{2c^2}[2 + (\mathbf{n}_\nu\boldsymbol{\xi}) - \\ & - (\mathbf{n}_\nu\mathbf{n})(\boldsymbol{\xi}\mathbf{n})] \Big\}. \end{aligned} \tag{9.70}$$

Here $\mathbf{n}_\nu$ is the initial direction of motion of the neutrino, $\mathbf{n} = \mathbf{k}/k$, $\boldsymbol{\xi}$ is the unit vector in the direction of the spin of the atomic electron before the collision, and $p' = p - \omega/c$.

The presence of terms with $\boldsymbol{\xi}\mathbf{n}_\nu$ in Eq. (9.70) is the result of the nonconservation of parity during bremsstrahlung, since these terms change sign on spatial inversion. In addition, the terms proportional to $\boldsymbol{\xi}\mathbf{n}_\nu$ determine the intensity of the emission as a function of the mutual orientation of the electron spin and the direction of motion of the neutrino.

We now consider two special cases of Eq. (9.70) in more detail. For $\boldsymbol{\xi}\mathbf{n}_\nu = -1$ (direction of motion of the neutrino opposite to the polarization of the atomic electron), Eq. (9.70) yields

$$d\sigma_{\text{el}}(\omega, \Omega_{\mathbf{k}}, \uparrow\downarrow) = \frac{G^2\omega^5 d\omega}{2\pi^3 c^5} d\Omega_{\mathbf{k}} \, |\alpha_d(\omega)|^2 p'^2 (1 + \cos^2\theta). \tag{9.71}$$

In the case of vectors $\mathbf{n}_\nu$ and $\boldsymbol{\xi}$ in the same direction $[(\boldsymbol{\xi}\mathbf{n}) = 1]$, we have

$$d\sigma_{\text{el}}(\omega, \Omega_{\mathbf{k}}, \uparrow\uparrow) = \frac{G^2\omega^3 p'^2}{2\pi^3 c^3} d\omega d\Omega_{\mathbf{k}} \, |\alpha_d(\omega)|^2 \times \{(10pp'/3 + 3\omega^2/2c^2) - (2pp' + \omega^2/2c^2)\cos^2\theta\}. \tag{9.72}$$

Equations (9.71) and (9.72) show that the angular distribution and, along with it, the emission intensity, depend strongly on the mutual orientation of $\mathbf{n}_\nu$ and $\boldsymbol{\xi}$. This dependence shows up most strongly for photon frequencies that are low compared to the energy of the neutrino. Comparing Eqs. (9.71) and (9.72) in this limit, we obtain the estimate

$$\eta = \frac{d\sigma_{\text{el}}(\uparrow\downarrow)}{d\sigma_{\text{el}}(\uparrow\uparrow)} \approx \frac{\omega^2}{E^2}, \quad \omega \ll E. \tag{9.73}$$

For frequencies $\omega \sim 1$ Ry and energies $E \sim 1$keV, the ratio $\eta \sim 10^{-4}$. Thus, there is a fundamental possibility of determining the direction of a beam of incident neutrinos from the angular distribution of their bremsstrahlung on a polarized target (in speaking of the polarization, we have the electron component in mind).

Integrating Eq. (9.70) over the emission angles of the photons, we obtain the spectrum of the radiation on a polarized target,

$$d\sigma_{\text{el}}(\omega) = \frac{16}{3\pi^2} \frac{G^2 p^4 d\omega}{c^4 \omega} \, |\omega^2 \alpha_d(\omega)|^2 (1 + \boldsymbol{\xi}\mathbf{n}_\nu). \tag{9.74}$$

It is clear from this formula that the spectrum has a maximum when $\boldsymbol{\xi}\uparrow\uparrow\mathbf{n}_\nu$ and is parametrically small ($\sim\omega^2/E^2$) for $\boldsymbol{\xi}\uparrow\downarrow\mathbf{n}_\nu$.

When the substitution $\mathbf{n}_\nu \to -\mathbf{n}_\nu$ is made in Eqs. (9.70)–(9.74), they describe the bremsstrahlung of antineutrinos. In this case the maximum emission occurs when $\boldsymbol{\xi}\uparrow\downarrow\mathbf{n}_\nu$.

In conclusion, we note that although Eqs. (9.70)–(9.74) are exact only for $E/c < R_a^{-1}$, it is nevertheless easy to obtain the analogous formulas for high-energy neutrinos. A formula for the bremsstrahlung amplitude that is valid over the entire range of variation in q (for $E/c > R_a^{-1}$) is obtained, as before, on going to the nonrelativistic limit in Eq. (9.65), but without expanding $e^{-i\mathbf{qr}}$ in the corresponding matrix elements.

Here some more complicated functions that also depend on q appear instead of the polarizabilities $\alpha_d(\omega)$ and $\beta_d(\omega)$. For example, the contribution of the vector part of the operator (9.64) in Eq. (9.66) is replaced by the expression in brackets in Eq. (9.5) when $|m\rangle = |0\rangle$; i.e., the generalized polarizabilities $\alpha_{00}(\omega, q)$ and $\beta_{00}^{ij}(\omega, q)$ appear here. We shall not write down the corresponding expressions for the contribution of the pseudovector part of $\hat{V}_{e,\nu}$ in the amplitude F_{el} because of their complexity. It is important only to note the following. When $q \gg R_a^{-1}$, all the functions behave in the same way: they decrease as $(qR_a)^{-4}$. It is easy to confirm this for the example of $\alpha_{00}(\omega, q)$ [see Eq. (9.6a)], where when $q \gg R_a^{-1}$ one must choose the states $|n\rangle$ in the form $e^{i\mathbf{pr}}$ and calculate the resulting integrals under the assumption that $|0\rangle = |1s\rangle$. Thus, in taking the integral over q, the region of large momentum transfers makes a small contribution to the bremsstrahlung spectrum, which we obtain by integrating Eq. (9.69) from $q_{\min} = \omega/c$ to some $q \sim R_a^{-1}$ using Eq. (9.66). As a result, we arrive at an expression analogous to Eq. (9.74), with the sole difference that the factor of $16p^4$ is replaced by κR_a^{-1}, where $\kappa \sim 1$. Thus, for $E/c > R_a^{-1}$ the spectrum of neutrino bremsstrahlung is independent of the energy E, and the parameter η introduced above is replaced by the quantity $(\omega^2/c^2)R_a^2 \ll 1$ for frequencies of order I (where I is the ionization potential of an H atom).

To summarize, the bremsstrahlung produced in collisions of relativistic particles with atoms has a number of specific features which are absent in the nonrelativistic case. These include singularities in the angular distribution of the scattered particles, in the total emission spectrum at high photon energies, in the angular distribution of the emission during collisions of complicated atomic particles, etc. In this regard, numerical calculations of bremsstrahlung processes should be performed, which will free us from the limitations of the Born approximation and yield reliable data on the magnitudes of the predicted effects. Although bremsstrahlung in atomic collisions was the main topic in the above discussion, it is evident that the internal structure of the scattered particles also plays an important role in the emission from nuclear relativistic collisions, such as those of nucleons with nuclei and of nuclei with nuclei, i.e., in collisions of elementary particles at high energies. Thus, studies in these areas are extremely topical.

Chapter 10

POLARIZATION BREMSSTRAHLUNG IN COLLISIONS OF ELECTRONS WITH ATOMS AND LASER BREAKDOWN IN GASES

B. A. Zon

10.1. The Limit of Low-Frequency Bremsstrahlung Photons

The cross section for the bremsstrahlung effect in collisions of electrons with neutral atoms controls a number of important physical parameters in low temperature plasmas. It can be shown from the most general considerations based on charge conservation and gauge invariance that the bremsstrahlung amplitude for nonrelativistic electrons can be written in the following form for low frequencies of the emitted photons [1]:

$$M = + \frac{e\hbar^2}{m}\sqrt{\frac{2\pi}{\omega}}\left[M_0(\varepsilon)\frac{\mathbf{qe}}{\hbar\omega} + (\mathbf{pe})\frac{\partial M_0(\varepsilon)}{\partial\varepsilon}\right]. \tag{10.1}$$

Here $M_0(\varepsilon)$ is the amplitude for elastic scattering of an electron with energy ε; ω is the frequency, $\mathbf{k}$ is the wave vector, and $\mathbf{e}$ is the polarization of the emitted photon; $\mathbf{p}$ and $\mathbf{p}'$ are the momenta of the electron before and after scattering; $p^2 = (p')^2 + 2m\hbar\omega$; and $\mathbf{q} = \mathbf{p}' - \mathbf{p}$ is the momentum transfer.

A low frequency of the bremsstrahlung photon corresponds to the condition

$$\omega \ll \varepsilon/\hbar. \tag{10.2}$$

Equation (10.1) is an expansion of the exact bremsstrahlung amplitude for $\omega \to 0$ that does not depend on the form of the interaction between the electron and the scatterer, in this case an atom, which, however, we shall assume to be quite heavy here. The second term in this expansion was obtained by Low [2]. Its asymmetry with respect to the momenta **p** and **p**' is a consequence of choosing the initial energy of the electron as an independent variable. The first term in Eq. (10.1) contains a pole at $\omega \to 0$ which is associated with the well-known "infrared catastrophe."

It is obvious that Eq. (10.1) will not be valid for frequencies ω lying in the region of the characteristic frequencies of the atom, since expanding the amplitude in terms of ω is clearly illegitimate there. For neutral atoms this corresponds to the visible and ultraviolet spectral ranges. In the case of bremsstrahlung on negative ions, this may correspond to still lower frequencies. Thus, it is desirable to obtain a generalization of Eq. (10.1) to higher frequencies.

10.2. The Born Approximation

We begin by considering the bremsstrahlung effect in the Born limit. The Born approximation is applicable for scattering of an electron on a neutral atom when the velocity of the electron before and after the scattering event is much greater than the characteristic velocities of the atomic electrons.

The Feynman diagrams for bremsstrahlung in this case are shown in Fig. 10.1. Here a thin continuous line denotes the electron propagator, a double line denotes the propagator of the atom, and a dashed line denotes the interaction between them, which need be taken into account only once (in the first order of perturbation theory) when the Born approximation is applicable. A wavy line denotes the emission or absorption of a photon.

The diagrams of Fig. 10.1a give the traditional description of bremsstrahlung in which the atom is treated merely as the source of the force field that perturbs the trajectory of the scattered electron. Bremsstrahlung or inverse bremsstrahlung (absorption) takes place as a result of this perturbation (in the latter case an external electromagnetic field must also be present).

The diagrams of Fig. 10.1b correspond to emission or absorption of a bremsstrahlung photon by atomic electrons, i.e., polarization bremsstrahlung. In this case the scattered electron is required to ensure conservation of energy and momentum.

Here we must point out an analogy between the contribution from the diagrams of Fig. 10.1b to bremsstrahlung and the Bersuker–Veselov correction to the oscillator strength of an atom [3, 3a]. The Bersuker–Veselov correction can also be represented by the diagrams of Fig. 10.1b, but the thin continuous line will correspond to a valence electron and the double line to the atomic core.

In the nonrelativistic approximation the interaction of the scattered electron with the atom is given by Coulomb's law,

$$U = -\frac{Ze^2}{r} + e^2 \sum_{j=1}^{Z} \frac{1}{|\mathbf{r} - \mathbf{r}_j|}. \tag{10.3}$$

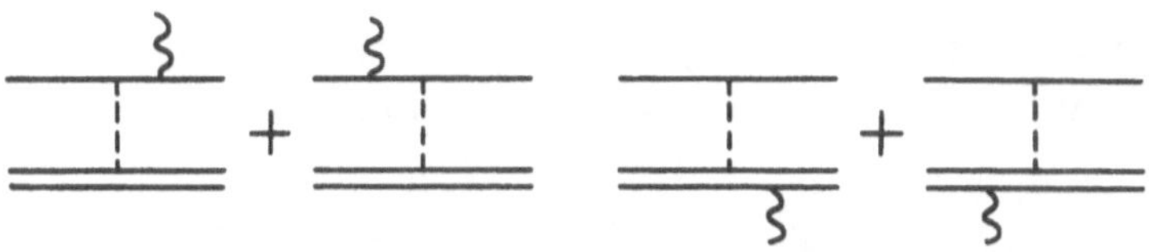

Fig. 10.1. Feynman diagrams for the bremsstrahlung effect during scattering of an electron on an atom.

Here Z is the nuclear charge, which is located at the coordinate origin, $\mathbf{r}_j$ are the coordinates of the atomic electrons, and $\mathbf{r}$ are the coordinates of the scattered electron.

We now write down the interactions of the atom V_a and the scattered electron V_e with the electromagnetic fields in the dipole approximation:

$$V_{a,e} = i \sum_{\mathbf{k},\sigma} (2\pi\hbar\omega)^{1/2} (\mathbf{d}^{(a,e)}\mathbf{e}_{\mathbf{k},\sigma}) (a^+_{\mathbf{k},\sigma} - a_{\mathbf{k},\sigma}), \tag{10.4}$$

$$d^{(e)} = -e\mathbf{r}, \quad \mathbf{d}^{(a)} = -e \sum_{j=1}^{Z} \mathbf{r}_j,$$

where $a_{\mathbf{k},\sigma}{}^+$ and $a_{\mathbf{k},\sigma}$ are the creation and annihilation operators for a photon with wave vector $\mathbf{k}$ and polarization σ, and $\mathbf{e}_{\mathbf{k},\sigma}$ is the real polarization vector (the quantization volume is taken to be unity).

The analytic expression for the bremsstrahlung amplitude has the following form:

$$M = \left\langle 0, \mathbf{p}' \left| V^{(e)} G^{(e)} \left(\frac{p^2}{2m}\right) U + U G^{(e)} \left(\frac{p'^2}{2m}\right) V^{(e)} \right| 0, \mathbf{p} \right\rangle +$$
$$+ \langle 0, \mathbf{p}' | V^{(a)} G^{(a)} (E_0 + \hbar\omega) U + U G^{(a)} (E_0 - \hbar\omega) V^{(a)} | 0, \mathbf{p} \rangle, \tag{10.5}$$
$$| 0, \mathbf{p} \rangle = | 0 \rangle | \mathbf{p} \rangle, \quad \langle 0, \mathbf{p}' | = \langle 0 | \langle \mathbf{p}' |.$$

Here $|0\rangle$ is the wave function of the ground state of an atom with energy E_0; $|\mathbf{p}\rangle$ and $\langle \mathbf{p}'|$ are the wave functions of the electron in the initial and final states; and $G^{(e,a)}$ are the Green functions of the electron and atom, which are given by

$$G^{(e)} (\varepsilon, \mathbf{r}, \mathbf{r}') = \int \frac{d\mathbf{p}}{(2\pi\hbar)^3} \frac{\exp\{i\mathbf{p}(\mathbf{r} - \mathbf{r}')/\hbar\}}{[p^2/2m - \varepsilon + i0]},$$
$$G^{(a)} (E, \mathbf{r}, \mathbf{r}') = \sum_n \frac{\langle \mathbf{r} | n \rangle \langle n | \mathbf{r}' \rangle}{E_n - E + i0}. \tag{10.6}$$

The sum over n in Eq. (10.6) is taken over the entire set of intermediate states of the atom, including the continuum. Exchange effects between the scattered electron and the atomic electrons are small when the Born approximation is applicable and are neglected here. These effects have been analyzed numerically for negative ions in [4].

Substituting Eqs. (10.3), (10.4), and (10.6) in the first matrix element of Eq. (10.5), we obtain the contribution to the amplitude from the diagrams of Fig. 10.1a:

$$M_e = 4\pi e^3 \sqrt{2\pi\hbar\omega\,(N_{\mathbf{k},\sigma}+1)}\,\frac{(\mathbf{e}_{\mathbf{k},\sigma}\mathbf{q})}{m\omega^2\mathbf{q}^2}\,[Z - F(\mathbf{q})], \tag{10.7}$$

where $N_{\mathbf{k},\sigma}$ is the number of photons; $\{\mathbf{k},\sigma\}$ are the modes of the external electromagnetic field; and

$$F(\mathbf{q}) = \left\langle 0 \left| \sum_{j=1}^{Z} \exp(-i\mathbf{q}\mathbf{r}_j/\hbar) \right| 0 \right\rangle \tag{10.8}$$

is the atomic form factor that determines the "spreading out" of the electron shell of the atom in the state $|0\rangle$. For concreteness, Eq. (10.7) has been written down for bremsstrahlung. In the case of inverse bremsstrahlung (absorption), we must make the substitution $N_{\mathbf{k},\sigma} + 1 \rightarrow N_{\mathbf{k},\sigma}$.

In calculating Eq. (10.7), we have used the Fourier transform of the Coulomb interaction,

$$\int d\mathbf{r}\,\frac{e^{-i\mathbf{p}\mathbf{r}}}{|\mathbf{r}-\mathbf{r}'|} = \frac{4\pi}{p^2}\,e^{-i\mathbf{p}\mathbf{r}'}.$$

The contribution to the amplitude from the diagrams of Fig 10.1b is calculated in analogous fashion:

$$M_a = \frac{4\pi i e^2}{q^2}\sqrt{2\pi\hbar\omega\,(N_{\mathbf{k},\sigma}+1)}\sum_n \langle 0|\,\mathbf{d}^{(a)}\mathbf{e}_{k,\sigma}\,|n\rangle \times$$
$$\times\left[\frac{F_{n0}(\mathbf{q})}{\omega_{n0}-\omega} + \frac{F_{0n}(\mathbf{q})}{\omega_{n0}+\omega}\right]\langle n|\,\mathbf{d}^{(a)}\mathbf{e}_{k,\sigma}\,|0\rangle, \tag{10.9}$$

with

$$\omega_{n0} = (E_n - E_0)/\hbar,$$
$$F_{mn}(\mathbf{q}) = \left\langle m \left| \sum_{j=1}^{Z} \exp(-i\mathbf{q}\mathbf{r}_j/\hbar) \right| n \right\rangle. \tag{10.10}$$

We shall initially neglect stimulated (induced) bremsstrahlung and consider only spontaneous emission. This means that in Eqs. (10.7) and (10.9), we can take $N_{\mathbf{k},\sigma} \ll 1$.

The amplitude of spontaneous electron bremsstrahlung satisfies Eq. (10.1) if we note that the amplitude of elastic electron scattering on an atom,

$$M_0 = -(4\pi e^2/q^2)[Z - F(\mathbf{q})] \tag{10.11}$$

is independent of the energy of the electron in the Born approximation and is determined solely by the momentum transfer. For this reason, the second term in Eq. (10.1) vanishes in this case.

It can be seen from Eq. (10.9) that the bremsstrahlung amplitude has a resonant maximum at $\omega = \omega_{n0}$ that corresponds to the characteristic atomic frequencies. Attention was first brought to this circumstance in [5]. The maxima have a simple physical interpretation: they correspond to excitation of the atom during inelastic electron scattering followed by radiative decay of the excitation. At these frequencies, the diagrams of Fig. 10.1b have poles. As already noted, this part of the amplitude cannot be found in the form of a series expansion in ω. For a long time, however, it was not known how wide the resonance peaks are or how noticeable they might be against the general spectrum of bremsstrahlung. Evidently, if the width of the resonances were determined just by the natural width of the spectrum lines, then the effect would be entirely trivial.

It has been shown [6] by a numerical analysis of the quantities (10.7) and (10.9) for the ground state of hydrogen that including the polarization (atomic) amplitude changes the bremsstrahlung spectrum substantially over an extremely wide range of photon energies, on the order of several electron volts (see Fig. 4.1). Further study is needed in order to clarify how general this result may be and to acquire the ability to calculate the bremsstrahlung amplitude for arbitrary atoms.

10.3. The Bethe Approximation

We note first that, depending on the scattering angle of the electron, the momentum transfer $\mathbf{q}$ [which determines the magnitude of the nondiagonal form factor (10.10)] varies within certain limits specified by the conservation of energy. If R_a denotes the size of an atom, then

$$q_{\max} R_a \approx 2\hbar\sqrt{\varepsilon/E_0} \gg \hbar. \tag{10.12}$$

The scattering angle at which $q \sim h/R_a$ is given by

$$\theta_a \approx v_a/v_0, \tag{10.13}$$

where v_a is the velocity of the atomic electrons and v_0 is the velocity of the scattered electron.

The same nondiagonal form factor (10.10) determines the amplitude for inelastic scattering of the electron with atomic excitation in the first Born approximation. It is well known from the theory of atomic collisions that F_{mn} is exponentially small for scattering at angles $\theta \gg \theta_a$. Thus, the main contribution to the integral cross section is from angles $\theta < \theta_a$ [7] and in calculating F_{mn} we can restrict ourselves to the lowest-order terms in the expansion of the exponent in the parameter $|\mathbf{q}\mathbf{r}_j|/\hbar \ll 1$ which yield a nonzero contribution because of the selection rules for the angular momentum. We emphasize that this sort of expansion is used for all scattering angles, despite the existence of a region, bounded by the condition (10.12), in which this expansion is not formally valid. This is the well-known Bethe (Born–Bethe) approximation in the theory of inelastic collisions. It has been applied to the bremsstrahlung problem by Zon [8].

Because the wave functions of the atom have a definite parity, it follows that a nonzero result will be obtained when $\exp(-i\mathbf{q}\mathbf{r}_j/\hbar)$ is replaced by $-i\mathbf{q}\mathbf{r}_j/\hbar$ in Eq. (10.10). We now introduce the dynamic polarizability tensor of the atom at the frequency ω,

$$\alpha_{ij}(\omega) = \frac{1}{\hbar}\sum_n \left[\frac{\langle 0|d_i^{(a)}|n\rangle\langle n|d_j^{(a)}|0\rangle}{\omega_{n0}+\omega} + \frac{\langle 0|d_j^{(a)}|n\rangle\langle n|d_i^{(a)}|0\rangle}{\omega_{n0}-\omega}\right]. \tag{10.14}$$

For spherically symmetric states of the atom, which are all that we are considering here, the polarization tensor is diagonal, i.e.,

$$\alpha_{ij}(\omega) = \delta_{ij}\alpha(\omega).$$

Given these remarks, Eq. (10.9) can be written in the form

$$M_a = \frac{4\pi e}{q^2}\sqrt{2\pi\hbar\omega}\,(\mathbf{q}\mathbf{e}_{\mathbf{k},\sigma})\,\alpha(\omega)\,\Theta(\theta_a-\theta), \tag{10.15}$$

where

$$\Theta(x) = \begin{cases}1, & x>0,\\ 0, & x<0\end{cases}$$

is the unit Heaviside function. The Θ-function in Eq. (10.15) is merely related to the use of the Bethe approximation. In this approximation the differential cross section for bremsstrahlung has the form

$$d\sigma(\omega) = \frac{8e^2 d\omega}{3\pi q^2\omega\hbar c^3}\,\frac{p'}{p}\,|e^2[Z-F(\mathbf{q})] - m\omega^2\alpha(\omega)\,\Theta(\theta_a-\theta)|^2\,d\Omega_{\mathbf{p}'}. \tag{10.16}$$

In Eq. (10.16) we have taken the sum over the two possible polarizations of the photons and integrated over the directions of emission of the photon. In doing this we have used the formula

$$\int d\Omega_{\mathbf{k}}\sum_\sigma(\mathbf{q}\mathbf{e}_{\mathbf{k},\sigma})^2 = \int d\Omega_{\mathbf{k}}\,\frac{[\mathbf{q}\mathbf{k}]^2}{\mathbf{k}^2} = \frac{8\pi}{3}\,q^2.$$

At first glance it seems that we must know the dependence of the form factor on the momentum transfer in order to integrate over the angles of p' in Eq. (10.16). For neutral atoms at $\theta < \theta_a$, however, we have $F(q) \to Z$ so that the main contribution to the bremsstrahlung cross section from the diagrams of Fig. 10.1a is from angles $\theta \gg \theta_a$ (cf. [7]). Therefore, the interference term associated with the presence of two terms under the absolute value sign in Eq. (10.16) vanishes upon integration over $d\Omega_{p'}$, since these terms are nonzero in different ranges of the scattering angle.

When the condition (10.2) is satisfied, the integral of the square of the first term can be expressed in terms of the transport cross section for elastic scattering in the form

$$\sigma_{tr} = (2e^4/v_0^2) \int (1/q^2)\,(Z - F(\mathbf{q})\,]^2 d\Omega_{p'}. \tag{10.17}$$

Then, we finally obtain [8]

$$d\sigma(\omega) = \frac{4\alpha v_0^2}{3\pi c^2}\frac{d\omega}{\omega}\sigma_{tr} + \frac{16\alpha m^2}{3c^2}\frac{d\omega}{\omega}\,|\,\omega^2\alpha(\omega)\,|^2 \times$$

$$\times \int_0^{\theta_a} \frac{\sin\theta}{q^2}\,d\theta = \frac{4\alpha v_0^2}{3\pi c^2}\frac{d\omega}{\omega}\sigma_{tr} + \frac{8\alpha}{3c^2 v_0^2}\frac{d\omega}{\omega}\,|\,\omega^2\alpha(\omega)\,|^2 \ln\left(1 + \gamma\frac{v_0^2}{\omega^2 R_a^2}\right), \tag{10.18}$$

where $\alpha = e^2/\hbar c \sim 1/137$ is the fine structure constant and $\gamma = \theta_a p R_a/\hbar$ is a factor of order unity whose exact value is not known since the exact value of the angle θ_a is not known. It is easy to see, however, that the result depends weakly on γ, since γ appears in the logarithm with a large coefficient $(v_0/\omega R_a)^2$. For the same reason, the 1 can be left out of the argument of the logarithm.

The absence of interference between the amplitudes corresponding to the diagrams of Figs. 10.1a and 10.1b in the Bethe approximation allows us to speak of distinct contributions to the total cross section from traditional bremsstrahlung by free electrons and bremsstrahlung by the bound electrons in the atom. The latter is usually referred to as polarization bremsstrahlung, since it is determined by the dynamic polarizability of the atom.

It is not difficult to obtain numerical values of the bremsstrahlung cross section from Eq. (10.18). Both the transport cross section for elastic scattering and the dynamic polarizability of different atoms have been measured and calculated by many authors. The calculated bremsstrahlung during scattering of an electron with momentum $3\hbar/a_B$ (where a_B is the Bohr radius) for the ground state of a hydrogen atom according to Eq. (10.18) is almost exactly the same as the result shown in Fig. 10.2 and obtained without using the Bethe approximation.

It is also apparent from Eq. (10.18) when the bremsstrahlung spectrum has a strong frequency dependence. If the atom has a transition with a large (~1) oscillator strength, then in the corresponding frequency range the dynamic polarizability takes on values considerably larger than a_B^3. Then polarization bremsstrahlung is dominant in Eq. (10.18) and the dispersion curve for bremsstrahlung has a broad maximum. These considerations provide an answer to the question posed at the end of the preceding section.

10.4. Applications to the Theory of Optical Breakdown and Multiphoton Bremsstrahlung

The equations obtained in the previous section can be used to explain experimental data on the optical breakdown of alkali metal vapors [8]. The phenomenon of optical breakdown manifests itself as a spark, i.e., rapid ionization of the medium in focussed laser light. The threshold intensity of the optical radiation at which breakdown begins has been measured as a function of the pressure in gaseous media. Two mechanisms for the development of optical breakdown have been considered in the literature [9]. For the avalanche mechanism one assumes that a group of free electrons acquires energy through inverse bremsstrahlung in collisions with neutral atoms or molecules of the gas. When the energy of an electron begins to exceed the ionization potential of the gas particles, electron multiplication takes place and an avalanche develops.

Whereas the avalanche theory of optical breakdown does not differ fundamentally from the theory of gas breakdown in the microwave region, the multiphoton theory of breakdown is vitally related to the high energy of optical photons, which is far higher than the energy of microwave photons. Although the energy of an optical photon is too small for single-photon ionization of the gas particles, ionization can occur as a result of multiphoton ionization. Thus, the multiphoton theory can be employed to explain gas breakdown at extremely low pressures.

The thresholds for breakdown of Rb and Cs vapor by ruby laser light have been measured [10]. Breakdown is observed at densities roughly four orders of magnitude lower than for He, Ar, and Hg vapor. It was pointed out [10] that the very low breakdown thresholds for Rb and Cs vapor cannot be explained by either the avalanche or the multiphoton theories of breakdown. If Eq. (10.18) is used in this case, it is possible to reduce the very large discrepancy between theory and experiment. In fact, the polarizabilities of Rb and Cs at the ruby laser wavelength are roughly 1200 and 800 a.u., respectively [11]. For electron velocities on the order of 10^8 cm/s and theoretically calculated values of the transport cross sections on the order of $4 \cdot 10^{-15}$ cm^2 [12], we find that polarization bremsstrahlung is greater than the traditional (electron) bremsstrahlung by roughly a factor of 130. The polarization term does not change the total bremsstrahlung cross sections for He, Ar, and Hg because of the small polarizabilities of these atoms at this frequency. Using the experimentally known transport cross sections of order $6 \cdot 10^{-16}$ cm^2 for He, Ar, and Hg, it is easy to see that the probability of the bremsstrahlung effect is roughly three orders of magnitude greater for Rb and Cs than for He, Ar, and Hg. According to the avalanche theory of optical breakdown, this difference in the probabilities of bremsstrahlung for the two groups of atoms leads to a difference of three orders of magnitude in the breakdown densities. The remaining difference of roughly an order of magnitude in the breakdown thresholds for the optical fluxes may be related to the difference in the ionization potentials of the two groups of atoms.

Thus, including polarization absorption makes it possible to match experimental data on the optical breakdown threshold for the alkali metal vapors with the avalanche theory. The resulting formulas for the cross section for the bremsstrahlung effect can be generalized to the multiphoton case when several photons are absorbed or emitted in a single elementary collision event between an electron and an atom [13]. This effect has been studied experimentally in a number of papers, which have been reviewed by Weingartshofer and Jung [14]. A review of the theory of this effect that does not include polarization bremsstrahlung and absorption has been written by Bunkin et al. [15]. In this theory there is a characteristic sum rule: in the Born approximation the total cross section for scattering of an electron on an atom or molecule in the presence of an external electromagnetic field is the same as for scattering in the absence of a field. The field causes a redistribution of the electron energies because the scattering may be accompanied by the emission or absorption of several photons, but the overall number of scattered electrons remains constant. It has been shown [13] that including polarization bremsstrahlung and absorption leads to an increase in the total cross section, i.e., to an increase in the number of scattered electrons, regardless of their final energy, compared to the case of scattering without a field. Referring to [13] for the details, we give just one numerical example here: during scattering of 5 eV electrons on lithium atoms in a radiation field of up to $5 \cdot 10^6$ V/cm produced by a ruby laser, the total scattering cross section increases to a value on the order of 10^{-15} cm^2.

10.5. The High-Frequency Limit

An analysis of the formulas obtained above is of special interest in the limit of high bremsstrahlung frequencies,

$$\hbar\omega \gg I, \tag{10.19}$$

where I is the energy required to detach all the electrons from an atom. It is clear from physical considerations that when condition (10.19) is satisfied, virtual excitation of the atomic electrons becomes inefficient so that polarization bremsstrahlung and absorption should be reduced.

This statement follows formally from Eq. (10.18). When condition (10.19) is satisfied, we can use the asymptotic value for the polarizability [16],

$$\alpha(\omega) = -Ze^2/m\omega^2, \tag{10.20}$$

and the second term in Eq. (10.18) falls off as $1/\omega$. The first term in Eq. (10.18) also contains a factor of $1/\omega$; thus, the two terms partially cancel out and the Bethe approximation cannot be used to find the asymptotic behavior of the bremsstrahlung spectrum.

When condition (10.19) is satisfied, it is possible to make a formal expansion in the parameter ω_{n0}/ω in Eq. (10.9). Limiting ourselves to two terms, we obtain [17]

$$M_a \approx \frac{4\pi e^3}{mq^2\omega^2} (\mathbf{e}_{\mathbf{k},\sigma}\mathbf{q}) \sqrt{2\pi\hbar\omega (N_{\mathbf{k},\sigma} + 1)}\, F(\mathbf{q}).$$

Comparing this expression with Eq. (10.7), we see that in the limit (10.19), the total bremsstrahlung amplitude does not contain the atomic form factor and is determined solely by scattering of an electron on the nucleus. This result has a clear physical significance: for large energies of the incident particle and bremsstrahlung photon, the atomic electrons can be regarded as free, while in the dipole approximation there is no bremsstrahlung during collisions of two electrons [18].

It should, however, be noted that if the incident particle is not an electron, but we are considering collisions of atoms with positrons, mesons, or protons, then the atomic form factors do not cancel out and the polarization amplitude does contribute to the bremsstrahlung cross section in the high-frequency limit.

10.6. Bremsstrahlung during Scattering of Electrons on Negative Ions

In view of the fact that negative ions have a large polarizability owing to the small binding energy of the extra electron, bremsstrahlung during scattering of electrons on negative ions should be substantially different from the emission during scattering on positive ions, although it is usually assumed that the corresponding cross sections are equal [19]. This process has been studied by Golovinskii and Zon [20]. One can obtain the following formula for the cross section for bremsstrahlung in collisions of electrons with negative ions in the s-state in the Bethe approximation:

$$d\sigma(\omega) = \frac{4\alpha v_0^2}{3\pi c^2}\frac{d\omega}{\omega}\sigma_{\mathrm{tr}} + \frac{16\alpha d\omega}{3\omega c^2 v^2}\left[\frac{e^2}{m} + \omega^2\alpha(\omega)\right]^2 \ln\frac{\gamma v_0}{R_a\omega}. \tag{10.21}$$

Here σ_{tr} is the transport cross section for scattering of electrons on neutral atoms and $\alpha(\omega)$ is the contribution of the weakly bound electron to the polarizability of the negative ion (the polarization of the neutral atom is neglected). The first term in Eq. (10.21) is related to scattering on the neutral atom and the second is related to scattering on the weakly bound electron, with the first part in the square brackets being the result of Coulomb scattering on the excess charge of the ion. When the energy of the photon is considerably above the energy for photodetachment, we can approximate $\alpha(\omega)$ in Eq. (10.21) by an asymptotic value analogous to Eq. (10.20): $\alpha(\omega) = -e^2/m\omega^2$. Here the second term in Eq. (10.21) vanishes as a result of the absence of bremsstrahlung during collisions between two free electrons in the dipole approximation, which was mentioned above.

In the other limiting case of $\omega \to 0$, the polarization term again does not contribute to the cross section. In the second term of Eq. (10.21) only the first part (in the square brackets), which is related to scattering on the extra negative charge of the ion, remains.

The second term of Eq. (10.21), therefore, reaches a maximum at some intermediate frequencies, specifically, those where the polarizability of the ion is greatest. This frequency range corresponds to the photodetachment limit, where the polarizability has a broad maximum (see [21], for example.)

Figure 10.2 shows the cross section for bremsstrahlung during scattering of an electron with an energy of 10 Ry on an H^- ion relative to the scattering cross section on a point Coulomb charge. Curve 1 corresponds to a calculation based on Eq. (10.21) for $\gamma = 1$ (the polarization was taken from [21]), curve 2, to a calculation without using the Bethe approximation, and curve 3, to a calculation neglecting the polarization amplitude. It can be seen that the Bethe approximation gives the correct location and shape of the maximum in the dispersion curve. The difference at large ω is related mainly to the violation of the condition (10.2) used in writing down the cross section for bremsstrahlung on a neutral atom in terms of the transport cross section. Neglecting the polarizability of the neutral atom means that the cross section for scattering on a negative ion does not approach (with increasing frequency) the cross section for scattering on an unshielded nucleus (in this case a unit charge).

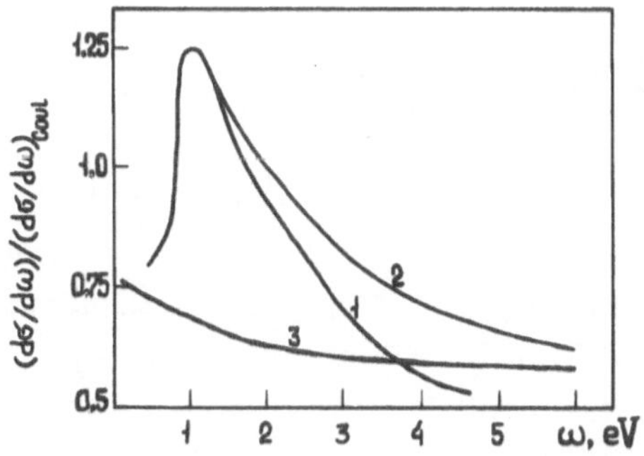

Fig. 10.2. The cross section for bremsstrahlung in collisions of electrons with H^- ions.

10.7. Bremsstrahlung during Scattering of Slow Electrons

The preceding results on the scattering of fast electrons can be obtained with the aid of elementary quantum mechanical perturbation theory if we describe the interaction of a free electron with a variable electric field $\mathbf{E}(t)$ by the effective Hamiltonian [8]

$$\widetilde{V}=e\,(\mathbf{rE}\,(t))-(e\alpha\,(\omega)/r^3)\,(\mathbf{rE}\,(t)). \tag{10.22}$$

Here the first term describes the direct effect of the field on the electron and the second describes the action on the electron of the dipole moment $\mathbf{p} = (\alpha(\omega)\mathbf{E}(t))$ which is induced in the atom by the field of frequency ω. According to perturbation theory, including the Hamiltonian (10.22) leads to the scattering amplitude (10.15) without the Θ-function. This is related to the fact that it is not correct to treat the atom as a point dipole for large momentum transfers.

Nevertheless, the Hamiltonian (10.22) is extremely useful, since it can be used to calculate the bremsstrahlung effect in the limit of low energies of the scattered electron. In this case, if we assume that the incident electron has a much lower energy than an atomic electron, then any momentum transfers will be small and the Hamiltonian (10.22) can be used for arbitrary scattering angles.

Inverse bremsstrahlung of photons by electrons with thermal energies during scattering on Ar and Kr atoms has been examined numerically with polarization effects taken into account [22]. An analytic formula for the cross section has been obtained in [23], which we shall follow in the subsequent discussion.

For a low-energy electron with

$$p\ll\hbar/R_a \tag{10.23}$$

the atomic potential can be viewed as a short-range potential. In this case, the wave function can be written in the form

$$\langle\mathbf{r}\,|\,\mathbf{p}\rangle=\sum_l i^l\,(2l+1)\exp\,[i\delta_l\,(p)]\,R_{pl}\,(r)\,P_l\left(\frac{\mathbf{pr}}{pr}\right), \tag{10.24}$$

where P_l are the Legendre polynomials and δ_l are the scattering phase shifts. The radial wave function of the s-state is taken to be equal to its asymptotic value

$$R_{p0}\,(r)=(\hbar/pr)\,\sin[\,pr/\hbar+\delta_0\,(p)\,]. \tag{10.25}$$

For angular momenta $l \geq 1$, the scattering phase shifts are assumed to be equal to zero, while the radial wave functions coincide with the wave functions for free motion,

$$R_{pl}\,(r)=J_l\,(pr/\hbar),\;\; l\geqslant1. \tag{10.26}$$

Here J_l are the spherical Bessel functions. Equations (10.25) and (10.26) have been used in [24, 25] for calculating the bremsstrahlung effect for a slow electron without the polarization corrections. The bremsstrahlung amplitude calculated with Eqs. (10.25) and (10.26) in the first order with respect to the interaction (10.22) has the form

$$M = -\frac{e\sqrt{2\pi\hbar\omega}}{m\omega^2}\left\{\left[\frac{(\mathbf{p}'\mathbf{e}_{\mathbf{k},\sigma})}{p}\sin\delta_0(p) - \frac{(\mathbf{p}\mathbf{e}_{\mathbf{k},\sigma})}{p'}\sin\delta_0(p')\right] + \frac{2m^2\omega^2}{\hbar}\frac{(\mathbf{q}\mathbf{e}_{\mathbf{k},\sigma})}{q^2}\alpha(\omega)\right\}. \tag{10.27}$$

The first term in Eq. (10.27) corresponds to the result in [24, 25] and the second is caused by the polarization interaction. If we neglect the latter term and the second term in the square brackets is expanded in a series in p' at the point $p' = p$, then the result is the same as Eq. (10.1) when it is noted that the elastic scattering amplitude is given by

$$M_0 = (\hbar/p)\sin\delta_0(p)$$

for a short-range potential.

The cross section for bremsstrahlung is easily obtained from the amplitude (10.27). For example, the cross section for absorption of a photon of frequency ω integrated over the directions of $\mathbf{p}'$ and averaged over the directions of $\mathbf{p}$ is given by

$$\begin{aligned} d\sigma(\omega, v) = {} & \frac{16\pi^2\alpha v'}{3v}\,d\omega\left\{\frac{\hbar^2}{2m^2\omega^3}\left[\frac{(v')^2}{v^2}\sin^2\delta_0(v) + \right.\right. \\ & + \left.\frac{v^2}{(v')^2}\sin^2\delta_0(v')\right] + \frac{(\alpha(\omega))^2\omega}{vv'}\ln\frac{v'+v}{v'-v} - \\ & - \frac{\hbar\alpha(\omega)}{m\omega}\left[\frac{1}{v}\sin\delta_0(v) + \frac{1}{v'}\sin\delta_0(v') + \right. \\ & + \left.\left.\frac{\hbar\omega}{mvv'}\left(\frac{1}{v}\sin\delta_0(v) - \frac{1}{v'}\sin\delta_0(v')\right)\ln\frac{v'+v}{v'-v}\right]\right\}. \end{aligned} \tag{10.28}$$

It is easy to confirm that all the quantities in Eq. (10.28) are of the same order for $v \sim 10^7$ cm/s and $\omega \sim 10^{15}$ Hz. It should be noted that the condition (10.23) is violated at such high frequencies (in the optical range). Nevertheless, a comparison with numerical calculations shows that the equations obtained under the assumption (10.23) can be extended into the region $pR_a \sim \hbar$. It is easy to see that the cross section (10.28) goes to zero for a negative scattering length L if the emitted frequency is

$$\omega_0 \approx \hbar|L|/m\alpha(\omega_0) \gg p^2/2m\hbar. \tag{10.29}$$

For this we must use the estimate

$$\sin\delta_0(p) \approx -pL/\hbar$$

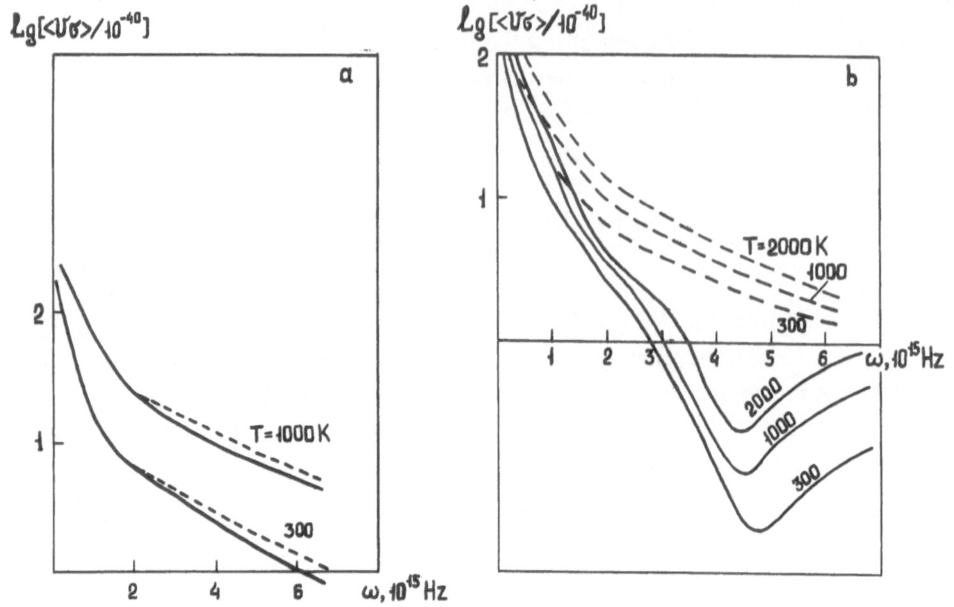

Fig. 10.3. The radiative absorption coefficient through free–free transitions of electrons in Ne (a) and Ar (b) ($v\sigma$ is in cm^3 s^{-1}).

for the zeroth-order phase shift. This behavior of the cross section is related to interference compensation of the two terms in the amplitude (10.27). For atoms with a positive scattering length the amplitude (and with it, the cross section) does not go to zero. From this fact, as well as from Eq. (10.28) for the cross section, it follows that for bremsstrahlung involving slow electrons the entire effect cannot be separated into parts associated with the interaction with the fields of the incident and bound electrons, as was done when the Born approximation was applicable. In particular, the bremsstrahlung photons are emitted by the incident and bound electrons within the same solid angle and in overlapping frequency intervals.

Figure 10.3 shows the behavior of the absorption coefficient for electrons in neon and argon, which is given in terms of the cross section $d\sigma(\omega)/d\omega$ by the formula

$$K_\omega = n_e n_a \langle v\, d\sigma(\omega, v)/d\omega \rangle,$$

where n_e and n_a are the densities of the electrons and atoms, respectively, and the average is taken over a Maxwellian electron distribution. The numbers on the curves denote the electron temperature. The phase shift δ_0 was calculated in the O'Malley–Spruch–Rosenberg approximation [26].

The continuous curves in the figure correspond to a calculation using Eq. (10.28) and the dashed curves, to the results of [24, 25]. It can be seen that with neon, for which the scattering length is positive, the polarization correction produces almost no change in the absorption coefficient, but that with argon the absorption has a deep minimum. We recall that a negative scattering length is also the reason for the appearance of the Ramsauer effect during elastic scattering of electrons on the corresponding atoms. However, the minimum in the photoabsorption curve is not related directly to the Ramsauer effect, since the frequency ω_0 given by Eq. (10.29) is considerably greater than the width of the Ramsauer dip. This fact is evident from the figure: the dashed curves for argon were calculated taking the Ramsauer effect into account in the scattering phase shifts, but they are still monotonic. We can, therefore, say that the Ramsauer dip and the minimum in the photoabsorption coefficient both have the same cause: a negative elastic scattering length.

Chapter 11

POLARIZATION RADIATION MECHANISMS IN ATOMIC COLLISIONS

A. B. Kukushkin and V. S. Lisitsa

11.1. Introductory Comments

The polarization radiation of atoms has been studied intensively in recent years [1–5] and belongs to a broad class of phenomena which can be characterized as the emission of a photon by a single composite "atom + incident particle" system. This general approach to bremsstrahlung processes was first formulated by Born. Recent papers [1–5] have clarified the significance of polarization effects in the bremsstrahlung spectrum of electrons on atoms at frequencies on the order of the atomic frequencies. The observed effects border upon Born's broader conception [6] of radiative processes in many particle systems. This approach has been used to study the polarization of radiation during collisions at resonant frequencies by Percival and Seaton [7] and for atomic spectral line broadening processes by Jablonski [8]. The importance of polarization effects outside the resonant frequency regions for bremsstrahlung processes has been demonstrated [1–5]. In the Born approach [6] one uses the total Hamiltonian of the system of particles, including the incident particle, the atomic electrons, and their interaction with the electromagnetic field. In later calculations, as opposed to [1–5], a static shielded potential has been used. Unlike in the customary treatment of bremsstrahlung, which attributes the emission to the dipole moment d_e of the incident electron alone, in this more general approach the total dipole moment $d_e + d_a$ of the system, which is the sum of the dipole moments of the electron and atom, is responsible for the radiation. The dipole moment d_a makes a much greater contribution to the emission in the spectral range on the order of the atomic frequencies than d_e. In the following we shall examine a number of effects, both resonant and nonresonant, where the emission of the atomic dipole moment excited by collisions with charged particles is important. This includes polarization of atomic emission owing to electron-impact excitation, broadening of atomic spectral lines, dielectronic and polarization recombination, and the emission of forbidden lines. In these processes a photon is emitted by an atom with a corresponding change in the energy of the scattered electron. Thus, these processes, like bremsstrahlung, can be characterized by a cross section for emission of a photon, $d\sigma(\omega)/d\omega$, that is analogous to the bremsstrahlung cross section.

We note that analogous processes are realized, in principle, for heavy particles (atoms and ions), as well as light particles (electrons). In fact, polarization radiation is independent of the mass of the incident particle, since the radiating object is always the electron shell of the atom. Only the spectral region of the emitted frequencies changes in accordance with the change in the characteristic frequencies of the relative motion of the heavy particles. One example of radiation during heavy particle collisions is emission during resonant charge exchange.

11.2. Polarization Radiation of Atoms during Electron-Impact Excitation. General Statement of the Problem of Radiation during Collisions

The formulation of the problem of polarization of atomic emission excited by an electron beam is extremely close to that of the polarization bremsstrahlung problem (here the term "polarization" has a different meaning for these two effects: for polarization radiation we are speaking of a deformation of the electronic shell of an atom, while for light we are speaking of a difference in the intensity at different observation angles). The essence of the polarization effect for radiation lies in the fact that the emitting atom excited by an electron "remembers" the direction of motion of the electron so that the radiation emitted by it has a different polarization along and perpendicular to the electron beam. In the following we shall not be interested in the details of the polarization effect, but only in its general physical formulation which allows us to trace the connection between the characteristics of the radiation and the scattering of the electron. This formulation was developed by Percival and Seaton [7] in complete accordance with the Born approach to bremsstrahlung processes mentioned above. They treated the emission of a photon during excitation of an atom as the emission from a composite "atom + incident electron" system with a single wave function $\psi(\mathbf{r}_e, \mathbf{r}_a)$. If E is the total energy of the "atom + electron" system and E_i is the energy of the initial state of the atom, then when the system emits a photon with energy $E - E'$ with a simultaneous transition of the atom into the state E_f, the initial p_i and final p_f momenta of the electron are related by the conservation of energy:

$$E - E' = \hbar\omega = p_i^2/2m_0 - p_f^2/2m_0 - (E_f - E_i). \tag{11.1}$$

Following [7] we denote the wave function of the colliding particles by $\psi(s_i, p_i|\mathbf{r}_e, \mathbf{r}_a)$, which describes the state of a system in which the incident electron has an initial momentum p_i and the atom is in state s_i. We now expand $\psi(s_i, p_i|\mathbf{r}_e, \mathbf{r}_a)$ in terms of the complete system of wave functions of the atom $\psi_s(\mathbf{r}_a)$ and the scattered electron $\chi_s(s_i, p_i|\mathbf{r}_e)$:

$$\psi(s_i, p_i \mid \mathbf{r}_e, \mathbf{r}_a) = \sum_s \psi_s(r_a)\, \chi_s(s_i, p_i \mid \mathbf{r}_e). \tag{11.2}$$

The wave function $\chi_s(s_i, p_i|\mathbf{r}_e)$ describes the scattering of an electron with momentum p_i on an atom in the initial state s_i accompanied by the transition of the atom into state s. At large distances the function $\chi_s(s_i, p_i|\mathbf{r})$ can be expressed in terms of the inelastic scattering amplitude $f_s(s_i, p_i|\mathbf{r}/r)$ as

$$\chi_s(s_i, p_i \mid \mathbf{r}) \approx \delta_{s,s_i} e^{i\mathbf{p}_i\mathbf{r}} + f_s(s_i, p_i \mid \mathbf{r}/r)\, e^{ip_s r}/r, \tag{11.3}$$

where p_s is the momentum of the electron in the final state s.

The total excitation cross section $\sigma^{\mathrm{ex}}(s_i, p_i \to s_f, p_f)$ is equal to

$$\sigma^{\mathrm{ex}}(s_i, p_i \to s_f, p_f) \equiv \sigma_{if}^{\mathrm{ex}} = \frac{p_f}{g_{s_i} p_i} \sum_{s_i} \int \left| f_s\left(s_i, p_i \left| \frac{\mathbf{r}}{r}\right.\right) \right|^2 d\Omega_r, \tag{11.4}$$

where g_{s_i} is the statistical weight of the initial state s_i ($s_i \in i$) and $d\Omega_r$ is the element of solid angle of the vector $\mathbf{r}$.

Using the representation (11.2) for the wave function $\psi(\mathbf{r}_e, \mathbf{r}_a)$, we write the general expression for the intensity Q_ω of photon emission by the combined "atom + electron" system in the form $[p_s^2 - p_i^2 = (E_s - E_i)2m,\ i = f]$

$$Q_\omega = Q_0 \frac{1}{g_{s_i}} \sum_{s_i, s} p_s \int |\langle s, p_s | \mathbf{e}(\mathbf{r}_a + \mathbf{r}_e) | s_i, p_i \rangle|^2\, d\Omega_{p_s}, \tag{11.5}$$

where $Q_0 = 4e^2\omega^3/3\hbar c^3$ and $\mathbf{e}$ is the unit polarization vector of the radiation.

Using the orthogonality properties of the atomic wave functions, we transform the matrix elements in Eq. (11.5) to the form

$$\langle s, p_s | \mathbf{r}_a + \mathbf{r}_e | s_i, p_i \rangle = \sum_{k,l} \langle k | \mathbf{r}_a | l \rangle \int \chi_l^+(s, p_s | \mathbf{r}_e)\, \chi_k(s_i, p_i | \mathbf{r}_e)\, d\mathbf{r}_e + \\ + \sum_{k,l} \int \chi_l^+(s, p_s | \mathbf{r}_e)\, \mathbf{r}_e \chi_k(s_i, p_i | \mathbf{r}_e)\, d\mathbf{r}_e. \tag{11.6}$$

The second term, associated with the dipole moment $\mathbf{r}_e$ of the scattered electron, is, strictly speaking, traditional bremsstrahlung. The first term, associated with the dipole moment $\mathbf{r}_a$ of the atom, describes polarization bremsstrahlung. The frequency dependence of the spectrum of ordinary and polarization bremsstrahlung is determined by the overlap integrals of the wave functions of the scattered electron, which contain the dependence on ω after substitution of Eq. (11.1).

Equations (11.5) and (11.6) give a general solution for the spectrum of the polarization radiation that does not involve the use of the Born approximation for scattering. They have the same degree of generality for polarization bremsstrahlung as the Sommerfeld formula has for ordinary bremsstrahlung in a Coulomb field.

The structure of the overlap integrals of the wave functions χ in Eq. (11.6) is extremely complicated. They obviously describe the interference of wave functions with different wave vectors, $p_s/\hbar$ and $p_k/\hbar$, which is small everywhere except in a resonance region where

$$\omega \approx \omega_{sk} = (E_s - E_k)/\hbar, \quad (p_s - p_k) \to 0. \tag{11.7}$$

For this resonance region the main contribution to the integrals is from large distances r_e, where the wave functions $\chi(\mathbf{r}_e)$ can be replaced by their asymptotic forms (11.3).

The resulting integrals reduce to integrals of the form

$$\int_0^\infty dr \cdot \exp\,[i\,(p_k - p_s)\,r/\hbar], \tag{11.8}$$

which are large just near a resonance. In order to obtain final expressions valid near a resonance as well, we must include the damping of the wave functions corresponding to motion of the electron in excited channels. Thus, for an atom in an excited state s with a finite lifetime Γ, the scattering amplitude of an electron moving at velocity v_s acquires a factor $\exp\,[-\Gamma r_e/2v_s]$. As a result, the overlap integrals of the wave functions are proportional to the resonance factor

$$\int_0^\infty \exp\,\{i\,[(p_s - p_k)/\hbar - \Gamma/2v_s]\,r\}\,dr = \frac{v_s}{iv_s\,(p_s - p_k)/\hbar - \Gamma/2}. \tag{11.9}$$

Substituting the overlap integrals calculated in this way in the general formula (11.5), we find an expression for the intensity of polarization bremsstrahlung in the resonance approximation:

$$Q_\omega = Q_0 \frac{v_f}{g_{si}} \int \sum_{s_i, s} \left| \sum_{s_f} \frac{\langle s \,|\, \mathbf{r}_a \mathbf{e} \,|\, s_f \rangle\, f_{sf}\,(s_i,\, p_i \,|\, \mathbf{r}/r)}{i\,(\omega - \omega_{s_f s}) - \Gamma/2} \right|^2 d\Omega_{\mathbf{k}}. \tag{11.10}$$

The structure of Eq. (11.10) coincides with the general formulas for polarization radiation given in [1–5] if we neglect the nonresonance terms containing the sum frequencies $\omega + \omega_{sf,s}$.

We note that the total scattering amplitude and not just its Born limit enter in Eq. (11.10). Furthermore, in the resonance approximation the main contribution to Eq. (11.10) is from the level with a frequency $\omega_{sf,s}$ closest to the frequency ω. In this case the sum under the absolute value sign in Eq. (11.10) extends to the degenerate states of this level. In is important to note (following [7]) that this kind of summing of the amplitudes reflects the "memory" of excitation during the subsequent emission by the atom. This means that during excitation by a directed beam of electrons, light from the polarization emission will have polarization properties. After summing over the polarizations, the interference of the amplitudes under the absolute value sign vanishes and the probability of the process is given in the resonance approximation simply by the product of the excitation probability and the spontaneous emission probability.

To summarize the above discussion, we note that the approach proposed by Percival and Seaton [7] is also the most general formulation for polarization emission. Here the main problem reduces to studying the complicated overlap integrals of the wave functions for electron scattering in the various channels.

11.3. Broadening of Atomic Spectrum Lines

The essence of spectral line broadening processes is that, as a result of collisions with other particles, an atom is able to emit light over a substantially wider spectral range than its radiative width. The reason for this is that an emitting atom can transfer part of its energy to the exciting particle. Thus, we are speaking again of the radiation from a composite "atom + incident particle" system. The result of collisional broadening is the emission by the atom of a photon with a frequency ω that differs significantly from the unperturbed transition frequency ω_0 and the energy difference $\hbar(\omega - \omega_0) = \hbar\Delta\omega$ is compensated by a change in the energy ε of the incident particle,

$$\varepsilon_i - \varepsilon_f = (p_i^2 - p_f^2)/2m = \hbar(\omega - \omega_0) \equiv \hbar\Delta\omega, \tag{11.11}$$

where $\varepsilon_{i,f}$ and $p_{i,f}$ are the initial and final energies and momenta of the incident particle and $\Delta\omega$ is the frequency detuning. It is clear that the broadening process, like ordinary bremsstrahlung, can be characterized by a cross section $d\sigma(\omega)/d\omega$ for emission of a photon of frequency ω during a collision of a broadening particle with the atom.

Let us consider the expression for the intensity $Q_{if}(\omega)$ of emission by an atom at frequency ω close to the frequency $\omega_{if} \equiv \omega_0$ of a transition between two levels i and f. The levels are regarded as nondegenerate in the momentum and the interaction of the radiating atom with the broadening electron in states i and f is determined by the spherically symmetric potentials $U_i(r)$ and $U_f(r)$.

Let us suppose that a photon $\hbar\omega$ is emitted by a unified system consisting of an atom and an electron broadening the emission line. The wave function of this system in the initial and final states is given by the product of the wave functions ψ_i and ψ_f of the atom with the wave functions $\chi_{p_{i,f}}^{(\pm)}$ that describe the scattering of a particle with momentum $p_{i,f}$ on the potentials U_i and U_f (the plus and minus signs correspond to the functions coincident with converging or diverging spherical waves at infinity).

We now write down the well-known expression for the probability dw of a transition of the entire system from state i to state f with emission of a photon of frequency ω and momentum $\mathbf{k}$:

$$dw = (2\pi/\hbar)\,|\langle i|V|f\rangle|^2\,\delta(\varepsilon_f - \varepsilon_i)\,d\mathbf{p}_f d\mathbf{k}/(2\pi)^6. \tag{11.12}$$

Here V is the operator for the interaction of the atom with the radiation field, and the wave functions of the system, $\langle i|$ and $|f\rangle$, have the forms

$$\langle i| = \psi_i \sqrt{\frac{m_0}{\hbar p_i}}\,\chi^{+}_{p_i}, \quad |f\rangle = \psi_f \chi^{-}_{p_f}, \tag{11.13}$$

while the energy $\varepsilon_{i,f}$ satisfies Eq. (11.11). The wave function $\langle i|$ is chosen to be normalized to a unit flux of broadening electrons. Thus, after averaging over the initial state and summing over the final states, Eq. (11.12) yields a differential cross section $d\sigma(\omega)/d\omega$ for emission of a photon at frequency ω. This cross section for emission of a photon, unlike that for ordinary bremsstrahlung, is determined by the dipole moment $e\mathbf{r}_a$ of the atom and, therefore, is independent of the mass of the incident particle, as is the cross section for polarization bremsstrahlung.

It is clear from these remarks that one can calculate the cross section $d\sigma(\omega)/d\omega$ for emission of a photon by an atom using the general formula for the cross section for ordinary bremsstrahlung (see [9], for example) on replacing the dipole moment of the electron $e\mathbf{r}_e$ by the dipole moment of the atom $e\mathbf{r}_a$ and using the wave functions (11.13) of the combined "atom + electron" system. As a result, we find that [10, 11]

$$\frac{d\sigma(\omega)}{d\omega} = \frac{4\omega^3\,|d_{if}|^2}{3c^3}\,\frac{\pi^2}{v_i\varepsilon_i p_f}\sum_l (2l+1)\,|A_l|^2. \tag{11.14}$$

Here v_i and ε_i are the initial velocity and energy of the incident electron. The matrix element of the dipole moment of the atom $e\mathbf{r}_a$ breaks up, as in Eq. (11.6), into the product of the matrix element of $e\mathbf{r}_a \equiv \mathbf{d}_a$ between states i and f of the atom and the overlap integral A_l of the wave functions of an electron with angular momentum l,

$$A_l = \int_0^\infty dr\,r^2 R_{p_i,l}(r)\,R_{p_f,l}(r), \tag{11.15}$$

where $R_{pi,l}$ and $R_{pf,l}$ are the radial wave functions of an electron with angular momentum l in the field $U_i(r)$ and $U_f(r)$. The power radiated from unit volume, $Q^{\mathrm{vol}}(\omega)d\omega$, is expressed, as in the case of ordinary bremsstrahlung, in terms of the cross section $d\sigma(\omega)/d\omega$ in the form

$$Q^{\mathrm{vol}}(\omega) = n_a n_e \hbar\omega \int dv_a v_a f(v_a)\,d\sigma(\omega)/d\omega, \tag{11.16}$$

where n_a and n_e are the densities of the atoms and electrons and $f(v_a)$ is the distribution function over the initial relative velocities. For simplicity in the following we shall not take the average over v_a (when necessary it can be done in the next stage). Dividing Eq. (11.16) by the total intensity of the emission from the atoms, $n_a 4\omega^4 |d_{if}|^2/3c^3$, we obtain an expression for the line profile I_ω of an individual atom,

$$I_\omega = n_e \frac{\pi^2\hbar}{\varepsilon_i p_f} \sum_l (2l+1)\,|A_l|^2. \tag{11.17}$$

Therefore, calculating the line profile reduces, as above (see Sec. 11.2), to calculating the overlap integrals A_l, whose dependence on the frequency shift $\Delta\omega$ is determined by the conservation of energy (11.11).

The theory of line broadening offers an interesting possibility for an analytic study of the structure of the overlap integrals. In the following we shall demonstrate the character of the spectral dependence of the profile I_ω using quasiclassical wave functions in Eq. (11.15).

The quasiclassical wave functions $\chi_{p_i}(r)$ in a spherically symmetric potential $U_i(r)$ have the form

$$\chi_{p_i}(r) \sim \frac{1}{\sqrt{p_i(r)}} \cos\left[\frac{1}{\hbar}\int_{r_0}^{r} p_i(r)\,dr\right], \tag{11.18}$$

where the radial component of the momentum $p_i(r)$ is given by

$$p_i(r) = \{2m[\bar{\varepsilon}_i - U_i(r)] - \hbar^2 l(l+1)/r^2\}^{1/2}. \tag{11.19}$$

Substituting Eqs. (11.18) and (11.19) in the overlap integrals (11.15) leads to integrals containing the sum $p_i(r) + p_f(r)$ and difference $p_i(r) - p_f(r)$ of the momenta in the arguments of trigonometric functions. The expressions in the integrands that contain the sum $p_i(r) + p_f(r)$ oscillate rapidly and therefore make only a small contribution to the overlap integrals. The remaining integrals, which contain the difference $p_i(r) - p_f(r)$, have the form [12]

$$A_l \propto \int_0^\infty \frac{dr}{\sqrt{p_i(r)\,p_f(r)}} \cos\left\{\frac{1}{\hbar}\int_{r_0}^{r} [p_i(r) - p_f(r)]\,dr\right\}. \tag{11.20}$$

The difference in the momenta under the integrals can be transformed with the aid of Eq. (11.19) to

$$p_i(r) - p_f(r) = \frac{2m}{(p_i(r) + p_f(r))} [\varepsilon_i - \varepsilon_f + U_i(r) - U_f(r)]. \tag{11.21}$$

In the quasiclassical region, the momenta of the particles in the initial (p_i) and final (p_f) channels differ little, so that we can set $p_i \approx p_f$ in the denominator of Eq. (11.21).

We note that the energy difference $\varepsilon_i - \varepsilon_f$ is equal to the frequency detuning $\hbar\Delta\omega$ according to Eq. (11.11). Proceeding further in Eq. (11.20) from an integral with respect to the coordinate dr to one with respect to the time $dt = dr/v$ and introducing the notation $\kappa(t) = (1/\hbar)[U_i(r) - U_f(r)]$ for the instantaneous frequency shift [where we assume that $r = r(t)$ is the classical trajectory of the particle], we find that

$$A_l(\Delta\omega) \propto \int_0^\infty dt \cos\left[\Delta\omega t - \int_0^t \varkappa(t')\,dt'\right]. \tag{11.22}$$

Equation (11.22) already has a purely classical structure and represents the spectrum of a classical oscillator that undergoes frequency shifts $\kappa(t)$ owing to collisions (phase modulation [9]). As a rule, the function $\kappa(t)$ is approximated by power law dependences of the form $\kappa(t) = c_n/r^n(t)$, where $r(t)$ is the classical trajectory of the unperturbed particle's motion.

Equation (11.22) is well known from studies of the classical theory of line broadening [9, 11]. Here we just limit ourselves to describing two limiting cases that are characteristic for the structure of the overlap integral (11.22). The first case (static) corresponds to slow collisions and fairly large frequency detunings $\Delta\omega$ that satisfy the condition $\Delta\omega r_{ef}/v \gg 1$, where r_{ef} is the effective collision radius. The main contribution to the integral (11.22) is from the stationary phase point t_k at which

$$\varkappa(t_k) = \Delta\omega,\quad U_i(r_k) - U_f(r_k) = \hbar\Delta\omega \qquad (r_k \equiv r(t_k)). \tag{11.23}$$

Equation (11.23) is an elementary expression of the Franck–Condon principle [13], which states that the main contribution to the intensity is from the points r_k where the system spends the most time. Calculating the integral (11.22) by the saddle point method and using Eq. (11.23) leads to a relationship between the intensity $I_{\Delta\omega}$ of the radiation and the time the system spends near the Franck–Condon points (11.23):

$$I_{\Delta\omega} \propto |\dot{\varkappa}(t_k)|^{-1} \propto \left| 1/\frac{\partial}{\partial r}[U_i(r) - U_f(r)] \right|_{r=r_k(\omega)}. \tag{11.24}$$

Substituting power law dependences for the potentials $U_{i,f}(r)$ leads to a power law drop in the radiative intensity (see [9, 11]).

The second case (impact) corresponds to rapid collisions with $\Delta\omega r_{ef}/v \ll 1$. In this case, the dependence on $\Delta\omega$ can be neglected in Eq. (11.22) and the main contribution to $\Delta\omega$ is from large distances $r \to \infty$, while the overlap integral can be expressed in terms of the quasiclassical scattering phase shifts δ_l. The resulting emission spectrum has a Lorentzian shape $\gamma/(\Delta\omega)^2$ in this case, where the collision frequency γ is determined by these phase shifts. A more detailed analysis can be found in [9, 11].

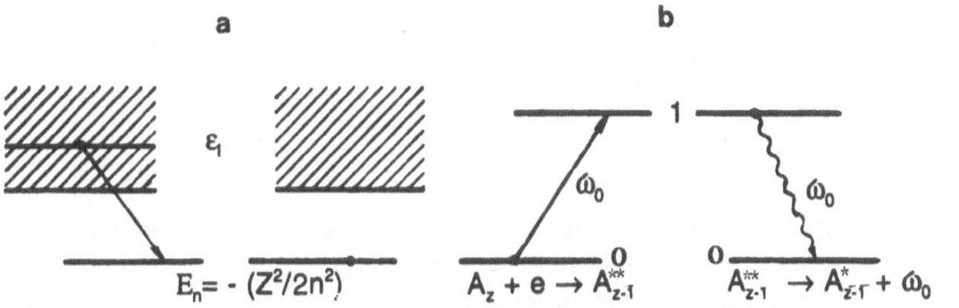

Fig. 11.1. Sketch of a dielectronic recombination process for the electron (a) and the core (b).

11.4. Dielectronic and Polarization Recombination

Photorecombination processes, as well as the recombination (photorecombination) radiation accompanying them, are a direct extension of bremsstrahlung processes. In fact, in a Coulomb field the matrix elements for the free–free transitions responsible for bremsstrahlung transform continuously into the matrix elements for the free–bound transitions responsible for recombination radiation. This is especially clear in the quasiclassical Kramers approximation, where the radiant intensity Q_ω is determined by the Fourier component of the radius vector of the classical trajectory, r_ω. For sufficiently high ω ($\omega \gg mv^3/Ze^2$), the intensity is determined by the most strongly curved portions of the electron trajectory that are closest to the nucleus. Then the intensities of both bremsstrahlung and recombination radiation are determined by the same formulas and these transform continuously from the former to the latter process with increasing ω. A difference arises only in the subsequent quasiclassical procedure of equating the emitted energy $\hbar\omega$ to the difference between the initial ε_i and final ε_f energies of the radiating particle: for bremsstrahlung both ε_i and ε_f belong to the continuum, while for recombination radiation the final state is bound with $\varepsilon_f = -1/n_f^2$ (ε_f in Ry).

It is clear from these considerations that polarization radiation processes can be accompanied by free–bound transitions leading to recombination, as well as by free–free particle scattering; that is, in principle, polarization recombination radiation is possible in addition to polarization bremsstrahlung.

Zon [3] has pointed out an analogy between polarization radiation and the scattering of light. During resonant scattering, as mentioned previously, polarization bremsstrahlung transforms simply into line emission by excited atoms (enters the mass surface). Analogously, under resonance conditions, polarization recombination transforms into the well-known process of dielectronic recombination [14].

The dielectronic recombination process essentially reduces to the following (Fig. 11.1): an incident electron with energy ε_i excites the ion core with an excitation energy of $\Delta\varepsilon = \hbar\omega_0$. If the energy ε_i of the electron is less than $\Delta\varepsilon$, then after exciting the core it will be captured into some level of the ion, $\varepsilon_f = -1/n_f^2$ (ε_f in Ry), which satisfies the condition

$$\varepsilon_i - \varepsilon_f = \varepsilon_i + 1/n_f^2 = \Delta\varepsilon = \hbar\omega_0. \tag{11.25}$$

As a result, a doubly excited state of the ion is formed: an electron in the core is excited by an amount $\Delta\varepsilon$ and the incident electron is captured into the state ε_f. Subsequently, this state can decay through two channels: a) autoionization, leading to the ejection of the captured electron back into the continuum with simultaneous relaxation of the core to the initial state, or b) radiative decay, in which the core returns to the ground state by emitting a photon of energy $\hbar\omega \approx \hbar\omega_0 = \Delta\varepsilon$ and the captured electron remains in the atom. The latter channel is dielectronic recombination. Thus, in terms of its result, dielectronic recombination, like radiative recombination, leads to capture of an electron and simultaneous emission of a photon. In the case of radiative recombination, the photon is emitted by the recombining electron itself and in the case of dielectronic recombination the photon is emitted by the excited core. Here there is an evident analogy with ordinary and polarization bremsstrahlung.

For a more complete clarification of the analogy between polarization radiation and the scattering of light, we now examine polarization recombination processes with the aid of Fermi's equivalent photon method [15] (for relativistic particles this idea of Fermi's later acquired the character of an exact theorem that forms the basis of the Weizsäcker–Williams method [8]). The basis of this method is the idea that the pulse of the electromagnetic field acting on an atom during a collision with a charged particle is fully equivalent to the action of a light pulse of the same intensity on the atom. Then a collision of an atom (or ion) with a charged particle can be viewed as the irradiation of the atom by a flux of equivalent photons whose intensity is determined by the Fourier components of the electric field produced by this particle. From the standpoint of Fermi's equivalent photon method, the dielectronic process can be viewed simply as the scattering of equivalent photons on an atom with their transformation into real photons. Then the rate of dielectronic recombination can clearly be related to the fluorescence probability for the equivalent photons.

We now find the intensity of the flux of equivalent photons. Since dielectronic recombination is the main recombination process for heavy ions in hot, rarefied plasmas, we shall assume that the recombining electron moves along a classical Coulomb trajectory in the field of a multiply charged ion with charge $Z \gg 1$. The dielectronic recombination rate is, as a rule, large for ions with a complicated core that has transitions within a given quantum number n (transitions with $\Delta n = 0$, such as $2s$–$2p$ transitions in lithiumlike and more complicated ions). The energy $\Delta\varepsilon = \hbar\omega_0$ of transitions with $\Delta n = 0$ in ions with charge $Z \gg 1$ is on the order of Z (in Ry) (see [14]), while their ionization energy is of order $Z^2 \gg \Delta\varepsilon \sim Z$. Since the energy of an incident recombining electron must be less than the excitation energy of the core $\Delta\varepsilon \sim Z$, it satisfies the condition

$$Ze^2/\hbar v \sim (Z^2/\varepsilon)^{1/2} \gg 1. \tag{11.26}$$

This condition justifies the use of the classical approach in the recombination processes of interest to us.

The magnitude of the electromagnetic field created by the external incident electron at the site $r_i = 0$ of the ion to be excited is equal to (in atomic units with $|e| = \hbar = m = 1$)

$$\mathbf{E}(0, t) = -\mathbf{r}_e(t)/r_e^3(t), \tag{11.27}$$

where $\mathbf{r}_e(t)$ is the electron trajectory. Using the equation of motion of the external electron in the field of the ion, $m\ddot{\mathbf{r}}_e = -Ze^2\mathbf{r}_e/r_e^3$, we can write Eq. (11.27) in the form

$$\mathbf{E}(t) = +\ddot{\mathbf{r}}_e(t)/Z. \tag{11.28}$$

The flux density I_ω of the equivalent photons produced by the electric field of the incident electron is given in terms of the Fourier component of the field (11.28) as

$$I_\omega = \frac{c}{8\pi^2}\frac{1}{\omega}\{|E_{x,\omega}|^2 + |E_{y,\omega}|^2\} = \frac{c\omega^3}{8\pi^2 Z^2}\{|x_\omega|^2 + |y_\omega|^2\}, \tag{11.29}$$

where x and y are the coordinates in the plane of motion of the incident electron. Using the standard expressions for the Fourier component of the coordinates of an electron moving in a Coulomb field [16], we obtain

$$I_\omega = \frac{c\omega}{8v_0^4}\left\{[H_{i\nu}^{(1)'}(i\nu e)]^2 - \frac{e^2-1}{e^2}[H_{i\nu}^{(1)}(i\nu e)]^2\right\}, \tag{11.30}$$

where v_0 is the initial velocity of the electron; $H_{i\nu}^{(1)}$ is the Hankel function of first type of order $i\nu$ (a prime denotes differentiation with respect to the argument);

$$e = 1 + 2\varepsilon M^2/Z^2; \quad \nu = \omega Z/v_0^3; \quad \varepsilon = v_0^2/2; \tag{11.31}$$

ε and M are the energy and angular momentum of the incident electron; and e is the eccentricity of the orbit.

In the low frequency limit for the equivalent photons, $\nu \ll 1$, the main contribution to the flux density of the photons averaged over the impact parameter ρ of the electron is from distant, almost linear trajectories ($\rho \gg a \equiv Z/2\varepsilon$) for which $e \gg 1$. In this case, Eq. (11.30) transforms to

$$I_\omega=(c\omega/2\pi^2v_0^4)\{K_0^2(\omega\rho/v_0)+K_1^2(\omega\rho/v_0)\}, \tag{11.32}$$

where K_0 and K_1 are the Macdonald functions. This expression was used by Fermi [15] to examine the excitation of atoms by a linearly moving particle.

In order to describe dielectronic (and polarization) recombination, in which the incident electron is captured by an excited ion, we must consider the high-frequency limit $\nu \gg 1$. The main contribution to a process with such a large energy transfer from the electron to the ion is from close, strongly curved trajectories with $\rho \ll a$. Fourier analysis of a classical Coulomb trajectory shows that a fairly narrow region of the trajectory near the turning point r_0 of the radial motion of the electron (located at its distance of closest approach to the ion) is responsible for the emission of such high-frequency photons (both real and equivalent). The effective distances from the ion that correspond to a given frequency ω are given by

$$r_\omega \sim (Z/\omega^2)^{1/3}. \tag{11.33}$$

The main contribution to the spectrum integrated over the angular momentum M is from trajectories with $M_{ef}(\omega) \sim (Z^2/\omega)^{1/3}$, which corresponds to the production of radiation at frequency ω during an (approximate) orbit of the electron around the ion at a distance r_0 with frequency $\omega \sim \omega_{rot}$:

$$\omega_{rot} \sim v_{max}/r_0 = M/r_0^2, \tag{11.34}$$

where v_{max} is the velocity of the electron at the turning point r_0.

At these short distances (11.33) the electron's trajectory and, because of the localization of the emission region, the resulting emission spectrum are independent of the initial energy ε of the electron and depend only on the angular momentum M. The loss of the dependence on ε ultimately means that the corresponding formula for the emission spectrum describes both the continuum transitions of the incident electron (bremsstrahlung) and the transitions from the continuum to discrete levels (recombination radiation):

$$I_\omega = \frac{c\omega M^4}{6\pi^2 Z^4}\left\{K_{1/3}^2\left(\frac{\omega M^3}{3Z^2}\right)+K_{2/3}^2\left(\frac{\omega M^3}{3Z^2}\right)\right\} \equiv \frac{cM}{2\pi^2 Z^2}\,G_0\left(\frac{\omega M^3}{3Z^2}\right). \tag{11.35}$$

The possibility of even describing the photorecombination spectrum with the aid of a classical formula is a result of the fact that when the spatial region responsible for emission at high frequencies is localized, the actual criterion for classical behavior in the spectrum integrated over M turns out to be a condition which can be written in the following equivalent forms [17]:

$$\lambda\!\!\!^{-}(r_\omega)/r_\omega \sim 1/M_{ef}(\omega) \sim \omega/\varepsilon_{kin}(r_0) \ll 1, \tag{11.36}$$

where r_ω is given by Eq. (11.33); $\varepsilon_{kin}(r)$ is the local kinetic energy of the electron; and

$$\lambdabar(r) \sim (r/Z)^{1/2} \tag{11.37}$$

is the local wavelength of the electron. The condition $\lambdabar(r_\omega) \ll r_\omega$ is the condition for quasiclassical motion of the electron, which is directly obtainable from the criterion

$$(1/p^2(\mathbf{r}))\,\mathrm{div}\,\mathbf{p}(\mathbf{r}) \ll 1, \tag{11.38}$$

where $\mathbf{p}(\mathbf{r})$ is the total local momentum of the electron. Equation (11.38) is a generalization of the standard criterion $|d\lambdabar(x)/dx| \ll 1$ to the three-dimensional case for the "rotational" segments of the trajectories (on these segments the momentum of the radial motion is small and the local radius of curvature of the trajectory is $\sim r_\omega$). It is clear from Eq. (11.36) that the often-introduced condition $\omega \ll \varepsilon$ is excessive in this case and can be replaced by the very much less rigid condition $\omega \ll \varepsilon_{kin}(r_\omega)$ [because $\varepsilon_{kin}(r_\omega) \gg \varepsilon$].

The criterion (11.36) for a classical spectrum follows from an analysis of the quantum mechanical corrections to the classical limit for the photon spectrum obtained in the high-frequency limit for a central field [17]. For the emission spectrum of an electron with a fixed momentum the criterion for classical behavior has the form $l = M \gg 1$ (at frequencies where the corresponding spectrum is not exponentially small). In fact, however, Eq. (11.35) is also applicable for small l (see [18]). This statement is consistent with the fact that the quantum mechanical correction to Eq. (11.35), obtained in [17] under the assumption that $l \gg 1$, is still small for the small values of the argument of the function G_0 which are realized for small l ($M = l + 1/2$) and sufficiently small ω/Z^2.

In this model description of dielectronic recombination it should be noted that it is resonance fluorescence with a complicated intermediate state which is formed when the incident electron is captured by the ion and provides an additional decay channel through autoionization. In resonance fluorescence we must consider three types of states: the initial (unexcited ion and an initial distribution I_0 of intensities of the equivalent photons) with energy E_1; an intermediate (excited ion with a plasma electron captured into one of its high-lying levels; i.e., a doubly excited ion whose ionization state is lowered by unity and the initial distribution I_0 with a single photon ω_{eq} absorbed from it) with energy E_2; and a final state (an ion whose ionization state is lowered by unity and is in a singly excited state, an emitted photon ω, and a distribution I_0 without any photons ω_{eq}) with energy E_3. The energies of these states are related by the conservation of energy:

$$E_3 - E_1 = \omega - \omega_{eq}, \quad E_3 - E_2 = \omega - \omega_0, \tag{11.39}$$

where ω_0 is the energy of the transition excited in the ion.

The probability distribution for resonance fluorescence has the form [19]

$$w_{RF}=\frac{|V_{21}|^2|V_{32}|^2}{[(\omega-\omega_{eq})^2+\Gamma^2/4][(\omega-\omega_0)^2+\gamma^2/4]}, \tag{11.40}$$

where V_{21} and V_{32} are the matrix elements for the absorption of a photon ω_{eq} by the ion and the emission of a photon ω by it. Γ and γ are the total probabilities (per unit time) of absorption and emission, respectively, of a photon by the ion:

$$\gamma(E)=2\pi\sum_{k}|V_{32}|^2\delta(E-E_3); \tag{11.41}$$

and

$$\Gamma(E)=\gamma\sum_{k_{eq}}|V_{21}|^2\frac{1}{[(E-E_2)^2+\gamma^2/4]}. \tag{11.42}$$

$\Gamma(E)$ and $\gamma(E)$ should be taken at the point $E = E_3$ in Eq. (11.40), but in practice they both depend only very weakly on the energy.

It is appropriate to mention a conclusion which follows from Eq. (11.40) to the effect that in the case of an individual absorption-emission process, the energies of the absorbed and emitted photons coincide (to within the width Γ which is very narrow in practice). The ion's "memory" of what kind of photon it absorbed is reflected in the fact that the probability (11.40) is not the same as the direct product of the probabilities of emission and absorption, since the first factor in the denominator relates (approximately, through the corresponding δ-function) the energies of the initial and final photons and not ω_{eq} and ω_0. And only when the ion is irradiated by light with a continuous spectrum will the shape of the absorbed line be as if two independent processes (absorption with subsequent emission) had taken place.

The width Γ_{DR} for dielectronic recombination will be given by the sum of the width γ (11.41) and the probability of autoionization Γ_A in the intermediate state,

$$\Gamma_{DR}=\gamma+\Gamma_A, \tag{11.43}$$

which includes the possibility that a recombined plasma electron may return to the continuum with reemission of a photon by the ion.

The matrix element V_{21} is determined by the oscillator strength of the radiative transition in the ion ($V_{21} \sim d_{21}$, d_{21} is the matrix element of the dipole moment of the radiating bound electron in the ion) and is directly proportional to the flux density of equivalent photons irradiating the ion. For the radiation with a continuous frequency spectrum incident on the ion that is realized in this case, the total absorption probability Γ_{RF} is related to $I_{0,\omega}$ by

$$\Gamma_{RF}=2\pi I_{0,\omega}\overline{|V_{21}|^2}, \tag{11.44}$$

where the bar denotes averaging over the angles of the wave vector of the absorbed photons. For the specific conditions of our dielectronic recombination model, the sum over k_{eq} in Eq. (11.42) should be supplemented by a sum over the final states of the captured electron, which, in combination with the conservation of the energy ε of the incident electron, $\omega_{eq} = \varepsilon_i + Z^2/2n^2$, gives

$$\Gamma_{DR} = \Gamma_{RF} Z^2/n^3. \tag{11.45}$$

Then, using the equation for the total probability of dielectronic recombination summed over all frequencies of the equivalent photons,

$$\sum_{\omega} w_{DR} = \frac{\gamma \Gamma_A}{\gamma + \Gamma_A}, \tag{11.46}$$

one can find an expression for the autoionization rate,

$$\Gamma_A = \frac{f_{12}}{\pi n^3} l G_0\left(\frac{\omega_0 M^3}{3Z^2}\right), \tag{11.47}$$

where f_{12} is the oscillator strength of the excited radiative transition in the ion and $G_0(x)$ is defined in Eq. (11.35). Equation (11.47) is the same as an exact quantum mechanical calculation [20] in the limit of classical motion by the incident electrons along quasiparabolic orbits which we are considering ($Ze^2/\hbar v \gg 1$, $\rho \ll a$).

The total rate of dielectronic recombination with capture of the incident electron into the n, l state is given by

$$\alpha_{DR} = \left(\frac{2\pi}{T}\right)^{3/2} \frac{g(2)}{g(1)} \frac{(2l+1)\gamma\Gamma_A}{(\gamma+\Gamma_A)} \exp\left[-\frac{\omega}{T} + \frac{Z^2}{2n^2 T}\right], \tag{11.48}$$

where T is the temperature of the Maxwellian plasma electrons and $g(2)$ and $g(1)$ are the statistical weights of the excited and ground states of the ion. We note that the above results for Γ_A and w_{DR} can also be obtained by using a method described in [14]. Thus, for Γ_A, when we include the mutual reversibility of autoionization for a weakly bound (freed) electron and electron-impact excitation of the ion near the excitation threshold (which is expressed, in turn, through the cross section $\sigma_{n,l\to\varepsilon}$ for photoionization of a highly excited electron with ejected electron energies on the order of ω_0, which is an analytic continuation of deexcitation to the discrete spectrum), we have

$$\Gamma_A = \frac{3}{10\pi} Z^2 \left(\frac{\omega_0}{Z^2}\right)^2 f_{12}\sigma_{n,l\to\varepsilon}. \tag{11.49}$$

Expressing $\sigma_{n,l\to\varepsilon}$ in terms of the cross section for excitation by a classical electron with $\rho \gg a$, we obtain Eq. (11.47).

All of these methods for obtaining Γ_A (and w_{DR}) are equivalent in the sense that, in one way or another, they use the dipole approximation for the interaction of the atomic and incident electrons, which makes it possible to express all the processes by which the incident electron loses energy (through radiation or slowing down at the ion and through inelastic Coulomb collisions) as the effective emission of photons (real or equivalent) with a probability that is determined by the dipole matrix element of the corresponding inelastic transition.

The above formulas (11.46) and (11.48) for dielectronic recombination describe a two-stage interaction of the recombining incident electron with the emitting atomic (or ionic) electron, where the intermediate state (a doubly excited ion) is real. This corresponds to having the energy of the intermediate state be close to the energy of the excited ion with the absorbed photon [ε' within the limits of the resonance in Eqs. (11.39) and (11.40)]. When the deviation of ε' from the resonance (i.e., the deviation of ω_{eq} from ω_0) is large, the resonance approximation is no longer applicable. It can be modified by simply replacing the resonant denominator $(\omega_0 - \omega - i\gamma/2)^{-1}$ in the resonance fluorescence amplitude by the sum

$$(\omega_0-\omega-i\gamma/2)^{-1}+(\omega_0+\omega-i\gamma/2)^{-1} \tag{11.50}$$

(see [19], for example). This also means that the spectral intensity of the polarization radiation (the emission from the atom owing to its dynamic polarization in the field of the incident electron) is given in terms of the elastic scattering cross section $\sigma_{\text{scat}}(\omega)$ for nonresonant (relative to ω_0) equivalent photons by

$$Q_\omega^{\text{R}} = I_\omega \sigma_{\text{scat}}(\omega), \tag{11.51}$$

where I_ω is the flux density of the equivalent photons (11.35).

The cross section σ_{scat} is related to the dynamic polarizability $\alpha(\omega)$ of the ion (atom) at a frequency of ω,

$$\sigma_{\text{scat}}(\omega) = \frac{8\pi\omega^4}{3c^4}\alpha^2(\omega), \tag{11.52}$$

where

$$\alpha(\omega) = \sum_n |d_{in}|^2 \{(\omega_{ni} - \omega - i0)^{-1} + (\omega_{ni} + \omega - i0)^{-1}\}, \tag{11.53}$$

in which (because of the off-resonance character) we must include the (virtual) transitions of the emitting electron over all the remaining levels of the ion [enumerated in Eq. (11.53) by the subscript n]. From Eq. (11.51) we obtain the ratio of the intensities of the polarization radiation of the ion and the bremsstrahlung of the electron which appear simultaneously when they collide. This is of some practical interest. It is expanded in terms of only the dynamic polarizability of the ion (in ordinary Gaussian units) by

$$\frac{Q_\omega^{\text{R}}}{Q_\omega^{\text{B}}} = \left[\frac{m_0\omega^2\alpha(\omega)}{e^2 Z}\right]^2. \tag{11.54}$$

It is important to note that this result, which we have derived for classical motion of the incident electron and high frequencies ($Ze^2/\hbar v_0 \gg 1$, $\omega \gg m_0 v_0^3/Ze^2$), is actually valid for arbitrary values of $Ze^2/\hbar v_0$ and ω; that is, it encompasses the regions of bremsstrahlung ($\hbar\omega \leq \varepsilon$) and recombination radiation ($\hbar\omega > \varepsilon$). The only condition for the validity of Eq. (11.54) is that the interaction between the emitting electron in the ion and the incident electron should be a dipole interaction. As pointed out above, this ensures that the intensity of the virtual photon emission can be separated in the final result (for dielectronic recombination and polarization radiation). Thus, in the Born and low-frequency cases $Ze^2/\hbar v_0 \ll 1$ and $\hbar\omega \ll m v_0^2$ (see [3], for example), which are alternatives to the case we have examined here, the result for Q_ω^R/Q_ω^t is the same as Eq. (11.54).

11.5. Polarization Mechanism for Emission of Forbidden Atomic Transitions

The polarization of atomic states during collisions with charged particles may make states that had previously been nonradiative become radiative. Lines then appear in the atomic spectra at transition frequencies that had been forbidden by the selection rules for dipole radiation in the absence of collisions. A clear example is the $4f$–$2p$ forbidden line of helium which shows up in the emission of this atom from plasmas. Another interesting example of the appearance of this type of forbidden transitions is the destruction of metastable states owing to collisions with charged particles. Here the $2p_{1/2}$ and $2s_{1/2}$ levels of hydrogen or a hydrogenlike ion are a classical example. The $2p_{1/2}$ level is radiative with a large rate constant for decay, γ, while the $2s_{1/2}$ level is metastable with a negligible rate of decay into the $1s$ state (owing to a two-photon transition). In a collision with a slow charged particle (such as a plasma ion), because of the strong polarization in the $2s$–$2p$ level system, a dipole moment develops on the $2s$–$1s$ transition and leads to the possibility of the emission of a photon at the frequency of the $2s$–$1s$ transition. This leads to a finite lifetime τ for the metastable level.

The two main mechanisms for the destruction of a metastable level should be noted [21–23]. The first is caused by direct collisional transitions from $2s$ into $2p$ and is determined by the cross section for inelastic flipping from the $2s$ to the $2p$ state followed by radiative decay of the $2p$ level. The second mechanism is caused by polarization during slow collisions and does not lead to an inelastic transition from $2s$ into $2p$, but only causes polarization mixing of these states which results in the appearance of a nonzero dipole moment for the $2s$–$1s$ transition. Thus, it is the second (polarization) mechanism for decay of a metastable level that is responsible for the production of forbidden lines. These lines can be considered in the spirit of our earlier discussion of the distinctive polarization radiation of atoms. We note that forbidden lines are formed even when the external perturbation is constant, i.e., during static polarization of atomic states. The study of forbidden line spectra is fairly complicated [24]. Thus, in the following we shall limit ourselves to an examination of an integral characteristic over the spectrum, namely the lifetime τ of the metastable level [22]. This time obviously determines the total intensity of the emission in a forbidden transition [22–24].

Let us consider the Schrödinger equations for the amplitudes of metastable a_0 and radiative a_1 levels with an energy interval $\hbar\omega$ between them:

$$i\dot{a}_0 = V_{01}(t)e^{i\omega t}a_1,$$
$$i\dot{a}_1 = -i\gamma a_1 + V_{10}(t)e^{-i\omega t}a_0, \tag{11.55}$$

where $\hbar V_{01}(t) = \mathbf{d}_{10}\mathbf{E}(t)$ is the perturbation of the levels by the external electric field $\mathbf{E}$ of the charged plasma particles.

We assume initially that the perturbation V_{10} is constant and then find the eigenfrequencies $\omega^{(1,2)}$ and the manner in which the amplitude $a_0(t)$ varies in this case:

$$\operatorname{Im}\omega^{(1,2)} = -\frac{\gamma}{2}\left[1 - \frac{\omega}{\sqrt{\omega^2 + 4|V_{01}|^2}}\right], \tag{11.56}$$

and

$$|a_0(t)|^2 = \exp\left\{-\gamma t\left[1 - \frac{\omega}{\sqrt{\omega^2 + 4|V_{01}|^2}}\right]\right\}. \tag{11.57}$$

Thus, applying a constant perturbation leads to radiative decay of the metastable state and for small $|V_{01}| \ll \omega$ this decay is entirely determined by the static polarizability of the $2s$–$2p$ level system:

$$\operatorname{Im}\omega = -\frac{\gamma}{\omega^2}|V_{10}|^2 = -\frac{\gamma}{\omega^2}\frac{|d_{01}|^2}{\hbar^2}E^2 = \frac{c}{e^2}E^2. \tag{11.58}$$

It can be seen from Eq. (11.58) that the decay of the state a_0 is determined (as are the level shifts) by a unique sort of quadratic Stark effect with a constant c that depends on γ.

We now consider how a metastable level decays during a slow collision between the atom and a charged particle accompanied by radiation in a forbidden transition. If the perturbing particle (say, an ion) is moving slowly, then at every time t the atom decays in the electric field $E(t)$ of the ion exactly as it would in a static field [Eq. (11.57)]. We can use Eq. (11.58) to find the steady-state transition probability per unit time:

$$w = -\frac{d}{dt}\ln|a_0(t)|^2 = \gamma\left[1 - \frac{\omega}{\sqrt{\omega^2 + 4|V_{01}|^2}}\right]. \tag{11.59}$$

Then the change in the total (up to time t) probability $P(t)$ of radiative decay of the state a_0 is given by the product of the transition rate $\omega(t)$ and the probability $(1 - P)$ of finding the system in state a_0:

$$dP/dt = -w(t)(1-P). \tag{11.60}$$

Given that the perturbation $V_{10}(t)$ is created by an incident charged particle, we integrate Eq. (11.60) subject to the initial condition $P(-\infty) = 0$ and obtain

$$P(\infty) = 1 - \exp\left\{-\gamma \int_{-\infty}^{\infty} \left[1 - \frac{\omega}{\sqrt{\omega^2 + 4|V_{01}(t)|^2}}\right] dt\right\}. \tag{11.61}$$

Equation (11.61) gives the probability of radiative decay of the metastable level in a single collision event. Assuming that the particle trajectory is rectilinear of the form $r^2(t) = \rho^2 + v_0^2 t^2$ (where ρ is the impact parameter of the collision and v_0 is the particle velocity), we find the emission probability to be

$$\sigma = 2\pi \int_0^{\infty} \rho d\rho P(+\infty). \tag{11.62}$$

Substituting the explicit form of the perturbation $V_{10} = \alpha R^{-2}(t)$, where $\alpha = (e/\hbar)|d_{10}|$, into Eq. (11.61) and introducing the dimensionless variable $\rho^2\omega/\alpha$ and the parameter $\xi = (\gamma/\omega)(2\omega\alpha/v_0^2)^{1/2}$, we find that [22, 23]

$$\sigma(\xi) = 4\pi(\alpha/\omega)\Lambda\,[(\gamma/\omega)\sqrt{2\omega\alpha/v_{0}^{2}}]. \tag{11.63}$$

Here the function Λ is given by the integral

$$\Lambda(\xi) = \int_0^{\infty} \frac{dz}{z^3}[1 - \exp(-\xi p(z))], \tag{11.64}$$

where the function $p(z)$ is equal to [cf. Eq. (11.61)]

$$p(z) = \frac{1}{z}\int_{-\infty}^{\infty} dx\left[1 - \frac{1}{\sqrt{1 + z^4/(1+x^2)^3}}\right] \approx \begin{cases} \pi z^3/4, & z \ll 1, \\ B\left(\frac{3}{4}, \frac{3}{4}\right) \approx 1{,}70, & z \gg 1 \end{cases} \tag{11.65}$$

(B is the beta function.)

The emission cross section has simple limiting forms for $\xi \ll 1$ and $\xi \gg 1$. Thus, for $\xi \gg 1$, it follows from Eq. (11.63) that

$$\sigma \simeq \pi^{5/3}\Gamma(1/3)\,\gamma^{2/3}\alpha^{4/3}/v_0^{2/3}\omega^{4/3} \equiv \sigma^{(1)}. \tag{11.66}$$

For $\xi \ll 1$, we obtain

$$\sigma \approx \frac{8\pi}{3} B\left(\frac{1}{4}, \frac{5}{4}\right)\frac{\gamma\alpha^{3/2}}{v_0\omega^{3/2}} \approx 31.1\frac{\gamma\alpha^{3/2}}{v_0\omega^{3/2}} \equiv \sigma^{(2)}. \tag{11.67}$$

The cross section $\sigma^{(1)}$ corresponds to the weak perturbations responsible for the ordinary polarization of an atom in a weak field E that is low compared to the separation ω between the $2s$–$2p$ levels. The cross section $\sigma^{(2)}$ corresponds to fairly strong perturbations where the distortion of the two-level system by the field $E(t)$ is large and cannot be written in terms of a power law. The cross sections (11.66) and (11.67) directly determine the emission rate

$$\gamma_0 = n v_0 \sigma \tag{11.68}$$

or the related lifetime $\tau = \gamma_0^{-1}$ of the atom in the metastable state. When it is multiplied by the transition energy $\hbar\omega_{2s\to 1s}$, Eq. (11.68) obviously gives the intensity of the forbidden line.

The mechanism for polarization excitation of forbidden spectral lines can serve as an important means for studying dense, hot plasmas that contain multiply charged ions with charges $Z \gg 1$. Estimates of the intensity and spectral distribution of these lines are given in [24]. The above scheme for quenching of the metastable $2s$-level by ions in a plasma has been used successfully for interpreting the intensity of lines of hydrogenlike ions in dense plasmas [23].

11.6. Interaction between Inelastic and Polarization Mechanisms for Quenching of Metastable Levels

The destruction (quenching) of metastable levels and the emission of forbidden lines is an interesting example in which one can follow the interaction between the two mechanisms for the destruction of metastable states fairly easily [22]. The mechanisms involve (1) direct inelastic transfer of the electron to a radiative level and (2) inducing a polarization dipole moment for a forbidden transition.

Our analysis of the transition between the inelastic and polarization mechanisms for quenching of metastable levels will be based on second-order perturbation theory. Assuming that

$$a_0(t) = e^{-i\varphi(t)}, \qquad a_0(0) = 1, \tag{11.69}$$

we write an expression for the phase $\varphi(t)$ of the metastable level,

$$\varphi(t) = -i \int_0^t dt' \cdot V_{01}(t') \exp(i\omega t' - \gamma t') \int_0^{t'} dt'' \; V_{10}(t'') \exp(-i\omega t'' + \gamma t''). \tag{11.70}$$

Since the phase shifts are small in this case, we shall assume that the resultant shift $\varphi(t)$ is the sum of the shifts from the individual collisions, i.e., $\varphi(t) = \sum_{k=1}^{N} \varphi_k(t)$, where N is the total number of particles. This makes it possible to express an average over the phase volume of N particles (denoted below by $\langle\ldots\rangle_N$) in terms of the average over the phase volume of a single particle (denoted by $\langle\ldots\rangle_1$). This procedure is carried out with the aid of the following chain of equations:

$$\langle|a_0(t)|^2\rangle_N = \langle\exp(2\,\mathrm{Im}\,\varphi(t))\rangle_N = \left\langle\prod_{k=1}^{N}\exp(2\,\mathrm{Im}\,\varphi_k(t))\right\rangle_N =$$
$$= \{\langle\exp(2\,\mathrm{Im}\,\varphi_1(t))\rangle_1\}^N = \left\{1 - \frac{1}{V}\int d\mathbf{r}\,\times\right.$$
$$\left.\times\,[1-\exp(2\,\mathrm{Im}\,\varphi_1(t))]\right\}^{nV} \approx \exp[-nV(t)], \qquad (11.71)$$

where the notation

$$V(t) = \int d\mathbf{r}\cdot[1-\exp(2\,\mathrm{Im}\,\varphi_1(t))], \qquad \varphi_1(t) = \varphi_1(t, \mathbf{r}). \qquad (11.72)$$

has been introduced for the "collision volume." Here and in the following the velocity v of the perturbing particles is assumed for simplicity to be fixed and equal to some characteristic velocity of a Maxwellian distribution. Thus, the average $\langle...\rangle_1$ reduces simply to integration over a normalization volume V containing N particles ($V \to \infty, N \to \infty, N/V = n = \mathrm{const}$).

Let us assume that the density n of perturbing particles is low enough that the evolution of the collision volume $V(t)$ is determined by sequential pairwise collisions of the emitting atom with perturbing particles. In this case the volume $V(t)$ is expressed in terms of the effective collision frequency Γ by

$$nV(t) = \Gamma t, \qquad (11.73)$$

where

$$\Gamma = nv\int_0^\infty 2\pi\rho d\rho\cdot[1-\exp(2\,\mathrm{Im}\,\varphi_1(\infty))] \equiv nv\sigma(v) \qquad (11.74)$$

and

$$\mathrm{Im}\,\varphi_1(\infty) = -\mathrm{Re}\int_0^\infty \exp(i\omega\tau - \gamma\tau)\,d\tau\int_{-\infty}^{\infty} V_{01}(t)\,V_{10}(t-\tau)\,dt. \qquad (11.75)$$

It is clear that the evolution of the amplitude of the metastable level $a_0(t)$ is determined by the cross section $\sigma(v)$ for pairwise collisions, which is related to the scattering phase $\varphi_1(\infty)$ in an individual collision.

We now calculate Eq. (11.75) using the simplest expression for the interaction potential $V_{10}(t)$ of the atom with a perturbing charged particle,

$$V_{10}(t) = \alpha/R^2(t), \qquad \alpha = d_{01}e/\hbar,\ R^2(t) = \rho^2 + v_0^2t^2, \qquad (11.76)$$

where the constant for the force interaction α is related to the matrix element of the dipole moment d_{01}, while ρ and v_0 are the impact parameter and velocity of the perturbing particle, whose motion is assumed to be rectilinear. Equation (11.75) is easily calculated if we use the condition $\gamma \ll \omega$. Then, we have

$$\operatorname{Im} \varphi_1(\infty) = -\frac{\pi}{2}\left(\frac{\alpha}{\rho v_0}\right)^2\left[\exp(-2\rho\omega/v_0) + \frac{\gamma}{\omega} F\left(2\frac{\rho}{v_0}\omega\right)\right], \tag{11.77}$$

where the function $F(x)$ is given in terms of the exponential integral as

$$F(x) = x\,[e^{-x}\operatorname{Ei}(x) + e^{x}\operatorname{Ei}(-x)] \approx \begin{cases} 2x^{-1}, & x \gg 1, \\ 2x\,[\ln x + C], & x \gg 1 \end{cases} \tag{11.78}$$

($C = 0.577$ is the Euler constant.)

According to Eq. (11.77), the phase $\varphi_1(\infty)$ consists of two parts. The first is determined by a direct inelastic transition from a metastable state [21] and the second is related to the phase shift in elastic scattering and originates in the possibility that the metastable level can be made to decay by inducing a polarization dipole moment, even in a purely elastic collision. We shall show that the first (inelastic) channel is predominant in fast collisions, while the second (polarization) channel predominates in slow collisions.

Going to the dimensionless variable $x = 2\rho\omega/v_0$ in Eq. (11.74) and introducing the parameter $\beta = \omega\alpha/v_0^2$, we obtain

$$\Gamma = \pi^{5/3}\Gamma(1/3)\, n\, \frac{v_0^{1/3}\alpha^{4/3}}{\omega^{2/3}}\, I_{\gamma/\omega}(\beta), \tag{11.79}$$

where

$$I_{\gamma/\omega}(\beta) = \frac{\beta^{-4/3}}{2\pi^{2/3}\Gamma(1/3)} \int_0^\infty dx \cdot x\,\{1 - \exp[-\beta^2 \chi_{\gamma/\omega}(x)]\} \tag{11.80}$$

and

$$\chi_{\gamma/\omega}(x) = 4\pi\left(\frac{\pi e^{-x}}{x^2} + \frac{\gamma}{\omega}\frac{F(x)}{x^2}\right) \approx 4\pi \begin{cases} (\gamma/\omega)\, 2x^{-3}, & x \gg 1, \\ \pi x^{-2}, & x \ll 1. \end{cases} \tag{11.81}$$

For $\beta \to 0$ (fast collisions), we have

$$I_{\gamma/\omega}(\beta) \approx \frac{\pi^{4/3}\beta^{2/3}}{\Gamma(1/3)} \ln(1/\beta), \tag{11.82}$$

which gives the inelastic collision frequency

$$\Gamma_{\text{inel}} = \pi^3 n(\alpha^2/v_0)\ \ln(1/\beta). \tag{11.83}$$

after substitution in Eq. (11.79). A rigorous calculation of the inelastic collision frequency Γ differs from Eq. (11.83) by a factor of $\pi^2/4$ because of our choice of the simplified potential (11.77). For $\beta \to \infty$ (slow collisions), we have $I_{\gamma/\omega}(\beta) \to (\gamma/\omega)^{2/3}$ and substitution into Eq. (11.83) yields the elastic collision frequency

$$\Gamma_{el}=\pi^{5/3}\Gamma(1/3)\,nv^{1/3}(\gamma\alpha^2/\omega^2)^{2/3}. \tag{11.84}$$

This frequency is obviously related to the emission cross section $\sigma^{(1)}$ (11.66) found above.

As the velocity of the perturbing particles increases, therefore, the polarization mechanism for quenching of a metastable state is supplanted by the inelastic mechanism. The transition from one mechanism to the other is determined by the quantity $\beta = \omega a/v_0^2$, the so-called Massey parameter [6], which marks the boundary between slow (adiabatic) and fast collisions.

11.7. Emission Spectra of Thermonuclear Plasmas (Experimental Aspects)

We now consider some basic experimental data from observations of atomic spectra in the hot, rarefied plasmas of tokamaks, a type of controlled thermonuclear fusion experiment. These spectra are produced by free–free (bremsstrahlung), free–bound (radiative recombination), and bound–bound (line emission) transitions in the field of the multiply charged ions of the medium and heavy elements present as impurities in the hydrogen plasma. Observations of these spectra form the basis of the modern spectroscopic diagnostics of hot, rarefied plasmas. These plasmas are characterized by electron temperatures $T_e \sim 1$ keV and densities $n_e \sim 10^{13}$ cm^{-3}. Note that these conditions are also typical of the plasma in the sun's corona.

In spectroscopic diagnostics one studies the intensity and spectrum of the radiation leaving the plasma, i.e., the radiative losses from the plasma. The radiation leaving the plasma carries energy and information on the plasma properties. Both the total (over the entire energy range $\Delta\varepsilon_{ph} = \hbar\Delta\omega$) losses Q^{tot} and their energy (spectral) distribution $\Delta Q/\Delta\varepsilon_{ph}$ are studied.

The total (integrated over all emitted wavelengths) radiative losses provide information on the overall amount of impurities in the plasma, their variation with time, and their distribution over the radius of the plasma column without any kind of detail as to the mechanisms for the radiation, types of emitting ions, etc.

Studies of the spectral distribution of the radiative losses ($\Delta Q/\Delta\varepsilon_{ph}$) make it possible to obtain very much more detailed information. Here we are interested in measurements of $\Delta Q/\Delta\varepsilon_{ph}$ with low resolution $\varepsilon_{ph}/\Delta\varepsilon_{ph} = \omega/\Delta\omega$ (or $\lambda/\Delta\lambda$) over a wide range of energies ε_{ph} (or wavelengths λ) and with high resolution over a narrower range of ε_{ph} (or λ). It is important to note that the radiation in a given interval $\Delta\varepsilon_{ph}$ generally originates from different emission mechanisms (bremsstrahlung, radiative recombination, line emission) and ions of different chemical elements in different stages of ionization. Thus, observed spectra must be interpreted on the basis of knowledge of both the individual mechanisms for emission and the kinetics of the radiating ions, especially their ionization equilibrium. We shall dwell on a few concrete examples related to studies of continuum and discrete (line) emission.

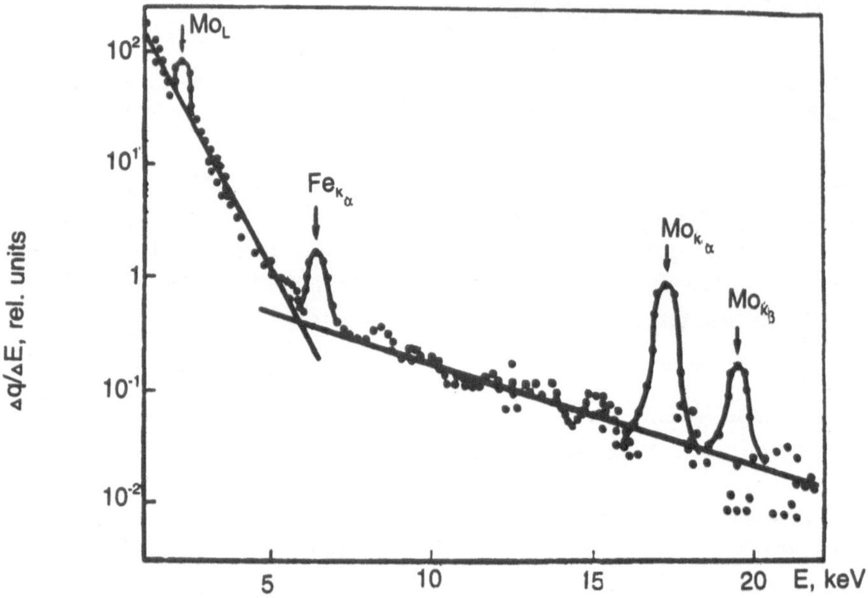

Fig. 11.2. The continuum emission spectrum of a plasma in the TM-4 tokamak ($T_e \sim 0.7$ keV, $n_e = 1.6 \cdot 10^{13}$ cm^{-3}) [25] ($E = \varepsilon_{ph}$, $q = Q$).

Figure 11.2 shows the continuum emission spectrum of a plasma in the TM-4 tokamak [25]. The spectral resolution $\varepsilon_{ph}/\Delta\varepsilon_{ph}$ was of order 10^1–10^2. The continuum spectrum in Fig. 11.2 was formed from bremsstrahlung and radiative recombination of electrons on different kinds of impurity ions. The break in the spectrum around 5–6 keV is apparently caused by highly energetic ("runaway") electrons. Thus, the observed plasma continuum provides information on the electron energy distribution. Furthermore, a comparison of the measured intensity of the continuum with the calculated intensity of bremsstrahlung for a pure hydrogen (electron–proton) plasma reveals that the observed level is far higher than the calculated level [26, 27]. This obviously is evidence of the presence of a fairly large amount of impurities in the plasma. If we assume that the emission of electrons on impurity ions is purely bremsstrahlung, then the degree of "contamination" of the plasma by impurities can be characterized conveniently by the parameter

$$Z_{\mathrm{eff}} = \sum_i n_i Z_i^2 / n_e, \tag{11.85}$$

where n_i is the density of ions with charge Z_i (including protons, for which Z_p = 1). The parameter Z_{eff} also shows up in calculations of plasma conductivity and is an important characteristic of the plasma.

It is difficult to determine Z_{eff} from spectroscopic measurements of the type shown in Fig. 11.2 for two reasons. First, the intensity of bremsstrahlung from electrons on on ions with a core not only includes the ionic charge Z_i, but also an

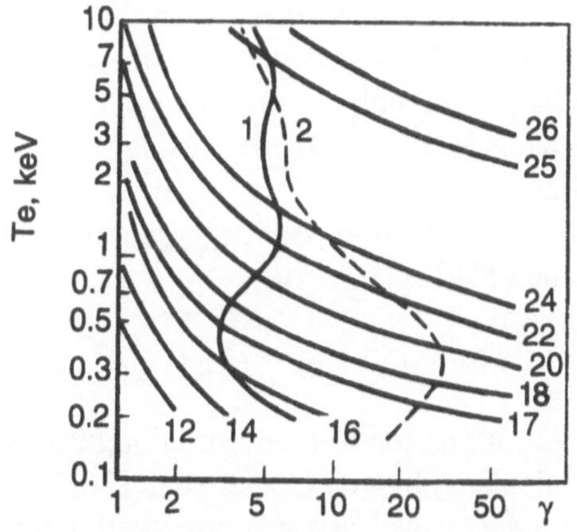

Fig. 11.3. The parameter γ, which determines the amount by which the total emission exceeds bremsstrahlung, for iron ions as a function of T_e [26]. The figures on the curves represent the ionic charge. Curves 1 and 2 are corona model calculations that include and neglect, respectively, dielectronic recombination.

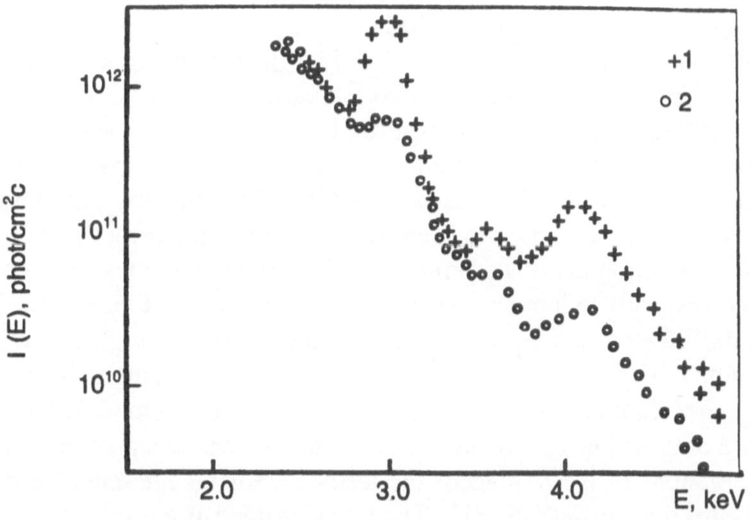

Fig. 11.4. The overall shape of the x-ray spectrum from T-10 [28] with (1) and without (2) added argon ($E = \varepsilon_{ph}$).

effective charge $Z_{eff}(\omega)$, whose value depends on the photon energy and ranges between Z_i and the nuclear charge Z. Second, the observed continuum is made up (as noted above) of radiative recombination, as well as bremsstrahlung, so that in order to determine Z_{eff} using Eq. (11.85), one must know the ratio of the intensities of the two types of radiation.

The excess of the total emission $(\Delta Q/\Delta\varepsilon_{ph})^{tot}$ above bremsstrahlung is characterized by the parameter [25]

$$\gamma = (\Delta Q/\Delta\varepsilon_{ph})^{tot}/(\Delta Q/\Delta\varepsilon_{ph})_{t\to B}. \tag{11.86}$$

The ratio of radiative recombination to bremsstrahlung is obviously equal to $\gamma - 1$. The factor γ has been calculated by von Goeler et al. [26] The intensity of radiative recombination is given in the quasiclassical approximation by

$$\hbar c\,(\Delta Q/\Delta\varepsilon_{ph})_R = 3\cdot 10^{11}\,\frac{n_e n_i}{10^{26}}\,Z_i^2\,(T_e)^{-1/2}\times$$

$$\times\, e^{-\varepsilon_{ph}/T_e}\left[\frac{\xi}{n^3}\frac{I_i}{T_e}e^{I_i/T_e}+\sum_{\nu}\frac{0.027}{T_e}\,\frac{Z_i^2}{(n+\nu)^3}\,\exp\left(\frac{Z_i^2}{(n+\nu)^2}\,\frac{0.0136}{T_e}\right)\right], \tag{11.87}$$

where n is the principal quantum number, ξ is the statistical weight, I_i is the ionization potential of the ground state (keV), and ε_{ph} and T_e are given in keV.

It is clear that the intensity of radiative recombination and, therefore, the value of γ depend strongly on the ionization state of the impurity (or impurities) that determines the values of n and ξ in Eq. (11.87). Figure 11.3 shows γ as a function of T for different ions of iron, as well as its value for corona ionization equilibrium with (curve 1) and without (curve 2) dielectronic recombination [26]. Including the additional recombination channel can obviously change γ by almost an order of magnitude. The data shown in Fig. 11.3 can be used to calculate Z_{eff} which, for example, is close to two for the conditions of Fig. 11.2. The features of the individual emission mechanisms can also be used for diagnostic purposes. Figure 11.4 shows spectra from a plasma with an argon impurity [28]. These spectra contain a jump in the radiative recombination that can be used to follow the dynamics of the He-like Ar ion. In Figs. 11.2 and 11.4 one can also see distinct isolated peaks corresponding to the K- and L-lines of Fe and Mo. Here the very possibility of detecting the K-line of molybdenum is entirely a result of the presence of a high-energy tail in the electron distribution function. It is important to note that each intensity peak corresponds to hundreds (or even thousands) of individual lines emitted by ions with different degrees of ionization, rather than to a single transition. The low resolution of Fig. 11.2 means that it is not possible to separate these masses of lines and higher resolution would be required to observed them. We now consider the structure of these lines using the K-lines of the ferrous group as an example.

Observations of the K-lines of the ferrous elements are one of the bases of modern plasma diagnostics [29–31]. The line is formed as a result of $2p$–$1s$ transitions when an electron is removed from the $1s$-shell. There are three mechanisms for forming a hole in the $1s$-shell:

- direct excitation of an electron from the $1s$ to the $2p$ or (if the latter states are filled) into higher shells;
- dielectronic recombination accompanied by excitation of the $1s$-shell of the electron core; and
- direct ionization from the $1s$-shell.

A complicated excited configuration of the type $1s$ consisting of a large number of individual sublevels is formed as a result of the creation of a hole in the $1s2s^22p^knl$ -shell. Radiative decay of this configuration into the initial $1s^2$-state involves the emission of a large number of individual lines that form the "mass" of lines within a definite spectral interval $\Delta\varepsilon_{ph}$. Another excited configuration radiates a mass of lines into a neighboring spectral interval. The resulting shape of the spectral radiative losses ($\Delta Q/\Delta\varepsilon_{ph}$) near the K-line develops as the result of the addition of different masses of lines from ions with different degrees of ionization emitting from different excited configurations. These types of spectra have been calculated in [32–34].

Figures 11.5 and 11.6 show K-line spectra of iron observed [30, 32] in the ST and T-10 tokamaks ($T_e \sim 1.2$ keV, $n_e \sim 7\cdot10^{13}$ cm^{-3}) and calculated in [34]. The spectral resolution $\Delta\varepsilon_{ph}$ was 20 eV ($\lambda/\Delta\lambda \sim 3\cdot10^2$).

It is clear that the experimentally observed intensity peaks (Fig. 11.5, curve 2) for ions with a low degree of ionization are anomalously high compared with the corona model calculations (Fig. 11.5, curve 3). This indicates that the ionization equilibrium differs from a corona equilibrium. This may be related to the diffusion of impurity ions. In fact, calculations of spectra with a diffusion coefficient typical of tokamaks, $D \sim 10^4$ cm^2/s (Fig. 11.5 ; 11.6, curve 1), give much better agreement with the experiments. Nevertheless, as can be seen from Fig. 11.6, even when diffusion is taken into account, some discrepancies remain between the calculated and observed intensities for the lowest ionization states (Fe XIX). These ions exist only in the edge of the plasma column, where the temperature of the bulk of the electrons is fairly low ($\leq$0.5 keV). Nevertheless, they make a contribution to the intensity of the K-line that is fully comparable to that from the hot central region of the plasma. The explanation for this effect is the existence at the plasma edge of a non-Maxwellian tail in the electron distribution function of the sort shown above in Fig. 11.2. The presence of a small number of runaway electrons does not have a significant influence on the ionization equilibrium, but can make a large contribution to the intensity of the K-lines from the ions at the edge. Calculations [34] based on the experimentally measured (from the continuum spectrum) electron distribution function [35] yield good agreement with the experimentally observed [30] intensities, even in those regions where the emission from ions with a low degree of ionization is important (see Fig. 11.6). In this way, the experimental data [35] on the

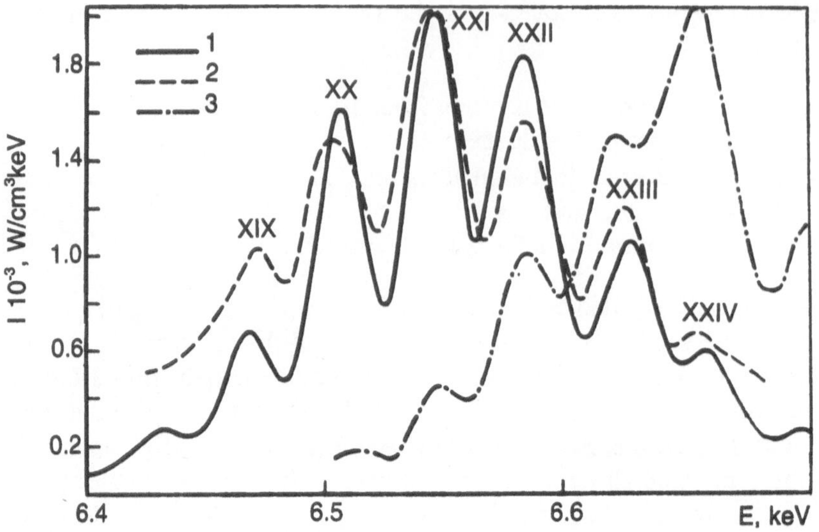

Fig. 11.5. The spectrum of the K-line of iron in a tokamak plasma: (1) calculated for a temperature of $T_e(0) = 1.2$ keV at the plasma center, $D/a^2 = 80\ s^{-1}$ (D is the diffusion coefficient of iron and a is the minor radius of the tokamak), an electron density of $n_e = 7 \cdot 10^{13}\ cm^{-3}$, and a density of iron ions of $n_{Fe} = 10^{12}\ cm^{-3}$; (2) measured in the ST tokamak at Princeton [32]; and, (3) calculated for coronal equilibrium at $T_e = 1.2$ keV. The Roman numerals denote the spectroscopic symbols for the ions ($E = \varepsilon_{ph}$).

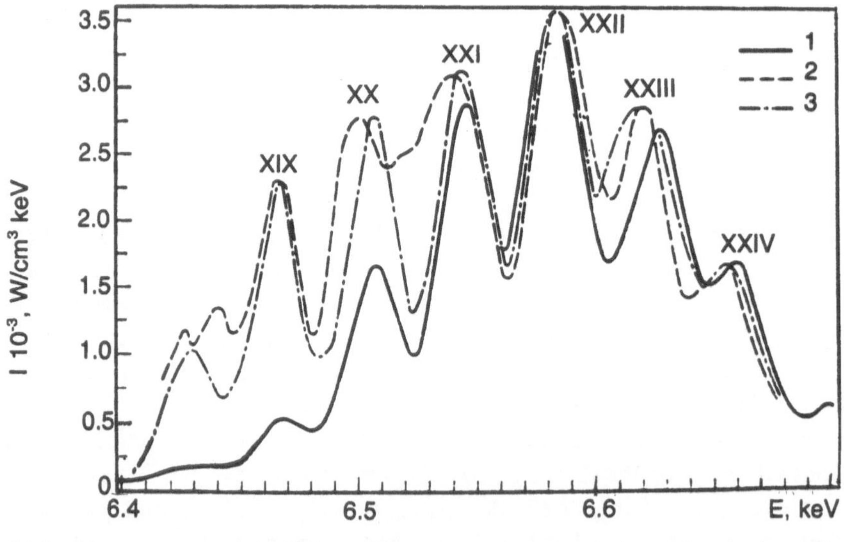

Fig. 11.6. The spectrum of the K-line of iron in a tokamak plasma: (1) calculated including diffusion $D/a^2 = 20\ s^{-1}$ with $n_e = 7 \cdot 10^{13}\ cm^{-3}$ and $T_e(0) = 1.2$ keV; (2) with additional inclusion of a non-Maxwellian distribution of the exciting electrons; and, (3) measured on the T-10 tokamak [30] ($E = \varepsilon_{ph}$).

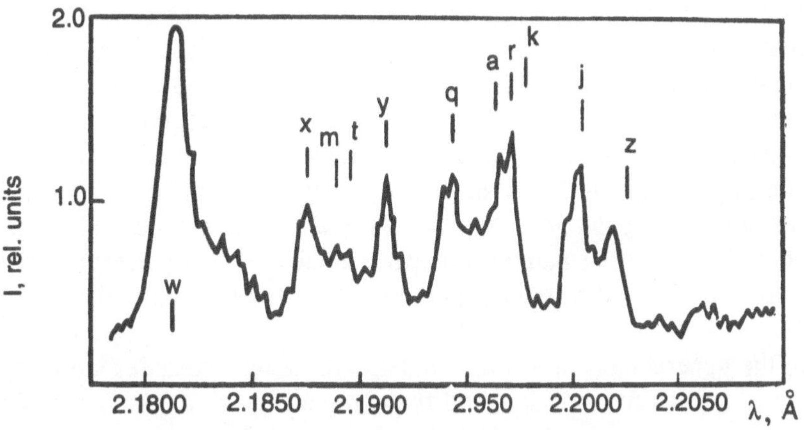

Fig. 11.7. The emission spectrum near the K-line of chromium observed on the T-10 tokamak [31].

continuum emission (bremsstrahlung + recombination) are coupled to experimental observations [30] of the intensity distribution in the K-lines (line emission spectra).

The latest advances in x-ray spectroscopy involve observations of the K-lines of elements in the ferrous group with extremely high resolution $\lambda/\Delta\lambda \geq 10^4$ [29, 31, 33]. These observations apply to a considerably narrower spectral range than in Figs. 11.5 and 11.6 and correspond to the resonance w-line of heliumlike ions with adjacent satellites. Figure 11.7 shows the spectrum $I(\lambda)$ near the resonant w-line of the heliumlike chromium ion observed in the T-10 tokamak [31]. The satellites are formed either as a result of the excitation of ions of the next ionization state (e.g., lithiumlike) or because of dielectronic recombination processes involving the formation of a hole in the $1s^2$-shell (for details, see [33]). It is noteworthy that, despite the noticeable difference in the radiative transition probabilities, all the satellites have comparable intensities. This is explained by the fact that under coronal excitation conditions, when all the radiative decay rates exceed the electron-impact excitation rates $n_e\langle v_e\sigma^{ex}\rangle \equiv k^{ex}$, the intensities of the transitions are proportional to the k^{ex}, which are of the same order for all the satellites. The high spectral resolution in Fig. 11.7 makes it possible to determine the ion temperature of the plasma from the Doppler broadening of the individual lines (the w-line). The ratio of the intensities of different satellites can be used to determine the electron temperature and the ionization equilibrium of the ions in the plasma [31, 33]. Studies of the quasicontinuum spectrum near the w-line (its "pedestal") are of some interest (see Fig. 11.7). It is usually assumed that the pedestal is formed by dielectronic satellites originating in $2p$–$1s$ transitions in Li-like Cr ions with a very highly excited outer electron (a configuration of the type $1s2pnl$ with $n \gg 1$). At the same time, because of the closeness of these segments of the spectrum to the resonance line, polarization effects could also make a certain contribution to this region. No quantitative estimates of this contribution, however, are available.

The close relationship of polarization radiation processes to other processes involving radiation during collisions is thus evident. One characteristic they have in common is a broad spectrum of the radiation emitted by an atom. In fact, in the customary approach to the radiation from atomic electrons the collision and radiative processes can be divided into two stages: a discrete level is initially excited and then (after the passage of the extremely long radiative lifetime $\tau \sim 1/\gamma$) the excited level radiates. The spontaneous emission spectrum that is produced has a narrow (natural) width γ, whose smallness compared to all the other frequencies of interest in the problem makes the above separation of collisional and radiative processes possible.

In the general case, the collisional and radiative processes are inseparable. This is well known for free–free and free–bound transitions (bremsstrahlung and radiative recombination), where the effective collision radii are related to the frequency range of the emission [36]. For line emission, the inseparability of the collision and emission events shows up primarily in the static theory of broadening. Here the range of observed frequency shifts $\Delta\omega$ exceeds the effective frequencies of the particle motion v_0/ρ, so that the emission spectrum of the atom in this region is related only to the static potential for the interaction between the broadening particle and the atom. An example of the inseparability of the emission and collision events is line broadening in a strong (laser) field [37]. Here the collision time ρ/v_0 is comparable to the period of the oscillations of the atom in the laser field $T \sim (d_{12}E_0)^{-1}$, where E_0 is the strength of the laser field and d_{12} is the matrix element of the transition being considered. The interrelation between the collision and emission events during polarization radiation follows especially clearly from the scattering of equivalent photons considered above. In fact, in the case of emission near the resonance frequencies of an atom, we are dealing with the resonance fluorescence of equivalent photons. Here a memory of the collision is retained in the polarization effects during the subsequent emission. After averaging over the polarizations, we can regard the two processes as sequential events. During the nonresonant scattering of equivalent photons responsible for polarization bremsstrahlung, the collisional and radiative interactions form a unified amplitude for a two-photon process that is analogous to the nonresonant scattering of light.

The overwhelming majority of papers on polarization bremsstrahlung are presently devoted to the scattering of fast particles (usually electrons) on atoms. If an atom has many electrons, then the characteristics of this radiation must be expressed in terms of the statistical characteristics of the atom, just as the characteristics of ordinary bremsstrahlung are expressed in terms of the parameters of the Fermi–Thomas potential [17]. This question is of a fundamental character, since it allows us to establish the relationship between polarization bremsstrahlung effects in plasmas and in the statistical theory of the atom. The extent to which the picture of an atom as a plasma cloud will be useful for describing these processes is currently still unclear.

Polarization bremsstrahlung processes involving slow heavy particles represent an interesting problem. In fact, any long-range (say, a van der Waals) interaction is caused by the mutual polarization of the atoms. When the atoms have thermal motion, a variable dipole moment should develop because of this polarization (for different atoms, at least) and lead to polarization emission from these atoms. Although the intensity of this emission is low because of the low frequencies of the relative motion of slow atoms, the mechanism for polarization bremsstrahlung of the atoms is of some interest since this type of radiation does not require excitation of an atomic medium.

Chapter 12

PHOTON–PLASMON TRANSITIONS AS A MECHANISM FOR POLARIZATION QUENCHING OF METASTABLE LEVELS OF ATOMS IN PLASMAS

E. B. Kleiman and I. M. Oiringel

12.1. Introductory Comments

We have given a detailed analysis of the polarization mechanisms for emission during atomic collisions in Chapter 11. The general physical cause of this mechanism is polarization of an atom by the electric field of an incident charged particle. In referring to processes in plasmas, one usually speaks of the action of an individual component of the plasma microfield on the atom. This situation holds in equilibrium plasmas and when the deviations from equilibrium are small [1, 2]. In highly nonequilibrium (turbulent) plasmas, the collective component of the plasma microfield can play an important role [1–3]. Processes for which this component of the field is responsible include spectral line broadening, forbidden line emission, the quenching of metastable states, and recombination. Spectral line broadening by turbulent fields in plasmas has been discussed in detail in [2, 3] and recombination in turbulent plasmas has been discussed in [4].

Polarization of an atom by turbulent fields can cause emission from states that previously were not radiative. An example of this kind of emission (which is especially important in space plasmas) is the $2s$–$1s$ transition in hydrogenlike atoms. Polarization mixing of the $2s$ and $2p$ states by the collective component of the field in a plasma leads to the appearance of a dipole moment for the $2s$–$1s$ transition. This mechanism is realized if the characteristic frequencies of the turbulent fields are not close to the frequency of the $2s$–$2p$ transition. When these frequencies are close, the metastable $2s$-state is quenched by another mechanism: $2s$–$2p$ transitions under the influence of the turbulent field with subsequent radiative decay of the $2p$

level. These mechanisms are competitive with the corresponding mechanisms for quenching of the $2s$ level in collisions of atoms with charged particles [2].

In the preceding discussion we have examined polarization emission of incident neutral particles. In this chapter we examine a realistic case where the incident particle is a plasmon which polarizes the atom and causes emission from metastable levels.

When considering radiative processes in turbulent plasmas, it is convenient to use the concept of plasmons as quantizing the long-wavelength part of the field in the medium. This approach to various problems in the theory of radiation in media has been developed by various authors [5–8]. It makes it possible to use the methods of quantum electrodynamics to account for all the types of waves which can be excited in a plasma. In the framework of this approach, polarization emission of forbidden lines can be treated as a two-quantum process involving the emission of a photon and a plasma wave quantum (plasmon) [8, 9].

12.2. Probabilities of Two-Quantum Transitions in Plasmas

In order to find the probabilities of two-quantum atomic transitions in plasmas it is convenient to use the S-matrix method, which has been developed for application to quantum electrodynamical processes in media [5–7]. As opposed to the standard S-matrix apparatus in vacuum [10], where only photons contribute to the operator 4-potential of the electromagnetic field A_μ ($\mu = 1, 2, 3, 4$), in our case we must include all the kinds of waves that can propagate in the medium in A_μ. The operator 4-potential in an arbitrary anisotropic medium (in the Schrödinger representation) has the form [7]

$$\mathbf{A}(\mathbf{r}) = \mathbf{A}^{(\sigma)}(\mathbf{r}) + \mathbf{A}^{(l)}(\mathbf{r}), \qquad A_4 = iA_0 = 0, \tag{12.1}$$

where

$$\mathbf{A}^{(\sigma)}(\mathbf{r}) = \sum_{\mathbf{k},\sigma} \left(\frac{2\pi\hbar c^2 k\, \partial\omega_{\mathbf{k},\sigma}/\partial k}{L^3 \omega^2_{\mathbf{k},\sigma} e_{\mathbf{k},\sigma,\alpha} \varepsilon_{\alpha\beta} e^*_{\mathbf{k},\sigma,\beta}} \right)^{1/2} (c_{\mathbf{k},\sigma} e^{i\mathbf{kr}} + c^+_{\mathbf{k},\sigma} e^{-i\mathbf{kr}})\, e_{\mathbf{k},\sigma} \tag{12.2}$$

and

$$\mathbf{A}^{(l)}(\mathbf{r}) = \sum_{\mathbf{k},l} \left(\frac{4\pi\hbar c^2 k^2}{L^3 \omega^2_{\mathbf{k},l} k_\alpha k_\beta \partial\varepsilon_{\alpha\beta}/\partial\omega_{\mathbf{k},l}} \right)^{1/2} (c_{\mathbf{k},l} e^{i\mathbf{kr}} + c^+_{\mathbf{k},l} e^{-i\mathbf{kr}}) \frac{\mathbf{k}}{k}. \tag{12.3}$$

Here $\mathbf{A}^{(\sigma)}(\mathbf{r})$ corresponds to ordinary waves and $\mathbf{A}^{(l)}(\mathbf{r})$ to longitudinal waves, L^3 is the volume occupied by the electromagnetic field, and $\varepsilon_{\alpha\beta} = \varepsilon_{\alpha\beta}(\mathbf{k}, \omega)$ is the dielectric permittivity tensor of the medium. In Eq. (12.2) the frequencies of the ordinary waves $\omega_{\mathbf{k},\sigma}$ and the directions $\mathbf{e}$ of the polarization vectors are found by solving the following system of algebraic equations:

$$[c^2 (k_\alpha k_\beta - k^2\delta_{\alpha\beta}) + \omega^2_{\mathbf{k},\sigma} \varepsilon_{\alpha\beta}(\mathbf{k}, \omega_{\mathbf{k},\sigma})]\, e_{\mathbf{k},\sigma,\beta} = 0, \tag{12.4}$$

where the subscripts α and β run over the values 1, 2, 3. Equating the determinant of the system (12.4) to zero, we arrive at the Fresnel equation

$$|c^2 (k_\alpha k_\beta - k^2 \delta_{\alpha\beta}) + \omega^2_{\mathbf{k},\sigma} \varepsilon_{\alpha\beta} (\mathbf{k}, \omega_{\mathbf{k},\sigma})| = 0, \qquad (12.5)$$

which specifies the frequency $\omega_{\mathbf{k},\sigma}$ as a function of $\mathbf{k}$ for every fixed value of the wave vector $\mathbf{k}$:

$$\omega_{\mathbf{k},\sigma} = \omega_{\mathbf{k},\sigma} (k, \mathbf{k}/k). \qquad (12.6)$$

The subscript σ takes integer values $\sigma = 1, 2, \ldots$ which denote the different roots $\omega_{k,1}, \omega_{k,2}, \ldots$ of Eq. (12.6). For longitudinal waves the frequencies $\omega_{\mathbf{k},l}$ ($l = 1, 2, \ldots$) are determined as functions of $\mathbf{k}$ by solving the equation

$$\varepsilon^l = \frac{k_\alpha k_\beta}{k^2} \varepsilon_{\alpha\beta} (\mathbf{k}, \omega_{\mathbf{k},l}) = 0. \qquad (12.7)$$

The creation and annihilation operators for ordinary $c_{\mathbf{k},\sigma}{}^+$, $c_{\mathbf{k},\sigma}$ and longitudinal $c_{\mathbf{k},l}{}^+$, $c_{\mathbf{k},l}$ quanta satisfy the standard commutation relations,

$$[c_{\mathbf{k}\sigma} c^+_{\mathbf{k}'\sigma'}] = \delta_{\mathbf{k},\mathbf{k}'} \delta_{\sigma,\sigma'}, \ [c_{\mathbf{k},l} c^+_{\mathbf{k}',l'}] = \delta_{\mathbf{k},\mathbf{k}'} \delta_{l,l'}, \qquad (12.8)$$

where the square brackets denote the commutators of the corresponding operators. In solving specific problems it is convenient to transform to the interaction representation by replacing the factor $e^{i\mathbf{kr}}$ in Eqs. (12.2) and (12.3) by $\exp[i(\mathbf{kr} - \omega_{k,\sigma} t)]$ and $\exp[i(\mathbf{kr} - \omega_{k,l} t)]$, respectively. It should be noted that the gauge chosen here for the potentials, where the scalar potential $A_0 = 0$, is more convenient for the problems to be considered below than the other gauges used in quantum electrodynamics [10].

In our studies of the simultaneous emission of two different quanta of the field in a medium, we shall follow Akhiezer and Berestetskii [11] and proceed from the second-order scattering matrix

$$S^{(2)} = \frac{e^2}{\hbar^2 c^2} N \int \{\hat{\bar{\psi}} (x_1) \hat{A} (x_1) S_c^{(e)} (x_1, x_2) \hat{A} (x_2) \psi (x_2)\} \, dx_1 dx_2, \qquad (12.9)$$

where $\hat{\psi}(x_2)$ and $\hat{\bar{\psi}}(x_1) = \hat{\psi}^+(x_1)\beta$ are the operators of the electron–positron field, $\hat{A}(x) = \gamma_\mu A_\mu(x)$; $\gamma_\mu = \{\gamma, i\beta\}$ are the Dirac matrices, $x = \{\mathbf{r}, ict\}$, N is the normal product of the operators, and $S_c{}^{(e)}(x_1, x_2)$ is the Green function for the Dirac equation for an electron in an external field, which is given by

$$S_c^{(e)} (x_1, x_2) = \begin{cases} \sum_s \psi_s^{(+)} (x_1) \psi_s^{(+)} (x_2), & t_1 > t_2, \\ -\sum_s \psi_s^{(-)} (x_1) \psi_s^{(-)} (x_2), & t_1 < t_2. \end{cases} \qquad (12.10)$$

$$\psi_s^{(+)}(x) = \psi_s^{(+)}(\mathbf{r})\, e^{-i\omega_s^{(+)}(t)}, \quad \omega_s^{(+)} > 0;$$

Here

$$\psi_s^{(-)}(x) = \psi_s^{(-)}(\mathbf{r})\, e^{-i\omega_s^{(-)}t}, \quad \omega_s^{(-)} < 0 \tag{12.11}$$

is the solution of the Dirac equation with positive and negative frequencies in an external field.

The differential probability for the simultaneous emission of two quanta ($\mathbf{k}_1$, ω_1 and $\mathbf{k}_2$, ω_2) has the form [10]

$$dw = \frac{2\pi}{\hbar} |M|^2 \delta(E_i - E_f - \hbar\omega_1 - \hbar\omega_2) \frac{d\mathbf{k}_1 d\mathbf{k}_2}{(2\pi)^6}, \tag{12.12}$$

where M is the matrix element of this process and E_i and E_f are the energies of the initial and final electronic states, respectively. With the aid of the scattering matrix S, it is easy to obtain an expression for the matrix elements $M_{\sigma\sigma'}$, $M_{l\sigma}$, and $M_{ll'}$ of different two-quantum processes in a medium:

(a) σσ'-radiation,

$$dw_{\sigma,\sigma'} = \frac{k_1^3 k_2^3 dk_1 d\Omega_{\mathbf{k}_1,\sigma} d\Omega_{\mathbf{k}_2,\sigma'} \dfrac{\partial\omega_{\mathbf{k}_1,\sigma}}{\partial k_1} (N_{\mathbf{k}_1,\sigma}+1)(N_{\mathbf{k}_2,\sigma}+1) D^{\sigma\sigma'}}{(2\pi)^3 (\varepsilon_{\alpha\beta}(\mathbf{k}_1,\omega_{k_1,\sigma})\, \overset{*}{e}_{\mathbf{k},\sigma,\alpha} e_{\mathbf{k},\sigma,\beta})(\varepsilon_{\alpha'\beta'}(\mathbf{k}_2,\omega_{k_2,\sigma})\, \overset{*}{e}_{\mathbf{k}_2,\sigma',\alpha'} e_{\mathbf{k}_2,\sigma',\beta'})},$$

$$D^{\sigma,\sigma'} = |(a_{\alpha\beta})_{2,1} \overset{*}{e}_{\mathbf{k}_1,\sigma,\alpha} e_{\mathbf{k}_2,\sigma',\beta}|^2,$$

$$(a_{\alpha\beta})_{21} = \sum_s \left[\frac{(d_\alpha)_{fs}(d_\beta)_{s1}}{E_l - E_s - \hbar\omega_{\mathbf{k}_1,\sigma}} + \frac{(d_\beta)_{fs}(d_\alpha)_{s1}}{E_l - E_s - \hbar\omega_{\mathbf{k}_2,\sigma'}} \right], \tag{12.13}$$

where the sum is taken over all states of an atomic electron, including the continuum,

(b) σl-radiation,

$$dw_{\sigma,l} = k_1^3 k_2^4 dk_1 d\Omega_{\mathbf{k}_1,\sigma} d\Omega_{\mathbf{k}_2,l} \frac{\partial\omega_{\mathbf{k}_1\sigma}}{\partial k_1} (N_{\mathbf{k}_1\sigma}+1)(N_{\mathbf{k}_2 l}+1) \times$$

$$\times D^{\sigma l} \left\{ 4\pi^3 (\varepsilon_{\alpha\beta}(\mathbf{k}_1\omega_{\mathbf{k}_1\sigma})\, \overset{*}{e}_{\mathbf{k}_1,\sigma,\alpha} e_{\mathbf{k}_1,\sigma,\beta}) \left(k_{2\alpha'} k_{2\beta'} \frac{\partial\varepsilon_{\alpha\beta}(\mathbf{k}_2,\omega_{\mathbf{k}_2,l})}{\partial\omega_{\mathbf{k}_2 l}} \right) \frac{\partial\omega_{\mathbf{k}_2 l}}{\partial k_2} \right\}^{-1},$$

$$D^{\sigma l} = \left| (a_{\alpha\beta})_{21} \frac{k_{2\alpha}}{k_2} \overset{*}{e}_{\mathbf{k}_1\sigma,\beta} \right|^2, \tag{12.14}$$

where $(a_{\alpha\beta})_{21}$ is given by that in Eq. (12.13) with the substitution $\omega_{\mathbf{k}\sigma} \to \omega_{\mathbf{k}l}$, and

(c) ll'-radiation,

$$dw_{ll'} = k_1^4 k_2^4 dk_1 d\Omega_{\mathbf{k}_1 l} d\Omega_{\mathbf{k}_2 l'} (N_{\mathbf{k}_1,l} + 1)(N_{\mathbf{k}_2,l'} + 1) D^{ll'} \times$$
$$\times \left\{ 2\pi^3 \left(k_{1\alpha} k_{1\beta} \frac{\partial \varepsilon_{\alpha\beta}(\mathbf{k}_1, \omega_{\mathbf{k}_1 l})}{\partial \omega_{\mathbf{k}_1 l}} \right) \left(k_{2\alpha} k_{2\beta} \frac{\partial \varepsilon_{\alpha\beta}(\mathbf{k}_2, \omega_{\mathbf{k}_2 l'})}{\partial \omega_{\mathbf{k}_2 l'}} \right) \frac{\partial \omega_{\mathbf{k}_2 l'}}{\partial k_2} \right\}^{-1}, \tag{12.15}$$
$$D^{ll'} = |(a_{\alpha\beta})_{21} k_{2\beta} k_{1\alpha} / k_2 k_1|^2,$$

where $(a_{\alpha\beta})_{21}$ is given by that in Eq. (12.13) with the substitutions $\omega_{\mathbf{k}\sigma} \to \omega_{\mathbf{k}1l}$ and $\omega_{\mathbf{k}2\sigma} \to \omega_{\mathbf{k}2l}$. In Eqs. (12.13)–(12.15), $N_{\mathbf{k},\sigma}$ and $N_{\mathbf{k},l}$ are the numbers of ordinary and longitudinal quanta, respectively, $d\Omega_{\mathbf{k}\sigma}$ and $d\Omega_{\mathbf{k}l}$ are the solid angles of the vector $\mathbf{k}$ for the corresponding quanta, and $(\mathbf{d}_\alpha)_{mn}$ are the matrix elements of the components of the dipole moment of the atom corresponding to the $n \to m$ transition.

Equations (12.13)–(12.15) are the solution in the dipole approximation to the problem of the simultaneous emission of two atomic quanta in an arbitrary anisotropic medium. In order to find the probabilities of specific two-quantum processes in plasmas, one must substitute expressions for the dielectric permeability of the plasma in these formulas as well as the ω_{kl} and polarization vectors for the waves. General expressions for the probabilities of two-quantum radiation processes in isotropic plasmas are given in [9].

Besides the processes discussed above, processes involving the scattering of quanta also take place. These processes are also second-order processes, so they are described by the scattering matrix $S^{(2)}$. The differential effective cross sections for scattering of different quanta by atoms in a medium are given in [9]. We shall not examine these processes further here.

Two-quantum processes in a medium have a number of features which distinguish them from vacuum processes. First, in a medium new types of processes, σl and ll', take place. Second, the characteristic frequencies of the quanta of a medium are generally low compared to optical transition frequencies in atoms, so that nearby levels may play an important role in the sums over intermediate states in Eqs. (12.13)–(12.15). Finally, stimulated processes, in which the numbers of quanta (especially $N_{\mathbf{k},l}$) are large, play an important role in a medium.

12.3. The 2s–1s Photon–Plasmon Transition in Hydrogenlike Atoms

As pointed out above, the 2s–1s transition in hydrogenlike atoms is an important example of a forbidden transition. Quenching of the metastable 2s level in a plasma is determined by the following processes: (a) two-photon 2s–1s transitions, (b) collisions with charged plasma particles, and (c) 2s–1s photon–plasmon processes.

The first of these processes has been thoroughly studied and its probability is 8.226 s^{-1} [10]. Collisional processes have been studied in detail (see [12], for example, and the literature cited there.) Their probabilities depend substantially on the velocities and densities of the particles in the plasma. Emission of a photon and a plasma wave in a 2*s*–1*s* transition is of interest in turbulent plasmas. The probability of this process can become large as a result of two circumstances. Since the plasma temperature is usually considerably higher than the energy of the quanta corresponding to the plasma waves, this process is stimulated. This increases its probability far above that of a spontaneous transition. Furthermore, the hydrogen atom has close-lying levels and this also increases the probability of this photon–plasmon process.

Let us consider a two-quantum 2*s*–1*s* transition in which a photon and Langmuir plasmon (quantum) are emitted [13]. Using Eq. (12.14) for the dielectric permeability tensor of an isotropic plasma and the dispersion relation for the transverse and Langmuir waves [5], one can easily obtain the probability of the two-quantum transition. Here we use the fact that the main contribution is from a virtual transition through the intermediate $2p_{1/2}$ state. As a result, we obtain [13]

$$dw_{tl} = \frac{2e^4 (\omega^t)^2 (\omega^l)^2 f_1 \tilde{f}_2 N_{k_l} dk_t}{\pi m^2 c^2 v_{Te}^2 v_{\rm ph}^l \left[(\omega_* - \omega^l)^2 + \Gamma^2/4\right]} . \qquad (12.16)$$

Here e^2/mc^2 is the classical electron radius, v_{Te} is the electron thermal speed, $v_{\rm ph}{}^l = \omega^l/k_l$ is the phase velocity of the Langmuir waves, ω^t and ω^l are the frequencies of the photon and Langmuir plasmon, respectively, ω_* is the Lamb frequency, f_1 is the oscillator strength of the $1s$–$2p_{1/2}$ transition, $\tilde{f}_2 = f_2\omega^t/\omega_*$ (where f_2 is the oscillator strength of the $2s$–$2p_{1/2}$ transition), and Γ is the width of the $2p_{1/2}$ level. Expressing dk_t in terms of dk_l using the conservation of energy,

$$\omega_{21} = \omega_{k_t}^t + \omega_{k_l}^l, \qquad (12.17)$$

where ω_{21} is the frequency of the 2*s*–1*s* transition, while k_t and k_l are the wave numbers of the photon and plasmon, and taking the integral of Eq. (12.16), we obtain

$$w_{tl} = \frac{12\pi e^4 \omega_{21}^2 f_1 \tilde{f}_2}{m^2 c^3 \left[(\omega_* - \omega^l)^2 + \Gamma^2/4\right]} \frac{U^l}{\hbar\omega_*} . \qquad (12.18)$$

In this equation U^l is the energy density of the Langmuir waves, which is given by

$$U^l = \frac{1}{2\pi^2} \int \hbar\omega_{k_l}^l N_{k_l} k_l^2 dk_l . \qquad (12.19)$$

We shall now illustrate the possible role of this transition with the aid of some examples. It appears to be of greatest interest in space plasmas. Thus, for a plasma with an electron density $n_e \sim 10^4$ cm^{-3} and an electron temperature $T_e \sim 10^4$ K (planetary nebulae), Eq. (12.18) yields

$$w_{tl} \approx 2 \cdot 10^{10} U^l, \qquad (12.20)$$

where w_{tt} is in seconds.

It is known [14] that under these conditions the competing process for quenching of the $2s$ level is two-photon emission. On comparing the probabilities of these processes, we find that the photon–plasmon transition is predominant for energy densities $U^l > 4 \cdot 10^{-10}$ erg/cm^3. Since the thermal energy in nebulae is

$$U_T^l \approx n_e T_e \approx 10^{-8}\ \text{erg/cm}^3, \qquad (12.21)$$

these densities U^l appear to be attainable. The predominance of the photon–phonon transition leads to a reduction in the population of the $2s$ level and to an attenuation of the continuum spectrum of nebulae. Thus, an interesting prospect has arisen for diagnostics of the plasma turbulence from the continuum emission.

Let us estimate the reduction in the intensity of the $2s$–$1s$ two-photon emission. Following [14], we shall assume that the hydrogen atoms are excited during photoionization and subsequent recombination events. We include the $2s$–$1s$ photon–plasmon transition in addition to the usual radiative and collisional transitions. The system of balance equations for the level populations has the form [11, 15]

$$\begin{aligned} n_{2s}(w_{2s,1s} + c_{2s,2p}) &= x_1 R + n_{2p_{1/2}} c_{2p_{1/2},2s} + n_{2p_{3/2}} c_{2p_{3/2},2s}, \\ n_{2p_{3/2}}\left(\frac{w_{2p_{3/2},1s}}{N_1} + c_{2p_{3/2},2s}\right) &= x_2 R + n_{2s} c_{2s,2p_{3/2}}, \\ n_{2p_{1/2}}\left(\frac{w_{2p_{1/2},1s}}{N_2} + c_{2p_{1/2},2s}\right) &= x_3 R + n_{2s} c_{2s,2p_{1/2}}. \end{aligned} \qquad (12.22)$$

Here n_{2s}, $n_{2p3/2}$, and $n_{2p1/2}$ are the populations of the corresponding sublevels of the second level, $w_{2s,1s} = w_{2s,1s}{}^{tt'} + w_{2s,1s}{}^{tl}$ is the probability of the $2s$–$1s$ transition, $c_{2s,2p} = c_{2s,2p1/2} + c_{2s,2p3/2}$ is the probability of the $2s$–$2p$ transition owing to collisions with charged particles, and $w_{2p3/2,1s}$ and $w_{2p1/2,1s}$ are the conventional probabilities of optical transitions with emission of photons. In addition, x_1R, x_2R, and x_3R are the numbers of atoms falling into the $2s$, $2p_{3/2}$, and $2p_{1/2}$ states, respectively, after recombination and cascading transitions per cubic centimeter per second (R is the number of recombinations), while N_1 and N_2 are the average number of scatterings of the corresponding photons. Introducing these numbers allows us to take multiple scattering of photons prior to their departure from the medium into account. It is known [14] that the average number of scattering events is determined by the optical thickness of the emitting region. For the close-lying $2p_{1/2}$ and $2p_{3/2}$ levels being considered here, N_1 and N_2 are roughly the same, so for simplicity we shall assume that $N_1 = N_2 = N$. On solving the system of Eqs. (12.22), it is easy to obtain an expression for the population of the $2s$ level [9]. It is fairly complicated in general. Here we give an expression for the numbers of two-photon transitions $Z_{2s,1s} = n_{2s} w_{2s,1s}{}^{tt'}$ in some limiting cases. Thus, when the conditions $w_{2p3/2,1s} \gg Nc_{2p3/2,2s}$ and $w_{2p1/2,1s} \gg Nc_{2p1/2,2s}$ are satisfied, we have [9]

$$Z_{2s,1s} = \left(1 + \frac{w^{tl}_{2s,1s} + c_{2s,2p}}{w^{tt'}_{2s,1s}}\right)^{-1} x_1 R. \tag{12.23}$$

Let us introduce the attenuation factor α for the intensity of the two-photon emission. Then, under these conditions we have

$$\alpha = \left(1 + \frac{w^{tl}_{2s,1s} + c_{2s,2p}}{w^{tt'}_{2s,1s}}\right)\left(1 + \frac{c_{2s,2p}}{w^{tt'}_{2s,1s}}\right)^{-1}. \tag{12.24}$$

In the opposite limit of $w_{2p3/2,1s} \ll Nc_{2p3/2,2s}$ and $w_{2p1/2,1s} \ll Nc_{2p1/2,1s}$, we find α to be given by

$$\alpha = 1 + w^{tl}_{2s,1s}/w^{tt'}_{2s,1s}. \tag{12.25}$$

Photon–plasmon transitions can even be important in stellar chromospheres (including that of the sun), where we might also expect intense plasma turbulence. Taking $n_e \sim 10^{10}$ cm^{-3} and $T_e \sim 10^4$ K for purposes of an estimate and using Eq. (12.18), we find that

$$w_{tl} \approx 7 \cdot 10^{11} U^l \ \mathrm{s}^{-1}. \tag{12.26}$$

It is easy to see that under these conditions, the probability of photon–plasmon processes is substantially higher than that of two-photon processes. The latter, however, are unimportant under chromospheric conditions. Here the competing process is a $2s$–$2p$ transition owing to collisions with protons [12, 16], which has a probability of

$$c_{2s,2p} = 5{,}3 \cdot 10^{-4}\, n_e \ \mathrm{s}^{-1}.$$

It follows from this that if $U^l > 8 \cdot 10^{-6}$ erg/cm^3, then the photon–plasmon transition is more efficient than collisions and it controls the population of the metastable $2s$-level.

12.4. The 2^3s_1–1^3s_1 Photon–Plasmon Transition in Positronium

Positronium is a structural analog of a hydrogenlike atom, although there are some substantial differences between them. First, the reduced mass of a positron is $1/2m$, so that the energy of its nonrelativistic levels is equal to half that for a hydrogen atom and the characteristic atomic distances are doubled. The equality of the masses of the electron and positron also has an effect on the fine structure of the levels of positronium [10]. Second, an electron and positron can annihilate one another and emit photons. For states with nonzero orbital angular momentum, the probability of this process is low. But, in the *s*-state it may exceed the probability of ordinary radiative transitions.

The properties of the radiation produced when an electron–positron pair is annihilated depend to a substantial degree on the spin states of the colliding particles [10]. There are two types of states for positronium: singlet and triplet. According to the symmetry conditions, annihilation from a singlet state is possible only with the emission of an even number of photons, and from a triplet state, only with an odd number of photons (at least three). As noted above, the cross section and character of the annihilation depend strongly on the mutual orientation of the spins of the annihilating particles; thus, the lifetimes are also different. Two-photon annihilation from singlet states of positronium is characterized by a lifetime of

$$\tau_{2\gamma}=1.25\cdot 10^{-10}\, n^3\ \mathrm{s}^{-1}, \tag{12.27}$$

and three-photon annihilation from triplet states has a lifetime of

$$\tau_{3\gamma}=1.4\cdot 10^{-7}\, n^3\ \mathrm{s}^{-1}, \tag{12.28}$$

where n is the principal quantum number.

As a rule, the probabilities of cascade radiative transitions are greater than the probability of annihilation, but the orthopositronium atom has a metastable 2^3s_1 state. The probability of a two-photon radiative transition from this state into the 1^3s_1 states, which equals $1.8\cdot 10^{-3}$ s^{-1}, is very much less than the probability of three-photon annihilation directly from the 2^3s_1 level, which equals $8.9\cdot 10^5$ s^{-1}. Therefore, it is usually assumed that orthopositronium is annihilated predominantly from the 2^3s_1 state. In plasmas, however, there is a competing process for quenching of the metastable level, namely, the 2^3s_1–1^3s_1 photon–plasmon transition. We shall now find an expression for the probability of this transition with emission of a photon and a Langmuir plasmon. Proceeding as in our derivation of Eq. (12.18) and noting that the main contribution to this probability is from virtual transitions through the $2^3p_{0,1,2}$ states (fine structure levels), we obtain

$$w_{tl}=\sum_{i=0}^{2} w_{tl}^{(i)}, \tag{12.29}$$

where

$$w_{tl}^{(i)}=\frac{2\pi e^4\omega^2 g_i f_i f_i'}{m^2c^3\,[(\omega_i-\omega^l)^2+\Gamma_i^2/4]}\,\frac{U^l}{\hbar\omega_i}. \tag{12.30}$$

Here $g_i = (2J_i + 1)/2(2J + 1)$; J and J_i are the angular momenta of the 2^3s_1 and $2^3p_{0,1,2}$ states, respectively; ω is the frequency of the 2^3s_1–1^3s_1 transition; ω_i are the frequencies of the 2^3s_1–$2^3p_{0,1,2}$ transitions; f_i and f_i' are the oscillator strengths of the $2^3p_{0,1,2}$–1^3s_1 and 2^3s_1–$2^3p_{0,1,2}$ transitions; and Γ_i are the widths of the $2^3p_{0,1,2}$ levels. Referring to [17] for the details, we list here the numerical values of the different quantities in Eq. (12.30):

$$\omega=0.75\cdot10^{16}\ \mathrm{s}^{-1},\quad \omega_0=4\cdot10^{11}\ \mathrm{s}^{-1},\quad \omega_1=3\cdot10^{11}\ \mathrm{s}^{-1},$$
$$\omega_2=2\cdot10^{11}\ \mathrm{s}^{-1},\quad f_1'\approx f_2'=5\cdot10^{-5},\quad f_0'=2\cdot10^{-5}.$$

Using these values, for $\omega_i \gg \omega^l$ we obtain

$$w_{tl}\approx6\cdot10^{6}U^l\ \mathrm{s}^{-1}. \tag{12.31}$$

A comparison of Eqs. (12.28) and (12.31) yields the conditions under which the photon–plasmon transition will predominate over three-photon annihilation:

$$U^l>0.1\ \mathrm{erg/cm^3} \tag{12.32}$$

In the opposite case of $\omega^l \gg \omega_i$, we have

$$U^l\gg0.1\,(\omega^l/\omega_2)^2\ \mathrm{erg/cm^3}. \tag{12.33}$$

The inequalities (12.32) and (12.33) show that the 2^3s_1–1^3s_1 photon–plasmon transition can control the population of the 2^3s_1 level. Thus, searches for the longer-lived state of positronium may be used as a diagnostic of plasma turbulence.

The importance of the process described above is evident in the following circumstances. Electron–positron annihilation in plasmas proceeds most often through a stage in which positronium atoms are formed [18]. Thus, only 25% of positrons with an initial energy of 100 MeV are annihilated in a "free" state. The rest are slowed down to thermal energies and form positronium atoms with plasma electrons, of which 75% are in the ortho-state and 25% are in the para-state. More than 50% of the positronium atoms are formed in an excited state. Thus, research on the photon–plasmon decay channel for the metastable 2^3s_1 state is of considerable interest.

In conclusion, we emphasize again that there are a large number of atomic processes, besides those discussed in this chapter, in which the collective component of the plasma microfield plays an important part. Two areas for research in this group of problems can be distinguished.

The first area [19] involves the development of a general theory of processes in which the collective interactions of atomic systems and charged particles play a dominant role in a kinetic theory approach. In the second area, specific atom–plasma processes are to be studied. Thus, the influence of collective effects on spectral line broadening [2] and on electron–ion recombination in plasmas [4, 20] has been studied. The role of polarization plasma effects in dielectronic recombination [21] and the influence of these effects on charge exchange processes in plasmas [22] have also been studied.

It should be noted that we have not considered a number of questions in the framework of the theory of photon–plasmon processes discussed above. First of all, we must mention the formation of satellite lines that are shifted relative to the transition frequencies by the Langmuir frequency [1] and the related diagnostics for superdense plasmas [23]. The scattering of quanta by metastable atoms is of considerable interest. Thus, a situation in which hydrogen atoms are acted on by fluxes of different kinds of radio emission is extremely widespread in astrophysical plasmas [24]. The effect of these radio frequency quanta on the polarization quenching of the metastable level of the hydrogen atom [1] is not fully clear at present.

One must assume that further progress in research in this area will ultimately provide the additional information needed to interpret the radiative energy losses from laboratory and space plasmas and will aid in resolving some fundamental questions about the kinetics of partially ionized, nonequilibrium plasmas.

Chapter 13

NONPOLARIZATION BREMSSTRAHLUNG OF ELECTRONS IN STATIC POTENTIALS

V. I. Gervids and V. I. Kogan

13.1. Introduction

This chapter is devoted to a discussion of the basic analytic and a few numerical results on the effective cross sections for bremsstrahlung, first in Coulomb (Secs. 13.2–13.4) and then in non-Coulomb static fields (the field of a rigid many-electron atom or ion; Secs. 13.5–13.7). In Sec. 13.6 we analyze the interrelation of the quantum mechanical and classical descriptions of bremsstrahlung, and discuss the possibility of developing an intermediate, semiclassical approach.

The considerable attention given to the Coulomb field reflects both the analytic solubility of this problem (on the classical and quantum mechanical levels) and the real physical conditions that prevail, for example, in hot plasmas. The discussion of the theory of bremsstrahlung in a Coulomb field begins with an examination of the purely classical (and nonrelativistic) approach (Sec. 13.2), continues with an examination of Sommerfeld's exact (but also nonrelativistic) quantum mechanical theory (Sec. 13.3), and concludes with a generalization of the quantum mechanical results to the relativistic case (Sec. 13.4). In our opinion, this "concentric" presentation of the theory of bremsstrahlung in a Coulomb field is justified by the fact that there is no exact quantum mechanical result for an arbitrary degree of relativistic motion of the electron, but also, primarily, by the fact that the classical theory presented in Sec. 13.2, which originates in the pioneering work of Kramers [41], still retains an independent general physical and practical value. In fact, this theory has great clarity and physical intuitiveness, can be applied over a wide range of conditions ($Ze^2/\hbar v \gg 1$), and its most important limiting case (the high frequency asymptote for the bremsstrahlung spectrum at quasiclassical energies) serves as a natural standard (a frequency-independent scaling factor) for descriptions of bremsstrahlung spectra in all the more complicated cases.

We note that the independent role of the classical description emphasized in the above remarks is one of the most distinctive illustrations of the "special character of the interrelation between quantum and classical mechanics": the first not only contains the second "as a limiting case, but simultaneously requires this limiting case for its own justification" [1].

On the other hand, the "nonanalytic" nature of the non-Coulomb case (Secs. 13.5–13.7) objectively forces us into a more attentive analysis of the entire physics of the bremsstrahlung process [33–36]. This, in turn, allows us to remove a transparent gap [37, 32, 8] in those branches of the theory concerned with describing the emission at high frequencies in terms of the classical theory, with quantum mechanical restrictions on the classical nature of the spectrum, etc. At the same time, this analysis leads us to a new, simple, and practically useful conception – "Kramers electrodynamics" (see Sec. 13.6 and Fig. 13.5.)

A large number of articles and reviews have been devoted to bremsstrahlung in collisions of electrons with nuclei and neutral atoms [2–5] (see [6–14], as well). Bremsstrahlung in Coulomb fields [15–18], in collisions of low-energy electrons with neutral atoms [4, 24, 25, 62], and in collisions of relativistic electrons with electrons, nuclei, and atoms [2, 7–9, 19–23] has been studied in detail. Recently, considerable attention has been devoted to bremsstrahlung effects related to the polarization of an atom and (or) the medium, especially plasmas (see, for example, [28–31], as well as the preceding chapters of this book).

13.2. Classical Theory of Bremsstrahlung of Nonrelativistic Electrons in a Coulomb Field. "Switch-off" of the Energy Integral

In the framework of the classical theory, the role of an effective spectral cross section $d\sigma(\omega)$* is played [32] by the effective spectral emission (radiation) $d\kappa(\omega)$, which is given by the product of $d\sigma(\omega)$ and the photon energy $\hbar\omega$,

$$d\kappa(\omega) = \hbar\omega \cdot d\sigma(\omega). \tag{13.1}$$

The classical effective emission $d\kappa$, of course, does not contain Planck's constant $\hbar$. Expressing $d\kappa(\omega)$ in terms of $d\mathscr{E}(\rho, \omega)$ (the energy emitted over the entire duration of a collision with an impact parameter ρ within the frequency interval $d\omega$ [32]), we obtain

$$d\sigma(\omega) = \frac{d\omega}{\hbar\omega} \int_0^\infty 2\pi\rho\, d\rho\, (d\mathscr{E}(\rho, \omega)/d\omega). \tag{13.2}$$

*Everywhere in this chapter $d\sigma = d\sigma' \equiv d\sigma^{st}$ of the preceding chapters.

At nonrelativistic energies the main contribution to the bremsstrahlung intensity is from dipole radiation. For purposes of convenience in calculating the bremsstrahlung cross section, we shall write $d\mathscr{E}(\rho, \omega)$ in the dipole approximation in the form [32]

$$d\mathscr{E}(\rho, \omega) = \frac{2e^2}{3\pi c^3}\omega^2(|\dot{x}(\rho, \omega)|^2 + |\dot{y}(\rho, \omega)|^2)d\omega, \tag{13.3}$$

where $\dot{x}(\rho, \omega)$ and $\dot{y}(\rho, \omega)$ are the Fourier transforms of the velocity components $\dot{x}(\rho, t)$ and $\dot{y}(\rho, t)$

Equations (13.1)–(13.9) are general in the sense that they can be used to calculate the dipole emission $d\kappa$ (or $d\sigma$) for motion in an arbitrary central field.

In the classical theory the total effective emission

$$\kappa = \int \hbar\omega d\sigma(\omega) = \int_0^\infty (d\kappa(\omega)/d\omega)d\omega \tag{13.4}$$

can be calculated independently without first calculating the spectrum:

$$\kappa = \int_0^\infty 2\pi\rho d\rho \Delta\mathscr{E}(\rho) \tag{13.5}$$

and

$$\Delta\mathscr{E}(\rho) = \frac{2e^2}{3c^3}\int_{-\infty}^{\infty}[w(\rho, t)]^2 dt, \tag{13.6}$$

where $w(\rho, t)$ is the acceleration of the electron as a function of time for scattering with an impact parameter ρ and $\Delta\mathscr{E}(\rho)$ is the total (over the entire frequency range from 0 to ∞) energy emitted throughout the entire duration of the collision. For motion in an arbitrary central field with a potential $U(r)$, Eq. (13.6) can easily be rewritten in the form

$$d\mathscr{E}(\rho) = \frac{4e^2}{3m^2c^3}\int_{r_0(\rho)}^{\infty}\left(\frac{dU}{dr}\right)^2\frac{dr}{v_r}. \tag{13.7}$$

Here $v_r(\rho, r)$ is the radial component of the electron velocity and $r_0(\rho)$ is the classical turning radius. $r_0(\rho)$ satisfies the equation

$$1 = (\rho^2/r_0^2) + U(r_0)/E, \tag{13.8}$$

where $E = mv^2/2$ is the energy of the electron and v is its velocity at infinity. When Eqs. (13.7) and (13.8) are substituted into Eq. (13.5), the integral with respect to the impact parameter ρ can be evaluated in a general form, which yields the following final expression for the total dipole emission of an electron in a central field:

$$\kappa = \frac{8\pi e^2}{3m^2 vc^3} \int_{r_{00}}^{\infty} \left(\frac{dU}{dr}\right)^2 \sqrt{1 - \frac{U(r)}{E}}\, r^2 dr, \tag{13.9}$$

where $r_{00} = r_0(0)$ is the distance of closest approach to the center of the field for $\rho = 0$.

Here we note the fundamental inability of classical electrodynamics to describe the total effective emission κ in an attractive field. In such a field $r_{00} = 0$ and Eq. (13.9) diverges at the lower limit for any potential with a singularity stronger than $r^{-\alpha}$ at the origin, where $\alpha > 2/5$. For a Coulomb field ($\alpha = 1$), Eq. (13.9) shows that κ goes to infinity as $\int_0 dr/r^{5/2}$ [the corresponding integral of Eq. (13.4) diverges as $\int^{\infty} d\omega$; see below]. The physically obligatory finiteness of κ can be ensured by cutting the integrals (13.4) and (13.9) off at some finite values of ω_{max} and r_{min}, respectively, that are determined by the quantum mechanical limitations on the applicability of the classical approach. This is, in rough outline, the essence of one variant of the semiclassical approach to the description of bremsstrahlung ([13, 33, 34], and Sec. 13.6).

Calculating the bremsstrahlung spectrum of nonrelativistic electrons in a Coulomb field $U = -Ze^2/r$ yields the following expression for $d\mathscr{E}(\rho, \omega)$ (13.3) [32, 39]:

$$d\mathscr{E}(\rho, \omega) = \frac{2\pi}{3}\left(Z^2 e^6 \omega^2/\left(m^2 v^4 c^3\right)\right)$$

$$\times \{[H_{i\nu}^{(1)\prime}(i\nu\tilde{\epsilon})]^2 + (1/\tilde{\epsilon}^2 - 1)\,[H_{i\nu}^{(1)}(i\nu\tilde{\epsilon})]^2\}\; d\omega. \tag{13.10}$$

Here $H_{i\nu}{}^{(1)}(i\nu\epsilon)$ and $H_{i\nu}{}^{(1)\prime}(i\nu\epsilon)$ are the Hankel function of first order and its derivative with respect to the argument, $\nu = \omega/\tilde{\omega}$ is the dimensionless "classical" frequency, $a = Ze^2/mv^2$ is the Coulomb scale length, $\tilde{\varepsilon} = (1 + \rho^2/a^2)^{1/2}$ is the eccentricity of the hyperbolic trajectory, and $\tilde{\omega} = v/a = mv^3/Ze^2$ is the Coulomb frequency. Intuitively, the latter represents the orbital rotational velocity of an electron with an impact parameter of $\rho \sim a$ at the turning point r_0 (13.8).

We shall refer to the spectral distribution (13.10) of the bremsstrahlung from an individual collision of an electron with a force center as a "subline." According to Eq. (13.2), it serves as the basis for calculating the bremsstrahlung spectrum of an electron beam undergoing scattering on a single center. The form of the subline is also (see Sec. 13.5) of major independent interest for clarifying the "physics" of the interrelation between the classical and quantum mechanical descriptions of bremsstrahlung (Sec. 13.6), as well as for generalizing the theory of the non-Coulomb case (Sec. 13.7). For this reason, we shall study Eq. (13.10), as such, below.

The final result [41, 32] for the classical bremsstrahlung spectrum in a Coulomb field ($\alpha = e^2/\hbar c = 1/137$) has the form

$$d\sigma_{cl}(\omega) = \frac{16\pi}{3\sqrt{3}}\alpha\frac{v^2}{c^2}a^2\frac{d\nu}{\nu}\left\{\frac{\pi\sqrt{3}}{4}i\nu H_{i\nu}^{(1)}(i\nu)H_{i\nu}^{(1)\prime}(i\nu)\right\}. \tag{13.11}$$

The scale of the bremsstrahlung cross section is determined by the factor $\alpha a^2(v^2/c^2)$ that contains the square of the Coulomb distance (the scale of the Rutherford cross section for elastic scattering in a Coulomb field), which shows up in Eq. (13.2) in the form $2\pi\rho d\rho$. The remaining factor $\alpha(v^2/c^2)$ emphasizes the nonrelativistic character of the cross section formula. The fine structure constant $\alpha = e^2/\hbar c$ has been effectively "smuggled" into the classical formula as a result of dividing $d\kappa$ by $\hbar\omega$. Its presence reflects the fact that this is a cross section for a single-photon radiative process.

The frequency dependence of the cross section is a universal function of a single dimensionless parameter, $\nu = \omega a/v$. We now examine the limiting cases of high ($\nu \gg 1$) and low ($\nu \ll 1$) frequencies.

When $\nu \gg 1$, the standard asymptotic expansions of the Hankel functions [38] can be used to reduce the expression in the curly brackets of Eq. (13.11) to the form [34]

$$\{\cdots\} = 1 + \frac{3^{2/3}\Gamma^2(1/3)}{10\pi\sqrt{3}\,2^{1/3}}\left(\frac{1}{\nu}\right)^{2/3} - \frac{1}{420}\frac{\Gamma(2/3)}{\Gamma(1/3)}\left(\frac{6}{\nu}\right)^{4/3} + \cdots \tag{13.12}$$

Thus, for $\nu \gg 1$, the classical bremsstrahlung cross section is approximately equal to

$$d\sigma_{cl}(\omega) \sim \frac{16\pi}{3\sqrt{3}}\alpha\frac{v^2}{c^2}a^2\frac{d\nu}{\nu} = \frac{16\pi}{3\sqrt{3}}\frac{Z^2e^6}{m^2v^2c^3\hbar}\frac{d\omega}{\omega} \equiv d\sigma_{Kr}, \tag{13.13}$$

where $d\sigma_{Kr}$ is the "Kramers" bremsstrahlung cross section.

Equation (13.13) shows that at high frequencies ($\nu \gg 1$), the effective emission $d\kappa = \hbar\omega d\sigma$ is independent of the frequency. In the classical theory the bremsstrahlung spectrum has no high-frequency limit, i.e., $\omega_{max}{}^{cl} = (mv^2/2\hbar)_{\hbar=0} = \infty$, and this is the origin of the divergence in the total effective emission (13.4) and (13.9).

In order for the bremsstrahlung spectrum to be classical (see [33–36] and Secs. 13.3 and 13.6), it is sufficient that the *motion* of the electron be classical, i.e., in the case of the Coulomb field being examined here, the condition

$$\eta \equiv Ze^2/\hbar v \gg 1 \tag{13.14}$$

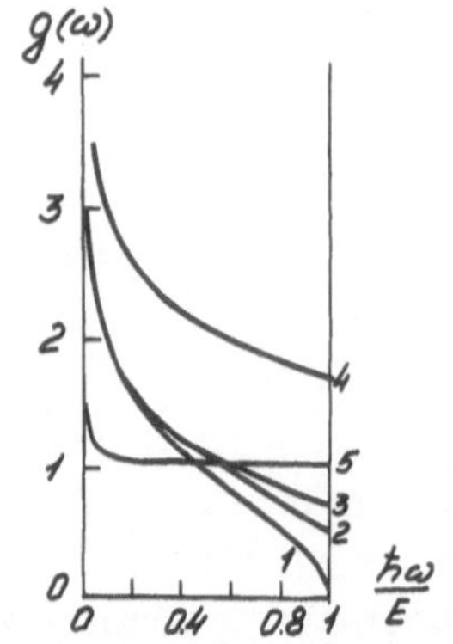

Fig. 13.1. Frequency dependence of the Coulomb Gaunt factor: 1) Born approximation; 2) Sommerfeld's quantum mechanical theory [Eq. (13.31)] for $\eta = 0.1$; 3) semiclassical spectrum [Eq. (13.70)] for $\eta \ll 1$; 4) classical spectrum [Eqs. (13.15) and (13.16)] for $\eta = 0.1$; 5) classical spectrum [Eqs. (13.15) and (13.12)] for $\eta = 50$.

should be satisfied. Since $\nu = 1/2\eta y$, where $y = \hbar\omega/E$ is the dimensionless "quantum mechanical" frequency, it is easy to see that the region of high classical frequencies, $\nu \geq 1$, where the Kramers formula (13.13) is valid, covers almost all of the real $(0 \leq y \leq 1)$ bremsstrahlung spectrum, $y = 2\nu/\eta \geq 2/\eta \ll 1$, except for the low quantum mechanical frequency range $y \leq 2/\eta \ll 1$.

It is clear from the above remarks why the Kramers bremsstrahlung cross section (13.13) is (as noted in Sec. 13.1) a natural standard for the description of bremsstrahlung spectra in all the more complicated cases. The ratio $d\sigma/d\sigma_{Kr}$ is known as the Gaunt factor. The expression in the curly brackets of Eq. (13.11) is the classical Gaunt factor

$$g_{cl}(\omega) = \frac{\pi\sqrt{3}}{4}\, i\nu H_{i\nu}^{(1)}(i\nu) H_{i\nu}^{(1)'}(i\nu), \tag{13.15}$$

which takes into account the deviation of the spectrum from its high-frequency limit when $\nu \leq 1$.

When $\nu \ll 1$, $g_{cl}(\omega)$ can be written in the form [42, 34]

$$g_{cl}(\omega) = \frac{\sqrt{3}}{\pi}\left[(1 + \pi\cdot\nu)\ln\frac{2}{\gamma\nu} - \frac{2}{3}\nu^2\left(\ln\frac{2}{\gamma\nu}\right)^3 + \cdots\right], \tag{13.16}$$

where $\gamma = 1.78$ ... is the Euler constant.

The classical Gaunt factor is plotted as a function of frequency in Fig. 13.1 (curves 4 and 5). Equation (13.15) and its limiting forms (13.12) and (13.16) will be compared in detail with with the corresponding quantum mechanical expressions (Sec. 13.3) below.

We now return to the form of the subline (13.10). The asymptotic expansions for the Hankel functions [51, 55] can be used to reduce it to a simpler, universal form in two physically opposite ranges of the parameters ω and ρ, specifically: for low frequencies ($\nu = \omega/\bar{\omega} \ll 1$) and distant, almost linear trajectories ($\rho \gg a$) and for high frequencies ($\nu \gg 1$) and close, strongly curved, almost parabolic trajectories ($\rho \ll a$). We shall write down the corresponding limiting forms for the subline normalized to unity, i.e., for $d\mathscr{E}(\rho, \omega)$ (13.10) divided by the integral $\int_0^\infty [d\mathscr{E}(\rho, \omega)/d\omega]d\omega = \Delta\mathscr{E}(\rho)$ of Eqs. (13.6) and (13.7) and given explicitly in the Coulomb case by

$$\Delta\mathscr{E}(\rho) = \frac{mv^5}{3Zc^3}\left(\frac{a}{\rho}\right)^3 \left\{\left(\pi + 2\arctan\frac{a}{\rho}\right)\left[1 + 3\left(\frac{a}{\rho}\right)^2\right] + 6\frac{a}{\rho}\right\}. \tag{13.17}$$

The above limiting forms of the normalized subline are universal functions of two different dimensionless arguments, s and u, and have the forms

$\nu \ll 1, \rho \gg a$:

$$\frac{d\mathscr{E}(\rho, \omega)}{\Delta\mathscr{E}(\rho)} \approx \frac{8}{\pi^2}\{[sK_0(s)]^2 + [sK_1(s)]^2\}\, ds, \tag{13.18}$$

$s = M\omega/2E$

and

$\nu \gg 1, \rho \ll a$

$$\frac{d\mathscr{E}(\rho, \omega)}{\Delta\mathscr{E}(\rho)} \approx \frac{12}{\pi^2}\{[uK_{1/3}(u)]^2 + [uK_{2/3}(u)]^2\}\, du, \tag{13.19}$$

$u = M^3\omega/3Z^2me^4$,

where $M = mv\rho$ is the angular momentum for a given trajectory (ρ, v) and the $K_\nu(x)$ are Macdonald functions. The $\Delta\mathscr{E}(\rho)$ in the denominators are given by the corresponding limits of the general dependence (13.17):

$\rho \gg a$:

$$\Delta\mathscr{E}(\rho) \approx \frac{\pi}{3}\frac{Z^2e^6}{c^3m^2v}\frac{1}{\rho^3} \propto \frac{Z^2E}{M^3}, \tag{13.20}$$

and

$\rho \ll a$:

$$\Delta\mathscr{E}(\rho) \sim 2\pi \frac{Z^4 e^{10}}{c^3 m^4 v^5} \frac{1}{\rho^5} \propto \frac{Z^4}{M^5}, \tag{13.21}$$

(The scaling laws for the dependence on Z and on the integrals E and M of the motion are also shown here.)

The form (13.19) for the subline and the corresponding $\Delta\mathscr{E}(\rho)$ (13.21) are characterized by a single important property (which, we note, has not been examined in the literature [32, 39, 12]) which subsequently will play a key role, both in generalizing the classical theory of bremsstrahlung to the non-Coulomb case and in identifying the region in which the classical theory itself is valid (Secs. 13.5–13.7). Specifically, both these quantities depend on only one of the two integrals of motion that characterize the trajectory of the electron (the angular momentum M), while they have lost their dependence on the energy E [35, 36]. This is true for the entire class of potentials that will be examined, not just for the Coulomb case, and this means that the part of the trajectory responsible for emission at frequencies $\omega \gg \tilde{\omega}$ (simply referred to as the *radiating part of the trajectory* in the following) is almost independent of E and, therefore, lies in that region of r where the electron has already been significantly *accelerated* in the central (especially in a Coulomb) attractive potential, i.e., it lies at distances close to the center where $|U(r)| \gg E$.

We shall refer to this property (which we emphasize exists only for an attractive field, since this region of r is classically inaccessible for a repulsive field) as the "*energy integral switch-off*" (EIS) effect.

The fact that the Coulomb subline (13.19) is independent of the energy E and, therefore, of its sign, means that to an electron that radiates practically only on the quasiparabolic section of the trajectory (for details, see below), it is a matter of "indifference" whether the nonradiative parts of the trajectory are hyperbolic ($E > 0$), or elliptical ($E < 0$), or even, as happens in the quantum mechanical case of radiative recombination (photorecombination), involve a transition with a change in the sign of E.

It should be emphasized that the inequality $\omega \gg \tilde{\omega}$ and the closely related EIS effect essentially encompass a region where the frequency is not so much "high" as, in general, "not low." In fact, in purely classical terms, according to Eq. (13.13) this part of the bremsstrahlung spectrum extends without any sort of "dip" to $\omega = \infty$. In the real, quantum mechanical case, the bremsstrahlung spectrum is limited to frequencies $\omega_{max} = E/\hbar$; however, when the quasiclassical condition $Ze^2/\hbar v \gg 1$ (necessary for the validity of all our classical analysis) is satisfied, we have $\omega_{max}/\tilde{\omega} \sim Ze^2/\hbar v \gg 1$, so that the frequencies $\omega \gg \tilde{\omega}$ encompass the *overwhelming* bulk of the total quasiclassical bremsstrahlung spectrum.

Using the analytical solubility of the Coulomb case, it is useful to analyze the concept of the "radiating portion of the trajectory" in somewhat more detail, especially as it relates to the EIS effect. In the course of this analysis we shall introduce the angular velocity of the orbital rotation of the electron about the center of the field at the turning point r_0 (v_{max} is the velocity of the electron at the turning point),

$$\omega_{rot}(r_0) = \frac{M}{mr_0^2} = \frac{v_{max}}{r_0} = \sqrt{\frac{2(E + |U(r_0)|)}{mr_0^2}}, \tag{13.22}$$

a quantity which is of great importance for the subsequent discussion. For now we leave the form of the attractive potential $U(r)$ unspecified under the assumption that it is not necessary to require that it fall at the center, i.e., $\lim_{r_0 \to 0} r_0^2 U(r_0) = 0$

The effective angular extent φ_{eff} of the radiating portion of the trajectory is coupled naturally by the formula $\varphi_{eff} = \omega_{rot}(r_0)\Delta t_{eff}$ to the effective duration of the bremsstrahlung process, Δt_{eff}, which, in turn, can be extracted from Eq. (13.6) by writing it in the form

$$\Delta\mathscr{E}(\rho) = \frac{2e^2}{3c^3} w_{max}^2 \Delta t_{eff}, \tag{13.23}$$

where $w_{max} = U'(r_0)/m$ is the acceleration of the electron at the turning point.

In the Coulomb case we have $w_{max} = Ze^2/mr_0^2$ and $r_0 = -a + (a^2 + \rho^2)^{1/2}$ [from Eq. (13.8)], while $\Delta\mathscr{E}(\rho)$ is given by Eq. (13.17) and it is easy to obtain the following expression for $w_{rot}(r_0)$:

$$\omega_{rot}(r_0) = \frac{Z^2me^4}{M^3}\left[1 + \sqrt{\left(Mv/Ze^2\right)^2}\right]^2. \tag{13.24}$$

As a result, we obtain the following universal formula for φ_{eff} as a function of the parameter ρ/a:

$$\varphi_{eff} = \frac{1}{2}\left(\frac{a}{\rho}\right)^2\left[\sqrt{1 + \left(\frac{\rho}{a}\right)^2} - 1\right]^2 \left\{\left(\pi + 2\arctan\frac{a}{\rho}\right)\left(1 + 3\frac{a^2}{\rho^2}\right) + 6\frac{a}{\rho}\right\}. \tag{13.25}$$

In the limiting cases of $\rho \gg a$ (almost a straight-line, or rectilinear trajectory) and $\rho \ll a$ (almost parabolic trajectory), Eq. (13.25) yields $\varphi_{eff}^{str} = \pi/2$ and $\varphi_{eff}^{parab} = 3\pi/4$, respectively.

We now turn to the equation for a hyperbolic trajectory [37],

$$\frac{M^2}{mZe^2}\frac{1}{r} = 1 + \sqrt{1 + (\rho/a)^2}\cos\varphi = 1 + \sqrt{1 + \left(\frac{Mv}{Ze^2}\right)^2}\cos\varphi. \tag{13.26}$$

It is clear that in the quasiparabolic case ($\rho \ll a$ or $M \ll Ze^2/v$), $r(\varphi)$ is controlled solely by the the integral M (the EIS effect) over the range of φ within which the contribution of the small term $(\rho/a)^2 = (Mv/Ze^2)^2$ to the right-hand side of Eq. (13.26) is small. This happens, in particular, up to the "angular amplitude" of the radiating part of a quasiparabolic trajectory, which equals $\pm(1/2)\varphi_{eff}^{parab} = \pm 3\pi/8$.

The EIS effect begins to fail only when the two terms on the right-hand side of Eq. (13.26) compensate one another strongly, i.e., as $r \to \infty$, near the asymptotes of the hyperbola $\varphi_\infty = \pm\arccos\left(-1/\sqrt{1 + (\rho/a)^2}\right) \sim \pm\pi$ with an angular deviation from each of these on the order of $\rho/a \ll 1$, that is, clearly beyond the confines of the radiating part.

As for the radial coordinate r of the radiating part, when $\rho \ll a$, according to Eq. (13.26) within the same limits of $\pm(1/2)\varphi_{eff}^{parab}$, it cannot be greater than M^2/mZe^2 in order of magnitude, and the latter, to the same accuracy, cannot exceed $2r_0$. Thus, for bremsstrahlung at frequencies $\omega \geq \tilde{\omega}$ there is a unique sort of localization or "compression" of the spatial region responsible for the emission that is characterized by the relations

$$r_{eff} \sim r_0 \ll \rho \ll a. \tag{13.27}$$

The above discussion makes it possible to obtain a physically intuitive, qualitative derivation of the fundamental, "standard" Kramers formula (13.13) for the entire theory of bremsstrahlung [13, 14]. In fact, in this case the frequencies of the motion responsible for the emission at the high frequencies ω of interest to us can only be the angular rotation velocities $\omega_{rot}(r_0)$ of electrons in the turning regions of highly curved, nearly parabolic trajectories.

In view of this order of magnitude equality, $(\omega_{rot})_{eff} \sim \omega$, we can expect that the shape of the $d\kappa/d\omega$ *spectrum* as a characteristic of the bremsstrahlung integrated over ρ (and, therefore, a cruder characteristic than the ***subline***) will be correctly conveyed by the distribution of the "effective emission" from a flux of electrons, $d\kappa(\rho) = \Delta\mathscr{E}(\rho)2\pi\rho d\rho$ [see Eqs. (13.5) and (13.6)], over such a purely mechanical quantity as $\omega_{rot}(r_0)$. For $\Delta\mathscr{E}(\rho)$ (13.6), we have Eq. (13.23), in which

$$\Delta t_{eff} \sim \frac{r_0}{v_{max}} = \sqrt{\left(\frac{mr_0^2}{2(E + |U(r_0)|)}\right)} \tag{13.28}$$

is the effective duration of the (classical) bremsstrahlung process for a given trajectory. Switching to the more convenient variable r_0 using the formula $\rho = M/mv$ and Eqs. (13.23) and (13.28), we find

$$d\kappa(\rho) = d\kappa(r_0) \sim \frac{e^2}{m^3v^2c^3E} \cdot \frac{|U'(r_0)|^2}{\sqrt{E + |U(r_0)|}} \cdot \frac{d}{dr_0}\left[r_0^2(E + U(r_0))\right]r_0 dr_0. \tag{13.29}$$

Now let $\omega_{rot}(r_0)$ be large enough so that $|U(r_0)| \gg E$ (EIS). Then, for the Coulomb case $U(r) = -Ze^2/r$ of interest to us here, Eq. (13.22) gives $\omega_{rot} \sim (Ze^2/mr_0^3)^{1/2}$, so that Eq. (13.29) reduces to the form

$$d\kappa(r_0) = d\kappa(\omega_{\text{rot}}) \sim \frac{e^2}{c^3 v^2}\left(\frac{Ze^2}{mr_0}\right)^{5/2} dr_0 \sim \frac{Z^2 e^6}{c^3 m^2 v^2}\, d\omega_{\text{rot}}, \tag{13.30}$$

which indeed coincides with the Kramers formula (13.13) to within a numerical factor of order unity. A criterion for the applicability of Eq. (13.30) can be extracted from the inequality $Ze^2/r_0 \gg E$ used in this derivation by expressing r_0 in terms of $\omega_{\text{rot}} \sim \omega$, which, as should happen, leads to the well-known condition $\omega \gg \tilde{\omega}$.

13.3. Quantum Mechanical Theory of Bremsstrahlung of Nonrelativistic Electrons in a Coulomb Field

At nonrelativistic energies ($\beta_2 < \beta_1 \ll 1$) an exact analytic calculation of the bremsstrahlung cross section is possible. This was first done by Sommerfeld [6, 47], Sommerfeld and Maue [48], and Menzel and Pekeris [49] (see [3, 7, 8], as well). For the spectral effective bremsstrahlung cross section in a Coulomb field we have the Sommerfeld formula [3, 6-8] (the corresponding calculations are done in [6-8])

$$d\sigma(\omega) = d\sigma_{\text{Kr}} \cdot g_{\text{Som}}(\omega);$$

$$g_{\text{Som}}(\omega) = \frac{\pi\sqrt{3}}{(\exp(2\pi\eta_1) - 1)(1 - \exp(-2\pi\eta_2))}\, x_0 \frac{d}{dx_0} |{}_2F_1(i\eta_1, i\eta_2, 1; x_0)|^2. \tag{13.31}$$

Here ${}_2F_1(a, b, c; z)$ is the hypergeometric function and $x_0 = -4\eta_1\eta_2/(\eta_2 - \eta_1)^2$.

Equation (13.31) is complicated and difficult to grasp, so we shall discuss its limiting forms. A detailed analysis of over 20 different cases has been made in the review by Brussard and van de Hulst [3] and the results of a numerical tabulation of Eq. (13.31) are given in [18].

The frequency dependence of the quantum mechanical Gaunt factors (13.31), as opposed to that of the classical (13.15), is determined by two independent parameters. It is convenient to take the Coulomb parameter η_1 and the "quantum mechanical" frequency $y = \hbar\omega/E_1$ as the "primary" parameters. (Here $\hbar\omega = E_1 - E_2$, where E_1 and E_2 are the initial and final energies of the electron.) In the analysis we shall find some dimensionless parameters derived from them to be useful: the "classical" frequency $\nu = Ze^2\omega/mv_1^3 = \eta_1 y/2$, the variable

$$\xi = \eta_2 - \eta_1 = \eta_1\,[(1/(1-y)^{1/2}) - 1] \tag{13.32}$$

and the argument x_0 of the hypergeometric function in Eq. (13.31).

The existence of a number of dimensionless combinations of the two initial physically different parameters (η_1 related to the "motion" and y related the "spectrum") means that there are a substantial number of regions on the (η_1, y) plane in which Eqs. (13.31) can be simplified significantly. Figure 13.2 is a diagram of the regions of applicability of the different approximations in the theory of bremsstrahlung in a Coulomb field.

The conditions for the validity of the different approximations in the theory of bremsstrahlung in a Coulomb field have been discussed in almost of all the papers devoted to studies of the limiting expressions for the Sommerfeld formula (13.31) or to developing the various approximate versions of the theory. Of the many such papers we list just a few here: [2-5, 7, 8, 11, 13, 16, 17, 19, 34, 39]. In most cases the regions of validity of the individual approximations are discussed without any attempt to construct a general picture. The review by Brussard and van de Hulst [3] is an exception in this regard; however, their analysis is based on a numerical criterion (a 1% deviation of a given approximation from the exact Sommerfeld formula) which is useful for practical applications of these formulas but does overshadow the physical side of the question. The latter aspect was the subject of a diagram contained in their review [3], but that does not answer the question in the best possible way, since it was not based on the "natural" dimensionless parameters of the problem. The natural dimensionless parameters are

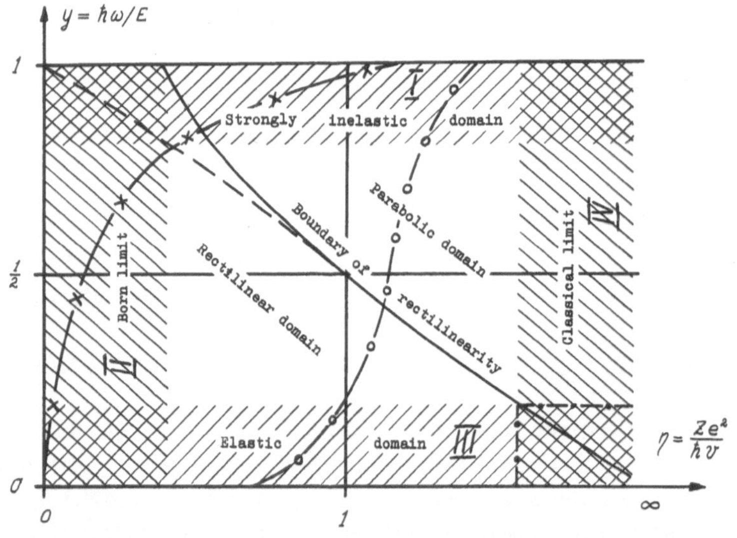

Fig. 13.2. Diagram of the regions of validity of the different approximations for the theory of bremsstrahlung in a Coulomb field: I) $y \sim 1$; II) $\eta_1 \ll 1$; III) $y \ll 1$; IV) $\eta_2 > \eta_1 \gg 1$; the "straight-line region" $\nu \ll 1$; the "parabolic" region $\nu \gg 1$; – – – limit for the "large argument" approximation [Eq. (13.47)]; –×–×– the limit of validity of the semiclassical spectrum (13.70); –o–o– the actual region of validity of the classical spectrum (13.15); and, –●–●– the conventional [8, 39, 43] region of validity of the classical spectrum: $\eta_1 \gg 1$ and $y \ll 1$.

$$\eta_1, y, \eta_2, \nu, \xi, x_0. \tag{13.33}$$

Comparing the quantities (13.33) with unity yields the following seven major regions:

$\eta_1 \ll 1, \eta_2 \ll 1$, the Born region (II),
$\eta_1 \gg 1$ $(\eta_2 > \eta_1)$, the quasiclassical region (IV),
$y \ll 1$, the elastic region (III),
$1 - y \ll 1$, the highly inelastic region (I),
$\nu \ll 1$, the region of straight-line trajectories,
$\nu \gg 1$, the region of parabolic trajectories, and
$|x_0| \gg 1$ $(\xi \ll 1)$, the region where the "large argument" approximation holds.

The first four regions are indicated by shading in Fig. 13.2. The Roman numerals in parentheses correspond to the classification employed by Brussard and van de Hulst in their review [3]. The origins of the names of regions I-IV are obvious. We refer to Sec. 13.2 of this book for a discussion of the regions with straight-line and parabolic trajectories [see Eqs. (13.18) and (13.19) and the associated discussion]. The "large argument" approximation and the other regions of the diagram in Fig. 13.2 will be discussed below.

Let us begin with the quasiclassical region (IV), where $\eta_1 \gg 1$ (then we also have $\eta_2 > \eta_1 \gg 1$). Here the Gaunt factor (13.31) reduces approximately to the classical limit (13.15) over almost the entire spectrum. The limit $\hbar \to 0$ in the Sommerfeld formula (13.31) was first derived by Biedenharn [50], but not directly. Instead this was done with the aid of an expansion of the bremsstrahlung cross section in a series in l (the orbital angular momentum) with a subsequent transition from a difference equation for the matrix element to a differential equation for its Fourier transform with a corresponding replacement of the sum over l by an integral with respect to the impact parameter. Gervids and Kogan [34] have obtained this limit directly by letting $\hbar$ approach zero in the Sommerfeld formula:

$$\lim_{\hbar \to 0} g_{\text{Som}} \equiv g_{\text{cl}} = \frac{\pi\sqrt{3}}{4} i\nu H_{i\nu}^{(1)}(i\nu) H_{i\nu}^{(1)\prime}(i\nu), \tag{13.34}$$

where the prime in the last factor denotes differentiation with respect to the argument.

As is to be expected, Eq. (13.34) is exactly equal to the classical expression for the g factor [32, 41] [see Eq. (13.15)]. Thus, the formulas for the classical theory of bremsstrahlung in a Coulomb field are the zeroth-order terms in the expansions of the corresponding exact quantum mechanical expressions in a power series in $\hbar$. The agreement between the limit of the quantum mechanical Gaunt factor as $\hbar \to 0$ and the classical formula is essentially a special (for the example of bremsstrahlung in a Coulomb field) demonstration of the correspondence principle for inelastic transitions in the continuum. Expanding directly in powers of $\hbar$ makes it possible, in principle, to calculate the quantum mechanical correction to the classical spectrum at arbitrary frequencies. Evidently, the result of such a calculation will yield an unambiguous answer to the question of the limits of applicability of the classical theory; however, these calculations are extremely cumbersome, so that here we shall limit ourselves to studying the quantum mechanical corrections in the limiting cases (see below). This question has been examined elsewhere, with an explicit calculation and analysis of the quantum mechanical corrections to the classical limit for the matrix element of an inelastic transition [35, 36] (in [35], the question was examined for other processes in addition to bremsstrahlung).

For $\eta_1 \gg 1$ and $\nu \gg 1$, i.e., for almost the entire spectrum except a small region near the low-frequency limit $\omega = 0$ (see curve 5 of Fig. 13.1, for example), the Sommerfeld Gaunt factor (13.31) can be written in the form [42, 56]

$$g_{\text{Som}} \approx 1 + \frac{3^{2/3}\Gamma^2(1/3)}{10\pi\sqrt{3}\,2^{1/3}}\left(\frac{1}{\nu}\right)^{2/3}(1 - y/2) + \cdots \tag{13.35}$$

For $\eta_1 \gg 1$ and $\nu \ll 1$, the Elwert formula [57] (also, see the "large argument" approximation below) yields the following form for the Sommerfeld Gaunt factor [34]

$$g_{\text{Som}} \approx \frac{\sqrt{3}}{\pi}\left((1 + \pi\nu)\ln\frac{2}{\gamma\nu} - \frac{1}{12\eta_1^2} - \frac{3}{4}y - \cdots\right). \tag{13.36}$$

Equations (13.35) and (13.36) [cf. Eqs. (13.12) and (13.16)] yield explicit forms for the quantum mechanical corrections to the bremsstrahlung spectrum at quasiclassical energies ($\eta_2 > \eta_1 \gg 1$) for the cases of high ($\nu \gg 1$) and low ($\nu \ll 1$) frequencies, respectively. It is natural to take the inequality

$$|(g - g_{\text{cl}})/g_{\text{cl}}| \ll 1 \tag{13.37}$$

as the condition for classical behavior of the spectrum. Substituting Eqs. (13.12) and (13.35) or (13.16) and (13.36) in Eq. (13.37) gives

$\nu \gg 1$:

$$\frac{y}{\nu^{2/3}} \sim \left(\frac{y}{\eta_1^2}\right)^{1/3} \ll 1 \quad \text{or } y \ll \eta_1^2, \tag{13.38}$$

and

$\nu \ll 1$:

$$12\eta_1^2 \ln(2/\gamma\nu) \gg 1. \tag{13.39}$$

Since $\eta_1 \gg 1$, while $y = \hbar\omega/E_1$ does not exceed unity in the bremsstrahlung spectrum, both conditions are automatically satisfied. We note that the condition (13.39) for low frequencies is, as might be expected, weaker than condition (13.38) for high frequencies.

It is clear from Eqs. (13.38) and (13.39) that the condition for quasiclassical motion $\eta_2 > \eta_1 \gg 1$ is a sufficient condition for classical behavior in the spectrum of bremsstrahlung in a Coulomb field at *all* frequencies. The condition $\hbar\omega \ll E_1$ for a classical bremsstrahlung spectrum, which is widespread in the literature [8, 39, 43, 79], is superfluous when the motion is quasiclassical. This was demonstrated above for the example of a Coulomb field, but it has been shown [35, 36] to be true for the entire class of potentials under consideration. The physical reasons for this important fact have been analyzed in [35 and 36] and will be discussed in Sec. 13.6.

We note that the condition (13.38) can be rewritten in the form

$$\hbar\omega \ll Z^2(me^4/2\hbar^2) = Z^2Ry, \tag{13.40}$$

which means that not only the entire bremsstrahlung spectrum $\hbar\omega \leq mv_1^2/2$, but also the bulk of the radiative recombination spectrum, is classical, provided only that the motion of the electron is classical in the initial state. This circumstance, in combination with the favorable outcome of the numerical factors (see the footnote on page 332), is the reason for the good accuracy of the essentially classical formulas of Kramers [41, 13, 14] for the radiative recombination cross section, even when $n \sim 1$ (where n is the principal quantum number).

We now consider the opposite, "anticlassical" case of energies in the Born approximation. If $\eta_1 \ll 1$ (only the initial velocity of the electron satisfies the Born approximation), then the bremsstrahlung spectrum is given by the Born–Elwert approximation,

$$g_{\text{B.E.}}(\eta_1, \eta_2) = \frac{\sqrt{3}}{\pi}\frac{\eta_2}{\eta_1}\frac{1-\exp(-2\pi\eta_1)}{1-\exp(-2\pi\eta_2)}\ln\frac{\eta_2+\eta_1}{\eta_2-\eta_1}. \tag{13.41}$$

If the final velocity of the electron also satisfies the Born approximation and, additionally, $2\pi\eta_2 \ll 1$, then Eq. (13.41) becomes the ordinary Born approximation (region II of Fig. 13.2, curve 1 of Fig. 13.1),

$$g_{\text{B}}(\eta_1, \eta_2) = \frac{\sqrt{3}}{\pi}\ln\frac{\eta_2+\eta_1}{\eta_2-\eta_1} = \frac{\sqrt{3}}{\pi}\ln\frac{1+\sqrt{1-y}}{1-\sqrt{1-y}}. \tag{13.42}$$

When $\omega \ll \omega_{max}$, i.e., when $\eta_2 \sim \eta_1$, Eqs. (13.41) and (13.42) give [cf. Eq. (13.16)]

$$g \sim ((3^{1/2})/\pi)\ln(4/y). \tag{13.43}$$

An accurate treatment of the limit $\omega \to 0$ or, more precisely, of the range $\omega \ll Ze^2mv_1/\hbar^2$ ($y \ll \eta_1 \ll 1$), requires that we return to the original exact formula (13.31). This leads to the classical low-frequency logarithmic term (13.16) instead of the low-frequency logarithmic term of the Born approximation (13.43) [42].

In the highly inelastic region (region I of Fig. 13.2, where $\omega \sim \omega_{max}$ or $1 - y \ll 1$, so that $2\pi\eta_2 \gg 1$), the Born approximation does not work, and as a result g_B of Eq. (13.42) goes to zero at $\omega = \omega_{max}$ (see curve 1 of Fig. 13.1). The fact that g_B goes to zero at the short wavelength limit is a specific feature of the Born approximation (with any potential and not just a Coulomb potential). The physics of its origin can be understood from the following considerations. The short-wavelength limit is the place where the bremsstrahlung spectrum transforms to the radiative recombination spectrum. In recombination the electron ends up in a bound state. The condition for applicability of the Born approximation at low energies is essentially (see [1], for example) that there should be no bound states; that is, in this case, there should be no recombination, which shows up in the fact that the corresponding Gaunt factor is equal to zero, regardless of the character of the potential. The fact that $g_B(\omega_{max}) = 0$ always, means that the simple Born approximation cannot be used for calculating the effect of shielding of the nucleus on the bremsstrahlung spectrum in the field of an atom or ion at nonrelativistic energies and at energies that are not low ($\hbar\omega \sim E_1$) (see Sec. 13.7*). A modified Born approximation with the Elwert factors (see the development of the results of [53] in [11]) can be used to obtain agreement with calculations of the shielding and penetration effects by other methods (numerical quantum mechanical and semi- and quasiclassical).

It is possible to find an exact expression for the Gaunt factor g at the short-wavelength limit for arbitrary η_1 by analyzing Eq. (13.31). The result has the form [54]

$$g_{Som}(\eta_1, \eta_2 = \infty) = 2(3)^{1/2} F_0(-\eta_1, 2\eta_1)F_0'(-\eta_1, 2\eta_1). \qquad (13.44)$$

Here $F_0'(-\eta_1, 2\eta_1) = [dF_0(-\eta_1, \rho)/d\rho]|_{\rho=2\eta_1}$, and $F_L(\eta, \rho)$ is the Coulomb wave function, which is a regular (for $\rho \to 0$) solution of a Schrödinger equation of the form

$$\frac{d^2F_L}{d\rho^2} + \left[1 - \frac{2\eta}{\rho} - \frac{L(L+1)}{\rho^2}\right]F_L = 0.$$

(The normalization and other properties of $F_L(\eta, \rho)$ are given in, for example, [55].) For $\eta_1 \ll 1$ the expression for the Gaunt factor at the short-wavelength limit has the form

*We stress here that the standard Bethe–Heitler calculations [7, 9, 52] refer to relativistic energies.

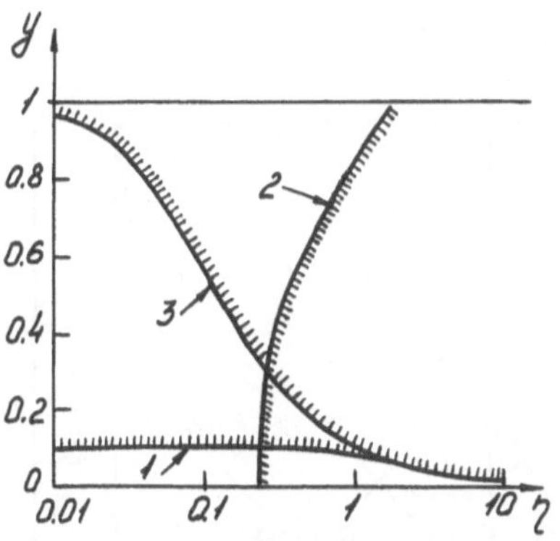

Fig. 13.3. The regions with a 1% error in the Elwert formula (1), in the Born–Elwert formula (2), and in Eq. (13.47) (the "large argument" approximation) (3). The curves are shaded on the side where the error exceeds 1%. Curves 1 and 2 were taken from [3] and curve 3 from [34].

$$g_{\rm Som} \approx \frac{8\pi\sqrt{3}\,\eta_1^2}{\exp(2\pi\eta_1)-1}\left(1+\frac{10}{3}\eta_1^2+\frac{196}{45}\eta_1^4+\cdots\right). \tag{13.45}$$

At quasiclassical energies ($\eta_1 \gg 1$), an approximate expression for the Gaunt factor at the short-wavelength limit can be obtained from Eq. (13.35) ($y = 1, \nu = 1/2\eta_1$) or directly from Eq. (13.44). We have [54]

$$g_{\rm Som} \approx 1+\frac{0.1728}{\eta_1^{2/3}}-\frac{0.0496}{\eta_1^{4/3}}-\frac{0.0172}{\eta_1^{6/3}}-\cdots \tag{13.46}$$

Over almost the entire extent of the bremsstrahlung spectrum the argument x_0 [see Eq. (13.31)] is large (regardless of the value of η_1). This means that, following [57–59] (see [60, 34] as well), we can develop a "large argument" approximation with $|x_0| \gg 1$ that will be valid for $\xi \equiv \eta_2 - \eta_1 \ll 1$. When $\xi \ll 1$, Eq. (13.81) reduces to

$$g_{\rm Som}(\eta_1, \eta_2) \approx \frac{2\sqrt{3}\,(\eta_2-\eta_1)}{1-\exp(-2\pi(\eta_2-\eta_1))}\left\{\ln\frac{\eta_2+\eta_1}{\eta_2-\eta_1} + \psi(1) - \operatorname{Re}\psi(i\eta_1) + \frac{1}{2}(\eta_2-\eta_1)\operatorname{Im}\psi'(i\eta_1)\right\}, \tag{13.47}$$

where $\psi(x) = d\ln\Gamma(x)/dx$ is the logarithmic derivative of the Γ-function.

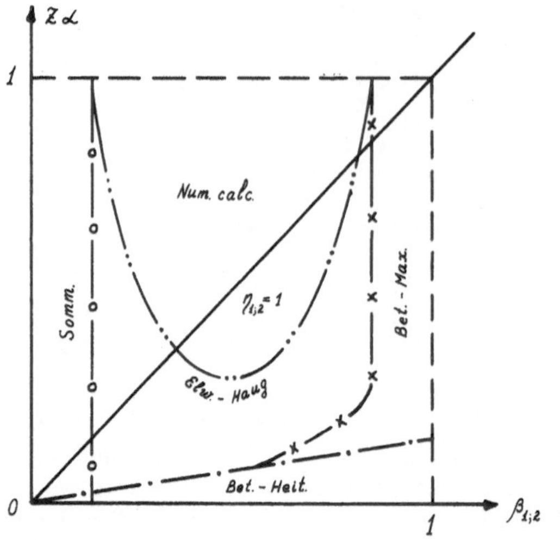

Fig. 13.4. A diagram of the regions of validity of the different approximations in the theory of bremsstrahlung of relativistic electrons in a Coulomb field; $\alpha = e^2/\hbar c = 1/137$, $\beta_{1,2} = v_{1,2}/c$. The bisector $\eta_{1,2} = 1$ separates the regions with $Z\alpha > \beta_{1,2}$ (above the bisector) and $Z\alpha < \beta_{1,2}$ (below it). The Sommerfeld formula ($\beta_{1,2} \ll 1$, $Z\alpha \le 1$) applies to the left of —o—; the Bethe–Heitler formulas ($\eta_{1,2} \ll 1$, i.e., $Z\alpha \ll \beta_{1,2}$) below —•—; the Bethe–Maximon formulas [$E_{1,2} \gg 1/2(Z\alpha)^2mc^2$] to the right of -×-×-, including below —•—; and, the Elwert–Haug formula below -••-. Only numerical calculations [20, 22, 23] have been carried out above the last curve.

When $\eta_1 \ll 1$ and $\eta_2 \ll 1$, Eq. (13.47) transforms to Eq. (13.42). When $y \ll 1$ [in the region $\eta_1 \ge 1$, because of Eq. (13.32) and the fact that $\xi \ll 1$, this restriction is obligatory], Eq. (13.47) converts to the Elwert formula [57]

$$g_E = \frac{\sqrt{3}}{\pi}\left[\ln\frac{4}{y} + \psi(1) - \operatorname{Re}\psi(1 + i\eta_1)\right]. \tag{13.48}$$

The Elwert formula (13.48) describes the long-wavelength Gaunt factor for arbitrary η_1. In the limit $\eta_1 \ll 1$ it transforms to the Born logarithm (13.43) and in the limit $\eta_1 \gg 1$, to the classical logarithm (13.16). Figure 13.3 shows the regions in the (η_1, y) plane where the errors in the Elwert (13.48) and Born–Elwert (13.41) equations and Eq. (13.47) are less than 1%. It is clear that Eq. (13.47) "grafts" a high-frequency region onto the Elwert formula for $\eta_1 \le 1$ and a low ($\nu \ll 1$)

frequency region onto the Born–Elwert formula for $\eta_1 \geq 1$. In some sense Eq. (13.47) "synthesizes" the Born and long-wavelength forms of the Sommerfeld formula. It can be shown that in terms of its physical significance, Eq. (13.47) corresponds to an approximation of straight-line trajectories (for $\hbar \to 0$ and $\xi \to \nu$; also, see the diagram showing the regions of applicability of the different approximations in Fig. 13.2). Equation (13.47) has been tabulated by Gervids and Kogan [34].

13.4. Bremsstrahlung of Relativistic Electrons in a Coulomb Field

An exact accounting for the interaction of an electron with a Coulomb field requires exact solutions of the Dirac equation. Solutions of this sort were obtained in 1928 by Darwin [61]. The resulting expressions are extremely complicated and we shall not list them here. They have been obtained and analyzed by, among others, Akhiezer and Berestetskii [7].

Simplifications leading to analytic expressions for the bremsstrahlung cross section can be made in the following basic cases (see the diagram of Fig. 13.4): 1) nonrelativistic energies β_1 and $\beta_2 \ll 1$ (where $\beta = v/c$) with arbitrary η_1 and η_2; this is the Sommerfeld theory; 2) sufficiently small Z that $\eta_1 < \eta_2 \ll 1$ for arbitrary β_1 and β_2 ; this is the relativistic Born approximation and the Bethe–Heitler formula [2, 7, 8, 52]; 3) ultrarelativistic energies $E_{1,2} \gg (Z\alpha)^2 mc^2$ ($\alpha = e^2/\hbar c = 1/137$) and arbitrary $Z\alpha$ (in the sense that $Z\alpha \leq 1$); this is referred to by Akhiezer and Berestetskii [7] as the Furry–Sommerfeld–Maue approximation [2, 7, 8, 63]; 4) arbitrary $Z\alpha$ at nonrelativistic and ultrarelativistic energies and small $Z\alpha$ at intermediate energies; this is an approximation that is, like case (3), based on using the Furry–Sommerfeld–Maue wave functions, but in which a number of terms that are small in the ultrarelativistic limit are retained. This leads to expressions, the Elwert–Haug [19] formulas, that contain the Sommerfeld, Bethe–Heitler, and Bethe–Maximon formulas as limiting cases. As can be seen from the diagram of Fig. 13.4, the Elwert–Haug formulas have the widest range of applicability.

It is clear from Fig. 13.4 that a substantial portion of the $(Z\alpha, \beta)$ plane is covered only by numerical calculations. As an example we turn to a series of papers [20, 22, 23] which contain numerical results for energies of E_1 = 1–500 keV with Z = 1–92, as well as a fairly detailed analysis of the accuracy of the different approximations, including the formulas of Sommerfeld, Bethe–Heitler, Bethe–Maximon, and Elwert–Haug mentioned above.*

*Here we note also a recent paper by Seltzer and Berger [78] that contains calculations of bremsstrahlung for energies of 1 keV to 10 GeV and atoms with Z = 1–100.

We now list for reference the Bethe–Heitler ($d\sigma_{\rm B.H.}$, Born approximation with $\eta_1, \eta_2 \ll 1$) and Bethe–Maximon [$d\sigma_{\rm B.M.}$, ultrarelativistic energies with $E_{1,2} \gg 1/2(Z\alpha)^2mc^2$] formulas for the bremsstrahlung spectrum. Here $\hbar = c = 1$ everywhere. The Bethe–Heitler formulas have the form ($r_0 = e^2/mc^2$)

$$d\sigma_{\rm B.H.}(\omega) = Z^2\alpha r_0^2 \frac{d\omega}{\omega}\frac{p_2}{p_1}\{\frac{4}{3} - 2\epsilon_1\epsilon_2\frac{p_1^2+p_2^2}{p_1^2p_2^2}$$
$$+ m^2(\frac{\mu_1\epsilon_2}{p_1^3} + \frac{\mu_2\epsilon_1}{p_2^3} - \frac{\mu_1\mu_2}{p_1p_2}) + L\,[\frac{8}{3}\frac{\epsilon_1\epsilon_2}{p_1p_2} + \frac{\omega^2}{p_1^3p_2^3}\left(\epsilon_1^2\epsilon_2^2 + p_1^2p_2^2\right)$$
$$+ \frac{m^2\omega}{2p_1p_2}(\mu_1\frac{\epsilon_1\epsilon_2+p_1^2}{p_1^3} - \mu_2\frac{\epsilon_1\epsilon_2+p_2^2}{p_2^3} + \frac{2\omega\epsilon_1\epsilon_2}{p_1^2p_2^2})]\}, \tag{13.49}$$

where

$$L = \ln\frac{p_1^2 + p_1p_2 - \epsilon_1\omega}{p_1^2 - p_1p_2 - \epsilon_1\omega} = 2\ln\frac{\epsilon_1\epsilon_2 + p_1p_2 - m^2}{m\omega},$$

$$\mu_1 = \ln\frac{\epsilon_1 + p_1}{\epsilon_1 - p_1} = 2\ln\frac{\epsilon_1 + p_1}{m},$$

$$\mu_2 = 2\ln\frac{\epsilon_2 + p_2}{m}.$$

At low energies Eq. (13.49) becomes much simpler and transforms to the nonrelativistic Born formula corresponding to the Gaunt factor (13.42). In the ultrarelativistic region $\varepsilon_{1,2} \gg m$, Eq. (13.49) takes the form

$$d\sigma_{\rm B.H.} \approx 4Z^2\alpha r_0^2\frac{d\omega}{\omega}\frac{\epsilon_2}{\epsilon_1}(\frac{\epsilon_1}{\epsilon_2} + \frac{\epsilon_2}{\epsilon_1} - \frac{2}{3})(\ln\frac{2\epsilon_1\epsilon_2}{m\omega} - \frac{1}{2})$$
$$= 4Z^2\alpha r_0^2\frac{d\omega}{\omega}\left[1 + \left(1 - \frac{\omega}{\epsilon_1}\right)^2 - \frac{2}{3}\left(1 - \frac{\omega}{\epsilon_1}\right)\right]\left[\ln\left(\frac{2\epsilon_1}{m}\frac{\epsilon_1 - \omega}{\omega}\right) - \frac{1}{2}\right]. \tag{13.50}$$

We note that the relativistic Born formula for the bremsstrahlung spectrum (13.49) has the same shortcoming as the nonrelativistic Born formula: the cross section goes to zero at the short-wavelength limit; i.e., $d\sigma_{\rm B.H.}$ $(\omega = \omega_{\max} = \epsilon_1 - m) = 0$. This deficiency is eliminated by multiplying $d\sigma_{\rm B.H.}$ by the Elwert factor [see Eq. (13.41)].

The bremsstrahlung spectrum in the Bethe–Maximon approximation has the form

$$d\sigma_{\text{B.M.}}(\omega) = d\sigma_{\text{B.H.}}(\omega) - 4Z^2\alpha^2 \frac{1}{m^2\omega p_1^2}\left(\epsilon_1^2 + \epsilon_2^2 - \frac{2}{3}\epsilon_1\epsilon_2\right) f(Z)\, d\omega, \quad (13.51)$$

where the function $f(Z)$ is given by

$$f(Z) = (Z\alpha)^2 \sum_{n=1}^{\infty} \frac{1}{n(n^2 + Z^2\alpha^2)}.$$

It is clear from Eq. (13.51) that the Born approximation gives excessive values of the bremsstrahlung cross section. The shape of the bremsstrahlung spectrum in the Bethe–Maximon approximation is, on the other hand, practically the same as that in the Born approximation.

13.5. Dipole Bremsstrahlung Spectrum of Classical Electrons in a Central Attractive Potential

In the framework of this chapter we still have, first of all, to clarify the "physics" of the interrelation between the quantum mechanical and classical descriptions of bremsstrahlung and, second, to generalize some of the results of Secs. 13.2 and 13.3 to the non-Coulomb case. In general, the key to attaining both objectives is to study the classical Fourier coefficient of the acceleration (and the corresponding quasiclassical matrix element of the dipole radiative transition) of an electron in a central attractive potential $U(r)$ of a certain class. The role of this duality for the classical–quantum mechanical aspect is evident, while for the Coulomb–non-Coulomb aspect it means that for this class of $U(r)$ the motion of the electrons is quasiclassical at *low* energies, which are simultaneously more sensitive to the non-Coulomb character of the atomic potential (shielding), including when the Born approximation [an "anticlassical" approximation for these $U(r)$] is clearly not valid.

Thus, we shall examine the class of monotonic central attractive potentials for which $U(r) \to -a/r^n$ as $r \to 0$, where $a > 0$ and $0 < n < 2$ (the condition that the electron does not fall into the center). This class includes, in particular, all atomic potentials. For this class of potentials, the motion of the electron ceases to be quasiclassical at *small* r [1].

Let an electron with a positive energy move in a potential $U(r)$ along a trajectory that is characterized by the integrals of motion E and M. The spectral distribution of the energy emitted by this electron as dipole bremsstrahlung over the entire duration of the collision (subline) can be written in the form

$$\frac{d\mathcal{E}_\omega}{d\omega} = \frac{2e^2m\omega^4}{3\pi c^3}\left[F_+^2(\omega) + F_-^2(\omega)\right], \quad (13.52)$$

where the functions

$$F_{\pm}(\omega) = \int\limits_{r_0(E,M)}^{\infty} \frac{r dr}{\sqrt{E + \left| U\left(\sqrt{\frac{M}{m\omega_{rot}(r)}}\right) \right| - \frac{M\omega_{rot}(r)}{2}}} \times$$

$$\times \cos\left\{ \int\limits_{r_0(E,M)}^{r} [\omega \pm \omega_{rot}(r')] \frac{dr'}{\sqrt{\frac{2}{m}\left[E + \left| U\left(\frac{M}{m\omega_{rot}(r')}\right)\right| - \frac{M\omega_{rot}(r')}{2}\right]}} \right\} \quad (13.53)$$

depend both on the parameters and on the two integrals of motion E and M. Here $\omega_{rot}(r) = M/(mr^2)$, the "local" angular rotation velocity of the electron around the center of force, is of key importance for the entire analysis that follows.

The bremsstrahlung spectrum of a homogeneous electron beam has the form

$$\frac{d\kappa(\omega)}{d\omega} = \int\limits_0^{\infty} \frac{d\mathcal{E}_\omega}{d\omega} 2\pi\rho d\rho = \frac{4e^2\omega^4}{3mc^3v^2} \int\limits_0^{\infty} \left[F^2_{+(E,M)}(\omega) + F^2_{-(E,M)}(\omega)\right] M dM. \quad (13.54)$$

The subline (13.52) is of major independent interest, since in the limit $\hbar \to 0$ the square of the radial matrix element of the corresponding radial transition (see Sec. 13.6) transforms into it. In the limit $\omega \to 0$, the form of Eq. (13.52) ceases to depend on ω [32]. This accordingly simplifies the spectrum (13.54) at low frequencies $\omega \ll \tilde{\omega}$ [the meaning of $\tilde{\omega}$ will be pointed out later; see Eq. (13.56)]: for a potential $U(r)$ that falls off faster than a Coulomb potential as $r \to \infty$, the spectrum will also approach a constant, while for a Coulomb potential it varies as $\ln(\tilde{\omega}/\omega)$, in accordance with Sec. 13.2. These results (in particular, the proportionality between the low-frequency limit of the spectrum and the transport cross section for elastic scattering, $\sigma_{tr} = \int_0^{\infty} [1 - \cos\theta(\rho)] 2\pi\rho d\rho$, in the non-Coulomb limit) have been adequately covered in the literature [39, 67] and we shall not dwell on them here (see Sec. 13.7).

The behavior of the subline and bremsstrahlung spectrum is of greatest interest to us at the rather high frequencies $\omega \gg \tilde{\omega}$ (where $\tilde{\omega}$ is the characteristic scale of the angular rotation frequency and emission frequency) corresponding to the characteristic scale length a of the problem defined by

$$|u(a)| = E = mv^2/2. \quad (13.55)$$

On introducing the angular momentum $M = \tilde{M}$ corresponding to this length, we then have

$$\tilde{\omega} \equiv v/a, \quad \tilde{M} \equiv mva. \tag{13.56}$$

In this range of ω a key role is played by the rapidity of the oscillations in the structure cos $\{\int_{r_0}^{r} [\omega \pm \omega_{rot}(r')] \ldots\}$ as a function of r under the outer integral of Eq. (13.53). When these oscillations are fast enough, this integral and, with it, the intensity of the subline at a given frequency are exponentially small (the asymptotic behavior of the functions $K_{1/3}$ and $K_{2/3}$ that describe the "Coulomb" subline in Sec. 13.2 is a concrete example of this). Thanks to this, when $\omega \gg \tilde{\omega}$ the two regions of the radial integral are effectively "localized" near the lower limit (the turning point r_0) and

$$r_{eff}' \sim r_{eff} \sim r_0(E, M). \tag{13.57}$$

The obvious meaning of this result is that with increasing ω, the radiating portion of the electron trajectory is shortened with its midpoint at r_0.

The main contribution to bremsstrahlung (13.54) at a given frequency $\omega \gg \omega_0$ is from an interval of "effective" values of $M \sim M_{eff}$ which lead precisely to quenching of the ω-oscillations in Eq. (13.53) when the order of magnitude equation

$$\omega_{rot}(r_0) \sim \omega \tag{13.58}$$

is satisfied. In accordance with the general properties of Fourier analysis, this equation indicates that a significant amount of components with frequencies ω on the order of the (required) emission frequency are present in the motion of the electrons. Crudely speaking, according to Eq. (13.58) each of the frequencies $\omega \gg \tilde{\omega}$ in the bremsstrahlung spectrum somehow selects an electron trajectory of "its own" with $M \sim M_{eff}(\omega)$ from the incident beam that corresponds to this value of ω in the integral with respect to M (13.54).

Equation (13.58) determines the form of the function $M_{eff}(\omega, E)$, but, thus far, only through r_0. It is possible to eliminate r_0 and, thereby, express the desired function solely in terms of quantities that also retain their strict significance in the quantum mechanical case (Sec. 13.6) with the aid of Eqs. (13.57), according to which the effective values of both the radicals under the integrals of Eq. (13.53) are close to zero and can simultaneously be expressed in terms of ω by using Eq. (13.58).

In this region of $\omega \gg \tilde{\omega}$, we can everywhere neglect the terms in E compared to the others. This is the energy intergral "switch-off" (EIS) effect that was identified and discussed in Sec. 13.2 for the special case of a Coulomb subline. When the EIS effect is included, the desired function $M_{eff}(\omega)$ is obtained by solving the approximate equation

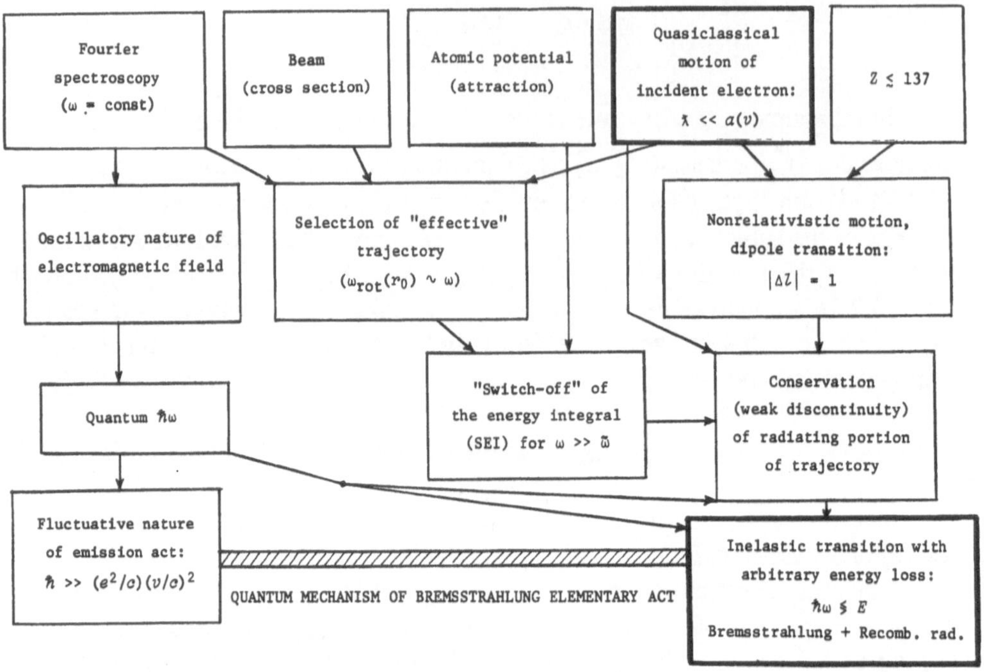

Fig. 13.5. Diagram illustrating the "origin" of Kramers electrodynamics, or the maintenance of classical spectra of electron bremsstrahlung in an atomic potential despite the quantum mechanical nature of the elementary emission mechanism. The range of validity of Kramers electrodynamics for a Coulomb potential is $v/c \ll Z/137$.

$$\left|U\left(\sqrt{\frac{M_{\rm eff}}{m\omega}}\right)\right| \approx \frac{1}{2}M_{\rm eff}\omega, \ \omega \gg \tilde{\omega}. \tag{13.59}$$

The function $M_{\rm eff}(\omega)$ decreases monotonically.

As is to be expected from the physical meaning of the EIS conditions and $\omega \gg \tilde{\omega}$, the corresponding electron trajectories are highly curved:

$$r_0 \ll \rho \ll a.$$

It follows from the above remarks that when $\omega \gg \tilde{\omega}$, the subline (13.52) in the portion where it is significant in an integral sense [i.e., is nonexponential or, more precisely, satisfies Eq. (13.58)] does not contain E at all and the bremsstrahlung spectrum (13.54) depends on E only through the universal, purely scale factor $1/v^2$.

In the special case of a Coulomb potential $U(r) = -Ze^2/r$ the above formulas yield all the main results of Sec. 13.2 on a qualitative level. Thus, the solutions of Eqs. (13.58) and (13.59) have the form [cf. the argument of the functions $K_{1/3}$ and $K_{2/3}$ in Eq. (13.19)]

$$r_0^{\rm eff}(\omega) \sim \left(\frac{Ze^2}{m\omega^2}\right)^{1/3}, \quad M_{\rm eff}(\omega) \sim \left(\frac{Z^2me^4}{\omega}\right)^{1/3} \ll \widetilde{M}, \quad \omega \gg \widetilde{\omega}.$$

13.6. Interrelation between the Quantum Mechanical (Quasiclassical) and Classical Descriptions of Bremsstrahlung. The Semiclassical Approach

In Sec. 13.3 we pointed out that when the condition for quasiclassical motion $Ze^2/\hbar v \gg 1$ is satisfied, the quantum mechanical spectrum of dipole bremsstrahlung from a beam of electrons in an attractive Coulomb field, $\hbar\omega d\sigma/d\omega$, coincides approximately with the classical spectrum $(d\kappa/d\omega)_{\rm cl}$ up to its short-wavelength limit $\omega_{\rm max} = E/\hbar$. The following question arises: what is the *physical* reason for the unexpectedly wide range of applicability of the classical theory? And, in particular, are the specific features of the Coulomb interaction important in this regard?

From a formal standpoint, any agreement between a quantum mechanical result and a classical one is related to the *loss* of $\hbar$ in the former. In a strict quantum mechanical formulation of the bremsstrahlung problem, $\hbar$ shows up in a total of three dimensionless parameters: $Ze^2/\hbar v$ (or its non-Coulomb generalization $a/\bar{\lambda}$; see below), $\hbar\omega/E$, and $e^2/\hbar c \sim 1/137$.

It is appropriate to organize this situation according to the following logical scheme (Fig. 13.5) in anticipation of the subsequent analysis:

The shaded horizontal "connecting bar" represents the unique interrelation of the two nonclassical factors in the *mechanism* for a single-photon radiative transition that is expressed by quantum electrodynamics and the arrows denote individual cause-and-effect relationships which together ultimately form the "channels for the suppression of $\hbar$." The upper row of the diagram includes the *initial* physical assumptions, among which the only "controlling" parameter (the change in E for a given potential) is the "quasiclassicality of the initial electron" [see Eq. (13.66)]. Here we shall touch upon the left side of the diagram only casually.

We now follow the passage between the quantum mechanical and classical descriptions of bremsstrahlung in terms of the diagram in Fig. 13.5. It is more instructive to do this at the level of the matrix element for the radiative transition and the corresponding classical subline (which corresponds, in turn, to a definite trajectory of the electron), rather than at the less detailed level of the spectrum $d\kappa/d\omega$. We now illustrate the passage from the quantum mechanical intensity to the classical subline (13.52) in the limit $\hbar \to 0$.*

*See the paper by Kogan and Kukushkin [36]. Unlike that paper, here we limit ourselves to deriving only the zeroth-order approximation in $\hbar$, but without the restriction to $\omega \gg \widetilde{\omega}$, i.e., for arbitrary ω.

A necessary intermediate stage in this passage is deriving the *quasiclassical* (qc) form of the subline $(d\mathcal{E}_\omega/d\omega)_{qc}$. The initial quantum mechanical analog of the classical subline for bremsstrahlung is the partial cross section corresponding to a given orbital angular momentum l. We begin with the *exact* formula [12]

$$\frac{d\sigma}{d\omega} = \frac{32m\omega^3 q'}{3\hbar c^3 a_0 q} \sum_{l=0}^{\infty} (l+1)(R_{l+1,l}^2 + R_{l,l+1}^2) \tag{13.60}$$

where $a_0 = \hbar^2/me^2$ is the Bohr radius, $\mathbf{q} = m\mathbf{v}/\hbar$ and $\mathbf{q}' = m\mathbf{v}'/\hbar$ are the wave vectors of the incident and scattered electrons, and $R_{ll'}$ is the radial matrix element of the $l \to l'$ transition, given by

$$R_{ll'} = \int_0^{\infty} u_l(q, r) u_{l'}(q', r) r dr, \tag{13.61}$$

where $u_l(q, r)$ $[\equiv rR_l(q, r)]$ is the radial wave function of the electron with the asymptote

$$u_l(q, r) = \frac{1}{q} \sin\left(qr - \frac{\pi l}{2} + \delta_l\right) \tag{13.62}$$

$[\delta_l(q)$ denotes the scattering phase shift].

In the quasiclassical case of interest to us ($l_{\text{eff}} \gg 1$), on introducing the angular momentum $mv\rho = M = \hbar(l + 1/2) \approx \hbar l$, we obtain [cf. Eq. (13.54)]

$$\frac{d\sigma}{d\omega} = \frac{1}{\hbar\omega} \cdot \frac{d\kappa}{d\omega} = \frac{1}{\hbar\omega} \int_0^{\infty} \frac{d\mathcal{E}_\omega}{d\omega} 2\pi\rho d\rho \approx \frac{2\pi\hbar}{m^2 v^2 \omega} \int_0^{\infty} \left(\frac{d\mathcal{E}_\omega}{d\omega}\right)_{qc} l dl. \tag{13.63}$$

Given that $l_{\text{eff}} \gg 1$ and taking $dl = 1$, a comparison of Eqs. (13.60) and (13.63) yields the desired expression for the quasiclassical subline:

$$\left(\frac{d\mathcal{E}_\omega}{d\omega}\right)_{qc} \approx \frac{16 m^4 e^2 \omega^4 v v'}{3\pi\hbar^4 c^3} \left[\tilde{R}_{l+1,l}^2(\omega) + \tilde{R}_{l,l+1}^2(\omega)\right], \tag{13.64}$$

where $\tilde{R}_{ll'}$ is the matrix element (13.61), but now taken between *quasiclassical* wave functions of the electron with the asymptotic behavior of Eq. (13.62).

Using the fact that the quasiclassical wave functions oscillate rapidly, we obtain

$$\tilde{R}_{l+1,l}(\omega) = \frac{\hbar^2}{\sqrt{8m^3 vv'}} \int_{r_0(E,M)}^{\infty} \frac{rdr}{\sqrt{E + |U(r)| - \dfrac{M^2}{2mr^2}}}$$

$$\times \cos\left\{\sqrt{m/2} \int_{r_0(E,M)}^{r} \frac{\omega - \omega_{\rm rot}(r')}{\sqrt{E + |U(r')| - (M/2mr'^2)}} dr'\right\}. \qquad (13.65)$$

$\tilde{R}_{l,l+1}(\omega)$ differs from Eq. (13.65) only in replacing $[\omega - \omega_{\rm rot}(r')]$ by $[\omega + \omega_{\rm rot}(r')]$.

The full agreement between the structures of Eqs. (13.65) and (13.53) and, when the remaining factors are taken into account, between the final expressions for $d\mathscr{E}_\omega/d\omega$, Eqs. (13.64) and (13.52), is now evident, as we set out to demonstrate.

Since the matrix element $\tilde{R}_{l+1\to l}(\omega)$ (13.65) transforms to the classical function $F_-(\omega)$,which makes the dominant contribution to the intensity of the subline (13.52) for $\omega \gg \tilde{\omega}$, at these frequencies it predominates over the matrix element $\tilde{R}_{l\to l+1}$, in the subline (13.64). In other words, here the angular momentum changes during a bremsstrahlung event in the same direction as the energy with an overwhelming probability. The last result is consistent both with the classical (strictly) [32] and the empirical Bethe selection rule (approximately) for dipole Coulomb matrix elements [68], which is theoretically justified for the quasiclassical case [69, 70]. This inequality between the matrix elements $R_{ll'}(\omega)$ can be regarded as a generalization of the Bethe selection rule to the non-Coulomb case, while its direct reduction to an analogous inequality between the classical quantities $F_\pm(\omega)$ can be regarded as evidence of the purely classical nature of this rule.

Formally, we have only demonstrated the limiting transition $\hbar \to 0$ from the quantum mechanical intensity to the classical subline (13.52). In fact, however, an analysis of the approximations used in the course of this demonstration also allows us to establish, on a qualitative level, the *criteria* for approximate classical behavior of the bremsstrahlung spectrum, taking the fact that $\hbar \neq 0$ into account, both in terms of the parameters of the electron's motion and in terms of ω. Satisfaction of the first of these criteria determines the presence of a wide range of realizations of the second (see the vertical coupling between the two "main" boxes on the right of Fig. 13.5).

Since the classical subline is the result of a Fourier analysis of the acceleration of an electron moving along a given, unchanged trajectory (Sec. 13.5), the above criteria for classical behavior of an actual, quantum mechanical subline mean physically that the *radiating portion of the trajectory is approximately conserved during a bremsstrahlung event.*

In this case the angular momentum l is approximately conserved in a bremsstrahlung event, since $l \gg 1$ in the quasiclassical approximation used here, while the dipole selection rule states that $|l - l'| = 1$. As for the energy E, on the contrary, it changes very strongly (Sec. 13.3) in the main part of the spectrum during a bremsstrahlung event. Here, however, the EIS effect comes in to "rescue" the conservation of the radiating part of the trajectory and, thereby, maintain the classical behavior (Secs. 13.2 and 13.5). As a result, the conservation of the radiating part at low frequencies is ensured by two conditions, $E' \sim E$ and $l' \sim l$, and for frequencies $\omega \gg \tilde{\omega}$ it is ensured just by the condition $l' \sim l$. The condition for conservation of the radiating part for the *continuous* region $\omega \ll \ldots$ (for specifics, see below) is evidently that the two subregions $E' \sim E$ and $\omega \gg \tilde{\omega}$ should overlap. For this to be so, it is necessary that the EIS effect set in even when the bremsstrahlung act is only slightly inelastic; i.e., when $\hbar\tilde{\omega} \ll E$ or, given Eq. (13.56), when

$$\tilde{M}/\hbar \equiv \tilde{l} \gg 1, \quad \text{or } \hbar/mv \equiv \lambda\!\!\!^{-} \ll a. \tag{13.66}$$

In the Coulomb case ($\tilde{M} \sim Ze^2/v$, $a \sim Ze^2/mv^2$), this inequality takes the standard form $Ze^2/\hbar v \gg 1$ of the condition for quasiclassical motion of the electron (in the region $r \gg \hbar^2/(Zme^2)$ [1]. It is clear that the physical significance of the criterion (13.66) is the same in the general, non-Coulomb case.

In sum, the criterion for quasiclassical motion of the "initial" electron and for approximate conservation of the radiating part of the trajectory during a bremsstrahlung event is the same inequality (13.66).

We now establish the high-frequency limit of the region in which the subline has a classical form: $\omega \ll \omega^*$. We define ω^* as the frequency produced by an angular momentum of $l = l'$ that violates the condition $l' \sim l$ for conservation of the radiating part. Since $|l - l'| = 1$, we have $l^* \sim 1$ or $M^* = \hbar l^* \sim \hbar$. Substituting M^* in Eq. (13.59), we arrive at the following equation for ω^*:

$$|U(\sqrt{\hbar/m\omega^*})| \sim \hbar\omega^*. \tag{13.67}$$

For the Coulomb case this yields $\omega_{Coul}^* \sim Z^2me^4/\hbar^3$, i.e., a frequency on the order of the Bohr frequency, which corresponds to the short-wavelength limit for the *combined* bremsstrahlung and radiative recombination spectra (brems.+ RR). This means that, when the favorable numerical factors are included, the *entire* bremsstrahlung + radiative recombination spectrum, $d\kappa(\omega)/d\omega$, is approximately classical when the single condition $Ze^2/\hbar v \gg 1$ is met.*

The above remarks emphasize the *physical* unity of bremsstrahlung and radiative recombination. On a practical level, we must warn of the need to "correct" this initially classical and purely bremsstrahlung spectrum in the radiative recombination region ($\hbar\omega > E$) by taking the discreteness of the final states of the recom-

*Thus, for radiative recombination, even into the nonquasiclassical ground state ($n = 1$), the difference between the exact quantum mechanical spectrum and the classical (Kramers) spectrum is characterized by a gaunt factor of $8\pi\sqrt{3}/e^4 \sim 0.80$ [12, 14]. The most recent work known to us concerning an approximate analytic theory of radiative recombination of electrons with nonhydrogen-like ions is the paper by Kim and Pratt [81].

bined electron into account. This is usually done by constricting parts of this region of the continuum into the corresponding lines [12, 14].

From all this discussion it is clear why the classical behavior of the entire bremsstrahlung *spectrum* (and essentially the radiative recombination spectrum, well) is ensured by the quasiclassical *motion* of the electron, i.e., $\lambda\!\!\!^{-} \ll a$, and the additional condition $\hbar\omega \ll E$ is unnecessary. This result can be justified rigorously by a direct calculation that demonstrates the smallness of the quantum mechanical correction to the classical limit for the matrix element of a dipole radiative transition in a central attractive potential [35, 36]. The relative magnitude of the corresponding quantum mechanical correction to the bremsstrahlung spectrum $d\kappa(\omega)/d\omega$ can be written in the following mutually equivalent forms, each of which clearly characterizes one or another aspect of the deviation from the classical *motion* of the electron near the turning point $r_0[M_{\text{eff}}(\omega)] \equiv r_\omega$ of the trajectory responsible for emission at frequencies $\omega \gg \tilde{\omega}$:

$$\left[\frac{d\kappa/d\omega}{(d\kappa/d\omega)_{\text{cl}}} - 1\right] \sim \frac{\lambda\!\!\!^{-}(r_\omega)}{r_\omega} \sim \frac{1}{l_{\text{eff}}(\omega)} \sim \frac{\hbar\omega}{E_{\text{kin}}(r_\omega)}, \tag{13.68}$$

where $\lambda\!\!\!^{-}(r_\omega) = \hbar/(2mE_{\text{kin}}(r_\omega))^{1/2} \ll \hbar/mv\ (\equiv\lambda\!\!\!^{-})$ is the local wavelength of the electron and $E_{\text{kin}}(r_\omega)$ is its maximum kinetic energy. The last equality in Eq. (13.68) shows that the "kinematic" condition for a classical spectrum, $\hbar\omega \ll E$, must be replaced by a much weaker "dynamic" condition, $\hbar\omega \ll E_{\text{kin}}(r_\omega)$, in the quasiclassical case. It is clear that this last condition provides the most direct coupling between the breakdown of classical behavior in the bremsstrahlung spectrum and the distortion of the radiating part of the trajectory by means of the recoil effect in an actual bremsstrahlung event.

In connection with the controlling role of quasiclassical motion in determining classical behavior in the spectrum demonstrated here, it is important to note that the dipole selection rule $|l - l'| = 1$, itself, actually follows from the criterion (13.66) for quasiclassical behavior. Thus, for the Coulomb case, on rewriting the inequality $Ze^2/\hbar v \gg 1$ in the form $v/c \ll Ze^2/\hbar c \approx Z/137$, we conclude that for all real values of $Z \leq 137$, it expresses the *nonrelativistic* behavior of the electron and, therefore [32], the dipole nature of this radiative process.

Returning to the scheme of Fig. 13.5, we see that it sums up all the physics discussed thus far in Secs. 13.5 and 13.5 (and partially in Sec. 13.2) and demonstrates the many-branched "cooperation" between quantum mechanical ($\hbar\omega$), on one hand, and quasiclassical ($\lambda\!\!\!^{-} \ll a$) and purely classical (the block that is completed by the EIS cell) behavior in keeping the radiating part of the electron trajectory constant. In the quasiclassical case, even the emission of a large photon $\hbar\omega$ cannot disrupt (and, on the whole, distort) the part $\omega_{\text{rot}} \sim \omega$ of the trajectory that it "samples" from an electron beam which is responsible for bremsstrahlung at the frequency ω. This bremsstrahlung regime continues to *imitate the classical regime* up to those high frequencies ω for which the minimum classically allowable angular momenta

$l \geq 1$ are responsible. Thanks to this (as well as to the oscillatory behavior of the photon field) and despite the particularly quantum mechanical, fluctuating mechanism for an elementary bremsstrahlung event and its strongly "kinematic" inelasticity [up to $(\hbar\omega/E)_{max} \sim (Ze^2/\hbar v)^2 \gg 1$ in the Coulomb case], the spectral distribution of the bremsstrahlung (+ radiative recombination) intensity remains approximately classical.

This discussion tells us that, within the framework of quantum electrodynamics, there is a physical region within which even highly inelastic radiative processes involving collisions of low-energy, "plasma" electrons with atoms and ions can be described classically. As an acknowledgment of the pioneering work of Kramers [41] on the theory of electron bremsstrahlung in a Coulomb field, it is appropriate to refer to this region as the *Kramers* region, and the new, classical method for calculating radiative processes in quantum mechanical systems, as *Kramers electrodynamics.*

Semiclassical Approach to the Theory of Bremsstrahlung. A fairly general formula for the bremsstrahlung cross section that unifies a number of the limiting results of Sec. 13.3 for the Coulomb case can be obtained by a semiclassical method [33, 34]. The essence of this method is to use the results of the classical theory (Sec. 13.2) in combination with certain quantum mechanical limitations that require that the quantum mechanical uncertainties in the classical quantities of importance to the bremsstrahlung process should be small.* In practice, these limitations reduce to cutting off the classical trajectories at short distances.

We begin with the classical formula (13.13). Integrating its product with $2\pi\rho$ over ρ from 0 to ∞ yields the classical Gaunt factor (13.15). We now require that the quantum mechanical uncertainty in the distance of closest approach, r_0, should be relatively small; i.e., $\Delta r_0 \leq r_0$. We have $\Delta r_0 \geq \hbar/\Delta p_1 \geq \hbar/p_1$ $(p_1 = mv_1)$, from which $r_0 \geq \lambda\!\!\!^{-}_1 = \hbar/mv_1$. Setting $r_0 = B\lambda\!\!\!^{-}_1$, where B is a numerical coefficient of order unity, which will be refined later on, and using the relation between ρ and r_0, $\rho = r_0(1 + a/r_0)^{1/2}$, we find

$$\rho_{min} = B\lambda\!\!\!^{-}_1\sqrt{1 + 2\eta_1/B}. \tag{13.69}$$

Now, integrating over the limits from ρ_{min} from Eq. (13.69) to $\rho = \infty$, we obtain

$$g_{semicl}(\nu, \eta_1) = \frac{\pi\sqrt{3}}{4} i\nu\left(1 + \frac{1}{\eta_1'}\right) H_{i\nu}^{(1)}\left[i\nu\left(1 + \frac{1}{\eta_1'}\right)\right] H_{i\nu}^{(1)'}\left[i\nu\left(1 + \frac{1}{\eta_1'}\right)\right], \tag{13.70}$$

*We emphasize the difference between the *semiclassical* method considered here and the conventional *quasiclassical* method. The latter is a regular expansion for a limited region of variation in the parameter $Ze^2/\hbar v$, while the first pretends (although more or less as a set of "recipes") to encompass in an approximate way the entire range of variation in $Ze^2/\hbar v$, including the Born, anticlassical limit.

where $\eta_1' = \eta_1/B$. Comparing the limit (13.70) for $\eta_1 \ll 1$ and $\nu \ll 1$ with the logarithmic Born formula (13.43) shows that we must set $B = \gamma \equiv 1.781...$ in order for them to coincide.

When $\eta_1 \gg 1$, the semiclassical spectrum (13.70) transforms into the classical spectrum (13.15) with its limiting formulas (13.12) and (13.16), for $\eta_1 \ll 1$ and $y \ll 1$ into Eq. (13.43) and when $y \ll 1$ and $\nu \ll 1$ (with arbitrary $\eta_1 = 2\nu/y$), into

$$g \approx \frac{\sqrt{3}}{\pi}\ln\frac{2}{(\gamma + 1/\eta_1)\nu} = \frac{\sqrt{3}}{\pi}\ln\frac{4}{2\gamma\nu + y}. \tag{13.71}$$

This formula has an analogous significance and is numerically very close to the Elwert formula (13.48). In view of its "symmetry," Eq. (13.71) is an obvious interpolation between Eqs. (13.16) and (13.43).

It should be noted that the practical applicability of the semiclassical formula (13.70) as an approximation to the exact result (13.31) is wider than might seem at first glance (see Fig. 13.1). Formally, its most vulnerable point is its inability to give a correct description of the short-wavelength limit of the bremsstrahlung spectrum ($y \sim 1$) at Born energies $\eta_1 \ll 1$. In fact, however, the values of η_1 are fairly strongly limited from below in the nonrelativistic theory considered here. Thus, at the boundary between the Born and quasiclassical regions, which correspond to $\eta_1' \sim 1$ or $\eta_1 \sim 1/\gamma$, according to Eqs. (13.70) and (13.71), we have $v_1/c = Ze^2\gamma/\hbar c \sim Z/80$. Thus, for sufficiently large Z, the Born limit $\eta_1 \ll 1/\gamma$ can in general be realized only in the relativistic region. It is clear from Fig. 13.1 that even when $\eta_1 = 0.1$, the semiclassical formula (13.70) is significantly different from the exact Eq. (13.31) only in the neighborhood of the short-wavelength limit $y \sim 1$.

The possibilities of the quasiclassical approach can be extended with the aid of a "symmetrizing" procedure over the initial and final states that is widely used in the theory of the Coulomb excitation of nuclei [82].

13.7. Electron Bremsstrahlung in the Field of a Rigid Many-Electron Atom or Ion

In any study of the bremsstrahlung of electrons on many-electron atoms (ions) one must take the shielding of the Coulomb field of the nucleus into account. For electrons with sufficiently high energies, this can be done in the framework of the well-explored Born approximation [2, 8, 68], which we shall not discuss in detail here. For slower electrons, the shielding effect is clearly just as important, but the Born approximation is not applicable. Thus, for bremsstrahlung [71] it gives a qualitatively incorrect (falling with increasing ω) form for the spectrum. This is explained by the inability of the Born (essentially a "straight-line") approximation to account for the dominant contribution of highly curved trajectories to the emission at high frequencies. On the other hand, elastic scattering of electrons with

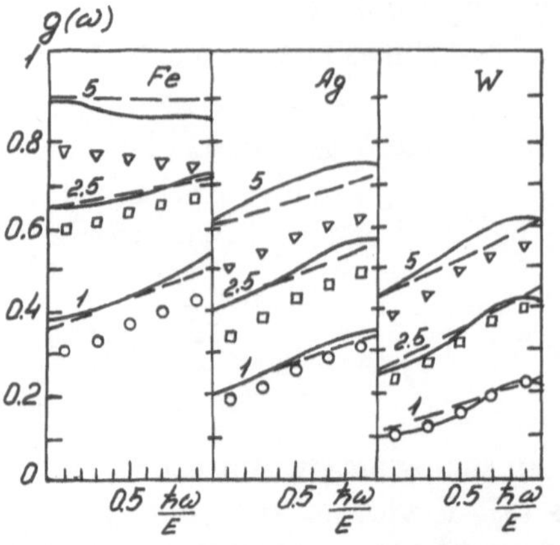

Fig. 13.6. Comparison of bremsstrahlung spectra on neutral Fe, Ag, and W atoms calculated with the aid of the "rotational" approximation (13.73) (dashed curve) with the results of numerical quantum mechanical calculations [22] (smooth curves) and [75] (circles, squares, and triangles). The numbers on the curves denote the energy of the electrons in keV.

the required energies (say, $E \leq 10$ keV) on heavy atoms is well described by the quasiclassical approximation [72, 73], which agrees with experiments. This is also natural: for all potentials of the class examined here, the degree of quasiclassical behavior in the motion increases at lower electron energies.Thus, in light of the results of Sec. 13.6, constructing a theoretical description of the bremsstrahlung spectra in this range of parameters (we now specify the atomic number as well to be $Z \geq 20$) on a purely classical basis is fully justified.

This approach was suggested by Gervids and Kogan [33] and subsequently realized by Kogan and Kukushkin [35, 36]. The corresponding region of radiative processes is physically interesting and of practical importance. Although these processes take place in quantum mechanical systems, they are essentially classical and obey "Kramers electrodynamics" (see Sec. 13.6 and Fig. 13.5).

The analytic structure of the spectrum (13.54) can be greatly simplified for arbitrary $U(r)$ over a wide range of $\omega \geq \tilde{\omega}$. It is natural to refer to this new approximation for the classical bremsstrahlung spectrum as the *rotational* approximation. It generalizes the Kramers approximation (13.13) to the non-Coulomb case and to lower frequencies.

This approach corresponds to replacing the actual subline with a model form $\delta[\omega - \omega_{\mathrm{rot}}(r_0)]$. In order to find the corresponding spectrum (the superposition of all the sublines), it is convenient to use the exact formula (13.9) for the integral effective emission κ. This yields

$$\frac{d\kappa}{d\omega} \sim \frac{8\pi e^2}{3c^3 m^2 v} \int_0^\infty \left(\frac{dU}{dr}\right)^2 \sqrt{1 - \frac{U(r)}{E}}\, r^2 \delta(\omega - \omega_{\rm rot}(r))\, dr. \tag{13.72}$$

The undetermined numerical coefficient of order unity is obtained by comparing the main term in the expansion of the spectrum (13.72) in terms of $1/\omega$ with the corresponding term in the exact asymptote of the spectrum (13.54). For the potentials of the class considered here which are of greatest practical importance, namely atomic potentials that transform to the Coulomb potential $U = -Ze^2/r$ when $r \to 0$, we finally obtain the following Gaunt factor from Eq. (13.72) [35, 36]:

$$g_{\rm rot}(\omega) \equiv \frac{(d\kappa/d\omega)_{\rm rot}}{(d\kappa/d\omega)_{\rm Kr}} = \frac{6}{Z^2 e^4} \frac{D_\omega^2}{2 + D_\omega} \frac{[E + |U(r_\omega)|]^3}{m\omega^2}, \tag{13.73}$$

where

$$D_\omega = -\frac{d \ln (E + |U(r_\omega)|)}{d \ln r_\omega}, \tag{13.74}$$

and the effective radius r_ω is the root of the equation

$$\frac{E + |U(r_\omega)|}{r_\omega^2} = \frac{m\omega^2}{2}. \tag{13.75}$$

For the purely Coulomb case, these formulas describe the exact classical spectrum [16], even at a frequency as low as $\omega = 1/2\tilde{\omega}$, with an accuracy of better than 5%. They cannot, however, describe the logarithmic growth in this spectrum for $\omega \ll \tilde{\omega}$.

Equations (13.73)–(13.75) can be used in an analytic description of the bremsstrahlung of electrons in the keV range on many-electron atoms [35, 36, 74]. This classical description can be used to "universalize" a large amount of related numerical quantum mechanical calculations [22, 75] in terms of simple, Fermi–Thomas similarity parameters. In the region with $Z \geq 20$ and $\varepsilon \leq 2$, agreement exists even for $\omega \sim \tilde{\omega}$ to within 10–20%. Here the dimensionless parameter $\varepsilon = 32.6E(\text{keV})/Z^{4/3}$ and $1/\varepsilon$ plays the role of the quasiclassicality parameter $a/\lambda\!\!\!^-$ (13.66).

It is useful to make an overall comparison of the quantum mechanical and classical results at the level of the "reference" functions $g_1(\varepsilon)$ and $g_0(\varepsilon)$ (Fig. 13.7), which are the Gaunt factors for bremsstrahlung at the low-frequency ($\omega \sim \tilde{\omega}$) limit of the region where the rotational approximation is valid and at a frequency of $\omega = 0$, where the bremsstrahlung intensity can be expressed [67] in terms of the transport cross section for elastic scattering of electrons in a given potential.

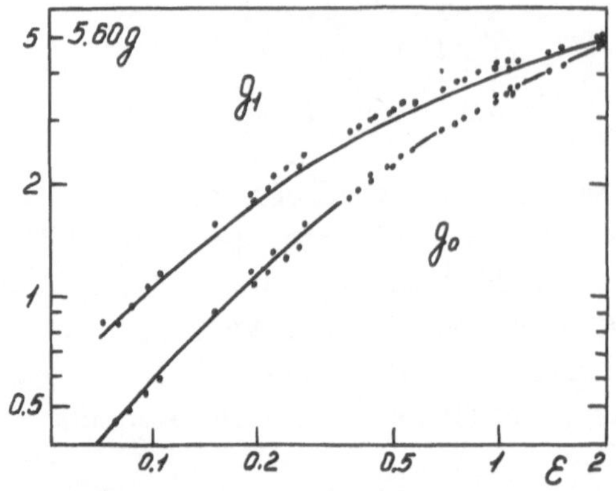

Fig. 13.7. The classical universal functions $g_0(\varepsilon)$ and $g_1(\varepsilon)$ (curves) compared with the corresponding (readjusted) results of some numerical quantum mechanical calculations [22].

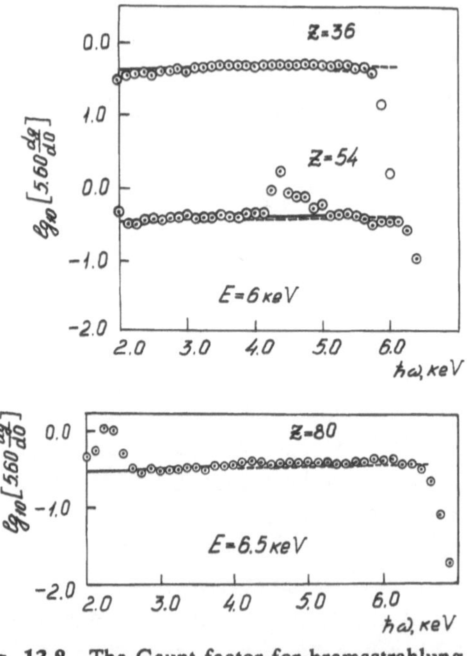

Fig. 13.8. The Gaunt factor for bremsstrahlung of electrons with energy E on neutral Kr, Xe, and Hg atoms for a photon exit angle of $\theta = 90°$ with $E_1 = 6$ keV (a) and $E_2 = 6.5$ keV (b). The points are the results of measurements [76], the smooth curves are quantum mechanical calculations [23], and the dashed curves are calculations using the classical formulas [36].

The rotational approximation can also be used to obtain the similarity law $g(E/Z)$ for the short-wavelength limit $\omega_{max} = E/\hbar$ of the bremsstrahlung spectrum. As ω is reduced to zero, there is a transition to the similarity law $g(E/Z^{4/3})$ [74].

In this region of its validity, therefore, the *classical* description simultaneously provides a simple physical explanation for the main *quantum mechanical* features of the bremsstrahlung spectra: the degree of shielding of the nucleus, the generally slow variation in the spectra, and their increase with ω that is supplanted by a decrease after $\varepsilon \sim 2$ (Figs. 13.6 and 13.7). This demonstrates the adequacy of the classical approach developed here as a method for describing bremsstrahlung, while the specific choice of a model for the atomic potential (Hartree–Fock [22] or Fermi–Thomas [75]) is of little importance.

The theory developed here has been compared [36] with experiment [76] (and with corresponding numerical quantum mechanical calculations [23]) for the case of bremsstrahlung spectra on many-electron atoms with values of E and Z that lie within a region $\varepsilon \leq 2$ of sufficiently quasiclassical behavior. The spectra were measured only for a fixed exit angle $\theta = 90°$ of the photons relative to the direction of the incident electrons. The formulas for the differential spectrum with respect to θ needed for the comparison with the experiments [76] were obtained with the aid of the present approach in [36].

The results of this comparison are shown in Fig. 13.8. The good agreement is evidence that this model of a static atomic potential for describing bremsstrahlung at moderate frequencies is realistic. The observed intensity peaks are from characteristic x-ray lines superimposed on the bremsstrahlung spectrum [76]. The dip in the spectrum at the short-wavelength limit (i.e., the breakdown of the rotational character of the spectrum on passing beyond bremsstrahlung) is related simply to the fact that here we are dealing with photoattachment, rather than radiative recombination (because of the absence of vacancies in the electron shell of a neutral atom).

Chapter 14

CONCLUSION

M. Ya. Amus'ya, V. M. Buimistrov, B. A. Zon, V. I. Kogan, and V. N. Tsytovich

In ending this presentation, we would like to note that we have come to our conclusion about the importance of the new polarization mechanism for bremsstrahlung by different methods and on the basis of different initial assumptions (as well as for different physical systems), which are mostly independent, but also somewhat interdependent. As almost always happens in these cases, when a certain clarity of physical understanding has been attained, the question arises of why this was not done long ago and right away. The answer to this question is essentially standard: the process of understanding the results of a calculation often takes longer and is more complicated than the calculations themselves (i.e., than the solution of specific problems). This is especially true when a physical picture develops that contradicts the conventional understanding. Only later, when the isolated fragments are combined into a complete whole, does it seem that the new point of view essentially has no opposition.

The main purpose of our effort has been to attain the required clarity or trivialization of the results. Ultimately, we descended (or, perhaps, in a certain sense, rose) from the "heights" of nonlinear plasma theory and quantum electrodynamics to almost doing arithmetic "on our fingers." Of course, this simplification of the results should not be overemphasized. A rigorous mathematical theory forms a reliable basis for the results and makes it possible to solve more complicated problems. And after all, once it has developed, a physical picture seemingly lives an independent life and stirs us to action. What sort of action should this be?

In this regard we may touch upon the prospects for new experimental research. We begin with polarization (transition) bremsstrahlung in plasmas. As usual, new experimental studies will probably require additional theoretical and computational work, since good agreement between theory and experiment can only be obtained this way. However, the general qualitative consequences of the theory and the general physical concepts concerning polarization bremsstrahlung allow us even now to see several new possibilities.

We shall first consider spontaneous emission and then briefly discuss the generation and amplification of radiation through the polarization bremsstrahlung mechanism.

The first set of problems which might be a subject for experimental investigations is radiation by heavy particles, such as protons. All of the variants of proton emission that were considered in this book are actually of interest. In a degenerate solid-state plasma, the frequency $\omega_{pe}c/v_F$, where v_F is the velocity on the Fermi surface, is quite high. The polarization (transition) bremsstrahlung of heavy particles extends up to these frequencies. The frequency $\omega_{pe}c/v_F$ may be as high as 10^{18}–10^{19} s^{-1} and correspond to the kilovolt x-ray range. Only polarization bremsstrahlung is still possible for heavy particles within a certain frequency range and this is a qualitatively new effect. At these frequencies, polarization bremsstrahlung on bound electrons of the atoms or ions that make up a solid will also be significant. That is, a combination of the two effects will be important, as follows from the results of Chapter 6. Naturally, more detailed theoretical calculations will be required for solid-state plasmas in order to take into account the specific features of the processes taking place in solids. (In particular, the momentum transferred during the emission process may be taken up to a certain extent by the lattice as a whole, etc.)

A second area for possible experiments is bremsstrahlung of ultrarelativistic electrons with $\gamma = \varepsilon/m_0c^2 > c/v_F$, in solid-state plasmas. This range of electron energies (on the order of 1 GeV) is quite attainable in modern accelerators and here we should expect a dip in the emission spectrum at $\omega \sim \omega_{pe}\varepsilon/mc^2$, i.e., in the x-ray range. At frequencies below $\omega_{pe}\varepsilon/mc^2$, the "density effect" shows up in ordinary bremsstrahlung. It has been observed experimentally by Arutyunyan et al. [1, 1a]. In those experiments, however, the electron energies were not high enough to reveal the dip in the frequency dependence of the emission. At frequencies below $\omega_{pe}c/v_F$ the radiation from the electrons becomes entirely polarization bremsstrahlung radiation (a growth in the intensity should be observed as the frequency is lowered, followed by saturation at frequencies around $\omega_{pe}c/v_F$). Again, the theoretical calculations should be refined for specific targets and to take the bound electrons into account.

Except for a few cases in recent years, experiments dealing with bremsstrahlung on bound electrons (of which there are quite a few) are compared with a theory based on the shielding approximation. At first glance, this may seem unimportant since experimental results hardly depend on theoretical formulas. In reality, of course, theory allows us to find the conditions under which a given effect shows up. As a rule, an experiment is much easier to set up when you know, essentially, what you are looking for. From the more precise theory developed in this book, it is clear that the shielding approximation works better and the polarization corrections are small in certain frequency ranges. As an example, for electron scattering on hydrogen atoms the exact theory yields a new physical picture and a new formula for the differential cross section, but the integral cross section is close to that obtained in the shielding approximation (because scattering on the nucleus is decisive here). For scattering of protons and other heavier nuclei, the situation is completely different. Here polarization bremsstrahlung is decisive at relatively low frequencies, in both the differential and total cross sections. In an experiment, however, matters are made more complicated by the fact that other processes, such as the emission of δ-electrons, operate to produce the x-ray background. Serious analyses of specific situations, such as the analysis by Ishii and Morita [2], are needed in order to isolate polarization bremsstrahlung from other effects. This sort of work is essentially only beginning, but this should not discourage us if we recall the history of the discovery of transition emission, which has a physically similar mechanism. There also, many years of steadfast work by theorists led to the construction of the decisive experiments.

Now we briefly consider the amplification of stimulated radiation. Absorption and emission are related by the Einstein coefficients. In the general case of nonequilibrium distributions of the incident particles and target particles, the absorption (emission) effect is obtained from the balance equation in terms of the bremsstrahlung probability (see [3, 4] and Chapter 3.) The probability, of course, includes the combined effect of traditional and polarization bremsstrahlung. The possibility of amplifying and generating bremsstrahlung arose almost simultaneously with the development of lasers. This problem is the subject of a thorough review [5] based on the traditional mechanism for bremsstrahlung. It is clear that including the polarization mechanism might introduce some new elements. Absorption always occurs with a Maxwellian distribution for the colliding particles (for a plasma with a Maxwellian distribution for all the particles, and for atoms with a Maxwellian distribution of the incident particles and Boltzmann thermal populations in the atomic levels). Amplification requires a nonequilibrium population inversion. Such inversions can be produced in atoms by a variety of means, especially through excitation by a beam of incident particles.

The advantage of the bremsstrahlung mechanism lies in the possibility of using a population inversion in the beam of incident particles when there is no population inversion in the atomic levels. For centers at rest (the particles labeled β in Chapter 3), the growth rate for pumping of bremsstrahlung photons (Chapter 3) is given by

$$2\gamma_{\mathbf{k}}^{\beta} = \int w_{\mathbf{p}_\alpha,\mathbf{p}_\beta}(\mathbf{q},\ \mathbf{k}) \left(\mathbf{q} \frac{\partial \Phi_{\mathbf{p}_\beta}^{\beta}}{\partial p_\beta} \right) \Phi_{\mathbf{p}_\alpha}^{\alpha} \frac{d\mathbf{p}_\alpha dp_\beta d\mathbf{q}}{(2\pi)^9} .$$

When $\partial\Phi/\partial p > 0$ (which can happen for a beam of incident particles), we have $\gamma > 0$, which is a necessary condition for amplification. The process responsible for the emission may be any of the mechanisms for polarization bremsstrahlung on free or bound electrons. Here two circumstances must be kept in mind. First, γ_k is proportional to both the density of incident particles and to the density of scattering centers ($\Phi_{p\alpha}{}^{\alpha}$); that is, for practical realization of amplification it is desirable that these densities be a bit higher. The density of electrons in a beam (provided the incident particles are electrons) cannot be very high and, if we are interested in gain at high frequencies, then the small parameter ω_{pe}/ω necessarily enters, where ω_{pe} is the plasma frequency of the beam. For small values of ω_{pe}/ω, the growth rate will also be small. Thus, we must choose a fairly dense target, such as a solid-state plasma (where ω_{pe} is somewhat higher). Here, however, the question becomes one of how far the beam can penetrate into the target (this is limited by scattering). Second, besides the interaction with the incident particles, photons in the medium experience other interactions that cause them to be damped. If ν is the characteristic reciprocal damping time, then for amplification the threshold $\gamma > \nu$ must be exceeded. This condition either requires a sufficiently dense or a sufficiently monoenergetic beam and the larger the density of the target is, the more rigid the requirements on the beam will be. Of course, at high densities the complicated physics of the different collective processes enters and they must be taken into account. In short, although the possibility of amplification is clear in principle, practical realization of gain will require a lot of work to optimize all the conditions in a specific target material. Polarization bremsstrahlung opens up new possibilities here.

The problem of wave amplification during bremsstrahlung was formulated early on by Bunkin et al. [5] and Tsytovich [6] and has been stated for arbitrary media [3] and for plasmas [7]. It turns out that the optimum frequency range in plasmas for relativistic incident particles is $c/v_F < \omega < \epsilon/mc^2$. Sufficiently powerful relativistic beams with these energies, however, do not yet exist. With respect to nonrelativistic beams, this process requires further studies in which the dynamic polarization of the atoms is taken into account. Akopyan and Tsytovich [7] have examined the effect of a magnetic field on the amplification for the polarization (transition) mechanism and shown that the conditions for amplification are substantially better when a strong magnetic field is present in a plasma. Solid-state plasmas are of great interest, although we can only make rough estimates for them now. Nevertheless, the gain problem is of such great interest that an attempt may be made to obtain a number of these effects experimentally without waiting for the refined theoretical calculations.

We note here, as mentioned above, that the obvious impediment to observing the amplification of radiation by the polarization effect on bound electrons is its small absolute magnitude. Besides bremsstrahlung and inverse bremsstrahlung (absorption), there are other forms of absorption, as well as scattering of bremsstrahlung photons. Clearly, if this background absorption is greater than the (bremsstrahlung) emission, then no gain has actually been obtained. It is also understandable that the situation changes if the frequency of the radiation is close to resonant, because in this case there is a sharp increase in the emission cross section. The resonant case has been examined elsewhere [8, 9]. The possibility of amplifying and generating electromagnetic radiation during scattering of relativistic electrons on atoms with a population inversion has been studied recently [5, 6].

At the present time, it is of key importance to follow the transition from the traditional "static" mechanism for bremsstrahlung of electrons to the polarization mechanism as the emission frequency ω decreases and the nucleus Ze is simultaneously "overgrown" with a many-electron shell. In this regard, an extension of the experiments of Semaan and Quarles [10] on the bremsstrahlung spectrum of electrons in the keV range to neutral many-electron atoms would be of great interest.

The numerical calculations of Pratt's group [11] and the analytic theory [12] are in good agreement with these experiments at fairly high frequencies. There have been no measurements at low frequencies. Thus, it is extremely desirable either to extend the experimental techniques of [10] directly to lower frequencies ω or to greatly expand the range of electron energies ε and atomic numbers Z.

The proposed research on the transition between the two bremsstrahlung mechanisms at keV energies also has a direct bearing on the physics of thermonuclear plasmas. This would allow us to reach a well-founded conclusion about the actual region within which the polarization mechanism operates in this important problem.

There is, therefore, a perceptible need for further experiments that have been specially designed for observing and studying polarization bremsstrahlung, as well as for new theoretical work aimed at proposing and analyzing these experiments. In fact, stirring experimentalists to action has been one of the main purposes of this book.

REFERENCES

Chapter 1

1. V. L. Ginzburg, *Theoretical Physics and Astrophysics* (2nd edn.) [in Russian], Nauka, Moscow (1981)[1st edn.: Pergamon, New York (1979)].
2. L. D. Landau and E. M. Lifshitz, *Classical Theory of Fields* [in Russian], Fizmatgiz, Moscow (1962) [Pergamon, New York (1976)].
3. G. S. Landsberg, *Optics* [in Russian], Gostekhizdat, Moscow (1947).
4. A. Sommerfeld, *Atomic Structure and Spectral Lines*, London (1924).
5. L. D. Landau and I. Ya. Pomeranchuk, "Limits of applicability of the theory of electron bremsstrahlung and pair formation at high energies," *Dokl. Akad. Nauk SSSR* **92**, 535 (1953).
6. A. B. Migdal, Effect of multiple scattering on bremsstrahlung at high energies *Dokl. Akad. Nauk SSSR* **96**, 49 (1955).
7. M. L. Ter-Mikaélyan, *Effect of a Medium on High-Energy Electromagnetic Processes* [in Russian], Izd. Akad. Nauk Arm. SSR, Erevan (1969) [*High-Energy Electromagnetic Processes in Condensed Media*, Wiley, New York (1972)].
8. W. Heitler, *The Quantum Theory of Radiation*, Oxford University Press, Oxford (1954).
9. F. Sauter, "Über die Bremsstrahlung schneller Elektronen," *Ann. Phys.* **20**, 404 (1934).

9a. F. Sauter, "Zur unrelativistischen Theorie des kontinuierlichen Röntgenspektrums," *Ann. Phys.* **18**, 486 (1933).

10. V. N. Tsytovich, "Bremsstrahlung of a relativistic plasma," *Tr. Fiz. Inst. Akad. Nauk SSSR* **66**, 191 (1973) [*Proc. P. N. Lebedev Phys. Inst.* **66** (1973)].
11. A. V. Akopyan and V. N. Tsytovich, "Bremsstrahlung in nonequilibrium plasmas," *Fiz. Plazmy* **1**, 673 (1975) [*Sov. J. Plasma Phys.* **1**, 371 (1975)].
12. A. V. Akopyan and V. N. Tsytovich, "Transition bremsstrahlung of relativistic particles," *Zh. Eksp. Teor. Fiz.* **71**, 166 (1976) [*Sov. Phys. – JETP* **44**, 87 (1976)].

12a. A. V. Akopyan and V. N. Tsytovich, "Bremsstrahlung of relativistic electrons in a strong magnetic field," *Zh. Eksp. Teor. Fiz.* **72**, 1824 (1977) [*Sov. Phys. – JETP* **45**, 957 (1977)].

13. V. L. Ginzburg and V. N. Tsytovich, *Transition Radiation and Transition Scattering*, Adam Hilger, New York–London (1990)].

14. V. L. Ginzburg and V. N. Tsytovich, "Transition scattering," *Zh. Eksp. Teor. Fiz.* **65**, 1818 (1973) [*Sov. Phys. – JETP* **38**, 909 (1973)].

15. V. M. Buimistrov, "Resonant bremsstrahlung and absorption of photons," *Ukr. Fiz. Zh.* **17**, 640 (1972).

16. V. M. Buimistrov and L. I. Trakhtenberg, "Cross section for bremsstrahlung during scattering of electrons on hydrogen atoms," *Zh. Eksp. Teor. Fiz.* **69**, 108 (1975) [*Sov. Phys. – JETP* **42**, 55 (1975)].

16a. V. M. Buimistrov and L. I. Trakhtenberg, "Role of atomic electrons in bremsstrahlung," *Zh. Eksp. Teor. Fiz.* **73**, 850 (1977) [*Sov. Phys. – JETP* **46**, 447 (1977)].

17. M. Ya. Amus'ya, A. S. Baltenkov, and A. A. Paiziev, "Bremsstrahlung of electrons on atoms including polarizability," *Pis'ma Zh. Eksp. Teor. Fiz.* **24**, 366 (1976) [*JETP Lett.* **24**, 332 (1976)].

18. B. A. Zon, "Bremsstrahlung effect in collisions of electrons with atoms," *Zh. Eksp. Teor. Fiz.* **73**, 128 (1977) [*Sov. Phys. – JETP* **46**, 65 (1977)].

18a. B. A. Zon, "Absorption of optical radiation by weakly ionized gases," *Zh. Eksp. Teor. Fiz.* **77**, 44 (1979) [*Sov. Phys. – JETP* **50**, 21 (1979)].

19. V. M. Buimistrov, Yu. A. Krotov, and L. I. Trakhtenberg, "Emission of a photon during a collision of a proton and positron with an atom," *Zh. Eksp. Teor. Fiz.* **79**, 808 (1980) [*Sov. Phys. – JETP* **52**, 411 (1980)].

20. M. Ya. Amus'ya, "Atomic bremsstrahlung spectrum," *Comments At. Mol. Phys.* **11**, 123 (1982).

21. M. Ya. Amus'ya, M. Yu. Kuchiev, and A. V. Solov'ev, "Bremsstrahlung in atom–atom collisions," *Pis'ma Zh. Tekh. Fiz.* **10**, 1025 (1984) [*Sov. Tech. Phys. Lett.* **10**(9), 433 (1984)].

22. V. A. Astapenko, V. M. Buimistrov, Yu. A. Krotov et al., "Dynamic bremsstrahlung of relativistic charged particles on atoms," *Zh. Eksp. Teor. Fiz.* **88**, 1560 (1985) [*Sov. Phys. – JETP* **61**, 930 (1985)].

23. M. Ya. Amusia, A. V. Korol, and A. V. Soloviev, "Bremsstrahlung in electron–positronium scattering," *Z. Phys. D* **1**, 347 (1986).

24. M. Ya. Amusia and A. V. Soloviev, "Inelastic scattering on muonic hydrogen," *J. Phys. B: At. Mol. Phys.* **18**, 3663 (1985).

25. V. A. Astapenko, "Emission and absorption of photons in many-particle interactions," Abstract of Candidate's Dissertation, Moscow Physicotechnical Institute, MFTI, Moscow (1985).

26. M. Ya. Amus'ya, M. Yu. Kuchiev, A. V. Korol', and A. V. Solov'ev, "Bremsstrahlung of relativistic particles including dynamic polarizability of the target atom," *Zh. Eksp. Teor. Fiz.* **88**, 383 (1985) [*Sov. Phys. – JETP* **61**, 224 (1985)].

27. Ya. B. Zel'dovich and Yu. P. Raizer, "Avalanche breakdown of gases by a light pulse," *Zh. Eksp. Teor. Fiz.* **47**, 1150 (1964) [*Sov. Phys. – JETP* **20**, 772 (1965)].
28. G. Wendin and K. Nuroch, "Bremsstrahlung resonances and appearance-potential spectroscopy near the 3*d* thresholds in metallic Ba, La, and Ce," *Phys. Rev. Lett.* **39**, 48 (1977).
29. M. Ya. Amus'ya, T. M. Zimkina, and M. Yu. Kuchiev, "Broad emission bands in atomic emission produced by fast electrons," *Zh. Tekh. Fiz.* **52**, 1424 (1982) [*Sov. Phys. Tech. Phys.* **27**, 866 (1982)].
30. E. T. Verkhovtseva, E. V. Gnatchenko, and P. S. Pogrebnjak, "Investigation of the connection between giant resonances and atomic bremsstrahlung," *J. Phys. B: At. Mol. Phys.* **16**, L613 (1983).
31. T. M. Zimkina, I. I. Lyakhovskaya, A. S. Shulakov, et al., "Structure of x-ray absorption and emission spectra of lanthanum and thorium in compounds with oxygen and fluorine at energies of 60 to 140 eV," *Fiz. Tverd. Tela* **25**, 26 (1983) [*Sov. Phys. Solid State* **25**, 13 (1983)].
32. M. Ya. Amusia, N. B. Avdonina, L. V. Chernysheva, and M. Yu. Kuchiev, "'Stripping' of the atom in bremsstrahlung," *J. Phys. B: At. Mol. Phys.* **18**, L791 (1985).
32a. V. B. Avdonina, M. Ya. Amus'ya, M. Yu. Kuchiev, and L. V. Chernysheva, "Bremsstrahlung of fast electrons on atoms," *Zh. Tekh. Fiz.* **56**, 246 (1986) [*Sov. Phys. Tech. Phys.* **31**(8), 137 (1986)].
33. F. Folkmann, C. Gaard, T. Huus, et al., "Proton-induced x-ray emission as a tool for trace element analysis," *Nucl. Instrum. Methods* **116**, 487 (1974).
34. V. M. Kolyada, A. N. Zaichenko, and R. V. Dmitrenko, *X-Ray Spectral Analysis with Ion Excitation* [in Russian], Atomizdat, Moscow (1987).
35. D. N. Jacubassa and M. Kleber, "Bremsstrahlung in heavy-ion reactions," *Z. Phys. A* **273**, 29 (1975).
36. R. Anholt and T. K. Sailor, "Radiative ionization in slow ion–atom collisions," *Phys. Rev. Lett. A* **56**, 455 (1976).
37. K. Ishii and S. Morita, "Continuum x rays produced by light ion–atom collisions," *Phys. Rev. A* **30**, 2278 (1984).
38. D. F. Hubbard and M. E. Rose, "Nuclear structure effects in bremsstrahlung," *Nucl Phys.* **84**, 377 (1966).
39. R. L. McGrath, D. Arbiola, J. Karp, et al., "Direct γ-transition in $^{12}C + ^{12}C$," *Phys. Rev. C* **24**, 2374 (1981).
40. V. Metag, A. Lazzarini, K. Lesko, et al., Search for γ-rays from quasimolecular $C^{12} + C^{12}$ system," *Phys. Rev. C* **25**, 682 (1982).
41. E. M. Nyman, "Bremsstrahlung in heavy-ion collisions," *Phys. Lett. B* **136**, 143 (1984).
42. L. Landau and G. Rumer, "The cascade theory of electron showers," *Proc. R. Soc. London A* **166**, 213 (1938).
43. I. C. Percival and M. J. Seaton, "The polarization of atomic line radiation excited by electron impact," *Philos. Trans. R. Soc. London A* **251**, 113 (1958).
44. N. Mott and H. Massey, *Theory of Atomic Collisions*, Oxford University Press, Oxford (1965).

45. I. L. Fabelinskii, *Molecular Scattering of Light* [in Russian], Nauka, Moscow (1965) [Plenum, New York (1967)].
46. G. Bekefi, *Radiation Processes in Plasmas*, Wiley (1965).
47. V. L. Ginzburg and V. N. Tsytovich, "Some questions in the theory of transition radiation and transition scattering," Materials of the 2nd Symp. on Transition Radiation of High-Energy Particles, Erevan Phys. Inst., Erevan (1984), pp. 48–96.
48. V. N. Tsytovich, *Theory of Turbulent Plasma*, Consultants Bureau, New York (1977).
48a. V. N. Tsytovich, *Nonlinear Effects in Plasma*, Consultants Bureau, New York (1970).
49. B. B. Kadomtsev, *Plasma Turbulence*, Academic Press, New York (1965).
49a. B. B. Kadomtsev, *Collective Phenomena in Plasmas* [in Russian], Nauka, Moscow (1976).
50. A. Gailitis and V. N. Tsytovich, "Emission of transverse electromagnetic waves during scattering of charged particles on plasma waves," *Zh. Eksp. Teor. Fiz.* **46**, 1726 (1964) [*Sov. Phys. – JETP* **19**, 1165 (1964)].
50a. A. Gailitis and V. N. Tsytovich, "Emission during scattering of charged particles on electromagnetic waves in an isotropic plasma," *Zh. Eksp. Teor. Fiz.* **47**, 1469 (1964) [*Sov. Phys. – JETP* (1964)].
50b. V. N. Tsytovich, "Nonlinear effects in plasmas," *Usp. Fiz. Nauk* **90**, 435 (1966) [*Sov. Phys. Usp.* **9**, 805 (1967)].
51. A. G. Sitenko, *Fluctuations and Nonlinear Wave Interactions in Plasmas* [in Russian], Naukova Dumka, Kiev (1977) [Pergamon, New York (1982)].
52. V. B. Berestetskii, E. M. Lifshitz, and L. P. Pitaevskii, *Quantum Electrodynamics* [in Russian], Nauka, Moscow (1980) [Pergamon, New York (1982)].
53. M. L. Ter-Mikaélyan, "Spectrum of bremsstrahlung in a medium," *Dokl. Akad. Nauk SSSR* **94**, 1033 (1954).
54. D. V. Melrose, "Tsytovich–Razin effect in bremsstrahlung," *Astrophys., Space Sci.* **18**, 267 (1972).
55. L. P. Rapoport, B. A. Zon, and N. L. Manakov, *The Theory of Multiphoton Processes in Atoms* [in Russian], Atomizdat, Moscow (1978).
56. S. I. Vetchinkin and S. V. Khristenko, "Use of the Coulomb Green function for calculating the interaction between a hydrogen atom and the field of a light wave in second-order perturbation theory," *Opt. Spektrosk.* **25**, 650 (1968).
57. M. Gavrila, "Elastic scattering of photons," *Phys. Rev.* **163**, 147 (1967).
58. V. L. Ginzburg and I. M. Frank, "Radiation from a uniformly moving electron as it passes from one medium to another," *Zh. Eksp. Teor. Fiz.* **16**, 15 (1946) [short version *J. Phys. USSR* **9**, 353 (1949)].
59. I. E. Tamm and I. M. Frank, "Coherent radiation from a fast electron in a medium," *Dokl. Akad. Nauk SSSR* **14**, 107 (1937).
60. S. P. Kapitsa, "Radiation from a charge moving in a nonuniform medium," *Zh. Eksp. Teor. Fiz.* **39**, 1367 (1960) [*Sov. Phys. – JETP* **12**, 954 (1961)].
61. M. L. Ter-Mikaélyan, "Radiation from fast particles in a nonuniform medium," *Dokl. Akad. Nauk SSSR* **134**, 318 (1960) [*Sov. Phys. Dokl.* **5**, 1015 (1961)].

Chapter 2

1. V. L. Ginzburg and V. N. Tsytovich, *Transition Radiation and Transition Scattering* [in Russian], Nauka, Moscow (1984) [Adam Hilger, New York–London (1990)].
2. I. E. Tamm and I. M. Frank, "Coherent radiation from fast electrons in a medium," *Dokl. Akad. Nauk SSSR* **14**, 107 (1937).
3. V. L. Ginzburg and V. N. Tsytovich, "Transition scattering," *Zh. Eksp. Teor. Fiz.* **65**, 1818 (1973) [*Sov. Phys. – JETP* **38**, 909 (1973)].
4. V. N. Tsytovich, "Bremsstrahlung of relativistic plasmas," *Tr. Fiz. Inst. Akad. Nauk SSSR* **66**, 191 (1973) [*Proc. P. N. Lebedev Phys. Inst.* **66** (1973)].
5. A. V. Akopyan and V. N. Tsytovich, "Bremsstrahlung in nonequilibrium plasmas," *Fiz. Plazmy* **1**, 673 (1975) [*Sov. J. Plasma Phys.* **1**, 371 (1975)].
6. A. V. Akopyan and V. N. Tsytovich, "Transition bremsstrahlung of relativistic particles," *Zh. Eksp. Teor. Fiz.* **71**, 166 (1976) [*Sov. Phys. – JETP* **44**, 87 (1976)].
7. A. V. Akopyan and V. N. Tsytovich, Bremsstrahlung of nonrelativistic suprathermal electrons in plasmas, Preprint No. 184, Fiz. Inst. Akad. Nauk SSSR, Moscow (1978).
8. M. L. Ter-Mikaélyan, "Spectrum of bremsstrahlung in a medium," *Dokl. Akad. Nauk SSSR* **94**, 1033 (1954).
9 D. V. Melrose, "Tsytovich–Razin effect in bremsstrahlung," *Astrophys. Space Sci.* **18**, 267 (1972).
10. V. M. Buimistrov and L. I. Trakhtenberg, "Role of atomic electrons in bremsstrahlung," *Zh. Eksp. Teor. Fiz.* **73**, 850 (1977) [*Sov. Phys. JETP* **46**, 447 (1977)].

Chapter 3

1. V. N. Tsytovich, "Bremsstrahlung of a relativistic plasma," *Tr. Fiz. Inst. Akad. Nauk SSSR* **66**, 191 (1973) [*Proc. P. N. Lebedev Phys. Inst.* (1973)].
2. A. V. Akopyan and V. N. Tsytovich, "Bremsstrahlung in nonequilibrium plasmas," *Fiz. Plazmy* **1**, 673 (1975) [*Sov. J. Plasma Phys.* **1**, 371 (1975)].
3. S. P. Kapitsa, "Radiation from a charge moving in a nonuniform medium," *Zh. Eksp. Teor. Fiz.* **39**, 1367 (1960) [*Sov. Phys. – JETP* **12**, 954 (1961)].
4. M. L. Ter-Mikaélyan, "Radiation from fast particles in a nonuniform medium," *Dokl. Akad. Nauk SSSR* **134**, 318 (1960) [*Sov. Phys. Dokl.* **5**, 1015 (1961)].
5. L. D. Landau, "The transport equation in the case of Coulomb interactions," *Zh. Eksp. Teor. Fiz.* **7**, 203 (1937) [*Phys. Z. Soviet* **10**, 154 (1936)].
6. R. Balescu, *Statistical Mechanics of Charged Particles*, Wiley, New York (1963).
7. A. G. Sitenko, *Fluctuations and Nonlinear Interactions of Waves in Plasmas* [in Russian], Naukova Dumka, Kiev (1977) [Pergamon, New York (1982)].
8. Yu. L. Klimontovich, *The Statistical Theory of Nonequilibrium Processes in a Plasma* [in Russian], Izd. Mosk. Gos. Univ., Moscow (1964) [Pergamon, New York (1967)].

9. V. L. Ginzburg and V. N. Tsytovich, "Transition radiation," *Zh. Eksp. Teor. Fiz.* **65**, 1818 (1973) [*Sov. Phys. – JETP* **38**, 909 (1973)].
10. G. Bekefi, *Radiation Processes in Plasmas*, Wiley, New York (1965).

Chapter 4

1. A. Sommerfeld, *Atomic Structure and Spectral Lines* , London (1924).
2. H. Bethe and E. Salpeter, *The Quantum Mechanics of One- and Two-Electron Systems*, Springer-Verlag, Berlin (1958).
3. W. Heitler, *The Quantum Theory of Radiation*, Oxford University Press, Oxford (1954).
4. F. Sauter, "Über die Bremsstrahlung schneller Elektronen," *Ann. Phys.* **20**, 404 (1934).
4a. F. Sauter, "Zur unrelativistischen Theorie des kontinuierlichen Röntgenspektrums," *Ann. Phys.* **18**, 486 (1933).
5. M. L. Ter-Mikaélyan, *Effect of a Medium on High-Energy Electromagnetic Processes* [in Russian], *Izd. Akad. Nauk Arm. SSR*, Erevan (1969) [*High-Energy Electromagnetic Processes in Condensed Media*, Wiley, New York (1972)].
6. I. C. Percival and M. J. Seaton, *Philos. Trans. R. Soc. London A* **251**, 113 (1958).
7. O. B. Firsov and M. I. Chibisov, "Bremsstrahlung of slow electrons on neutral atoms," *Zh. Eksp. Teor. Fiz.* **39**, 1770 (1960) [*Sov. Phys.– JETP* **12**, 1235 (1961)].
8. V. Kas'yanov and A. Starostin, "Theory of bremsstrahlung of slow electrons on neutral atoms," *Zh. Eksp. Teor. Fiz.* **48**, 295 (1965) [*Sov. Phys.– JETP* **21**, 193 (1965)].
9. V. M. Buimistrov, "Resonant bremsstrahlung and absorption of photons," *Ukr. Fiz. Zh.* **17**, 640 (1972).
10. V. M. Buimistrov and L. I. Trakhtenberg, "Cross section for bremsstrahlung during scattering of electrons on hydrogen atoms," *Zh. Eksp. Teor. Fiz.* **69**, 108 (1975) [*Sov. Phys. – JETP* **42**, 55 (1975)].
10a. V. M. Buimistrov and L. I. Trakhtenberg, "Role of atomic electrons in bremsstrahlung," *Zh. Eksp. Teor. Fiz.* **73**, 850 (1977) [*Sov. Phys. – JETP* **46**, 447 (1977)].
11. M. Ya. Amus'ya, A. S. Baltenkov, and A. A. Paiziev, "Bremsstrahlung of electrons on atoms including polarizability," *Pis'ma Zh. Eksp. Teor. Fiz.* **24**, 366 (1976) [*JETP Lett.* **24**, 332 (1976)].
12. B. A. Zon, "Bremsstrahlung effect in collisions of electrons with atoms," *Zh. Eksp. Teor. Fiz.* **73**, 128 (1977) [*Sov. Phys. – JETP* **46**, 65 (1977)].
12a. B. A. Zon, "Absorption of optical radiation by weakly ionized gases," *Zh. Eksp. Teor. Fiz.* **77**, 44 (1979) [*Sov. Phys. – JETP* **50**, 21 (1979)].
13. S. Hayakawa, "Cosmic background x rays produced by intergalactic inner-bremsstrahlung," *Prog. Theor. Phys.* **41**, 1592 (1969).
14. F. R. Arutyunyan and V. A. Tumanyan, "Bremsstrahlung of electrons in a medium as heavy charged particles pass through it," Sci. Commun. of the Kh. Abovyan Arm. State Pedagogical Inst., Erevan (1975).

15. F. A. Agaronyan, S. R. Keller, V. G. Kirillov-Ugryumov, and Yu. D. Kotov, "Cosmic x rays produced by bremsstrahlung of nonthermal protons and nuclei," *Izv. Akad. Nauk SSSR, Ser. Fiz.* **43**, 2499 (1979).

16. V. M. Buimistrov, Yu. A. Krotov, and L. I. Trakhtenberg, "Emission of a photon during a collision of a proton and positron with an atom," *Zh. Eksp. Teor. Fiz.* **79**, 808 (1980) [*Sov. Phys. – JETP* **52**, 411 (1980)].

17. A. V. Akopyan and V. N. Tsytovich, "Transition bremsstrahlung of relativistic particles," *Zh. Eksp. Teor. Fiz.* **71**, 166 (1976) [*Sov. Phys. – JETP* **44**, 87 (1976)].

17a. A. V. Akopyan and V. N. Tsytovich, "Bremsstrahlung in nonequilibrium plasmas," *Fiz. Plazmy* **1**, 673 (1975) [*Sov. J. Plasma Phys.* **1**, 371 (1975)].

18. F. Folkmann, C. Gaard, T. Huuns, et al., "Proton-induced x-ray emission as a tool for trace element analysis," *Nucl. Instrum. Methods* **116**, 487 (1974).

19. D. N. Jacubassa and M. Kleber, "Bremsstrahlung in heavy-ion reactions," *Z. Phys. A* **273**, 29 (1975).

20. K. Ishii and S. Morita, "Continuum x rays produced by light ion–atom collisions," *Phys. Rev. A* **30**, 2278 (1984).

21. L. P. Rapoport, B. A. Zon, and N. L. Manakov, *Theory of Multiphoton Processes in Atoms* [in Russian], Atomizdat, Moscow (1978).

22. S. I. Vetchinkin and S. V. Khristenko, "Use of the Coulomb Green function for calculating the interaction between a hydrogen atom and the field of a light wave in second-order perturbation theory," *Opt. Spektrosk.* **25**, 650 (1968).

23. M. Gavrila, "Elastic scattering of photons," *Phys. Rev.* **163**, 147 (1967).

24. V. M. Buimistrov, "Probabilities of electronic transitions of atoms in nonuniform electric fields with an arbitrary angular dependence. A new computational technique," *Lit. Fiz. Sb.* **3**, 93 (1963).

25. M. Abramowitz and I. Stegun, *Handbook of Mathematical Functions* , Dover, New York (1964).

26. L. I. Trakhtenberg, "Resonant emission and absorption of photons in three-body collisions of electrons, photons, and atoms," [*Opt. Spectr.* **44**, 510 (1978)].

27. L. I. Trakhtenberg, "Radiative processes in many-particle interactions," Author's abstract of candidate's dissertation, L. Ya. Karpov Physicochemical Research Institute, Moscow (1977).

28. L. D. Landau and E. M. Lifshitz, *Quantum Mechanics* [in Russian], Nauka, Moscow (1974) [Pergamon, New York (1975)].

29. A. Weingartshofer, S. Holmes, et al., "Direct observations of multiphoton process in laser-induced free–free transitions," *Phys. Rev. Lett.* **39**, 269 (1977).

30. F. V. Bunkin, A. E. Kazakov, and M. V. Fedorov, "Interaction of intense optical radiation with free electrons (nonrelativistic case)," *Usp. Fiz. Nauk* **107**, 559 (1972) [*Sov. Phys. Usp.* **15**, 416 (1973)].

31. F. Novak, G. Friedman, and R. Hochstrasser, "Resonant scattering of light by molecules. Time-dependent and coherent phenomena," in *Laser and Coherent Spectroscopy* [Russian translation], Mir, Moscow (1982), pp. 548–620.

32. V. N. Tsytovich, "Bremsstrahlung of a relativistic plasma," *Tr. Fiz. Inst. Akad. Nauk SSSR* **66**, 191 (1973) [*Proc. P. N. Lebedev Phys. Inst.* **66** (1973)].
33. V. L. Ginzburg, *Theoretical Physics and Astrophysics* [in Russian], Nauka, Moscow (1981) [Pergamon, New York (1979)].
34. V. L. Ginzburg and V. N. Tsytovich, *Transition Radiation and Transition Scattering* [in Russian], Nauka, Moscow (1984) [Adam Hilger, New York–London (1990)].
35. V. M. Buimistrov, "Excitation of an atom by electron–photon impact," *Phys. Lett. A* **30**, 136 (1969).
36. N. K. Rahman and F. H. M. Faisal, "High-energy resonant cross sections for simultaneous electron–photon excitation of the 3*s* state of hydrogen," *J. Phys. B: At. Mol. Phys.* **11**, 2003 (1978).

Chapter 5

1. H. Bethe and W. Heitler, "On stopping of fast particles and on creation of positive electrons," *Proc. R. Soc. London A* **146**, 83 (1934).
2. H. Bethe, "The screening influence on the creation and stopping of electrons," *Proc. Cambridge Philos. Soc.* **30**, 524 (1934).
3. L. Landau and G. Rumer, "The cascade theory of electron showers," *Proc. R. Soc. London A* **166**, 213 (1938).
4. J. Wheeler and W. Lamb, "Influence of atomic electrons on radiation and pair production," *Phys. Rev.* **55**, 858 (1939).
5. M. L. Ter-Mikaélyan, *Effect of a Medium on High-Energy Electromagnetic Processes* [in Russian], Izd. Akad. Nauk Arm. SSR, Erevan (1969) [*High-Energy Electromagnetic Processes in Condensed Media*, Wiley, New York (1972)].
6. A. I. Akhiezer and V. B. Berestetskii, *Quantum Electrodynamics* [in Russian], Nauka, Moscow (1981).
7. W. Heitler, *The Quantum Theory of Radiation*, Oxford University Press, Oxford (1954).
8. V. B. Berestetskii, E. M. Lifshitz, and L. P. Pitaevskii, *Quantum Electrodynamics* [in Russian], Nauka, Moscow (1980) [Pergamon, New York (1982)].
9. H. Bethe and E. Salpeter, The Quantum Mechanics of One- and Two-Electron Systems, Springer-Verlag, Berlin (1958).
10. F. Rohrlich and J. Joseph, "Contribution of the electron to pair production in hydrogen," *Phys. Rev.* **78**, 161 (1950).
11. Yu. A. Krotov, Bremsstrahlung and scattering of photons in many-particle collisions, Abstract of candidate's dissertation, Moscow Physicotechnical Institute, Moscow (1981).
12. V. A. Astapenko, V. M. Buimistrov, Yu. A. Krotov, et al., "Dynamic bremsstrahlung of relativistic charged particles on atoms," *Zh. Eksp. Teor. Fiz.* **88**, 1560 (1985) [*Sov. Phys. – JETP* **61**, 930 (1985)].
13. M. Ya. Amus'ya, M. Yu. Kuchiev, A. V. Korol', and A. V. Solov'ev, "Bremsstrahlung of relativistic particles including dynamic polarizability of the target atom," *Zh. Eksp. Teor. Fiz.* **88**, 383 (1985) [*Sov. Phys. – JETP* **61**, 224 (1985)].

14. V. N. Baier, V. M. Katkov, and V. S. Fadin, *Radiation from Relativistic Electrons* [in Russian], Atomizdat, Moscow (1973).
15. S. Hayakawa, "Cosmic background x rays produced by intergalactic inner-bremsstrahlung," *Prog. Theor. Phys.* **41**, 1592 (1969).
16. E. Boldt and P. Serlemitsos, "Cosmic x-ray bremsstrahlung associated with suprathermal protons," *Astrophys. J.* **157**, 557 (1969).
17. F. R. Arutyunyan and V. A. Tumanyan, "Bremsstrahlung of electrons in a medium as heavy charged particles pass through it," Sci. Commun. Kh. Abovyan Arm. State Pedagogical Institute, Erevan (1975).
18. F. A. Agaronyan, S. R. Kel'ner, V. G. Kirillov-Ugryumov, and Yu. D. Kotov, "Cosmic x rays produced by bremsstrahlung of nonthermal protons and nuclei," *Izv. Akad. Nauk SSSR, Ser. Fiz.* **43**, 2499 (1979).
19. E. T. Verkhovtseva, E. V. Gnatchenko, and P. S. Pogrebnjak, "Investigation of the connection between giant resonances and atomic bremsstrahlung," *J. Phys. B: At. Mol. Phys.* **16**, L613 (1983).
20. V. A. Astapenko, "Emission and absorption of photons in many-particle interactions," Abstract of Candidate's Dissertation, Moscow Physicotechnical Institute, Moscow (1985).

Chapter 6

1. A. V. Akopyan and V. N. Tsytovich, "Bremsstrahlung in nonequilibrium plasmas," *Fiz. Plazmy* **1**, 673 (1975) [*Sov. J. Plasma Phys.* **1**, 371 (1975)].
2. A. V. Akopyan and V. N. Tsytovich, "Transition bremsstrahlung of relativistic particles," *Zh. Eksp. Teor. Fiz.* **71**, 166 (1976) [*Sov. Phys. – JETP* **44**, 87 (1976)].
3. V. A. Astapenko, V. M. Buimistrov, Yu. A. Krotov, et al., "Dynamic bremsstrahlung of relativistic charged particles on atoms," *Zh. Eksp. Teor. Fiz.* **88**, 1560 (1985) [*Sov. Phys. – JETP* **61**, 930 (1985)].
4. V. L. Ginzburg and V. N. Tsytovich, *Transition Radiation and Transition Scattering* [in Russian], Nauka, Moscow, (1984) [Adam Hilger, New York–London (1990)].
5. E. M. Lifshitz and L. P. Pitaevskii, *Statistical Physics* [in Russian], Part 2, Nauka, Moscow (1978).
6. L. D. Landau and E. M. Lifshitz, *Electrodynamics of Continuous Media* [in Russian], Nauka, Moscow (1982).
7. V. B. Berestetskii, E. M. Lifshitz, and L. P. Pitaevskii, *Quantum Electrodynamics* [in Russian], Nauka, Moscow (1980) [Pergamon, New York (1982)].
8. D. Pines, Elementary Excitations in Solids, Benjamin, New York (1963).
9. F. Platzman and P. Wolf, Waves and Interactions in Solid State Plasmas, Academic Press, New York (1973).
10. D. F. Hubbard and M. E. Rose, "Nuclear structure effects in bremsstrahlung," *Nucl. Phys.* **84**, 377 (1966).
11. M. L. Ter-Mikaélyan, "Interference radiation of superfast electrons," *Zh. Eksp. Teor. Fiz.* **25**, 296 (1953).

Chapter 7

1. V. B. Berestetskii, E. M. Lifshitz, and L. P. Pitaevskii, Quantum Electrodynamics [in Russian], Nauka, Moscow (1980) [Pergamon, New York (1982)].
2. I. C. Percival and M. J. Seaton, "The polarization of atomic line radiation excited by electron impact," *Philos. Trans. R. Soc. London A* **251**, 113 (1958).
3. V. M. Buimistrov and L. I. Trakhtenburg, "Cross section for bremsstrahlung during scattering of electrons on hydrogen atoms," *Zh. Eksp. Teor. Fiz.* **69**, 108 (1975) [*Sov. Phys. – JETP* **42**, 55 (1975)].
4. M. Ya. Amus'ya, A. S. Baltenkov, and A. A. Paiziev, "Bremsstrahlung of electrons on atoms including polarizability," *Pis'ma Zh. Eksp. Teor. Fiz.* **24**, 366 (1976) [*JETP Lett.* **24**, 332 (1976)].
5. M. Pindzola and H. P. Kelly, "Free–free radiative absorption coefficients for the negative argon ion," *Phys. Rev. A* **14**, 204 (1976).
6. M. Ya. Amus'ya, A. S. Baltenkov, and V. B. Gilerson, "Bremsstrahlung of fast electrons on atoms," *Pis'ma Zh. Tekh. Fiz.* **3**, 1105 (1977) [*Sov. Tech. Phys. Lett.* **3**, 455 (1977)].
7. G. Wendin and K. Nuroch, "Bremsstrahlung resonances and appearance-potential spectroscopy near the 3*d* thresholds in metallic Ba, La, and Ce," *Phys. Rev. Lett.* **39**, 48 (1977).
8. B. A. Zon, "Bremsstrahlung effect in collisions of electrons with atoms," *Zh. Eksp. Teor. Fiz.* **73**, 128 (1977) [*Sov. Phys. – JETP* **46**, 65 (1977)].
9. M. Ya. Amus'ya, T. M. Zimkina, and M. Yu. Kuchiev, "Broad emission bands in atomic emission produced by fast electrons," *Zh. Tekh. Fiz.* **52**, 1424 (1982) [*Sov. Phys. Tech. Phys.* **27**, 866 (1982)].
10. T. M. Zimkina, I. I. Lyakhovskaya, A. S. Shulakov, et al., "Structure of x-ray absorption and emission spectra of lanthanum and thorium in compounds with oxygen and fluorine at energies of 60 to 140 eV," *Fiz. Tverd. Tela* **25**, 26 (1983) [*Sov. Phys. Solid State* **25**, 13 (1983)].
11. M. Ya. Amusia, N. B. Avdonina, L. V. Chernysheva, and M. Yu. Kuchiev, "'Stripping' of the atom in bremsstrahlung," *J. Phys. B: At. Mol. Phys.* **18**, L791 (1985).
12. V. B. Avdonina, M. Ya. Amus'ya, M. Yu. Kuchiev, and L. V. Chernysheva, "Bremsstrahlung of fast electrons on atoms," *Zh. Tekh. Fiz.* **56**, 246 (1986) [*Sov. Phys. Tech. Phys.* **31**(2), 137 (1986)].
13. E. T. Verkhovtseva, E. V. Gnatchenko, and P. S. Pogrebnjak, "Investigation of the connection between giant resonances and atomic bremsstrahlung," *J. Phys. B: At. Mol. Phys.* **16**, L613 (1983).
14. V. M. Buimistrov and L. I. Trakhtenberg, "Role of atomic electrons in bremsstrahlung," *Zh. Eksp. Teor. Fiz.* **73**, 850 (1977) [*Sov. Phys. – JETP* **46**, 447 (1977)].
15. M. Ya. Amus'ya, M. Yu. Kuchiev, A. V. Korol', and A. V. Solov'ev, "Bremsstrahlung of relativistic particles including dynamic polarizability of the target atom," *Zh. Eksp. Teor. Fiz.* **88**, 383 (1985) [*Sov. Phys. – JETP* **61**, 224 (1985)].

16. V. A. Astapenko, V. M. Buimistrov, Yu. A. Krotov, et al., "Dynamic bremsstrahlung of relativistic charged particles on atoms," *Zh. Eksp. Teor. Fiz.* **88**, 1560 (1985) [*Sov. Phys. – JETP* **61**, 930 (1985)].
17. D. F. Hubbard and M. E. Rose, "Nuclear structure effects in bremsstrahlung," *Nucl. Phys.* **84**, 337 (1966).
18. V. M. Buimistrov, Yu. A. Krotov, and L. I. Trakhtenberg, "Emission of a photon during a collision of a proton and positron with an atom," *Zh. Eksp. Teor. Fiz.* **79**, 808 (1980) [*Sov. Phys. – JETP* **52**, 411 (1980)].
19. M. Ya. Amus'ya, "Atomic bremsstrahlung spectrum," *Comments At. Mol. Phys.* **11**, 123 (1982).
20. M. Ya. Amus'ya, M. Yu. Kuchiev, and A. V. Solov'ev, "Bremsstrahlung in atom–atom collisions," *Pis'ma Zh. Tekh. Fiz.* **10**, 1025 (1984) [*Sov. Tech. Phys. Lett.* **10**, 431 (1984)].
21. M. Ya. Amus'ya, M. Yu. Kuchiev, and A. V. Solov'ev, "Bremsstrahlung in atom–atom collisions," *Zh. Eksp. Teor. Fiz.* **89**, 1512 (1985) [*Sov. Phys. – JETP* **62**, 876 (1985)].
22. V. L. Ginzburg and V. N. Tsytovich, *Transition Radiation and Transition Scattering* [in Russian], Nauka, Moscow (1984) [Adam Hilger, New York–London (1990)].
23. M. Ya. Amus'ya, M. Yu. Kuchiev, and A. V. Solov'ev, "Total emission spectrum in collisions of heavy charged particles with atoms," *Pis'ma Zh. Tekh. Fiz.* **11**, 1401 (1985) [*Sov. Tech. Phys. Lett.* **11**, 577 (1985)].
24. N. March, W. Yang, and S. Sampanchar, *The Many-Body Problem in Quantum Mechanics*, Cambridge University Press, Cambridge (1973).
25. M. Ya. Amusia and N. A. Cherepkov, "Many-electron correlations in scattering processes," *Case Stud. At. Phys.* **5**, 47 (1975).
26. M. Ya. Amus'ya, V. K. Ivanov, S. A. Sheinerman, and S. I. Sheftel', "Readjustment of electron shells during ionization processes," *Zh. Eksp. Teor. Fiz.* **78**, 910 (1980) [*Sov. Phys. – JETP* **51**, 458 (1980)].
27. M. Ya. Amusia, V. K. Ivanov, and V. A. Kupchenko, "Photoionization of inner shells," *J. Phys. B: At. Mol. Phys.* **14**, L667 (1981).
28. B. A. Zon, "Absorption of optical radiation by weakly ionized gases," *Zh. Eksp. Teor. Fiz.* **77**, 44 (1979) [*Sov. Phys. – JETP* **50**, 21 (1979)].
29. L. D. Landau and E. M. Lifshitz, *Quantum Mechanics* [in Russian], Nauka, Moscow (1974) [Pergamon, New York (1975)].
30. P. A. Golovinskii and B. A. Zon, "Bremsstrahlung effect in collisions of electrons with negative ions," *Zh. Tekh. Fiz.* **50**, 1847 (1980) [*Sov. Phys. Tech. Phys.* **25**, 1076 (1980)].
31. N. B. Avdonina, M. Ya. Amus'ya, M. Yu. Kuchiev, and L. V. Chernysheva, "Effect of atomic structure on the angular distribution and polarization of bremsstrahlung," *Izv. Akad. Nauk SSSR. Ser. Fiz.* **50**, 1261 (1986).
32. A. B. Migdal, *Qualitative Methods in Quantum Theory* [in Russian], Nauka, Moscow (1975) [Benjamin–Cummings, Menlo Park (1977)].

33. A. A. Radzig and B. M. Smirnov, *Reference Data on Atoms, Molecules, and Ions*, Springer-Verlag, Berlin (1985).
34. L. V. Chernysheva, N. B. Avdonina, M. Ya. Amus'ya, and M. Yu. Kuchiev, "System for mathematical programming of atomic calculations 'Atom' XIII," Preprint No. 865, FTI. Leningrad (1983).
35. K. Ishii and S. Morita, "Continuum x rays produced by light ion–atom collisions," *Phys. Rev. A* **30**, 2278 (1984).

Chapter 8

1. M. Ya. Amusia, A. V. Korol, and A. V. Soloviev, "Bremsstrahlung in electron–positronium scattering," *Z. Phys. D* **1**, 347 (1986).
2. M. Ya. Amusia and A. V. Soloviev, "Inelastic scattering on muonic hydrogen," *J. Phys. B: At. Mol. Phys.* **18**, 3663 (1985).
3. A. A. Varfolomeev, "Coherent inelastic scattering of neutrinos on electrons in a medium," *Yad. Fiz.* **28**, 1034 (1978) [*Sov. J. Nucl. Phys.* **28**, 531 (1978)].
4. A. A. Varfolomeev, "Coherent interaction of neutrinos with a dense medium," *Yad. Fiz.* **31**, 1268 (1980) [*Sov. J. Nucl. Phys.* **31**, 655 (1980)].
5. M. Ya. Amus'ya, I. S. Gul'karov, M. B. Zhalov, and T. M. Potapova, "Effect of nuclear structure on the bremsstrahlung spectra of fast nucleons," *Yad. Fiz.* **40**, 1321 (1984) [*Sov. J. Nucl. Phys.* **40**, 839 (1984)].
6. R. L. McGrath, D. Arbiola, J. Karp, et al., "Direct γ-transitions in $^{12}C + ^{12}C$," *Phys. Rev. C* **24**, 2374 (1981).
7. V. Metag, A. Lazzarini, K. Lesko, and R. Vandenbosch, "Search for γ-rays from quasimolecular $^{12}C + ^{12}C$ system," *Phys. Rev. C* **25**, 682 (1982).
8. E. M. Nyman, "Bremsstrahlung in heavy-ion collisions," *Phys. Lett. B* **136**, 143 (1984).
9. M. Ya. Amusia and N. A. Cherepkov, "Many-electron correlation in scattering processes," *Case Stud. At. Phys.* **5**, 47 (1975).
10. M. Ya. Amusia, N. A. Cherepkov, L. V. Chernysheva, and S. G. Shapiro, "The elastic scattering of slow positrons on He atoms," *J. Phys. B: At. Mol. Phys.* **9**, L531 (1976).
11. V. L. Ginzburg and V. N. Tsytovich, "Transition scattering," *Zh. Eksp. Teor. Fiz.* **65**, 1818 (1973) [*Sov. Phys. – JETP* **38**, 909 (1974)].
12. V. M. Buimistrov and L. I. Trakhtenberg, "Role of atomic electrons in bremsstrahlung," *Zh. Eksp. Teor. Fiz.* **73**, 850 (1977) [*Sov. Phys. – JETP* **46**, 447 (1977)].
13. V. B. Berestetskii, E. M. Lifshitz, and L. P. Pitaevskii, *Quantum Electrodynamics* [in Russian], Nauka, Moscow (1980) [Pergamon, New York (1982)].
14. M. A. Amusia and V. A. Kharchenko, "Electron scattering by mesic hydrogen," *J. Phys. B: Atom and Mol Phys.* **14**, L219 (1981).
15. M. Gavrila, "Elastic scattering of protons," *Phys. Rev.* **163** (1967).
16. A. Bohr and B. Mottelson, *Nuclear Structure*, Vol. 2, Benjamin–Cummings, Menlo Park, California (1976).
17. M. Ya. Amus'ya, M. B. Zhalov, and V. I. Ryazanov, "Elastic scattering of fast protons by magic nuclei," Preprint No. 368, LIYaF, Leningrad (1977).

Chapter 9

1. M. Ya. Amusia, "Atomic bremsstrahlung spectrum," *Comments At. Mol. Phys.* **11**, 123 (1982).
2. M. Ya. Amus'ya, M. Yu. Kuchiev, A. V. Korol', and A. V. Solov'ev, "Bremsstrahlung of relativistic particles including dynamic polarizability of the target atom," *Zh. Eksp. Teor. Fiz.* **88**, 383 (1985) [*Sov. Phys. – JETP* **61**, 224 (1985)].
3. V. A. Astapenko, V. M. Buimistrov, Yu. A. Krotov, et al., "Dynamic bremsstrahlung of relativistic charged particles on atoms," *Zh. Eksp. Teor. Fiz.* **88**, 1560 (1985) [*Sov. Phys. – JETP* **61**, 930 (1985)].
4. M. Ya. Amus'ya, A. V. Korol', and A. V. Solov'ev, "Coherent emission of target electrons in atomic collisions," *Pis'ma Zh. Tekh. Fiz.* **12**, 705 (1986) [*Sov. Tech. Phys. Lett.* **12**(6), 290 (1986)].
5. M. Ya. Amus'ya, M. Yu. Kuchiev, and A. V. Solov'ev, "Total emission spectrum in collisions of heavy charged particles with atoms," *Pis'ma Zh. Tekh. Fiz.* **11**, 1401 (1985) [*Sov. Tech. Phys. Lett.* **11**(11), 577 (1985)].
6. M. Ya. Amus'ya, M. Yu. Kuchiev, and A. V. Solov'ev, "Bremsstrahlung in atom–atom collisions," *Pis'ma Zh. Tekh. Fiz.* **10**, 1025 (1984) [*Sov. Tech. Phys. Lett.* **10**(9), 431 (1984)].

6a. M. Ya. Amus'ya, M. Yu. Kuchiev, and A. V. Solov'ev, "Bremsstrahlung in atom–atom collisions," *Zh. Eksp. Teor. Fiz.* **89**, 1512 (1985) [*Sov. Phys. – JETP* **62**, 876 (1985)].

7. V. B. Berestetskii, E. M. Lifshitz, and L. P. Pitaevskii, *Quantum Electrodynamics* [in Russian], Nauka, Moscow (1980) [Pergamon, New York (1982)].
8. A. I. Akhiezer and V. B. Berestetskii, *Quantum Electrodynamics* [in Russian], Nauka, Moscow (1981).
9. V. M. Buimistrov and L. I. Trakhtenberg, "Role of atomic electrons in bremsstrahlung," *Zh. Eksp. Teor. Fiz.* **73**, 850 (1977) [*Sov. Phys. – JETP* **46**, 447 (1977)].
10. V. M. Buimistrov, Yu. A. Krotov, and L. I. Trakhtenberg, "Emission of a photon during a collision of a proton and positron with an atom," *Zh. Eksp. Teor. Fiz.* **79**, 808 (1980) [*Sov. Phys. – JETP* **52**, 411 (1980)].
11. M. Ya. Amus'ya, M. Yu. Kuchiev, and A. V. Solov'ev, "Bremsstrahlung in atom–atom collisions at relativistic velocities," *Zh. Tekh. Fiz.* **57**, 820 (1987) [*Sov. Phys. Tech. Phys.* **32**(4), 499 (1987)].
12. E. M. Lifshitz and L. P. Pitaevskii, *Relativistic Quantum Theory*, Part 2 [in Russian], Nauka, Moscow (1972).
13. A. A. Varfolomeev, "Coherent inelastic scattering of neutrinos on electrons in a medium," *Yad. Fiz.* **28**, 1034 (1978) [*Sov. J. Nucl. Phys.* **28**, 531 (1978)].

Chapter 10

1. A. I. Akhiezer and V. B. Berestetskii, *Quantum Electrodynamics* [in Russian], Nauka, Moscow (1969).

2. F. E. Low, "Bremsstrahlung of very low–energy quanta in elementary particle collisions," *Phys. Rev.* **110**, 974 (1958).
3. I. B. Bersuker, "Influence of the core on transitions of optical electrons," *Opt. Spektrosk.* **2**, 97 (1957).
3a. M. G. Veselov and A. V. Shtoff, "Computation of the oscillator strengths of the principal series of lithium in the adiabatic approximation," *Opt. Spektrosk.* **26**, 321 (1969).
4. P. A. Golovinskii and B. A. Zon, "Multiphoton processes on negative ions," *Izv. Akad. Nauk SSSR, Ser. Fiz.* **45**, 2305 (1981).
5. I. C. Percival and M. J. Seaton, "The polarization of atomic line radiation excited by electron impact," *Philos. Trans. R. Soc. London A* **251**, No. 3, 113 (1958).
6. V. M. Buimistrov and L. I. Trakhtenberg, "Cross section for bremsstrahlung during scattering of electrons on hydrogen atoms," *Zh. Eksp. Teor. Fiz.* **69**, 108 (1975) [*Sov. Phys. – JETP* **42**, 55 (1975)].
7. L. D. Landau and E. M. Lifshitz, *Quantum Mechanics* [in Russian], Nauka, Moscow (1974) [Pergamon, New York (1975)].
8. B. A. Zon, "Bremsstrahlung effect during collisions of electrons with atoms," *Zh. Eksp. Teor. Fiz.* **73**, 128 (1977) [*Sov. Phys. – JETP* **46**, 65 (1977)].
9. Yu. P. Raizer, *Laser Sparks and the Propagation of Discharges* [in Russian], Nauka, Moscow (1974) [*Laser-Induced Discharge Phenomena*, Consultants Bureau, New York (1977)].
10. J. E. Rizzo and R. C. Klewe, "The breakdown in vapor of metals due to laser radiation," *Br. J. Appl. Phys.* **17**, 1137 (1966).
11. L. P. Rapoport, B. A. Zon, and N. L. Manakov, *The Theory of Multiphoton Processes in Atoms* [in Russian], Atomizdat, Moscow (1978).
12. É. M. Karule and R. K. Peterkop, *Cross Sections for Scattering of Slow Electrons on Alkali Metal Atoms. Effective Cross Sections for Collisions of Electrons with Atoms* [in Russian], Zinante, Riga (1965), pp. 3–27.
13. E. L. Beilin and B. A. Zon, "On the sum rule for multiphoton bremsstrahlung," *J. Phys. B: At. Mol. Phys.* **16**, L15 (1983).
14. A. Weingartshofer and C. Jung, *Multiphoton Free–Free Transitions. Multiphoton Ionization of Atoms*, S. L. Chin and P. Lambropoulos (eds.), Academic Press, Toronto (1984), pp. 155–158.
15. F. V. Bunkin, A. E. Kazakov, and M. V. Fedorov, "Interaction of intense optical radiation with free electrons (nonrelativistic case)," *Usp. Fiz. Nauk* **107**, 559 (1972) [*Sov. Phys.– Usp.* **15**, 416 (1973)].
16. L. D. Landau and E. M. Lifshitz, *Electrodynamics of Continuous Media* [in Russian], Nauka, Moscow (1982) [Pergamon, New York (1960)].
17. V. M. Buimistrov and L. I. Trakhtenberg, "Role of atomic electrons in bremsstrahlung," *Zh. Eksp. Teor. Fiz.* **73**, 850 (1977) [*Sov. Phys. – JETP* **46**, 447 (1977)].
18. L. D. Landau and E. M. Lifshitz, *Classical Theory of Fields* [in Russian], Nauka, Moscow (1973) [Pergamon, New York (1976)].
19. V. L. Ginzburg, *The Propagation of Electromagnetic Waves in Plasmas* [in Russian], Nauka, Moscow (1967) [Pergamon, New York (1971)].

20. P. A. Golovinskii and B. A. Zon, "Bremsstrahlung effect in collisions of electrons with negative ions," *Zh. Tekh. Fiz.* **50**, 1847 (1980) [*Sov. Phys. Tech. Phys.* **25**, 1076 (1980)].
21. P. A. Golovinskii and B. A. Zon, "Dynamic polarizability of the negative hydrogen ion," *Opt. Spektrosk.* **45**, 854 (1978) [*Opt. Spectr.* **45**, 733 (1978)].
22. M. Ya. Amus'ya, A. S. Baltenkov, and A. A. Paiziev, "Bremsstrahlung of electrons on atoms including polarizability," *Pis'ma Zh. Eksp. Teor. Fiz.* **24**, 366 (1976) [*JETP Lett.* **24**, 332 (1976)].
23. B. A. Zon, "Absorption of optical radiation by weakly ionized gases," *Zh. Eksp. Teor. Fiz.* **77**, 44 (1979) [*Sov. Phys. – JETP* **50**, 21 (1979)].
24. O. B. Firsov and M. I. Chibisov, "Bremsstrahlung of slow electrons on neutral atoms," *Zh. Eksp. Teor. Fiz.* **39**, 1770 (1960) [*Sov. Phys. – JETP* **12**, 1235 (1961)].
25. T. Ohmura and H. Ohmura, "Continuous absorption due to free–free transitions in hydrogen," *Phys. Rev.* **121**, 513 (1961).
26. N. Mott and H. Massey, *Theory of Atomic Collisions*, Oxford University Press, Oxford (1965).

Chapter 11

1. V. M. Buimistrov and L. I. Trakhtenberg, "Cross section for bremsstrahlung during scattering of electrons on hydrogen atoms," *Zh. Eksp. Teor. Fiz.* **69**, 108 (1975) [*Sov. Phys. – JETP* **42**, 55 (1975)].
2. M. Ya. Amus'ya, A. S. Baltenkov, and A. A. Paiziev, "Bremsstrahlung of electrons on atoms including polarizability," *Pis'ma Zh. Eksp. Teor. Fiz.* **24**, 366 (1976) [*JETP Lett.* **24**, 332 (1976)].
3. B. A. Zon, "Bremsstrahlung effect in collisions of electrons with atoms," *Zh. Eksp. Teor. Fiz.* **73**, 128 (1977) [*Sov. Phys. – JETP* **46**, 65 (1977)].
4. M. Ya. Amus'ya, M. Yu. Kuchiev, A. V. Korol', and A. V. Solov'ev, "Bremsstrahlung of relativistic particles including dynamic polarizability of the target atom," *Zh. Eksp. Teor. Fiz.* **88**, 383 (1985) [*Sov. Phys. – JETP* **61**, 224 (1985)].
5. V. A. Astapenko, V. M. Buimistrov, Yu. A. Krotov, et al., "Dynamic bremsstrahlung of relativistic charged particles on atoms," *Zh. Eksp. Teor. Fiz.* **88**, 1560 (1985) [*Sov. Phys. – JETP* **61**, 930 (1985)].
6. N. Mott and H. Massey, *Theory of Atomic Collisions*, Oxford University Press, Oxford (1965).
7. I. C. Percival and M. J. Seaton, "The polarization of atomic line radiation excited by electron impact," *Philos. Trans. R. Soc. London A* **251**, 113 (1958).
8. A. Jablonski, "General theory of pressure broadening of spectral lines," *Phys. Rev.* **68**, 78 (1945).
9. I. I. Sobel'man, *Introduction to the Theory of Atomic Spectra* [in Russian], Fizmatgiz, Moscow (1963) [Pergamon, New York (1972)].
10. V. S. Lisitsa, "Line broadening as light emission during a collision process," *Acta Phys. Pol. A* **55**, 87 (1979).

11. V. S. Lisitsa, "Stark broadening of hydrogen lines in plasmas," *Usp. Fiz. Nauk* **122**, 449 (1977) [*Sov. Phys. – Usp.* **20**, 603 (1977)].

12. J. Szudy, "The Franck–Condon principle and the broadening of isolated spectral lines in plasmas," *Acta Phys. Pol.* **40**, 361 (1971).

13. L. D. Landau and E. M. Lifshitz, *Quantum Mechanics* [in Russian], Nauka, Moscow (1974) [Pergamon, New York (1975)].

14. L. A. Vainshtein, I. I. Sobel'man, and E. A. Yukov, *Excitation of Atoms and Broadening of Spectral Lines* [in Russian], Nauka, Moscow (1979).

15. E. Fermi, "On the theory of atomic collisions with electrically charged particles," in: *Scientific Papers* [in Russian], Vol. 1, Nauka, Moscow (1971), pp. 166–177.

16. L. D. Landau and E. M. Lifshitz, *Classical Theory of Fields* [in Russian], Nauka, Moscow (1973) [Pergamon, New York (1976)].

17. V. I. Kogan and A. B. Kukushkin, "Radiation from quasiclassical electrons in an atomic potential," *Zh. Eksp. Teor. Fiz.* **87**, 1164 (1984) [*Sov. Phys. – JETP* **60**, 665 (1984)].

18. R. A. Gantsev, N. F. Kazakova, and V. P. Krainov, "Radiative transition rates in hydrogen-like plasmas," in: *Plasma Chemistry* [in Russian] (1985), Energoatomizdat, Moscow, pp. 96–118.

19. W. Heitler, *The Quantum Theory of Radiation*, Oxford University Press, Oxford (1954).

20. I. L. Beigman, L. A. Vainshtein, and B. N. Chichkov, "Dielectronic recombination," *Zh. Eksp. Teor. Fiz.* **80**, 964 (1981) [*Sov. Phys. – JETP* **53**, 490 (1981)].

21. E. M. Purcell, "The lifetime of the $2\,^2S_{1/2}$ state of hydrogen in an ionized atmosphere," *Astrophys. J.* **116**, 457 (1952).

22. V. I. Kogan, V. S. Lisitsa, and A. D. Selidkovkin, "Two-level system with quenching in a plasma," *Zh. Eksp. Teor. Fiz.* **65**, 152 (1973) [*Sov. Phys. – JETP* **38**, 75 (1974)].

23. I. L. Beigman, V. A. Boiko, S. A. Pikuz, and A. Ya. Faenov, "Collisional deexcitation of metastable levels and the intensity of the resonant doublet components of hydrogenlike ions in laser plasmas," *Zh. Eksp. Teor. Fiz.* **71**, 975 (1976) [*Sov. Phys. – JETP* **44**, 511 (1976)].

24. A. V. Vinogradov and E. A. Yukov, "Forbidden transitions induced by collisions in dense plasmas," *Fiz. Plazmy* **1**, 860 (1975) [*Sov. J. Plasma Phys.* **1**, (1975)].

25. V. I. Bugarya, A. V. Gorshkov, and S. A. Grashin, "Measurement of plasma column rotation and potential in the TM-4 tokamak," in: Proc. 9th International Conf. on Plasma Phys. and Contr. Nucl. Fusion Research, Baltimore (1982), IAEA, Vienna (1983), Vol. 3, pp. 263–272.

26. S. von Goeler, W. Stodiek, H. Eubank, et al., "Thermal x-ray spectra and impurities," *Nucl. Fusion* **15**, 301 (1975).

27. N. D. Vinogradova, Yu. V. Esipchuk, P. E. Kovrov, and K. A. Razumova, "Investigation of plasma x-ray emission in the T-10 tokamak," in: Proc. International Conf. on Plasma Phys. and Contr. Nucl. Fusion Research, Innsbruck (1978), IAEA, Vienna (1979), Vol. 1, pp. 257–267.

28. V. I. Bugarya, N. L. Vasin, V. A. Vershkov, et al., "Transport of multiply charged ions in the T-10 tokamak," *Fiz. Plazmy* **9**, 914 (1983) [*Sov. J. Plasma Phys.* **9**, 529 (1983)].

29. S. Suckewer, "Spectroscopic diagnostics of Tokamak plasmas," *Phys. Scr.* **23**, No. 2, 72 (1981).
30. E. V. Aglitskii, V. A. Rantsev-Kartinov, M. M. Stepanenko, and D. A. Shcheglov, "Observation of x-ray spectra of chromium and iron in plasmas in the Tokamak-10 machine," *Pis'ma Zh. Eksp. Teor. Fiz.* **26**, 544 (1977).
31. V. A. Bryzgunov, S. Yu. Luk'yanov, M. T. Pakhomov, et al., "X-ray spectrum of He-like chromium in a laboratory plasma," *Zh. Eksp. Teor. Fiz.* **82**, 1904 (1982) [*Sov. Phys. JETP* **55**, 1095 (1982)].
32. A. L. Merts, R. D. Cowan, and N. H. Magee, Jr., "The calculated power output from a thin iron-seeded plasma," Informal Report No. LA 6220-SN, Los Alamos (1976).
33. J. Durban and M. Lourergue, "Electron excitation cross-section and oscillator strength for highly ionized atoms," *Phys. Scr.* **23**, No. 2, 136 (1981).
34. V. G. Gontis and V. S. Lisitsa, "Inhomogeneity effects and the structure of x-ray lines in tokamak plasmas," *Fiz. Plazmy* **11**, 483 (1985) [*Sov. J. Plasma Phys.* **11**, 282 (1985)].
35. Yu. V. Esipchuk and P. E. Kovrov, "Measurement of soft x-ray emission in the T-10 tokamak," Preprint No. 3258/7, Institute of Atomic Energy, Moscow (1980).
36. V. I. Gervids, V. I. Kogan, and V. S. Lisitsa, "Multiply charged ions and plasma radiation," in: *Plasma Chemistry*, Vol. 10 [in Russian], Energoatomizdat, Moscow (1983), pp. 3–73.
37. V. S. Lisitsa and S. I. Yakovlenko, "Nonlinear theory of broadening and a generalization of the Karplus–Schwinger formula," *Zh. Eksp. Teor. Fiz.* **68**, 479 (1975) [*Sov. Phys. – JETP* **41**, 233 (1975)].

Chapter 12

1. H. Griem, *Spectral Line Broadening by Plasmas*, Academic Press, New York (1974).
2. V. I. Kogan, V. S. Lisitsa, and G. V. Sholin, "Broadening of spectral lines in plasmas," in: *Reviews of Plasma Physics*, Vol. 13, Consultants Bureau, New York (1987).
3. E. A. Oks and G. V. Sholin, "Stark profiles of hydrogen spectral lines in plasmas with Langmuir turbulence," *Zh. Eksp. Teor. Fiz.* **68**, 974 (1975) [*Sov. Phys. – JETP* **41**, 482 (1975)].
4. E. Ya. Kogan and E. V. Martysh, Recombination in turbulent plasmas, *Fiz. Plazmy* **9**, 881 (1983) [*Sov. J. Plasma Phys.* **9**, 515 (1983)].
5. V. L. Ginzburg, *Theoretical Physics and Astrophysics* [in Russian], Nauka, Moscow (1981) [Pergamon, New York (1979)].
6. A. I. Alekseev and Yu. P. Nikitin, "Radiation from atoms in an anisotropic medium," *Zh. Eksp. Teor. Fiz.* **48**, 1669 (1965) [*Sov. Phys. – JETP* **21**, 1121 (1965)].
7. A. I. Alekseev and Yu. P. Nikitin, Quantization of the electromagnetic field in a dispersive medium, *Zh. Eksp. Teor. Fiz.* **50**, 915 (1966) [*Sov. Phys. – JETP* **23**, 608 (1966)].
8. E. B. Kleiman and I. M. Oiringel , "Atomic systems in turbulent plasmas," in: *Problems in the Physics of Space Plasmas* [in Russian], Nauka, Moscow (1979), pp. 76–92.
9. I. M. Oiringel and E. B. Kleiman, *Atomic Emission in Space Plasmas* [in Russian], Nauka, Novosibirsk (1984).

10. V. B. Berestetskii, E. M. Lifshitz, and L. P. Pitaevskii, *Quantum Electrodynamics* [in Russian], Nauka, Moscow (1980) [Pergamon, New York (1982)].
11. A. I. Akhiezer and V. B. Berestetskii, *Quantum Electrodynamics* [in Russian], Nauka, Moscow (1981).
12. V. I. Kogan, V. S. Lisitsa, and A. D. Selidkovkin, "Two-level system with quenching in a plasma," *Zh. Eksp. Teor. Fiz.* **65**, 152 (1973) [*Sov. Phys. – JETP* **38**, 75 (1974)].
13. S. A. Kaplan, E. B. Kleiman, and I. M. Oiringel, Two-photon $2s_{1/2}$–$1s_{1/2}$ photon–plasma transition," *Astron. J.* **49**, 294 (1972) [*Sov. Astron.* **16**, 241 (1972)].
14. S. A. Kaplan and S. B. Pikel'ner, *The Physics of the Interplanetary Medium* [in Russian], Nauka, Moscow (1979).
15. E. B. Kleiman, "Populations of the second level of the hydrogen atom in a plasma medium," *Astron. Zh.* **49**, 892 (1972) [*Sov. Astron.* **16**, 724 (1973)].
16. V. V. Sobolev, *A Course in Theoretical Astrophysics* [in Russian], Nauka, Moscow (1975).
17. S. A. Kaplan, E. B. Kleiman, and I. M. Oiringel, "Annihilation of positronium in a plasma," *Astrofizika* **9**, 417 (1973) [*Astrophysics* **9** (1974)].
18. V. I. Gol'danskii, *The Physical Chemistry of Positrons and Positronium* [in Russian], Nauka, Moscow (1968).
19. Yu. L. Klimontovich, *The Kinetic Theory of Electromagnetic Processes* [in Russian], Nauka, Moscow (1980).
20. E. B. Kleiman and I. M. Oiringel, "Effect of polarization plasma effects on quenching of metastable atomic states," in: Material from the 9th All-union Conf. on the Theory of Atoms and Atomic Spectra (1985).
21. V. P. Zhdanov, "Dielectronic recombination," in: *Reviews of Plasma Physics*, Vol. 12, Consultants Bureau, New York (1987).
22. A. K. Grigoriadi and O. I. Fisun, "Charge exchange of ions in plasmas," *Zh. Eksp. Teor. Fiz.* **49**, 975 (1979) [*Sov. Phys. – JETP* **22**, 678 (1979)].
23. A. V. Vinogradov, I. I. Sobel'man, and E. A. Yukov, "Spectroscopic methods for diagnostics of extremely dense hot plasmas," *Kvantovaya Elektron.* **1**, 268 (1974) [*Sov. J. Quant. Electron.* **4**, 149 (1974)].
24. S. A. Kaplan and V. N. Tsytovich, *Plasma Astrophysics* [in Russian], Nauka, Moscow (1972) [Pergamon, New York (1973)].

Chapter 13

1. L. D. Landau and E. M. Lifshitz, *Quantum Mechanics. Nonrelativistic Theory* [in Russian], Nauka, Moscow (1974) [Pergamon, New York (1977)].
2. H. W. Koch and J. W. Motz, *Rev. Mod. Phys.* **31**, 920 (1959).
3. P. J. Brussaard and H. C. van de Hulst, *Rev. Mod. Phys.* **34**, 507 (1962).
4. R. R. Johnston, *J. Quant. Spectrosc. Radiat. Transfer* **7**, 815 (1967).
5. G. R. Blumenthal and R. J. Gould, *Rev. Mod. Phys.* **42**, 237 (1970).
6. A. Sommerfeld, *Atomic Structure and Spectral Lines*, London (1924).

7. A. I. Akhiezer and V. B. Berestetskii, *Quantum Electrodynamics* [in Russian], Nauka, Moscow (1981).
8. V. B. Berestetskii, E. M. Lifshitz, and L. P. Pitaevskii, *Quantum Electrodynamics* [in Russian], Nauka, Moscow (1980) [Pergamon, New York (1982)].
9. W. Heitler, *The Quantum Theory of Radiation*, Oxford University Press, Oxford (1954).
10. I. I. Sobel'man, *Introduction to the Theory of Atomic Spectra* [in Russian], Nauka, Moscow (1977) [Pergamon, New York (1972)].
11. V. I. Gervids, V. P. Zhdanov, V. I. Kogan, B. A. Trubnikov, and M. I. Chibisov, in: *Reviews of Plasma Physics*, M. A. Leontovich and B. B. Kadomtsev (eds.), Vol. 12, Consultants Bureau, New York (1987).
12. V. P. Krainov and B. M. Smirnov, *Radiative Processes in Atomic Physics* [in Russian], Vysshaya Shkola, Moscow (1983).
13. V. I. Gervids, V. I. Kogan, and V. S. Lisitsa, in: *Plasma Chemistry*, B. M. Smirnov (ed.) [in Russian], Vol. 10 (1983).
14. V. I. Kogan and V. S. Lisitsa, in: *Progress in Science and Technology, Series on Plasma Physics* [in Russian], V. D. Shafranov (ed.), Vol. 4, VINITI, Moscow (1983).
15. J. M. Barger, *Phys. Rev.* **105**, 35 (1957).
16. I. P. Grant, *Mon. Not. R. Astron. Soc.* **118**, 241 (1958).
17. J. Green, *Astrophys. J.* **130**, 693 (1959).
18. W. J. Karzas and R. Latter, *Astrophys. J., Suppl.* **6**, 167 (1961).
19. G. Elwert and E. Haug, *Phys. Rev.* **183**, 90 (1969).
20. H. K. Tsang and R. M. Pratt, *Phys. Rev. A* **3**, 100 (1971).
21. E. Haug, *Z. Naturforsch.* **30a**, 1099, 1546 (1975).
22. C. M. Lee, L. Kissei, R. M. Pratt, and H. K. Tsang, *Phys. Rev. A* **13**, 1714 (1976).
23. H. K. Tsang, R. H. Pratt, and C. M. Lee, *Phys. Rev. A* **19**, 187 (1979).
24. L. M. Biberman and G. É. Norman, *Usp. Fiz. Nauk* **91**, 193 (1967) [*Sov. Phys. – Usp.* **10**, 52 (1967)].
25. O. B. Firsov and M. I. Chibisov, *Zh. Eksp. Teor. Fiz.* **39**, 1770 (1960) [*Sov. Phys. – JETP* **12**, 1235 (1961)].
26. S. von Goeler, J. Stevens, S. Bernabei, et al., *Nucl. Fusion* **25**, 1515 (1985).
27. J. Stevens, S. von Goeler, S. Bernabei, et al., *Nucl. Fusion* **25**, 1529 (1985).
28. B. A. Zon, *Zh. Eksp. Teor. Fiz.* **73**, 128 (1977) [*Sov. Phys. – JETP* **46**, 65 (1977)].
29. M. Ya. Amus'ya, A. S. Baltenkov, and A. A. Paiznev, *Pis'ma Zh. Eksp. Teor. Fiz.* **24**, 366 (1976) [*JETP Lett.* **24**, 332 (1976)].
30. V. M. Buimistrov and L. I. Trakhtenberg, *Zh. Eksp. Teor. Fiz.* **69**, 108 (1975); **73**, 850 (1977) [*Sov. Phys. – JETP* **42**, 54 (1975); **46**, 447 (1977)].
31. V. N. Tsytovich and I. M. Oiringel, *Polarization Bremsstrahlung of Particles and Atoms* [in Russian], Nauka, Moscow (1987).
32. L. D. Landau and E. M. Lifshitz, *Classical Theory of Fields* [in Russian], Nauka, Moscow (1973) [Pergamon, New York (1976)].
33. V. I. Gervids and V. I. Kogan, *Pis'ma Zh. Eksp. Teor. Fiz.* **22**, 308 (1975) [*JETP Lett.* **22**, 142 (1975)].

34. V. I. Gervids and V. I. Kogan, Preprint No. 2720, Institute of Atomic Energy, Moscow (1976).
35. V. I. Kogan and A. B. Kukushkin, Preprint No. 3660/6, Institute of Atomic Energy, Moscow (1982).
36. V. I. Kogan and A. B. Kukushkin, *Zh. Eksp. Teor. Fiz.* **87**, 1164 (1984) [*Sov. Phys. – JETP* **60**, 665 (1984)].
37. L. D. Landau and E. M. Lifshitz, *Mechanics* [in Russian], Nauka, Moscow (1973) [Pergamon, New York (1976)].
38. H. Bateman and A. Erdelyi, *Higher Transcendental Functions*, Vol. 2, McGraw-Hill, New York (1954).
39. I. Shkarovsky, T. Johnston, and M. Bachynski, *The Particle Kinetics of Plasmas*, Addison-Wesley, Reading, Massachusetts (1966).
40. J. Jackson, *Classical Electrodynamics*, Wiley, New York (1962).
41. H. A. Kramers, *Philos. Mag.* **46**, 836 (1923).
42. V. V. Babikov, in: *Plasma Physics and the Problem of Controlled Thermonuclear Reactions* [in Russian], M. A. Leontovich (ed.), Vol. 2, Izd. AN SSSR (1958).
43. V. L. Ginzburg, *Theoretical Physics and Astrophysics*, 3rd edn. [in Russian], Nauka, Moscow (1987) [1st edn., Pergamon, New York (1979)].
44. P. I. Fomin, *Zh. Eksp. Teor. Fiz.* **35**, 707 (1958) [*Sov. Phys. – JETP* **8**, 491 (1959)].
45. H. W. Furry, *Phys. Rev.* **81**, 115 (1951).
46. F. J. Dyson, *Phys. Rev.* **75**, 486, 1736 (1949); R. P. Feynman, *Phys. Rev.* **76**, 749, 769 (1949).
47. A. Sommerfeld, *Ann. Phys.* **11**, 257 (1931).
48. A. Sommerfeld and A. W. Maue, *Ann. Phys.* **23**, 589 (1935).
49. D. H. Menzel and C. L. Pekeris, *Mon. Not. R. Astron. Soc.* **96**, 77 (1935).
50. L. C. Biedenharn, *Phys. Rev.* **102**, 262 (1956).
51. I. S. Gradshtein and I. M. Ryzhik, *Tables of Integrals, Sums, Series, and Products* [in Russian], Fizmatgiz, Moscow (1962) [Academic Press, New York (1966)].
52. H. Bethe and W. Heitler, *Proc. R. Soc. London* **146**, 83 (1934).
53. V. D. Kirillov, B. A. Trubnikov, and S. A. Trushin, *Fiz. Plazmy* **1**, 218 (1975) [*Sov. J. Plasma* **1**, 117 (1975)].
54. T. Guggenberger, *Z. Phys.* **149**, 523 (1957).
55. A. Abramowitz and I. Stegun, *Handbook of Mathematical Functions*, Dover, New York (1964).
56. I. I. Sobel'man, *Introduction to the Theory of Atomic Spectra* [in Russian], Fizmatgiz, Moscow (1963) [Pergamon, New York (1972)].
57. G. Elwert, *Z. Naturforsch.* **3A**, 477 (1948).
58. G. Elwert, *Ann. Phys.* **34**, 178 (1939).
59. K. Kummerer, *Z. Phys.* **147**, 373 (1957).
60. J. M. C. Scott, *J. Phys. A* **6**, 1853 (1973).
61. C. G. Darwin, *Proc. R. Soc. London* **118A**, 654 (1928).
62. V. A. Kas'yanov and A. N. Starostin, *Zh. Eksp. Teor. Fiz.* **48**, 295 (1965) [*Sov. Phys. – JETP* **21**, 193 (1965)].

63. H. Bethe and L. Maximon, *Phys. Rev.* **93**, 768 (1954).
64. H. W. Furry, *Phys. Rev.* **46**, 391 (1934).
65. W. Nakel, *Phys. Lett.* **22**, 614 (1966); **25A**, 569 (1967).
66. W. Nakel, *Z. Phys.* **214**, 618 (1968).
67. G. Bekefi, *Radiation Processes in Plasmas*, Wiley, New York (1965).
68. H. Bethe and E. Salpeter, *The Quantum Mechanics of One- and Two-Electron Systems*, Springer-Verlag, Berlin (1958).
69. S. P. Goreslavskii, N. B. Delone, and V. P. Krainov, *Zh. Eksp. Teor. Fiz.* **82**, 1789 (1982); Preprint No. 3, Fiz. Inst. Akad. Nauk SSSR (1982) [*Sov. Phys. - JETP* **55**, 1032 (1982)].
70. A. B. Kukushkin and V. S. Lisitsa, *Zh. Eksp. Teor. Fiz.* **88**, 1570 (1985) [*Sov. Phys. - JETP* **61**, 937 (1985)].
71. F. Sauter, *Ann. Phys.*, **18**, 486 (1933).
72. W. Henneberg, *Z. Phys.* **83**, 555 (1933).
73. P. Gambos, *The Statistical Theory of the Atom and Its Applications* [Russian translation], IL, Moscow (1951).
74. V. I. Kogan and A. B. Kukushkin, *Pis'ma Zh. Eksp. Teor. Fiz.* **37**, 272 (1983) [*JETP Lett.* **37**, 322 (1983)].
75. V. P. Zhdanov, *Fiz. Plazmy* **4**, 128 (1978) [*Sov. J. Plasma Phys.* **4**, 71 (1978)].
76. M. Semaan and C. Quarles, *Phys. Rev. A* **26**, 3152 (1982).
77. R. A. Gantsev, N. F. Kazakova, and V. P. Krainov, in: *Plasma Chemistry* [in Russian], Vol. 12, B. M. Smirnov (ed.), Energoatomizdat, Moscow (1985).
78. S. M. Seltzer and M. J. Berger, *At. Data Nucl. Data Tables* **35**, 345 (1986).
79. F. E. Low, in: *Problems in Theoretical Physics* (in Memory of I. E. Tamm), [in Russian], Nauka, Moscow (1972).
80. A. A. Sokolov, I. M. Ternov, V. Ch. Zhukovskii, and A. V. Borisov, *Quantum Electrodynamics* [in Russian], Izd. MGU, Moscow (1983).
81. Y. S. Kim and R. H. Pratt, *Phys. Rev. A* **27**, 2913 (1983).
82. K. Alder, A. Bohr, et al., in: *Deformation of Atomic Nuclei* [Russian translation], IL, Moscow (1958).

Chapter 14

1. F. R. Arutyunyan, K. A. Ispiryan, A. G. Oganesyan, et al., "Density effect in bremsstrahlung of electrons with energies of up to 600 MeV," *Dokl. Akad. Nauk Arm. SSR, Fizika* **44**, 65 (1967).

1a. F. R. Arutyunyan, A. A. Nazaryan, and A. A. Frangyan, "Effect of a medium on radiation from relativistic electrons," *Zh. Eksp. Teor. Fiz.* **62**, 2044 (1972) [*Sov. Phys. - JETP* **35**, 1067 (1972)].

2. K. Ishii and S. Morita, "Continuum x rays produced by light ion–atom collisions," *Phys. Rev. A* **30**, 2278 (1984).

3. V. L. Ginzburg, *Theoretical Physics and Astrophysics*, 2nd edn. [in Russian], Nauka, Moscow (1981) [1st edn., Pergamon, New York (1979)].

INDEX